I0063087

Toxins in Drug Discovery and Pharmacology

Special Issue Editor
Steve Peigneur

MDPI • Basel • Beijing • Wuhan • Barcelona • Belgrade

MDPI

Special Issue Editor
Steve Peigneur
Laboratorium for Toxicology & Pharmacology
Belgium

Editorial Office
MDPI
St. Alban-Anlage 66
Basel, Switzerland

This edition is a reprint of the Special Issue published online in the open access journal *Toxins* (ISSN 2072-6651) from 2017–2018 (available at: http://www.mdpi.com/journal/toxins/special_issues/ Discovery_and_Pharmacology).

For citation purposes, cite each article independently as indicated on the article page online and as indicated below:

Lastname, F.M.; Lastname, F.M. Article title. *Journal Name* **Year**, *Article number*, page range.

First Edition 2018

ISBN 978-3-03842-861-9 (Pbk)
ISBN 978-3-03842-862-6 (PDF)

Cover image courtesy of Maria Elena de Lima.

Articles in this volume are Open Access and distributed under the Creative Commons Attribution (CC BY) license, which allows users to download, copy and build upon published articles even for commercial purposes, as long as the author and publisher are properly credited, which ensures maximum dissemination and a wider impact of our publications. The book taken as a whole is © 2018 MDPI, Basel, Switzerland, distributed under the terms and conditions of the Creative Commons license CC BY-NC-ND (http://creativecommons.org/licenses/by-nc-nd/4.0/).

Table of Contents

About the Special Issue Editor

Steve Peigneur is a researcher in the lab of Toxicology and Pharmacology at the Catholic University of Leuven (KU Leuven) in Belgium. He has a BSc in Biomedical Lab Techniques. Since 2013, he holds a permanent position at the KU Leuven. His research in the field of drug discovery and development mainly focuses on the characterization of novel ligands for ion channels and receptors, starting from natural sources such as venomous animals and poisonous plants.

Preface to "Toxins in Drug Discovery and Pharmacology"

Venoms from marine and terrestrial animals (cone snails, scorpions, spiders, snakes, centipedes, cnidarian, etc.) can be seen as untapped cocktails of biologically active compounds that are being increasingly recognized as a new emerging source of peptide-based therapeutics. Venomous animals are considered specialized predators that have evolved the most sophisticated peptide chemistry and neuropharmacology for their own biological purposes by producing venoms that contain myriads of toxins with an amazing structural and functional diversity. These neurotoxins have been shown to be highly selective and potent ligands for a wide range of ion channels and receptors. Therefore, they represent interesting lead compounds for the development of novel medicines, for example, analgesics, anti-cancer drugs, and drugs for neurological disorders. A successful example of venom-derived peptides in use is the drug Prialt® or Ziconotide, which is a synthetic version of a peptide that is found in the venom of a marine snail, Conus magus. This molecule is a 25-amino acid peptide that blocks N-type voltage-gated calcium channels. Its FDA definition reads that ziconotide is intended for the treatment of chronic severe pain in patients that are intolerant or refractory to systemic analgesics or intrathecal morphine. The characterization of small, bradykinin-potentiating peptides from the venom of the South American snake Bothrops jararaca led to the development of Captopril, even prior to Prialt®. Captopril, and its follow-up compounds, are widely used as inhibitors of angiotensin-converting enzyme (ACE). Another example is Exenatide®, which is a synthetic version of a hormone called exendin-4, isolated from the saliva of the glia monster. Exendin-4 is a glucagon-like peptide-1 receptor agonist and hereby finds its use in the treatment of type 2 diabetes. Exenatide® was approved by the FDA in 2005.

In this Special Issue, "Toxins in Drug Discovery and Pharmacology", we have attempted to provide the reader with a comprehensive overview of new toxins and toxin-inspired leads. This issue focuses on the mechanisms of action, structure–function relationships, and the evolution of pharmacologically interesting venom components, including, but not limited to, recent developments related to the emergence of venoms as an underutilized source of highly evolved bioactive peptides with clinical potential. The following is a short synopsis of the six reviews and 15 research papers that constitute this Special Issue.

Agwa and colleagues have explored the pharmaceutical potential of spider venoms. Their work shows that spider-derived gating modifier toxins are undeniably among nature's more interesting pharmacological probes in the study of voltage-gated ion channels. The current work has provided additional insight into the potential of these ICK peptides as templates for drugs designed to target ailments linked to the voltage-gated ion channels. Freitas et al. also focused on spider venoms for the discovery of novel lead compounds with potential interesting pharmaceutical properties. PnPP-19 is a toxin-derived peptide from Phoneutria nigriventer that activates μ-opioid receptors without inducing β-arrestin2 recruitment. PnPP-19 is the first spider toxin derivative that, among opioid receptors, selectively activates μ-opioid receptors. The lack of β-arrestin2 recruitment highlights its potential for the design of new improved opioid agonists. Dong et al. describe a high-throughput way to discover AMPs from fish gastrointestinal microbiota, which can be developed as alternative pathogen antagonists or toxins for micro-ecologics or probiotic supplements.

The anti-cancer properties of snake venom have been investigated by Osipov et al. by verifying the anti-tumor effects of nerve growth factor from cobra venom. This work suggests that the antitumor effect of nerve growth factor in vivo depends critically on the normal status of the immune system. Nerve growth factor may exert an antitumor effect by causing an increase in lymphocytic infiltration in the tumor, a rise in the levels of IL-1β and TNF-α in the serum of tumor-bearing mice, and an increase in aerobic glycolysis. Sales et al. questioned whether inhibitors of snake venom phospholipases A2 can lead to new insights for the development of anti-inflammatory therapies in humans. This work reports a proof-of-principle study demonstrating that snake venom toxins, more specifically snake venom phospholipases A2, can be used as tools for studies of human phospholipases A2, based on a careful selection of specific snake venom phospholipases A2. Azemiopsin is a linear peptide from viper venom and it is a selective inhibitor of nicotinic acetylcholine receptors. Azemiopsin has good drug-like properties as a local muscle relaxant. Another source for venom-based drug discovery is bee venom. Bee venom given subcutaneously attenuates allodynia in mice models of CPIP without notable adverse effects. The anti-allodynic effects were found to be closely associated with a significant decrease in NK-1 receptor expression in DRG. These findings suggest that a repetitive bee venom-based therapy could be a useful modality for the treatment of complex regional pain syndrome type I. Shin and colleagues report that melittin and apamin can inhibit the fungi-induced production of chemical mediators and extracellular matrix (ECM) by nasal fibroblasts. This study suggests the possible role of melittin and apamin in the treatment of fungi-induced airway inflammatory diseases. Another study with bee venom showed that bee venom acupuncture has potent suppressive effects on paclitaxel-induced neuropathic pain, which is mediated by spinal α2-adrenergic receptor activity. In addition, bee venom may be a useful preventive and therapeutic agent in the treatment of obesity. It was found that bee venom mediates anti-obesity effects by suppressing obesity-related transcription factors. Lepiarczyk and colleagues describe a first study suggesting that both resiniferatoxin and tetrodotoxin can modify the number of noradrenergic and cholinergic NF supplying the porcine urinary bladder.

Mastoparan V1 is derived from the venom of the social wasp Vespula vulgaris and presents potent antimicrobial activity against Salmonella infection. However, there exist some limits for its practical application due to the loss of its activity in the presence of a high bacterial density and the difficulty of its efficient production. Ha et al. modulated successfully the antimicrobial activity of synthetic mastoparan V1 against an high-density Salmonella population using protease inhibitors. Furthermore, they developed an Escherichia coli secretion system efficiently producing active mastoparan V1. Short toxin-like proteins from insects are often overlooked in drug discovery. The study performed by Linial et al. indicates dozens of new candidates for peptide-based therapy and discusses their potential for drug design. It concludes that the overlooked endogenous toxin-like proteins from insects, characterized by structural stability and enhanced specificity, are attractive templates for drug design.

Two dipeptides with anticoagulant activity have been isolated from scorpion venom. This study shows for the first time the ability of short venom peptides to slow down blood coagulation. Using molecular dynamics simulations of complexes between scorpion toxins and the Kv1.2 channel, the authors identified hydrophobic patches, hydrogen bonds, and salt bridges as the three essential forces mediating the interactions between this channel and the toxins. This discovery might help design highly selective Kv1.2-channel inhibitors.

Park and Park provide a comprehensive review to summarize the experimental and clinical evidence of the mechanism by which Botulinum toxin acts on various types of neuropathic pain and describe why this molecule is a successful example in toxin-based drug discovery. Botulinum toxin has been used for approximately 40 years for the treatment of excessive muscle stiffness, spasticity, dystonia, and various types of neuropathic pain.

Royal and Motoba review the possible mechanisms behind cholera toxin B subunit anti-inflammatory activity and discuss how this protein could impact the treatment of mucosal inflammatory disease.

The anti-metastatic mechanisms of snake toxins can be classified on the basis of three molecular targets. These are the inhibition of extracellular matrix components-dependent cell adhesion and migration, the inhibition of epithelial–mesenchymal transition, and the inhibition of cell motility by alterations in the actin cytoskeleton network. The molecular mechanisms by which snake toxins target metastases are reviewed by Urra and Araya-Maturana.

The genus Conus has become an important genetic resource for conotoxin identification and drug development. The many challenges of drug discovery from cone snail venom are reviewed by Gao et al. Animal toxins are valuable tools to study ion channels such as TRPV1. A comprehensive summary of the advancements made in TRPV1 research in recent years by employing venom-derived peptide toxins is provided by Geron and colleagues. In this review, the authors describe the functional aspects, behavioral effects, and structural features of each studied toxin, all of which have contributed to our current knowledge of TRPV1.

Steve Peigneur
Special Issue Editor

toxins

MDPI

Article

The Peptide PnPP-19, a Spider Toxin Derivative, Activates μ-Opioid Receptors and Modulates Calcium Channels

Ana C. N. Freitas [1], Steve Peigneur [2] , Flávio H. P. Macedo [1], José E. Menezes-Filho [1], Paul Millns [3], Liciane F. Medeiros [4], Maria A. Arruda [4,5] , Jader Cruz [1], Nicholas D. Holliday [4], Jan Tytgat [2], Gareth Hathway [3] and Maria E. de Lima [1,*]

[1] Departamento de Bioquímica e Imunologia, Universidade Federal de Minas Gerais, Belo Horizonte 31270-901, Brazil; acnfreitas@gmail.0100com (A.C.N.F.); flavio.hpmacedo@gmail.com (F.H.P.M.); menezesfilho10@gmail.com (J.E.M.-F.); jadercruzytrio@gmail.com (J.C.)
[2] Toxicology and Pharmacology, KU Leuven, 3000 Leuven, Belgium; steve.peigneur@pharm.kuleuven.be (S.P.); jan.tytgat@pharm.kuleuven.be (J.T.)
[3] Arthritis Research UK Pain Centre, School of Life Sciences, Queen's Medical Centre, University of Nottingham, Nottingham NG7 2UH, UK; paul.millns@nottingham.ac.uk (P.M.); gareth.hathway@nottingham.ac.uk (G.H.)
[4] Cell Signaling Research Group, School of Life Sciences, Queen's Medical Centre, University of Nottingham, Nottingham NG7 2UH, UK; licimedeiros@gmail.com (L.F.M.); Maria.Arruda@nottingham.ac.uk (M.A.A.); nicholas.holliday@nottingham.ac.uk (N.D.H.)
[5] Farmanguinhos, Fiocruz, Brazilian Ministry of Health, Rio de Janeiro 22775-903, Brazil
* Correspondence: melpg@icb.ufmg.br; Tel.: +55-31-3409-2638

Received: 22 December 2017; Accepted: 12 January 2018; Published: 15 January 2018

Abstract: The synthetic peptide PnPP-19 comprehends 19 amino acid residues and it represents part of the primary structure of the toxin δ-CNTX-Pn1c (PnTx2-6), isolated from the venom of the spider *Phoneutria nigriventer*. Behavioural tests suggest that PnPP-19 induces antinociception by activation of CB1, μ and δ opioid receptors. Since the peripheral and central antinociception induced by PnPP-19 involves opioid activation, the aim of this work was to identify whether this synthetic peptide could directly activate opioid receptors and investigate the subtype selectivity for μ-, δ- and/or κ-opioid receptors. Furthermore, we also studied the modulation of calcium influx driven by PnPP-19 in dorsal root ganglion neurons, and analyzed whether this modulation was opioid-mediated. PnPP-19 selectively activates μ-opioid receptors inducing indirectly inhibition of calcium channels and hereby impairing calcium influx in dorsal root ganglion (DRG) neurons. Interestingly, notwithstanding the activation of opioid receptors, PnPP-19 does not induce β-arrestin2 recruitment. PnPP-19 is the first spider toxin derivative that, among opioid receptors, selectively activates μ-opioid receptors. The lack of β-arrestin2 recruitment highlights its potential for the design of new improved opioid agonists.

Keywords: *Phoneutria nigriventer*; opioid receptor; spider toxin; antinociception

Key Contribution: The spider toxin derivative PnPP-19 activates μ-opioid receptors and blocks calcium channels in DRG neurons. Our data highlights the possible use of PnPP-19 for the development of new drug candidates for pain treatment.

---

1. Introduction

The venom of *Phoneutria nigriventer* has been the focus of intensive research in recent years since it is of interest for discovering novel pharmaceutical bioactive peptides. This venom has a potent neurotoxic effect and many of its toxins have been already isolated and studied in detail [1]. One of the best characterized toxins, δ-CNTX-Pn1c, also known as PnTx2-6 [2], exerts interesting pharmacological effects and it was originally studied as a modulator of voltage-gated sodium channels [1]. Recently, this toxin has been studied as a potentiator of erectile function. δ-CNTX-Pn1c improves erectile function of normotensive and DOCA-salt hypertensive rats [3] and it also ameliorates the erectile function of rats with bilateral cavernous nerve crush injury [4].

The synthetic peptide PnPP-19 comprehends 19 amino acid residues and it represents part of the primary structure of the spider toxin δ-CNTX-Pn1c. This peptide has been suggested to be a promising drug candidate for the treatment of both erectile dysfunction and pain. Through histopathological experiments [5], it was shown that PnPP-19 does not induce any sign of toxicity in different tissues (brain, heart, lung, liver and kidney) and it no longer modulates Nav channels [5]. Furthermore, it does not cause death or hypersensitivity reactions and it induces only low immunogenicity in mice. However, similar to the native toxin δ-CNTX-Pn1c, PnPP-19 does potentiate erectile function. The exact molecular target through which PnPP-19 improves erectile function still awaits elucidation [5]. Regarding the pain pathway, PnPP-19 induces both peripheral and central antinociception. This antinociceptive effect elicited by the peptide seems to involve the activation of opioid and cannabinoid receptors along with the activation of the NO/cGMP/K_{ATP} pathway [6–8].

Millions of people suffer from acute or chronic pain every year, which makes pain a serious global public health problem. Chronic pain, for instance, may cause an enormous socioeconomic impact with associated costs in treatment and reduced levels of productivity [9]. Nowadays, there is an urge for the development of novel potent and more selective analgesic drugs that elicit less undesirable side effects [10].

The opioid receptors belong to the superfamily of GPCRs and they are coupled to G_i/G_o proteins. The activation of these receptors may contribute to cellular hyperpolarization, and might impair neurotransmitters release, by suppressing calcium influx and stimulating potassium channels. The activation of the three different opioid receptor subtypes (μ-, δ- and κ-) might inhibit different calcium channels in various mammalian tissues [11]. Therefore, measuring calcium currents could be a complementary way for verifying opioid activation. In addition, the direct inhibition of calcium channels by exogenous substances may also induce per se antinociception. This is the case, for example, of two antinociceptive *P. nigriventer* toxins, PnTx3-3 and PnTx3-6 [12,13], and the well-known Food and Drug Administration (FDA) approved analgesic drug Prialt® (Ziconotide) [14]. Regarding the potassium channels, it has been shown that opioid receptor activation leads to opening of different potassium channels, among which are inward rectifying potassium channels (GIRK) [11]. As such, measuring the alteration of potassium flux through the cell membrane might also be an alternative way of investigating opioid receptor activation.

Since the peripheral and central antinociception induced by PnPP-19 involves opioid activation [6,8], the aim of this work was to identify whether this synthetic peptide could directly activate opioid receptors and investigate the possible subtype selectivity for μ-, δ- and/or κ-opioid receptors co-expressed with GIRK1/GIRK2 and RGS4. Furthermore, we also studied the modulation of calcium influx driven by PnPP-19 in dorsal root ganglion (DRG) neurons, and analyzed whether this modulation was opioid-mediated. Our data show that PnPP-19 may selectively activate μ-opioid receptors, however with low potency. Interestingly, activation of opioid receptors induced by the PnPP-19 does not stimulate the recruitment of β-arrestin2. However, it does induce indirectly inhibition of calcium channels and, consequently, impairs calcium influx in DRG neurons.

2. Results

2.1. Electrophysiological Characterization of Direct Activation of Opioid Receptors Induced by PnPP-19 Using Two-Electrode Voltage-Clamp

Each receptor was individually co-expressed with GIRK1/GIRK2 channels and RGS4, mimicking the native neuronal G-protein-mediated pathway of K$^+$ channel activation. We used the two-microelectrode voltage-clamp technique to measure the opioid receptor-activated GIRK1/GIRK2 channel response as the increase of the inward K$^+$ current at -70 mV, evoked by the application of increasing concentrations of opioid ligands. The potency of PnPP-19 on human μ-opioid receptor (hMOR), human κ-opioid receptor (hKOR) and human δ-opioid receptor (hDOR) was investigated (Figure 1). Concentrations up to 10 μM could not evoke currents from oocytes expressing hKOR or hDOR. However, PnPP-19 could activate hMOR, albeit with low potency. Oocytes co-expressing only GIRK1/GIRK2 and RGS4 were used as a control to verify that PnPP-19 indeed interact with the opioid receptor and not the inward rectifying potassium channels. No activity was seen when PnPP-19 was applied to oocytes expressing only GIRK channels and RGS4 (Figure S1).

To confirm the interaction of PnPP-19 with the μ-opioid receptor, the activity of PnPP-19 in the presence of naloxone was investigated (Figure 2). First, expression of GIRK1/GIRK2/RGS4/hMOR was verified by applying 1 μM morphine as a control. Next, 1 μM PnPP-19 was applied as reference current. Application of 1 μM of the well characterized opioid antagonist naloxone was subsequently followed by another pulse of 1 μM PnPP-19. No PnPP-19 evoked current could be observed in the presence of naloxone (Figure 2). A similar experiment, investigating the activation of hMOR by 1 μM morphine in the presence of naloxone was performed as a control (Figure S2).

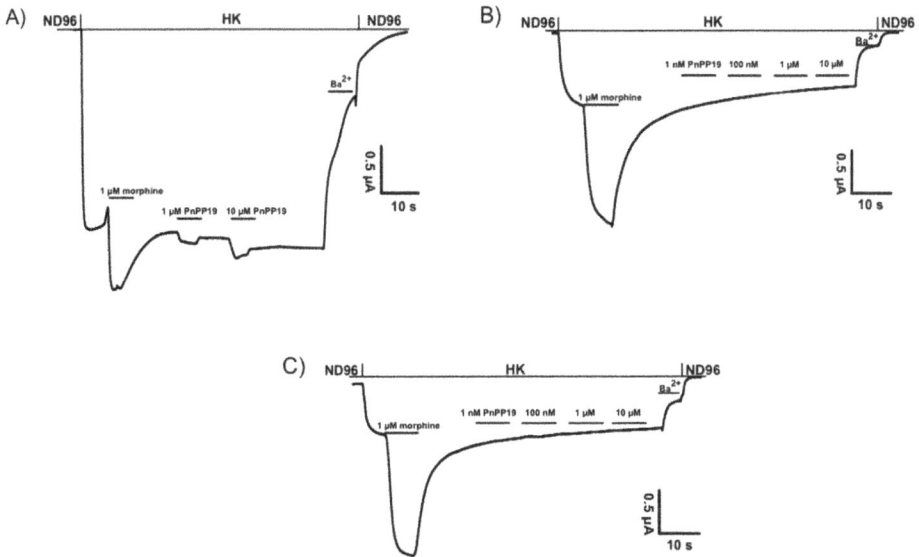

Figure 1. (**A**) Shows representative current traces of agonist-gated currents evoked from oocytes expressing human μ-opioid receptor (hMOR) by 1 μM morphine and 1 or 10 μM PnPP-19. PnPP-19 could not activate human δ-opioid receptor (hDOR) (**B**) or human κ-opioid receptor (hKOR) (**C**) up to a concentration of 10 μM.

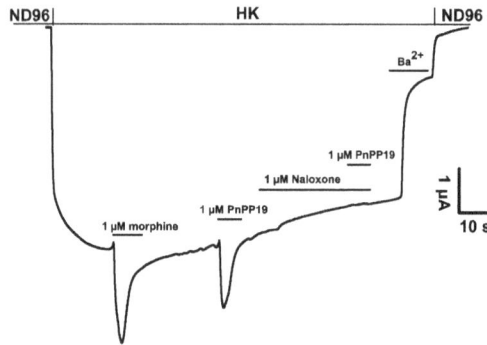

Figure 2. Representative current traces evoked from *X. laevis* oocytes co-expressing GIRK1/GIRK2 channels and RGS4 with hMOR. In addition, 1 µM naloxone inhibits the agonistic activity of PnPP-19.

2.2. Inhibition of Calcium Current Induced by PnPP-19

Intact neurons of rat dorsal root ganglia (DRG) were used for whole-cell patch-clamp recordings. PnPP-19, morphine or naloxone were added separately to the bath solution to give a final concentration of 1 µM, 1 µM and 10 µM, respectively. Figure 3 shows that PnPP-19 induced a reduction of the calcium evoked current density, with an efficacy comparable to morphine (MOR). Therefore, we could demonstrate that incubation of DRG neurons with the opioid agonist morphine (1 µM) or with PnPP-19 (1 µM) induced inhibition of calcium channels. Furthermore, pre-incubation of the cells with naloxone (10 µM) completely blocks the activity of PnPP-19, suggesting that the inhibition of calcium channels induced by the synthetic peptide is through activation of opioid receptors. Application of naloxone alone has no significant effect on the current density.

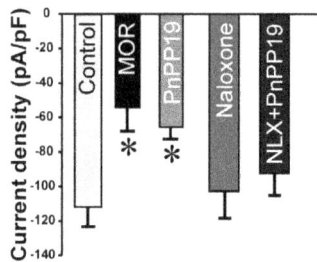

Figure 3. Effect of PnPP-19 and morphine on calcium current density evoked in dorsal root ganglion (DRG) neurons. Calcium currents were evoked by depolarizing pulses to 10 mV (200 ms) from a holding potential of −90 mV in DRG neurons incubated with 1 µM morphine, 1 µM PnPP-19 or 10 µM naloxone. Control group: cells incubated only with external/bath solution. Group NLX + PnPP-19: cells were previously incubated for 30 min with 10 µM naloxone and then PnPP-19 was added reaching a final concentration of 1 µM. MOR: morphine and NLX: naloxone. Data shown are the means ± SEM (n = 8 cells, 5 animals). * $p < 0.05$ compared with control (one-way ANOVA + Bonferroni's test).

2.3. Inhibition of Calcium Influx Induced by PnPP-19

DRG neurons were isolated and the calcium influx was evaluated using fluorescence microscopy. Cell stimulation with KCl (30 mM) induced calcium influx and consequently an increase of intracellular calcium concentration. The perfusion of the cells for 5 min with buffer did not alter the profile of calcium influx induced by KCl during the second course of stimulation if compared with the first set of stimuli (Figures 4 and 5). On the other hand, incubation of DRG neurons with PnPP-19 (1 μM) for 5 min induced a decrease of approximately 20% of calcium influx during the second course of stimulation with KCl. In addition, PnPP-19 (1 μM) did not cause any alteration of intracellular calcium concentration when the cells were not stimulated; therefore, no change of calcium concentration was observed during the 5 min period of cell perfusion with PnPP-19 (Figure 4).

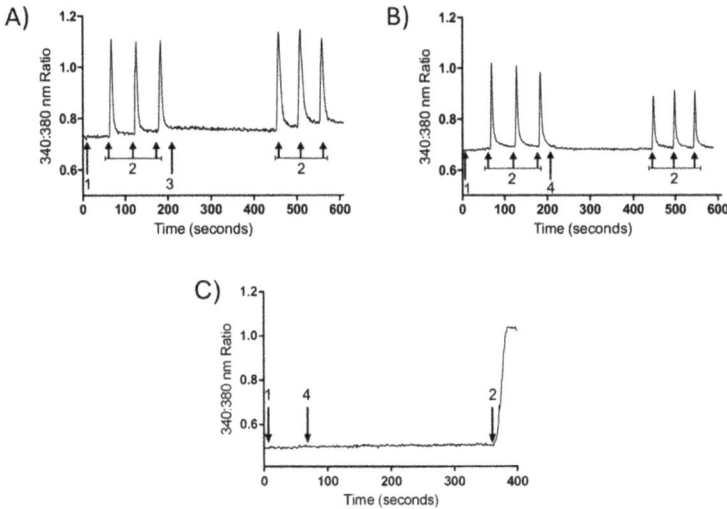

Figure 4. Representative trace showing calcium influx (changes in 340:380 nm ratios) in a single DRG neuron stimulated by KCl before and after 5 min incubation with buffer or PnPP-19. A perfusion system was used to incubate DRG neurons with buffer for 1 min (1) followed by consecutive KCl (30 mM) stimulations (2). After that, cells were perfused with buffer (3) or PnPP-19 (1 μM) (4) for 5 min, and again depolarized by KCl (30 mM) at three different time points (2). Control cells incubated only with buffer after first set of KCl stimulations do not show a significant difference in calcium influx during the second set of stimulations (**A**); However, cells incubated with PnPP-19 display a decrease in calcium influx during the second set of KCl stimulations (**B**); As a negative control, PnPP-19 does not influence calcium influx on its own (**C**).

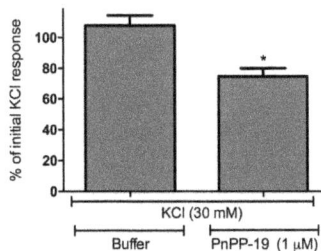

Figure 5. Effect of pre-incubation with PnPP-19 (1 μM) on KCl-evoked (30 mM) responses. The bars represent the percentage of the maximum amplitude response during the second set of KCl stimuli corresponding to the initial KCl stimulations (100%). The peak of response in each situation was calculated, and the amplitude was assessed by diminishing this value of the baseline. The baseline corresponds to the pre-incubation of cells with buffer, before any KCl stimulation. Data shown are the means ± SEM ($n = 5$). * $p < 0.05$ compared with KCl (30 mM) + Buffer (two-tailed *t*-test).

2.4. β-Arrestin2 Recruitment Induced by DAMGO and PnPP-19

Here, we intended to evaluate whether stimulation of HEK293T cells, coexpressing μ-opioid receptor-Yc and β-arrestin2-Yn, by PnPP-19 or the selective μ-opioid agonist [D-Ala^2, MePhe^4, Gly-ol^5] encephalin (DAMGO) would induce recruitment of β-arrestin2 by activating μ-opioid receptors. Incubation of the cells with DAMGO could clearly induce β-arrestin2 recruitment (Figure 6). However, different concentrations of PnPP-19 could not induce any μ-opioid receptor-β-arrestin2 association (Figure 6). In addition, pre-incubation of the cells with 10 μM of PnPP-19 was not able to prevent the binding and activation of μ-opioid receptors by the agonist DAMGO (Figure 6).

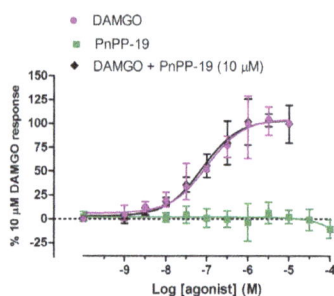

Figure 6. Recruitment of β-arrestin2 by activation of μ-opioid receptors. Stably transfected HEK293 cells coexpressing μ-opioid receptors and β-arrestin2 were pretreated for 60 min with DAMGO or PnPP-19 at the indicated concentrations. The group "DAMGO + PnPP-19" represents prior incubation of the cells with 10 μM of PnPP-19 for 30 min. β-arrestin2 recruitment was quantified by high content imaging complementation assay as described in Materials and Methods ($n = 5$).

3. Discussion

Previous literature data, obtained using behavioral tests, suggested that the peripheral and central antinociceptive effect induced by PnPP-19 is partially because of μ- and δ-opioid receptors activation [6–8]. Therefore, we verified whether this synthetic peptide could indeed directly bind and activate the different isoforms of opioid receptors (μ-, δ- and κ-). Moreover, since it was already described that activation of opioid receptors suppresses calcium influx through inhibition of different voltage-gated calcium channels, we also investigated the modulation of calcium influx induced by PnPP-19 in DRG neurons. Our data demonstrates that, among the μ-, δ- and κ-opioid receptor subtypes, PnPP-19 may selectively activate, with relatively low potency, the μ-opioid receptor subtype. Remarkably, it seems that the peptide does not induce the recruitment of β-arrestin2 by activating opioid receptors and, most likely, PnPP-19 binds to a different binding site of the opioid receptor than DAMGO (a selective opioid agonist). In addition, PnPP-19 induced an inhibition of calcium channels, very likely through activation of opioid receptors, in the whole-cell patch-clamp assay; it also diminished the calcium influx observed by fluorescence microscopy.

Among all the toxins isolated from the venom of *Phoneutria nigriventer* and its derivatives, only two of them are able to elicit antinociception via activation of opioid receptors. The antinociceptive effect of the toxin PnTx4(6-1), also known as δ-Ctenitoxin-Pn1a [2], is partially blocked when selective antagonists of both μ- and δ-opioid receptors are administered [15]. Likewise, antinociception of PnPP-19, a δ-CNTX-Pn1c derivative [5], occurs also through activation of those very same opioid receptors [6,8]. On the other hand, it was never investigated whether these peptides may directly bind and activate opioid receptors. Our present data show that PnPP-19 might selectively activate, with low potency, only the μ-opioid receptor subtype. However, through behavioral experiments, it was shown that the antinociception induced by PnPP-19 also involves activation of δ-opioid receptors. Here, we demonstrated that PnPP-19 is incapable of activating directly δ-opioid receptors. Therefore, activation of this specific subtype of receptor in vivo may occur via an indirectly pathway, as previously suggested by Freitas and collaborators [6].

PnPP-19 is the first synthetic peptide, derived from a spider toxin, proven to act directly on opioid receptors, and more specifically, on μ-opioid receptor subtype. Novel ligands of the μ-opioid receptor are of clinical and social importance since the common used analgesic drugs, such as morphine, fentanyl and oxymorphone, elicit both their beneficial pharmacological effect and undesirable side effects through activation of opioid receptors [16]. One of the very serious and life-threatening conditions developed following the use of the usual opioid agonist medicines is respiratory paralysis [16,17]. It has been demonstrated that the induction of respiratory paralysis, as well as other side effects, after the use of opioids may be linked with the recruitment of the β-arrestin pathway, which is stimulated downstream following activation of μ-opioid receptor [18–22]. Since opioid receptors are still one of the most relevant targets for pain treatment, great effort is being put in the development of new opioid agonists that elicit fewer negative side effects [22,23]. In this way, the lack of β-arrestin2 recruitment by PnPP-19 underlines the potential of this peptide as a possible lead compound in the development of improved opioid agonists. Recently, a very selective and potent μ-opioid agonist was developed and named PZM21. Despite its great potency and selectiveness against μ-opioid receptors, administration of PZM21 induced minimal β-arrestin2 recruitment. Therefore, the use of PZM21 induced a long-lasting analgesia along with decreased respiratory depression and constipation when compared to morphine [10]. For this reason, studies concerning the exact mechanism of action of PnPP-19 in the pain pathway are of interest since PnPP-19 showed no induction of β-arrestin2 recruitment in cell culture. It thus seems that the peptide has a site of interaction different from the DAMGO binding site, since in our experiments the presence of PnPP-19 has no influence on DAMGO-induced β-arrestin2 recruitment. However, one other hypothesis for the lack of β-arrestin2 recruitment by PnPP-19 could be the low potency of which PnPP-19 might bind to μ-opioid receptors. Therefore, further investigation is required in order to elucidate the exact mechanism of PnPP-19 interaction with the opioid receptors. Moreover, a better characterization of the target of this synthetic peptide in erectile function is required in order to develop a PnPP-19 derived drug without unwanted side effects.

The interaction between opioid receptors and ion channels has been a subject of much interest during decades. Various studies suggest that activation of opioid receptors causes hyperpolarization of the cell and consequently prevents neurotransmitter release by inducing an inhibition of calcium channels [11,24,25] and activation potassium channels [11,26–28]. According to in vitro studies, incubation of a selective agonist of μ-opioid receptors with HEK293 cells co-expressing μ-opioid receptors together with voltage-gated N-type calcium channels (Cav2.2) or R-type (Cav2.3) channels induced an inhibition of both calcium channels tested [29]. Moreover, experiments, conducted with primary culture of vestibular afferent neurons and DRG neurons, suggest that selective stimulation of μ-opioid receptors may inhibit T-, L- and N-type calcium channels supposedly through activation of a $G\alpha_{i/o}$ protein [30–32]. Accordingly, our results demonstrate that PnPP-19 inhibits calcium influx in DRG neurons and that this inhibition is suppressed by the unspecific opioid antagonist naloxone. These data show that inhibition of calcium influx induced by PnPP-19 is mediated by activation of opioid receptors. Recently, it was demonstrated that DAMGO (selective μ-opioid agonist) induces inhibition of calcium influx and action potential-evoked Ca^{2+} fluorescent transients in individual peripheral nociceptive fiber free nerve endings from trigeminal ganglion. The authors have shown that activation of "big conductance" Ca^{2+}-activated K^+ channels (BK_{Ca}) mediates this inhibition of calcium influx induced by DAMGO. Furthermore, the activation of this subtype of potassium channel plays a major role on μ-opioid induced antinociception in a behavioral test for trigeminal nociception. Therefore, it is likely that PnPP-19 might also modulate potassium channels, since this synthetic peptide may act as an opioid agonist. However, further experiments are needed to investigate whether PnPP-19 indeed interferes with potassium channel activity.

In conclusion, the data we present here shows for the first time a spider toxin derivative that may act as a selective μ-opioid agonist. PnPP-19 directly binds and activates, albeit with low potency, only the μ-opioid receptor subtype. In DRG neurons, the activation of μ-opioid receptors induced by PnPP-19 generates an inhibition of calcium channels, consequently reducing or even eliminating calcium influx. This modulation of calcium channels appears to follow activation of opioid receptors, confirming once again the role of PnPP-19 as an opioid agonist. Interestingly, notwithstanding the activation of opioid receptors, PnPP-19 does not induce β-arrestin2 recruitment. This could be due its low potency; however, it may also be a consequence of a differential opioid activation mechanism in which β-arrestin2 recruitment is not stimulated. Further studies with PnPP-19 could lead to the development of new and more potent opioid agonists that in turn could elicit antinociception with possibly less side effects by not inducing recruitment of β-arrestin2.

4. Materials and Methods

4.1. Expression of Voltage-Gated Potassium Channels in Xenopus laevis Oocytes

Xenopus laevis oocytes were isolated as previously described [33]. Oocytes were co-injected with 0.5 ng/50 nL of GIRK1, 0.5 ng/50 nL of GIRK2, and 10 ng/50 nL of RGS4 cRNA, with the addition of 10 ng/50 nL of either hMOR, hKOR, hDOR, hMORW318L, or hMORW318Y/H319Y cRNA. Injected oocytes were maintained in ND-96 solution (composition: 2 mM KCl, 96 mM NaCl, 1 mM MgCl$_2$, 1.8 mM CaCl$_2$, 5 mM HEPES, pH 7.5) supplemented with 50 μg/mL of gentamicin sulfate.

4.2. Electrophysiological Recordings: Xenopus laevis Oocytes

Whole-cell currents from oocytes were recorded from 1 to 2 days after injection using the two-microelectrode voltage-clamp technique. Resistances of voltage and current electrodes were kept between 0.7 and 1.5 MΩ and were filled with 3 M KCl. Currents were filtered at 20 Hz, using a 4-pole low-pass Bessel filter. To eliminate the effect of the voltage drop across the bath-grounding electrode, the bath potential was actively controlled. All experiments were performed at room temperature [19–23]. At the start and the end of each experiment, oocytes were superfused with low-potassium (ND-96) solution (composition: 2 mM KCl, 96 mM NaCl, 1 mM MgCl$_2$, 1.8 mM CaCl$_2$, 5 mM HEPES, pH 7.5). During application of increasing concentrations of ligands, oocytes were superfused with high-potassium (HK) solution (composition: 96 mM KCl, 2 mM NaCl, 1 mM MgCl$_2$, 1.8 mM CaCl$_2$, 5 mM HEPES, pH 7.5). In HK solution, the K$^+$ equilibrium potential is close to 0 mV and enables K$^+$ inward currents to flow through inwardly rectifying K$^+$ channels at negative holding potentials. A gravity-controlled fast perfusion system was used to ensure rapid solution exchanges. Analysis of un-injected cells ($n = 3$), under the same experimental conditions as injected oocytes, revealed an endogenous current that amounted maximally 1% as compared with the current measured in injected oocytes. Application of opioid ligands did not evoke an increase of the conductance in un-injected oocytes. In each experiment, oocytes were clamped at a holding potential of −70 mV and super-fused with ND-96 solution. Next, the super-fusion solution was switched from ND-96 to HK solution, after which increasing concentrations of morphine or peptide were applied. Each concentration was applied for as long as needed to achieve a steady-state GIRK1/GIRK2 current activation. Each ligand concentration was washed out by super-fusing it with an HK solution. During this washout period, the channels return to the control current level as a result of a deactivation process that is accelerated dramatically in the presence of RGS4, as previously described [34]. At the end of each experiment, the oocyte was super-fused with HK solution containing 300 μM BaCl$_2$, causing a blockage of the net GIRK1/2-gated inward current. Finally, the super-fusion was switched back to ND-96 solution to confirm complete reversibility.

4.3. Data Analysis of Two-Microelectrode Voltage-Clamp

The pCLAMP program (Axon Instruments, pCLAMP, Sunnyvale CA, USA) was used for data acquisition, and data files were directly imported, analyzed, and visualized with a custom-made add-in for Microsoft Excel (Redmond, WA, USA). The percentage-activated current was calculated using the equation: percentage activation = activated current amplitude control current amplitude $\times 100 - 100$ and 0% was taken as the control current level. Current percentages were then used for the calculation of concentration–response curves, using the Hill equation $I = I_{max}/[1 + (EC_{50}/A)^{n_H}]$, where I represents the current percentage, I_{max} the maximal current percentage, EC_{50} the concentration of the agonist that evokes the half-maximal response, A the concentration of agonist, and n_H the Hill coefficient. Averaged data are indicated as means $\pm$ SEM and were calculated using n experiments, where n indicates the number of oocytes tested. For each experiment, the number of oocytes tested was at least 6 ($n > 6$) For each experiment, averaged current percentages were normalized to 100%, and an averaged concentration–response curve was drawn using the average EC_{50} values and Hill coefficients of n experiments. Statistical analysis of differences between groups was carried out with Student's *t*-test, and a probability of 0.05 was taken as the level of statistical significance.

4.4. DRG Culture

DRGs were isolated from adult Wistar rats (200 ± 300 g) and neurons cultured as described by Lindsay (1988) [35] with minor modifications. The neurons were isolated and washed by gravity in phosphate-buffered saline (PBS). The cells were than incubated with collagenase type IV (sigma, St. Louis, MO, USA) solution (5 mL of Dulbecco's modified Eagle's medium—DMEM; 10% *v/v* fetal bovine serum; penicillin 200 units/mL—streptomycin 200 µg/mL; 12.5 mg of collagenase type IV) for 90 min at 37 °C. After that, ganglions were washed 3 times by gravity in PBS and trypsin solution (2500–6000 BAEE U/ML, sigma, St. Louis, MO, USA) was added. In order to dissociate the DRG neurons, the ganglions were taken up and down with the use of a fine tipped transfer pipette. Cells were then incubated with the trypsin solution for 10 min at 37 °C. After the incubation period, 1 mL of bovine serum albumin (BSA) solution (16% *v/v* in PBS) was added and cells were more firmly dissociated. The cell suspension was added on the top of 3 mL BSA solution and centrifuged at $500 \times g$ for 6 min. The supernatant was discarded and the pellet was resuspended in complete media (DMEM media; 10% *v/v* fetal bovine serum; 1% *v/v* penicillin/streptomycin; 0.1% *v/v* NGF). Cells were plated on poly-L-lysine and laminin coated cover slips and incubated at 37 °C with 5% CO_2 in a humidified incubator. The study was approved by the local Ethics Committee on Animal Experimentation (CETEA) of UFMG (Protocol number: 233/2013).

4.5. Whole-Cell Voltage-Clamp

DRG neurons were used for the measurements after 48 h of cell culture. The calcium current recordings were obtained by using the Patch Clamp amplifiers type EPC-9/EPC-10 (HEKA Instruments, Lambrecht/Pfalz, Germany) and the PULSE/PATCHMASTER data acquisition program (HEKA Instruments, Lambrecht/Pfalz, Germany) adjusted for the Whole Cell Voltage-Clamp configuration. Low resistance patch electrodes (3–4 MΩ) were filled with solution containing (in mM): 130 CsCl, 2.5 $MgCl_2$, 10 HEPES, 5 EGTA, 3 Na_2-ATP and 0.5 Li_3-GTP, pH 7.4 adjusted with 1 M CsOH. The external/bath solution contained (in mM): 125 CsCl, 10 $BaCl_2$, 1 $MgCl_2$, 10 HEPES and 60 Glucose, pH 7.4 adjusted with 1 M CsOH. An Ag-AgCl electrode was used as reference. The recordings were filtered with a Bessel low-pass filter set at 2.9 kHz and digitalized at a 10 kHz rate (100 µs interval) through an AD/DA interface (ITC 1600). Capacitive currents were electronically compensated and a P/4 protocol was used to correct the linear leakage current and to subtract residual capacity (BEZANILLA, ARMSTRONG, 1977) [36]. After establishing the Whole Cell configuration, the calcium current was evoked from negative holding potential of -90 mV to 10 mV (200 ms). Once the calcium current showed stable amplitude values, PnPP-19, morphine or naloxone were added separately to the bath solution to give a final concentration of 1 µM, 1 µM and 10 µM, respectively. To test whether the effect of PnPP-19 was through opioid receptors, cells were prior incubated with 10 µM of naloxone for 30 min. The experiments were performed on 35 mm diameter acrylic Petri dishes using inverted microscope (Axiovert 20, Carl Zeiss, Jena, Germany or Nikon TMF-100, Nikon, Chiyoda-Ku, Japan).

4.6. Calcium Imaging

The experiments were performed after 24 h of DRG neurons dissociation. On the day of the experiments, cells were incubated with Fura 2-AM (5 mM, 30 min, 37 °C). Intracellular Ca^{2+} concentrations ($[Ca^{2+}]i$) in individual neurons were estimated as the ratios of peak fluorescence intensities (measured at 500 nm) at excitation wavelengths of 340 and 380 nm, respectively (Bundey & Kendall, 1999) [37], using an Improvision imaging system. DRG neurons were superfused (2 mL min^{-1}) with buffer (NaCl 145 mM; KCl 5 mM; $CaCl_2$ 2 mM; $MgSO_4$ 1 mM; HEPES 10 mM; glucose 10 mM) for 1 min followed by three consecutive KCl (30 mM) stimulations. After that, cells were perfused with buffer (control group) or PnPP-19 (1 μM) dissolved in buffer for 5 min, and again depolarized by KCl (30 mM) at three different time points. Representative traces of calcium influx in a single DRG neuron are shown. Results are presented as means ± SEM and indicate the percentage of calcium influx related to the peak of calcium influx during the first course of activation with KCl (100%). Statistical analyses were carried out using GraphPrism software (version 7.0a, GraphPad Software, La Jolla, CA, USA, 2016). Our data were distributed normally and analyzed statistically by two-tailed *t*-test. Probabilities less than 5% ($p < 0.05$) were considered to be statistically significant.

4.7. Beta-Arrestin2 Recruitment

HEK293T were cultured in DMEM (Sigma-Aldrich) supplemented with 10% *v/v* fetal bovine serum. These cells were coexpressing μ-opioid receptor-Yc and β-arrestin2-Yn (Yc and Yn are complementary fragments of yellow fluorescent protein-YFP). To analyze whether activation of μ-opioid receptor would induce recruitment of β-arrestin2, the Bimolecular fluorescence complementation (BiFC) based detection of μ-opioid receptor-β-arrestin2 association was conducted. The cells were seeded at 33,000 cells/well onto poly (D-lysine)-coated Greiner 655,090 imaging plates. Plates were kept in a humidified incubator at 37 °C filled with 5% CO_2 for 24 h. HEK293T were stimulated with the selective opioid agonist DAMGO (Tocris, Minneapolis, MN, USA) or the synthetic peptide PnPP-19 in HEPES-buffered saline solution (HBSS) including 0.1% *v/v* BSA (10^{-10} M–10^{-4} M) for 60 min at 37 °C. In the experiment where we investigated whether PnPP-19 could impair the binding of DAMGO to μ-opioid receptors, cells were preincubated with PnPP-19 10 μM (30 min, 37 °C). After that, cells were fixated with 3% paraformaldehyde in PBS for 10 min at room temperature. Then, cells were washed once with PBS and the cell nuclei were stained for 15 min with H33342 (2 μg/mL in PBS, Sigma, St. Louis, MO, USA). H33342 was then removed by a final PBS wash. Images (4 central sites/well) were acquired automatically on the IX Ultra confocal plate reader, using 405 nm/488 nm laser lines for H33342 and complemented YFP excitation, respectively. Data was analyzed by the use of MetaXpress software (version 5.3, Sunnyvale, CA, USA, 2013) as described by Liu and co-authors [38] and normalized by 10 μM of DAMGO (100%).

Supplementary Materials: The following are available online at www.mdpi.com/2072-6651/10/1/43/s1, Figure S1: Representative current traces evoked from *X. laevis* oocytes co-expressing GIRK1/GIRK2 channels and RGS4. PnPP-19 does not interact with GIRK channels; Figure S2: Representative current traces evoked from *X. laevis* oocytes co-expressing GIRK1/GIRK2 channels and RGS4 with hMOR.

Acknowledgments: We would like to thank Nicholas D. Holliday, from the University of Nottingham (Nottingham, UK), for kindly donating the HEK293T cells coexpressing μ-opioid receptor-Yc and β-arrestin2-Yn. We also would like to thank Andrew Cooper, from the University of Nottingham (Nottingham, UK), for all the help with the calcium imaging studies. Fellowships and grants were awarded by the Brazilian Agencies FAPEMIG (Fundação de Amparo à Pesquisa do Estado de Minas Gerais), CAPES (Coordenação de Aperfeiçoamento de Pessoal de Nível Superior)-University of Nottingham Programme in Drug Discovery and CNPq (Conselho Nacional de Desenvolvimento Científico e Tecnológico).

Author Contributions: Ana C. N. Freitas, Jader Cruz, Nicholas D. Holliday, Jan Tytgat, Gareth Hathway and Maria E. de Lima conceived and designed the experiments; Ana C. N. Freitas, Steve Peigneur, Flávio H. P. Macedo, José E. Menezes-Filho, Paul Millns, Liciane F. Medeiros performed the experiments; Ana C. N. Freitas, Steve Peigneur, Maria A. Arruda, Jader Cruz analyzed the data; Jader Cruz, Nicholas D. Holliday, Jan Tytgat, Gareth Hathway, Maria E. de Lima contributed reagents/materials/analysis tools; Ana C. N. Freitas and Steve Peigneur wrote the paper.

Conflicts of Interest: The authors declare no conflict of interest.

References

1. De Lima, M.E.; Figueiredo, S.G.; Matavel, A.; Nunes, K.P.; da Silva, C.N.; de Marco Almeida, F.; Ribeiro, M.; Diniz, V.; do Cordeiro, M.N.; Stankiewicz, M.; et al. *Phoneutria nigriventer Venom and Toxins: A Review*; Springer: Amsterdam, The Netherlands, 2015; pp. 1–24.
2. King, G.F.; Gentz, M.C.; Escoubas, P.; Nicholson, G.M. A rational nomenclature for naming peptide toxins from spiders and other venomous animals. *Toxicon* **2008**, *52*, 264–276. [CrossRef] [PubMed]
3. Nunes, K.P.; Costa-Goncalves, A.; Lanza, L.F.; Côrtes, S.D.F.; Cordeiro, M.D.N.; Richardson, M.; Pimentad, A.M.C.; Webbe, R.C.; Leite, R.; De Lima, M.E. Tx2-6 toxin of the *Phoneutria nigriventer* spider potentiates rat erectile function. *Toxicon* **2008**, *51*, 1197–1206. [CrossRef] [PubMed]
4. Jung, A.R.; Choi, Y.S.; Piao, S.; Park, Y.H.; Shrestha, K.R.; Jeon, S.H.; Hong, S.H.; Kim, S.W.; Hwang, T.K.; Kim, K.H.; et al. The effect of PnTx2-6 protein from *Phoneutria nigriventer* spider toxin on improvement of erectile dysfunction in a rat model of cavernous nerve injury. *Urology* **2014**, *84*, 730. [CrossRef] [PubMed]
5. Silva, C.N.; Nunes, K.P.; Torres, F.S.; Cassoli, J.S.; Santos, D.M.; Almeida, F.D.M.; Matavel, A.; Cruza, J.S.; Santos-Miranda, A.; Nunes, A.D.C.; et al. PnPP-19, a synthetic and non toxic peptide designed from a *Phoneutria nigriventer* toxin, potentiates erectile function via NO/cGMP. *J. Urol.* **2015**, *194*, 1481–1490. [CrossRef] [PubMed]
6. Freitas, A.C.; Freitas, A.C.N.; Pacheco, D.F.; Machado, M.F.M.; Carmona, A.K.; Duarte, I.D.G.; Lima, M.E. PnPP-19, a spider toxin peptide, induces peripheral antinociception through opioid and cannabinoid receptors and inhibition of neutral endopeptidase. *Br. J. Pharmacol.* **2016**, *173*, 1491–1501. [CrossRef] [PubMed]
7. Freitas, A.C.; Silva, G.C.; Pacheco, D.F.; Pimenta, A.M.C.; Lemos, V.S.; Duarte, I.D.G.; de Lima, M.E. The synthetic peptide PnPP-19 induces peripheral antinociception via activation of NO/cGMP/K$_{ATP}$ pathway: Role of eNOS and nNOS. *Nitric Oxide* **2017**, *64*, 31–38. [CrossRef] [PubMed]
8. Da Fonseca Pacheco, D.; Freitas, A.C.N.; Pimenta, A.M.C.; Duarte, I.D.G.; de Lima, M.E. A spider derived peptide, PnPP-19, induces central antinociception mediated by opioid and cannabinoid systems. *J. Venom. Anim. Toxins Incl. Trop. Dis.* **2016**, *22*, 34. [CrossRef] [PubMed]
9. Phillips, C.J. The Cost and Burden of Chronic Pain. *Rev. Pain* **2009**, *3*, 2–5. [CrossRef] [PubMed]
10. Manglik, A.; Lin, H.; Aryal, D.K.; McCorvy, J.D.; Dengler, D.; Corder, G.; Levit, A.; Kling, R.C.; Bernat, V.; Hübner, H.; et al. Structure-based discovery of opioid analgesics with reduced side effects. *Nature* **2016**, *537*, 185–190. [CrossRef] [PubMed]
11. Law, P.Y.; Wong, Y.H.; Loh, H.H. Molecular mechanisms and regulation of opioid receptor signaling. *Annu. Rev. Pharmacol. Toxicol.* **2000**, *40*, 389–430. [CrossRef] [PubMed]
12. Souza, A.H.; Ferreira, J.; do Nascimento Cordeiro, M.; Vieira, L.B.; De Castro, C.J.; Trevisan, G.; Reis, H.; Souza, I.A.; Richardson, M.; Prado, M.A.M.; et al. Analgesic effect in rodents of native and recombinant Phα1β toxin, a high-voltage-activated calcium channel blocker isolated from armed spider venom. *Pain* **2008**, *140*, 115–126. [CrossRef] [PubMed]
13. Dalmolin, G.D.; Silva, C.R.; Rigo, F.K.; Gomes, G.M.; do Nascimento Cordeiro, M.; Richardson, M.; Silva, M.A.R.; Prado, A.M.; Gomez, M.V.; Ferreira, J. Antinociceptive effect of Brazilian armed spider venom toxin Tx3-3 in animal models of neuropathic pain. *Pain* **2011**, *152*, 2224–2232. [CrossRef] [PubMed]
14. McGivern, J.G. Ziconotide: A review of its pharmacology and use in the treatment of pain. *Neuropsychiatr. Dis. Treat.* **2007**, *3*, 69–85. [CrossRef] [PubMed]
15. Emerich, B.L.; Ferreira, R.; Cordeiro, M.N.; Borges, M.H.; Pimenta, A.; Figueiredo, S.G.; Duarte, I.D.G.; de Lima, M.E. δ-Ctenitoxin-Pn1a, a Peptide from *Phoneutria nigriventer* Spider Venom, Shows Antinociceptive Effect Involving Opioid and Cannabinoid Systems, in Rats. *Toxins* **2016**, *8*, 106. [CrossRef] [PubMed]
16. Benyamin, R.; Rajive Adlaka, M.; Nalini Sehgal, M. Opioid complications and side effects. *Pain Phys.* **2008**, *11*, S105–S120.
17. Wilson, K.C.; Saukkonen, J.J. Acute respiratory failure from abused substances. *J. Intensive Care Med.* **2004**, *19*, 183–193. [CrossRef] [PubMed]
18. Raehal, K.M.; Walker, J.K.; Bohn, L.M. Morphine side effects in beta-arrestin 2 knockout mice. *J. Pharmacol. Exp. Ther.* **2005**, *314*, 1195–1201. [CrossRef] [PubMed]
19. Bohn, L.M.; Lefkowitz, R.J.; Caron, M.G. Differential mechanisms of morphine antinociceptive tolerance revealed in (beta)arrestin-2 knock-out mice. *J. Neurosci.* **2002**, *22*, 10494–10500. [PubMed]
20. Bohn, L.M.; Lefkowitz, R.J.; Gainetdinov, R.R.; Peppel, K.; Caron, M.G.; Lin, F.T. Enhanced morphine analgesia in mice lacking beta-arrestin 2. *Science* **1999**, *286*, 2495–2498. [CrossRef] [PubMed]

21. Bohn, L.M.; Gainetdinov, R.R.; Lin, F.T.; Lefkowitz, R.J.; Caron, M.G. μ-opioid receptor desensitization by β-arrestin-2 determines morphine tolerance but not dependence. *Nature* **2000**, *408*, 720–723. [PubMed]

22. DeWire, S.M.; Yamashita, D.S.; Rominger, D.H.; Liu, G.; Cowan, C.L.; Graczyk, T.M.; Chen, X.; Pitis, P.M.; Gotchev, D.; Yuan, C.; et al. A G protein-biased ligand at the μ-opioid receptor is potently analgesic with reduced gastrointestinal and respiratory dysfunction compared with morphine. *J. Pharmacol. Exp. Ther.* **2013**, *344*, 708–717. [CrossRef] [PubMed]

23. Soergel, D.G.; Subach, R.A.; Burnham, N.; Lark, M.W.; James, I.E.; Sadler, B.M.; Skobieranda, F.; Violin, J.D.; Webster, L.R. Biased agonism of the μ-opioid receptor by TRV130 increases analgesia and reduces on-target adverse effects versus morphine: A randomized, double-blind, placebo-controlled, crossover study in healthy volunteers. *Pain* **2014**, *155*, 1829–1835. [CrossRef] [PubMed]

24. Piros, E.T.; Prather, P.L.; Law, P.Y.; Evans, C.J.; Hales, T.G. Voltage-dependent inhibition of Ca2+ channels in GH3 cells by cloned μ- and δ-opioid receptors. *Mol. Pharmacol.* **1996**, *50*, 947–956. [PubMed]

25. Rhim, H.; Miller, R.J. Opioid receptors modulate diverse types of calcium channels in the nucleus tractus solitarius of the rat. *J. Neurosci.* **1994**, *14*, 7608–7615. [PubMed]

26. North, R.A.; Williams, J.T.; Surprenant, A.; Christie, M.J. μ and δ receptors belong to a family of receptors that are coupled to potassium channels. *Proc. Natl. Acad. Sci. USA* **1987**, *84*, 5487–5491. [CrossRef] [PubMed]

27. Schneider, S.P.; Eckert, W.A.; Light, A.R. Opioid-activated postsynaptic, inward rectifying potassium currents in whole cell recordings in substantia gelatinosa neurons. *J. Neurophysiol.* **1998**, *80*, 2954–2962. [CrossRef] [PubMed]

28. Marker, C.L.; Luján, R.; Loh, H.H.; Wickman, K. Spinal G-protein-gated potassium channels contribute in a dose-dependent manner to the analgesic effect of μ- and δ- but not kappa-opioids. *J. Neurosci.* **2005**, *25*, 3551–3559. [CrossRef] [PubMed]

29. Berecki, G.; Motin, L.; Adams, D.J. Voltage-Gated R-Type Calcium Channel Inhibition via Human μ-, δ-, and κ-opioid Receptors Is Voltage-Independently Mediated by Gβγ Protein Subunits. *Mol. Pharmacol.* **2016**, *89*, 187–196. [CrossRef] [PubMed]

30. Seseña, E.; Vega, R.; Soto, E. Activation of μ-opioid receptors inhibits calcium-currents in the vestibular afferent neurons of the rat through a cAMP dependent mechanism. *Front. Cell. Neurosci.* **2014**, *8*, 90. [PubMed]

31. Schroeder, J.E.; Fischbach, P.S.; Zheng, D.; McCleskey, E.W. Activation of mu opioid receptors inhibits transient high- and low-threshold Ca²⁺ currents, but spares a sustained current. *Neuron* **1991**, *6*, 13–20. [CrossRef]

32. Rusin, K.I.; Moises, H.C. μ-Opioid receptor activation reduces multiple components of high-threshold calcium current in rat sensory neurons. *J. Neurosci.* **1995**, *15*, 4315–4327. [PubMed]

33. Liman, E.R.; Tytgat, J.; Hess, P. Subunit stoichiometry of a mammalian K⁺ channel determined by construction of multimeric cDNAs. *Neuron* **1992**, *9*, 861–871. [CrossRef]

34. Ulens, C.; Daenens, P.; Tytgat, J. Changes in GIRK1/GIRK2 deactivation kinetics and basal activity in the presence and absence of RGS4. *Life Sci.* **2000**, *67*, 2305–2317. [CrossRef]

35. Lindsay, R.M. Nerve growth factors (NGF, BDNF) enhance axonal regeneration but are not required for survival of adult sensory neurons. *J. Neurosci.* **1988**, *8*, 2394–2405. [PubMed]

36. Bezanilla, F.; Armstrong, C.M. A low-cost signal averager and data-acquisition device. *Am. J. Physiol.* **1977**, *232*, C211–C215. [CrossRef] [PubMed]

37. Bundey, R.A.; Kendall, D.A. Inhibition of receptor-mediated calcium responses by corticotrophin-releasing hormone in the CATH.a cell line. *Neuropharmacology* **1999**, *38*, 39–47. [CrossRef]

38. Liu, M.; Richardson, R.R.; Mountford, S.J.; Zhang, L.; Tempone, M.H.; Herzog, H.; Holliday, N.D.; Thompson, P.E. Identification of a Cyanine-Dye Labeled Peptidic Ligand for Y₁R and Y₄R, Based upon the Neuropeptide Y C-Terminal Analogue, BVD-15. *Bioconjug. Chem.* **2016**, *27*, 2166–2175. [CrossRef] [PubMed]

© 2018 by the authors. Licensee MDPI, Basel, Switzerland. This article is an open access article distributed under the terms and conditions of the Creative Commons Attribution (CC BY) license (http://creativecommons.org/licenses/by/4.0/).

Article

Azemiopsin, a Selective Peptide Antagonist of Muscle Nicotinic Acetylcholine Receptor: Preclinical Evaluation as a Local Muscle Relaxant

Irina V. Shelukhina [1], Maxim N. Zhmak [1], Alexander V. Lobanov [2], Igor A. Ivanov [1], Alexandra I. Garifulina [1], Irina N. Kravchenko [2], Ekaterina A. Rasskazova [2], Margarita A. Salmova [2], Elena A. Tukhovskaya [2], Vladimir A. Rykov [2], Gulsara A. Slashcheva [2], Natalya S. Egorova [1], Inessa S. Muzyka [1], Victor I. Tsetlin [1] and Yuri N. Utkin [1,*]

[1] Shemyakin-Ovchinnikov Institute of Bioorganic Chemistry, Russian Academy of Sciences, ul. Miklukho-Maklaya 16/10, Moscow 117997, Russia; shelukhina.iv@yandex.ru (I.V.S.); mzhmak@gmail.com (M.N.Z.); chai.mail0@gmail.com (I.A.I.); garifulinaai@gmail.com (A.I.G.); natalyegorov@yandex.ru (N.S.E.); mis_kou@mail.ru (I.S.M.); vits@mx.ibch.ru (V.I.T.)

[2] Branch of the Shemyakin-Ovchinnikov Institute of Bioorganic Chemistry, Russian Academy of Sciences, Pushchino 142290, Moscow Region, Russia; lobanov-av@yandex.ru (A.V.L); ikravchenko@bibch.ru (I.N.K.); katyarass@mail.ru (E.A.R.); mak401@gmail.com (M.A.S.); elentuk@mail.ru (E.A.T.); vladimirrykov@email.su (V.A.R.); slashcheva_ga@mail.ru (G.A.S.)

* Correspondence: Yutkin@yandex.ru; Tel.: +7-495-336-6522

Received: 3 November 2017; Accepted: 2 January 2018; Published: 7 January 2018

Abstract: Azemiopsin (Az), a linear peptide from the *Azemiops feae* viper venom, contains no disulfide bonds, is a high-affinity and selective inhibitor of nicotinic acetylcholine receptor (nAChR) of muscle type and may be considered as potentially applicable nondepolarizing muscle relaxant. In this study, we investigated its preclinical profile in regard to in vitro and in vivo efficacy, acute and chronic toxicity, pharmacokinetics, allergenic capacity, immunotoxicity and mutagenic potency. The peptide effectively inhibited (IC_{50} ~ 19 nM) calcium response of muscle nAChR evoked by 30 μM (EC_{100}) acetylcholine but was less potent (IC_{50} ~3 μM) at α7 nAChR activated by 10 μM (EC_{50}) acetylcholine and had a low affinity to α4β2 and α3-containing nAChR, as well as to $GABA_A$ or $5HT_3$ receptors. Its muscle relaxant effect was demonstrated at intramuscular injection to mice at doses of 30–300 μg/kg, 30 μg/kg being the initial effective dose and 90 μg/kg—the average effective dose. The maximal muscle relaxant effect of Az was achieved in 10 min after the administration and elimination half-life of Az in mice was calculated as 20–40 min. The longest period of Az action observed at a dose of 300 μg/kg was 55 min. The highest acute toxicity (LD_{50} 510 μg/kg) was observed at intravenous injection of Az, at intramuscular or intraperitoneal administration it was less toxic. The peptide showed practically no immunotoxic, allergenic or mutagenic capacity. Overall, the results demonstrate that Az has good drug-like properties for the application as local muscle relaxant and in its parameters, is not inferior to the relaxants currently used. However, some Az modification might be effective to extend its narrow therapeutic window, a typical characteristic and a weak point of all nondepolarizing myorelaxants.

Keywords: nicotinic acetylcholine receptor; azemiopsin; preclinical studies; toxicity; pharmacokinetics; myorelaxant

Key Contribution: Key Contribution: Investigation of the preclinical profile of azemiopsin demonstrated its high affinity and specificity for muscle type nicotinic acetylcholine receptor as well as good muscle relaxant capacity. Toxicology studies in mice indicated that azemiopsin was well tolerated during chronic dosing and showed no immunotoxicity, allergenic or mutagenic activity, which made it a good candidate for application as a local muscle relaxant.

1. Introduction

A linear peptide azemiopsin (Az) isolated from the *Azemiops feae* viper venom contains no disulfide bonds [1] and can be easily prepared by peptide synthesis. It is a high-affinity and selective inhibitor of muscle-type nicotinic acetylcholine receptor (nAChR) involved in fast synaptic signal transduction at nerve-muscle junction [2]. These receptors are well-known targets for muscle relaxant drugs (e.g., [3]). Muscle relaxants reduce the tone of the skeletal muscle with a decrease in motor activity up to complete immobilization. These drugs are generally classified into central muscle relaxants, which disrupt the transmission of excitation in the central nervous system and muscle relaxants of peripheral action that primarily and specifically disturb neuromuscular transmission.

Peripheral muscle relaxants can affect the signal transmission both at the presynaptic and postsynaptic membrane of the neuromuscular junction. The drugs acting at postsynaptic membrane are classified into depolarizing and nondepolarizing muscle relaxants (NMR). The action of depolarizing muscle relaxants (e.g., succinylcholine) is based on the persistent depolarization of the postsynaptic membrane, which makes impossible the propagation of the action potential and causes relaxation of the muscle fiber. NMRs (such as *d*-tubocurarine, atracurium, rocuronium) block the binding of acetylcholine to nAChR and disturb its function [4].

Nowadays, the muscle relaxants are used generally during large operations in order to achieve relaxation of the muscles (especially the abdomen) and thereby facilitate surgical manipulation. For most operations, the basic condition is a good relaxation of the striated muscles. NMRs are usually administered during anesthesia to facilitate endotracheal intubation and/or to improve surgical conditions. Currently, clinical practice cannot do without them. Muscle relaxants allowed reducing the depth of anesthesia and better controlling the conditions of the body's systems. In addition to anesthesiology, muscle relaxants have found application in traumatology and orthopedics for muscle relaxation in the treatments of dislocations, fractures, diseases of the back and ligament. Short-acting drugs in combination with general anesthetic agents are often used to facilitate laryngoscopy, bronchoscopy and esophagoscopy. NMRs are applied parenterally, almost always injected intravenously.

It should be noted that NMRs have undesirable side effects, primarily associated with their effects on the autonomic nervous system and with the release of histamine. The reason for these effects is the insufficient selectivity of low molecular NMR for muscle-type nAChR. Moreover, a number of side effects are due to blockade or activation of the muscarinic acetylcholine receptor [5,6]. Because of these side effects, *d*-tubocurarine is practically not applied today.

To treat conditions where muscles spasms and spasticity are a problem, muscle relaxants are also used. Muscle spasms are caused by an involuntary contraction of the muscles, which is often painful and causes difficulty in performing everyday tasks. Spasticity occurs when a muscle contracts and remains in this tight position, becoming very stiff and almost impossible to use. In cases like this, muscle relaxants are used to control stiffness and involuntary movements. They are used to treat so-called muscle dystonia. Dystonia is defined by Dystonia Medical Research Foundation as "a movement disorder characterized by sustained or intermittent muscle contractions causing abnormal, often repetitive, movements, postures, or both. Dystonic movements are typically patterned, twisting and may be tremulous. Dystonia is often initiated or worsened by voluntary action and associated with overflow muscle activation" [7]. Examples of muscle dystonias are blepharospasm (involuntary squinting), cervical dystonia (torticollis), spasticity (hypertonicity) of skeletal muscle, writer's cramp, foot dystonia, etc. The problem of constant high tone of particular groups of muscles also exists at spastic form of cerebral palsy. Currently, according to the European (http://dystonia-europe.org) and American (http://www.dystonia-foundation.org/) dystonia societies, the number of patients with various forms of dystonia is 500,000 in Europe and 300,000 in the United States. According to the Research Foundation for Cerebral Palsy Associations (UCPA: http://www.ucp.org/), there are approximately 760,000 patients with this disease in the United States. In Russia, the number of patients with muscular dystonia is estimated at 80,000–140,000 people and cerebral palsy—150,000–200,000.

Peripheral muscle relaxants have become medications for the treatment of muscle dystonia. Historically, the first applied in clinical practice peripheral relaxants were low-molecular alkaloids. After these first low-molecular cholinergic blockers, showing a number of side effects but being used up to the present time, the botulinum toxin has appeared. Currently, the main method of treatment for muscular dystonia and spastic form of cerebral palsy is the injection of botulinum toxin into the muscles involved in hyperkinesis. In 1989, the "BOTOX" (one of the drugs based on botulinum toxin) has been approved by the FDA for the treatment of blepharospasm, in 2000—cervical dystonia, in 2010—spasticity at the elbow, wrist and fingers. The clinical effect is achieved in 85–90% of cases and lasts 2–3 months, however patients are in need of repeated administration of the drug: for spastic torticollis—2 injections per year, blepharospasm—3–4, cerebral palsy—2 times a year. With the apparent effectiveness of "BOTOX", there are a number of disadvantages associated with side effects, which include itching, burning, swelling at the injection site, in some patients there is a general muscle weakness during the first two weeks after application of the preparation, antibody formation shows in 3–10% of patients. Even the duration of its effect is a disadvantage, since it does not allow promptly adjusting the dose of the drug in accordance with individual tolerability. A significant drawback of the drug is associated with the mechanism of action of botulinum toxin at the molecular level.

Negative moments of this mechanism of action (and especially its duration) are progressive atrophy of muscle fiber with a decrease in the average diameter of the fiber, scattering of nAChRs from the site of the synapse and a decrease in the activity of synaptic acetylcholinesterase. On rabbits, it was shown that after the injections of botulinum toxin that lasts for six months, the reduction in muscle mass could reach 76% and the contractile fibers could be replaced by fatty tissue elements. Currently, there are no medicines capable of replacing BOTOX in the treatment of muscular dystonia in the world. Therefore, the task of creating new effective drugs that do not have side effects, for local therapy of muscular dystonia is extremely urgent.

Muscle relaxants of peptide nature may be considered as alternatives to low-molecular alkaloids with a large number of side effects and to extremely toxic protein botulinum toxin. Peptides are not xenobiotics and, as a rule, have high selectivity to specific targets, which is due to the very nature of peptide-protein recognition. The natural source of such peptides has always been the animal venoms, especially the venoms of molluscs and snakes [8]. In the venoms of molluscs and snakes, polypeptide and peptide compounds acting on neuromuscular transmission have been identified and potentially can be regarded as agents for the treatment of muscular dystonia. In particular, the discovery and characterization of Az, blocking the neuromuscular transmission, open the possibility for the development of a novel muscle relaxant. As an inhibitor of muscle nAChR, Az may be regarded as a potentially applicable muscle relaxant itself [9]. This paper reports the results of preclinical studies, including single- and repeated-dose toxicities, immunotoxicity, pharmacokinetic and other studies. Overall, the results obtained demonstrate that Az has good drug-like properties.

2. Results

2.1. Az Synthesis

The Az peptide was prepared by a solid phase synthesis using a general Fmoc-strategy. The principal scheme was adopted from [10]. However, to minimize the consumption of materials for large-scale synthesis needs, all steps in the procedure were optimized and standardized. Thus, to determine the optimal excesses of amino acid derivatives, pilot experiments were performed. It was found, that for the first stage of the coupling of the C-terminal proline residue to a polymer, the 5-fold excess of the protected amino acid was necessary. For the coupling of the subsequent residues, the 4-fold excesses were enough. The couplings of the seventh and subsequent residues required the increase of reaction time from 1 to 2 h. The modified scheme of synthesis allowed to avoid unnecessary reagent consumption, in particular, of expensive protected amino acids. Optimization experiments showed that the repeated condensations were necessary only for coupling of residues His9 and Pro15.

The amounts of solvents and coupling reagents were optimized as well, reducing their consumption by several times. Large-scale purification of the peptide was performed in two steps. Ion exchange chromatography on a weak cation exchanger under moderately basic conditions was used at the first step. It allowed to obtain the peptide with purity of about 90% without any organic solvent consumption. To meet pharmacopeia specifications, the final purification step was carried out by a reversed-phase HPLC, increasing the substance purity to greater than 97% (Figure 1). Using the optimized procedure, the final product, Az, was obtained with 20% yield.

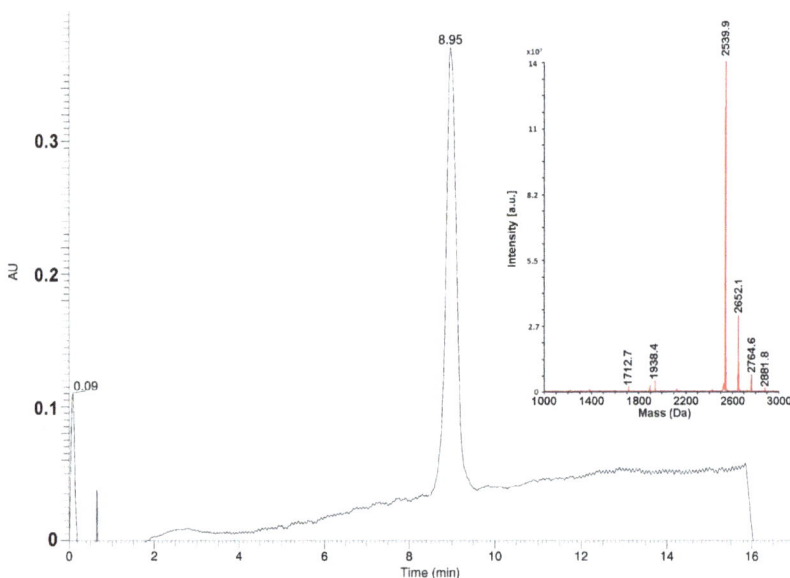

Figure 1. Analytical UPLC-MS on Phenomenex Aeris PEPTIDE XB-C18 column (1.7 µm, 2.1 × 150 mm) using a linear acetonitrile gradient from 10 to 35%. Inset. Deconvoluted mass-spectrum of Az sample obtained after the final purification step. a.u., arbitrary unit.

2.2. Efficacy and Specificity of Az In Vitro

To study a specific activity and selectivity of Az in vitro, two different methods were used: electrophysiological method of two-electrode voltage-clamp on *Xenopus* oocytes and calcium imaging using the genetically encoded calcium sensor Case12 or the low-molecular weight calcium indicator Fluo-4. In calcium imaging experiments on mouse muscle type $\alpha 1 \beta 1 \varepsilon \delta$ nAChR, Az showed a high inhibitory activity in nanomolar range ($IC_{50} = 19 \pm 8$ nM, Figure 2a). Az also manifested ability to interact with the human neuronal homopentameric $\alpha 7$ nAChR but with a much lower affinity ($IC_{50} = 2.67 \pm 0.02$ µM, Figure 2b). It should be mentioned, that the corresponding cellular calcium responses were provoked by acetylcholine (ACh) at concentrations of 30 µM (EC_{100} on muscle nAChR) and 10 µM (EC_{50} on $\alpha 7$ nAChR), respectively [11]. Electrophysiology experiments discovered no influence of Az on ion currents induced by 20 µM nicotine in rat neuronal heteromeric $\alpha 4 \beta 2$ nAChR at a concentration up to 50 µM (Figure 2c). Besides, no Az activity against human neuronal heteromeric $\alpha 3$-containing nAChRs ($\alpha 3 \beta 2$, $\alpha 3 \beta 4$, etc.) expressed in neuroblastoma SH-SY5Y cells was detected by calcium imaging at a concentration up to 100 µM (Figure 2f). In control experiments using the same cellular system, calcium responses induced by 100 µM nicotine ($EC_{50} = 22 \pm 2$ µM, Figure 2d) were successfully inhibited by α-conotoxin MII ($IC_{50} = 60 \pm 4$ nM, Figure 2e), a specific antagonist of $\alpha 3$-containing nAChRs. In sum, these data demonstrated high selectivity of the Az action on muscle nAChR.

For comparison, we have tested NMR rocuronium in some in vitro experiments and found that at muscle type receptor it was less effective than Az, IC_{50}s being 257.06 ± 95.54 nM and 19 ± 8 nM for rocuronium and Az, respectively (Figures 2a and 3a). At the human neuronal homopentameric $\alpha7$ nAChR rocuronium showed also lower affinity with IC_{50} of 25.69 ± 4.5 µM (Figure 3b) as compared to 2.67 ± 0.02 µM for Az (Figure 2b). In contrast to Az, at concentration up to 100 µM (Figure 2f) manifesting no activity against human neuronal heteromeric $\alpha3$-containing nAChRs ($\alpha3\beta2$, $\alpha3\beta4$, etc.) expressed in neuroblastoma SH-SY5Y cells, rocuronium dose-dependently inhibited these nAChR subtypes (Figure 3c). At 200 µM rocuronium inhibited Nic induced currents by about 60% (Figure 3c).

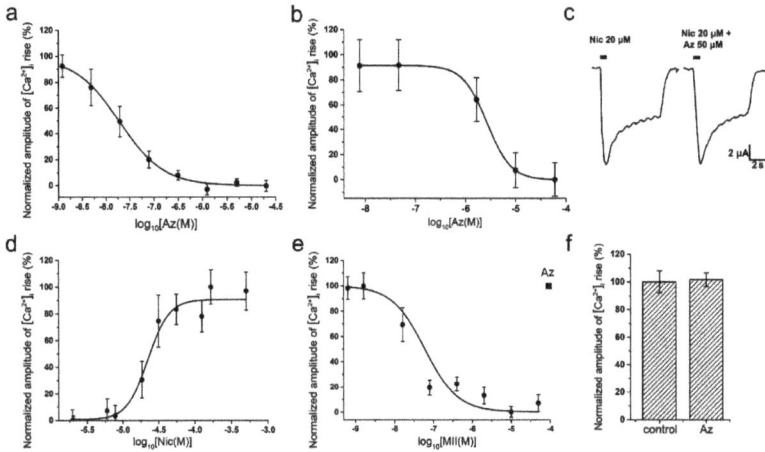

Figure 2. Interaction of Az with muscle and neuronal nAChRs. Inhibitory curves of Az action on ACh (30 and 10 µM)-evoked intracellular calcium concentration ($[Ca^{2+}]_i$) rises in neuroblastoma Neuro2a cells expressing (**a**) muscle $\alpha1\beta1\epsilon\delta$ and (**b**) $\alpha7$ nAChRs, respectively. (**c**) Representative nicotine (Nic)-induced current traces through $\alpha4\beta2$ nAChR and (**d**) dose-response curve of $[Ca^{2+}]_i$ amplitude rise in neuroblastoma SH-SY5Y cells expressing $\alpha3$-containing nAChRs in response to different concentrations of Nic. There are no inhibitory effects of Az on Nic-evoked (**c**) ion currents and (**f**) calcium responses mediated by $\alpha4\beta2$ and $\alpha3$-containing nAChRs, respectively ($p > 0.05$, Mann–Whitney U test). (**e**) Inhibitory curve of α-conotoxin MII action on Nic (100 µM)-evoked $[Ca^{2+}]_i$ rise in SH-SY5Y cells expressing $\alpha3$-containing nAChRs. Each point represents data obtained from 4 independent experiments (mean $\pm$ SEM).

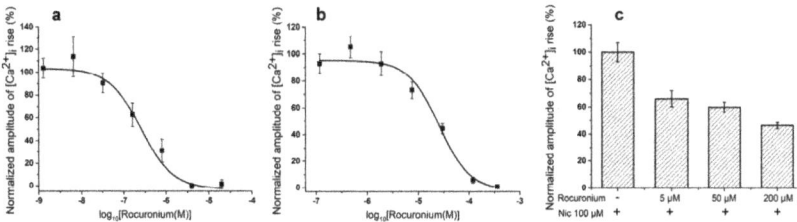

Figure 3. Interaction of rocuronium with muscle and neuronal nAChRs. Inhibitory curves of rocuronium action on ACh (30 and 10 µM)-evoked intracellular calcium concentration ($[Ca^{2+}]_i$) rises in neuroblastoma Neuro2a cells expressing (**a**) muscle $\alpha1\beta1\epsilon\delta$ and (**b**) $\alpha7$ nAChRs, respectively. Inhibition of $[Ca^{2+}]_i$ amplitude rise induced by Nic (100 µM) in neuroblastoma SH-SY5Y cells expressing $\alpha3$-containing nAChRs by different concentrations of rocuronium (**c**). Each point represents data obtained from 4 independent experiments (mean $\pm$ SEM).

2.3. *In Vivo Efficacy Tests*

2.3.1. In Vivo Az Efficacy

To study a specific activity of Az as an agent blocking neuromuscular transmission for the treatment of muscular dystonia, its effect on mouse muscular strength was estimated. Single administration of Az in the muscles of the forelimbs at doses of 0.03, 0.1 and 0.3 mg/kg caused a significant decrease in their muscular strength (Figure 4), while the dose of 0.01 mg/kg was not effective (Figure S1). The longest period of Az action was observed for a dose of 0.3 mg/kg and was maintained for 55 min from the 5th to the 60th minute after its administration. The maximal muscle relaxant effect of Az for all the doses studied was achieved 10 min after its administration and was preserved until the 30th minute. The dose of 0.03 mg/kg was considered as an initial effective dose. The average effective dose at the maximum response point (10 min after injection) was 0.09 mg/kg. These in vivo data showed the good muscle relaxing properties of Az.

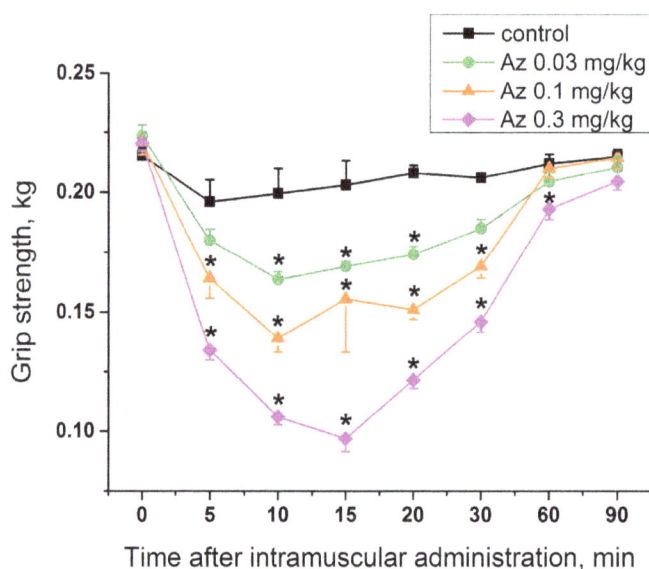

Figure 4. Muscle relaxant effect of Az. The time courses of a grip strength of mouse (ICR males) forelimbs at 0–90 min after Az (0.03, 0.1 and 0.3 mg/kg) or normal saline (control) intramuscular administration. The results are presented as mean values ± SEM, $n = 10$. Significant differences in the forelimb strength were revealed between control and experimental groups (one-way repeated measures ANOVA, * $p < 0.05$).

2.3.2. In Vivo Rocuronium Efficacy

The rocuronium effect was studied at doses of 0.13 mg/kg, 0.1 mg/kg and 0.08 mg/kg. The dose of 0.13 mg/kg was lethal and after 60 s the animal lost muscle tone (grip strength = 0 kg) and the ability to move. The introduction of rocuronium at doses of 0.08 and 0.1 mg/kg did not cause the death of animals and showed a dose-dependent decrease in muscle tone within the first 5 min after administration. A statistically significant effect compared to the control was observed 2 min after administration of rocuronium at a dose of 0.1 mg/kg (Figure 5). At the 3rd minute after the administration, muscle strength began to recover and did not significantly differ from the control. At a dose of 0.08 mg/kg no statistically significant difference from control was found and only a tendency to decrease the muscular strength was observed with the greatest effect at the 2nd and 3rd minutes (Figure 5).

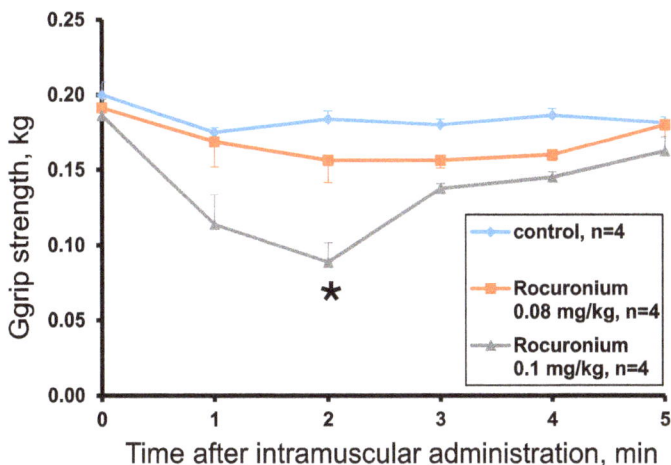

Figure 5. Muscle relaxant effect of rocuronium. The time courses of a grip strength of mouse (ICR males) forelimbs at 0–5 min after rocuronium (0.08, 0.1 mg/kg) or normal saline (control) intramuscular administration. The results are presented as mean values ± SEM, *n* = 4. Significant differences in the forelimb strength were revealed between control and experimental groups (one-way repeated measures ANOVA, * *p* < 0.05).

2.4. Pharmacokinetics of Az

To study a pharmacokinetics of Az, its radioiodinated ^{125}I-labeled analog ($[^{125}$I]-Az) was prepared.

2.4.1. Preparation of $[^{125}$I]-Az

In position 9, Az molecule contains a histidine residue which may be subjected to electrophilic iodination as published earlier [12]. However, two tryptophan residues (Trp3 and Trp4) might be oxidized under iodination conditions complicating the isolation of the target iodinated product. To overcome this problem, both tryptophan residues were protected by formylation. Diformyl-Az was iodinated following a standard chloramine protocol, optimized to obtain a better yield of iodinated peptide [13]. For the iodination reaction, a preliminary screening of the reaction conditions at which the pH and the substrate/chloramine ratio varied was carried out. It was not possible to obtain radioiodinated Az derivative containing only one iodine atom, therefore di-iodinated analogue was prepared. The complete iodination was achieved at pH 6.8 with 3 equivalents of iodide and 2.2 equivalents of chloramine T relative to peptide. The di-iodinated product was purified by HPLC and deprotected under alkaline conditions, followed by alkali neutralization. Analytical HPLC showed full removal of protecting groups, so the product was used further without additional purification.

2.4.2. Pharmacokinetics Studies

Single intravenous (iv) and intramuscular (im) administrations of $[^{125}$I]-Az at doses of 0.25 and 0.50 mg/kg to male ICR mice were performed. No lethality was observed after $[^{125}$I]-Az injections at these doses. The main pharmacokinetic parameters such as the area under a pharmacokinetic curve ($AUC(0 \rightarrow t)$), the maximum Az concentration in mouse blood (C_{max}) and its excretion half-life $T_{1/2}$ were determined (Table 1), allowing to evaluate the processes of excretion and elimination of the peptide.

Table 1. The main pharmacokinetic parameters estimated after a single intramuscular (im) or intravenous (iv) administration of Az to male ICR mice: the area under a pharmacokinetic curve (AUC(0 → t)), the maximum Az concentration in mouse blood (C_{max}) and its excretion half-life $T_{1/2}$.

Route/Dose	AUC(0 → t), h × ng/mL	C_{max}, ng/mL	$T_{1/2}$, h
im/0.25 mg/kg	328	278	0.30
im/0.50 mg/kg	622	257	0.68
iv/0.25 mg/kg	214	517	0.26
iv/0.50 mg/kg	542	745	0.29

For a single intravenous injection, the maximum concentration (C_{max}) of [^{125}I]-Az in mouse blood was observed 1 min after injection and its excretion half-life ($T_{1/2}$) was estimated as 15–20 min (Figure 6a). With intramuscular administration, the maximum [^{125}I]-Az concentration (C_{max}) was achieved within five minutes and the parameter $T_{1/2}$ was calculated as 20–40 min (Figure 6b). In both modes of administration, the drug was almost completely removed from the free blood flow during 24 h (Figure 6a,b). A greater maximum drug concentration (C_{max}) was observed at an intravenous route of administration than at intramuscular injection.

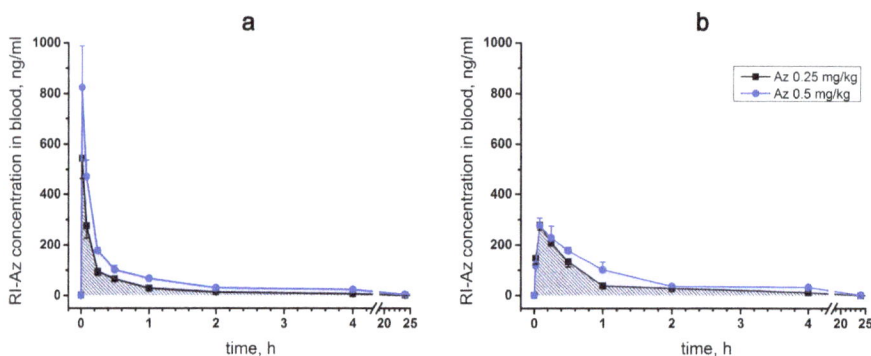

Figure 6. Pharmacokinetic curves of [^{125}I]-Az concentration in blood of male ICR mice 0–24 h after its (**a**) intravenous and (**b**) intramuscular administration at doses of 0.25 and 0.5 mg/kg. The results are presented as mean values ± SD, $n = 5$ (each plot point represents mean results for five animals).

2.5. Acute Toxicity Tests

2.5.1. Acute Toxicity of Az

The acute toxicity of Az after intraperitoneal administration to mice was determined earlier [1], the LD$_{50}$ value was 2.57 ± 0.27 mg/kg. In the present work, we studied Az acute toxicity to male ICR mice after intravenous and intramuscular administration. The LD$_{50}$ value was estimated as 0.51 ± 0.06 mg/kg after Az intravenous administration.

A single intramuscular Az injection at doses of 0.8, 0.775, 0.75 and 0.725 mg/kg resulted in a dose-dependent death of the mice. Death in 100% of cases was detected at a dose of 0.8 mg/kg. The doses of 0.775 and 0.75 mg/kg resulted in the death of 4 animals out of 5. At a dose of 0.725 mg/kg, 3 animals died out of 5 and at 0.7 mg/kg all mice were alive. The death of animals at doses of 0.725, 0.75, 0.775, 0.8 mg/kg was observed 19.7 ± 5.9, 28.8 ± 16.0, 15.8 ± 4.3 and 12.2 ± 1.8 min after Az administration, respectively. Based on these data, LD50 of 0.732 ± 0.13 mg/kg was calculated for Az at intramuscular injection. The maximal tolerant dose was 0.7 mg/kg after its intramuscular injection. No toxicity signs were observed at a dose of 0.3 mg/kg and lower used for grip strength tests.

Intramuscular injection of Az was accompanied by external signs of intoxication, the severity of which was dose-dependent. Visible toxic signs appeared 5–7 min after administration and were characterized by impaired coordination of movements and loss of muscle tone, decreased motor activity, impaired breathing, decreased response to external stimuli. Maximal manifestations of intoxication were noted between 10 and 20 min after administration and were characterized by loss of motor activity, a lacunar posture or posture on the side, loss of muscle tone, loss of response to external stimulation, delayed or intermittent breathing, coma. Mice almost completely recovered within 60 min after injection. After this period, the animals showed a decrease in motor activity and muscle tone. Complete recovery from the toxic effect of large doses occurred 24 h after administration. No function disturbances caused by Az large doses were observed 14 days after the administration.

At intramuscular administration way, a maximal tolerated dose (MTD) was determined. MTD is the highest dose of a drug that does not cause unacceptable side effects. In mice, MTD for Az was determined as 0.7 mg/kg; no lethality was observed at this dose. The doses up to 0.5 mg/mL used for the in vivo tests were lower than MTD of 0.7 mg/kg and induced no lethality as well.

2.5.2. Acute Toxicity of Rocuronium

Injection of rocuronium at a dose of 0.13 mg/kg caused the death of the test animal. The first signs of intoxication were detected 30 s after injection. The impaired coordination of movements, decreased muscle tone, decreased motor activity, increased respiratory movements, convulsions were observed. The severity of toxic disorders rapidly increased. After 60 s, the animal lost muscle tone (grip strength = 0 kg) and the ability to move, the frequency of respiratory movements decreased, the animal fell into a coma and then died 11.5 min after injection. Rocuronium at doses of 0.08 and 0.1 mg/kg did not cause the death of animals. A dose of 0.08 mg/kg did not induce visible signs of toxicity, while 0.1 mg/kg caused discoordination of movements, decreased muscle tone, gait disturbance, decrease or short-term loss of motor activity, increased respiratory movements, vocalization (in one animal). These toxicity signs disappeared in 3–4 min after drug administration.

2.6. Subchronic Toxicity of Az

For the study of the subchronic toxicity, the number of animals was increased in comparison with acute toxicity tests, because many biochemical, histological and other parameters needed to be measured. In order to reveal a statistically significant difference in these parameters, large groups of animals were used. For the studies, larger animals (rats) were used for experimenting with long-term administration of the drug and studying many blood parameters, this might be difficult for small animals such as mice.

On the basis of acute toxicity results a repeated dose 14-day intramuscular toxicity study was performed. During the 14-day period of Az administration at doses of 0.1 and 0.5 mg/kg, it did not cause any evident signs of toxicity in male or female Sprague Dawley rats. The increase in the mean body weight of animals and in the food intake did not differ significantly between the experimental and control groups.

A 14-day Az treatment at a dose of 0.1 mg/kg did not lead to any changes in the hemogram of experimental animals relative to control ones. However, in male rats at a dose of 0.5 mg/kg Az caused a statistically significant increase ($n = 6$, $p < 0.05$, Kruskall-Wallis ANOVA on ranks) in the number of platelets (810 ± 39 g/L) relative to the control level (729 ± 22 g/L). Two weeks after the 0.5 mg/kg Az administration the level of platelets was still slightly increased but non-significantly. There were no differences in the hemograms of female rats receiving Az at both doses and the control group.

A few biochemical parameters of rat blood serum were changed at the end of 28-day subchronic toxicity experiment (14 days of Az administration and 14 days of its withdrawal) in comparison to the control animal group. In the blood serum of males receiving Az at a dose of 0.5 mg/kg, a significant decrease in the mean level of triglycerides (0.97 mmol/L vs. 1.25 mmol/L in control, $n = 6$, $p < 0.05$, Kruskall-Wallis ANOVA on ranks) was observed. In a group of female rats treated with

0.1 mg/kg of Az, the levels of cholesterol (2.82 ± 0.36 mmol/L vs. 3.55 ± 0.39 mmol/L in control) and calcium (3.09 ± 0.06 vs. 3.25 ± 0.09 mmol/L in control) were reduced significantly ($n = 6$, $p < 0.05$, Kruskall-Wallis ANOVA on ranks).

In two weeks after the Az administration the rats were euthanized. Post-mortem necropsy of animals did not reveal any abnormalities in the anatomy or in the absolute weight of their internal organs. However, some differences in a relative heart weight of male rats were observed. Thus, the mean relative heart weight in a rat group treated with Az at a dose of 0.5 mg/kg was significantly ($n = 6$, $p < 0.05$, Kruskall-Wallis ANOVA on ranks) reduced ($0.352 \pm 0.033\%$) in comparison with the control group of animals ($0.428 \pm 0.086\%$). Histological analysis was performed for the following organs and tissues: liver, stomach, kidneys, adrenal glands, lungs, heart, spleen, thymus, submandibular lymph nodes, ovaries, testes, brain, femoral muscle of the right and left paws (the injection site). No pathological changes were observed in the organs examined.

All observed changes in biochemical, hematological and other tested parameters were among physiologically normal variations for Sprague Dawley rats and did not indicate Az toxicity [14]. Thus, during 14-day intramuscular administration of Az in two doses of 0.1 and 0.5 mg/kg, which are similar to the expected therapeutic doses, to female and male rats with a two-week withdrawal period the peptide showed no significant toxicity.

2.7. Immunotoxicity of Az

To study the possible immunotoxicity of Az, its effects on

(1)   cellular immunity (the delayed-type hypersensitivity reaction),
(2)   immune response to a standard antigen and
(3)   phagocytic activity of peritoneal macrophages was evaluated.

All three tests were carried out after 7-day intramuscular Az administration at doses of 0.15 and 0.5 mg/kg to male ICR mice. Control animals were treated with the sodium chloride physiological solution (normal saline).

To probe the Az influence on cellular immunity, the mice were primarily immunized subcutaneously (sbc) at the base of a tail with trinitrobenzenesulfonic acid (10 mM, 200 µL) and secondarily after 6 days with the same agent (50 µL) in the left hind paw. Simultaneously, the physiological solution (50 µL) was injected into the right control hind paw. Next day after the second immunization, the weight of left and right hind paws were compared and edema of the experimental paws was revealed in all animal groups (Figure 7a). However, Az treatment did not cause any significant changes in the degree of the observed edema, showing no influence on cellular immunity in mice (Figure 7a).

In the next test, the effect of a 7-day Az administration on mouse immune response to a bovine serum albumin (BSA) was determined. The mice from one control and two experimental groups were routinely immunized with BSA in two steps (1st and 10th days) and after a week the corresponding IgG titers were evaluated in mouse blood serum (Figure 7b). The obtained results did not demonstrate any significant difference in the immune response of mice treated with Az (0.15 and 0.5 mg/kg) or with the physiological solution (Figure 7b).

Az at doses of 0.15 and 0.5 mg/kg also did not significantly change the phagocytic activity (engulfing ink particles) of peritoneal macrophages, which were isolated from experimental animals, vs. control ones (Figure 7c). Thus, in all three tests no significant changes in the studied parameters of the immune system in animals receiving Az during a week were observed.

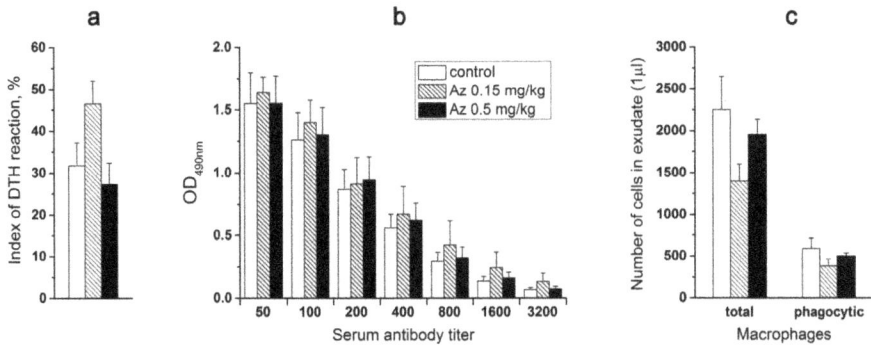

Figure 7. The influence of Az on immune system. The effect of 7-day intramuscular Az (0.15 and 0.5 mg/kg) administration to male ICR mice (**a**) on their delayed-type hypersensitivity (DTH) to a specific antigen (10 mM trinitrobenzenesulfonic acid) manifested by paw edema, (**b**) on their immune response to a bovine serum albumin (BSA) and (**c**) on phagocytic activity of their peritoneal macrophages. (**a**) The presented indexes of DTH reactions reflect the normalized difference in the weights of treated and control mouse hind paws. (**b**) The titers of IgG in blood serum of mice from experimental and control groups after their standard immunization with BSA are presented. (**c**) The number of total and phagocytic (engulfing ink particles) macrophages in 1 μL of peritoneal exudate isolated from experimental and control animals. In all three tests, there was no significant difference between control and experimental animals ($p > 0.05$, Kruskall-Wallis ANOVA on ranks) in the parameters studied. The results are presented as mean values $\pm$ SEM, $n = 5$–10.

2.8. Allergenicity of Az

To test the allergenicity of Az, its ability to induce a delayed-type hypersensitivity reaction in male and female ICR mice at a dose of 0.15 mg/kg was investigated. The scheme of animal immunization (the 1st day subcutaneously and the 5th day in a left hind paw) was similar to the test with trinitrobenzenesulfonic acid but Az was used as an immunogen. In the control groups, the animals were first given the sodium chloride physiological solution (normal saline) subcutaneously and after 5 days were similarly injected with Az in a left hind paw. The degree of left hind paw edema was estimated 6, 12 and 24 h after the second Az injection relatively to the control paw size (Figure 8). In all groups of animals, a small swelling of the experimental paws was observed 6 h after the second Az injection. After 12 h the revealed edema decreased and after 24 h there was practically no swelling. The size of the edema and its dynamics were not significantly different between the animal groups with preliminary Az sensitization and without it. In all tests carried out, no allergic reaction was detected. Thus, Az at a dose of 0.15 mg/kg demonstrated no allergic effect in the delayed-type hypersensitivity reaction test in ICR mice.

2.9. Mutagenicity of Az

To study Az mutagenicity in vitro, its ability to induce mutations in the hypoxanthine guanine phosphoribosyltransferase (hprt) gene of mammalian CHO-k1 cells was tested. This is the common assay for detection of gene mutations in mammalian cells [15]. For this purpose, Az was added to the cell growth medium at concentrations ranging from 2.8 to 2000 μg/mL for four hours in the presence or absence of a metabolic activation system (S9 mixture [16]) and after a cultivation period of 8 days, the cells were sub-cultured in the presence of a specific cytostatic agent 6-thioguanine. The inactivation of hprt gene due to any induced mutations led to the resistance of CHO-k1 cells to the cytostatic effect of this purine analog and allowed the selection and counting of the mutant cell colonies grown in its presence.

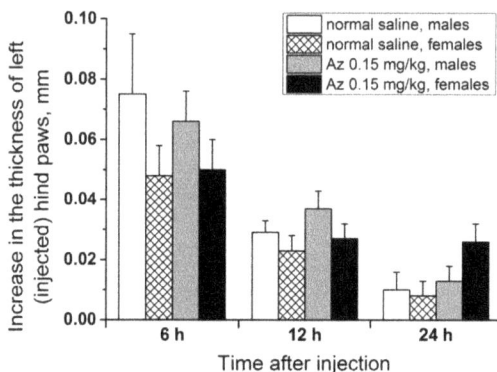

Figure 8. The capacity of Az (0.15 mg/kg subcutaneously) to provoke allergic (a delayed-type hypersensitivity) reaction in male and female ICR mice. The differences in the thickness of the experimental left (injected with Az) and the control right hind paws are presented for animal groups with preliminary Az sensitization and without it 6, 12 and 24 h after the second peptide administration. There was no significant difference in the increase in the thickness of left hind paws between all animal groups ($p > 0.05$, Kruskall-Wallis ANOVA on ranks). The results are presented as mean values $\pm$ SEM, $n = 10$.

As positive controls, two high-mutagenic agents were used: ethylnitrosurea (6.25 and 12.5 µg/mL) and methylcholanthrene (2.5 and 5 µg/mL) in the absence and in the presence of the metabolic activation system, respectively. To determine the basic level of spontaneous mutations in the hprt gene, CHO-k1 cells were cultivated in the intact growth medium before their sub-culturing in the presence of 6-thioguanine. The mean frequency of spontaneous mutations was evaluated as $25.2 \pm 1.6 \times 10^{-6}$. Although for all positive control conditions the statistically significant rise ($n = 6$, $p < 0.05$, Kruskall-Wallis ANOVA on ranks) in the frequency of mutations in the hprt gene was observed (33.9–39.5×10^{-6} for ethylnitrosurea and 35.7–45.7×10^{-6} for methylcholanthrene), Az in all tested concentrations (up to 2000 µg/mL) did not provoke any significant increase over a basic level of spontaneous mutations in this gene (23.5–30.2×10^{-6}, $p > 0.05$). Thus, in this cell system Az did not demonstrate any mutagenic capacity.

3. Discussion

As discussed above peptide and protein drugs in many cases possess higher efficacy and better specificity as compared to low molecular weight compounds. The Az, manifesting specific inhibiting activity against muscle nAChR, is a good candidate to be a local muscle relaxant. However, to claim Az as a perspective medicine, it is necessary to check a number of its biological characteristics, including muscle relaxing efficacy, acute and chronic toxicity, pharmacokinetics, mutagenicity, immunotoxicity and allergenicity. The current study was undertaken to determine if Az has a perspective to be used as a local muscle relaxant and it was carried out according to General requirements to conduct preclinical studies of drugs as stated in Appendix No. 7 to the Rules of Good Laboratory Practice of the Eurasian Economic Union in the field of drug circulation [17].

Basing on these requirements, we have studied the pharmacology and in vivo efficacy of the Az, its pharmacokinetics and toxicology, including toxicity after its single and repeated administration, specific toxicity and mutagenicity. Earlier in competition experiments with radioactive α-bungarotoxin, Az showed high affinity to muscle-type *Torpedo* nicotinic acetylcholine receptor (nAChR) (IC_{50} 0.18 ± 0.03 µM) and lower efficiency to human α7 nAChR (IC_{50} 22 ± 2 µM) [1]. In *Xenopus* oocytes heterologously expressing human muscle-type nAChR it was more potent against the adult form ($\alpha1\beta1\epsilon\delta$, IC_{50} 0.44 ± 0.1 µM) than the fetal form ($\alpha1\beta1\gamma\delta$, IC_{50} 1.56 ± 0.37 µM). In the present study, we have found that Az exhibited high affinity for mouse muscle $\alpha1\beta1\epsilon\delta$ nAChR (IC_{50} 19 ± 8 nM) but was less potent to human $\alpha7$ nAChR (IC_{50} 2.67 ± 0.02 μM) in calcium imaging assay. It was more active than rocuronium against both muscle and $\alpha7$ nAChR and manifested higher selectivity to muscle receptor (about 140 times) as compared to rocuronium (about 100 more active to muscle type). In general, these data are in agreement with earlier results. At concentrations up to 100 μM, Az had no effect on heteromeric rat $\alpha4\beta2$ or human $\alpha3$-containing nAChRs ($\alpha3\beta2$, $\alpha3\beta4$, etc.). It also showed no activity against 5-HT$_3$ receptors at concentration up to 10 μM and GABA$_A$ ($\alpha1\beta3\gamma2$ or $\alpha2\beta3\gamma2$) receptors at concentration up to 100 μM [1]. All these data show high Az selectivity to muscle type nAChR. It should be noted that the nondepolarizing neuromuscular blocking agents now used as muscle relaxants reversibly and concentration-dependently inhibited in the low micromolar range the neuronal nAChRs, including $\alpha3\beta2$, $\alpha3\beta4$, $\alpha4\beta2$ and $\alpha7$ subtypes [18]. In our experiments, rocuronium dose-dependently inhibited neuronal $\alpha3$-containing nAChRs. The mechanism (i.e., competitive vs. noncompetitive) of the block at the neuronal nAChRs was dependent both on the receptor subtype and the agent tested. Our data indicate that Az is more selective to muscle type nAChR than currently used relaxant, therefore it may produce less side effects in practice.

To determine the muscle relaxing capacity of Az, its influence on the forelimb grip strength of male ICR mice was studied. Grip strength test is a widely used non-invasive method to quantify objectively the muscular strength of mice and rats and to investigate the effects of neuromuscular disorders and drugs. It is based on the natural tendency of a rodent to grasp a bar or grid when it is suspended by the tail [19,20]. It was found that a single Az injection in mouse forelimb muscles resulted in the decrease of its grip strength. The effect was dose-dependent and the strongest decrease was observed at the highest dose used (0.3 mg/kg). The effect was evident 5 min after administration and maintained for 25–55 min depending on the dose. For comparison, we have tested nondepolarizing muscle relaxant rocuronium and found that it was very poor in grip strength test. Its effect was extremely fast and disappeared with 5 min after injection. Moreover, at a dose of 0.13 mg/kg it was lethal to mice and at 0.08 mg/kg produced no statistically significant relaxing effect as compared to control. Thus, Az demonstrated much better performance in grip strength test. At 0.1 mg/kg (the intermediate dose used) the Az effect was slightly more pronounced as compared to other nondepolarizing muscle relaxant pancuronium, the decrease being by 37% for Az and by 24.1% for pancuronium [20]. However, taking into consideration the 4.4-fold molecular mass difference, Az is much more active than pancuronium; 0.1 mg/kg corresponds 39 and 175 nmoles/kg, respectively. In grip strength test the relaxant activity of botulinum neurotoxins was assessed in rats [20]. In this test, the botulinum neurotoxin was more active and its effect was much more persistent. At 0.24 U of neurotoxin injected intramuscularly, the normal strength was not observed for more than 14 days [21]. This long muscle function disturbance might not be beneficial at some practical applications.

To study pharmacokinetics, the radioiodinated Az analogue was prepared. The only amino acid residue, which may be iodinated in the peptide, is histidine. Oxidative conditions used for iodination may result in oxidation of two tryptophan residues present in Az molecule. Therefore, to obtain the isotope-labeled derivative, a three-step procedure was chosen, including the introduction of formyl protective groups in the tryptophan residues, iodination of the protected Az using the iodide/chloramine T mixture and deprotection of the obtained derivative under alkaline conditions. The radioactive analogue possessing high radioactivity was used for the pharmacokinetics study. No lethality was observed at intravenous injection of [^{125}I]-Az at doses of 0.25 and 0.5 mg/kg. While the dose of 0.5 mg/kg is very close to LD_{50} of Az (0.51 mg/kg), it induced no death in injected mice. This fact may be explained by the lower toxicity of iodinated Az. Earlier we have shown that histidine residue is essential for Az activity [1]; its replacement by alanine resulted in strong decrease in capacity to bind to *Torpedo* nicotinic acetylcholine receptor. Thus, introduction of iodine in histidine residue may result in the decrease of [^{125}I]-Az toxicity.

Pharmacokinetics and pharmacodynamics are the empirical mathematical models that describe the time course of drug effect after administration [22]. Pharmacokinetics describes the disposition of drug in the organism, while pharmacodynamics relates the drug effects to their concentration in the plasma and at the site of action. They can be used to predict the drug action at different doses and by this way to optimize the safe and effective use of the drug. Information from pharmacokinetic studies can be used in the design and analysis of data from other toxicity studies. Several pharmacokinetics parameters for Az were determined including $AUC(0 \rightarrow t)$, C_{max} and $T_{1/2}$. It was found that a higher Az concentration was achieved in the blood after its intravenous injection. We were not able to find in the available literature pharmacokinetics parameters obtained for any peripherally acting muscle relaxant in mice. However, there are several studies published for humans [23–26]. It should be noted that the results obtained on different species with different administered doses and different analysis methods cannot be compared adequately. Nondepolarizing peripherally acting muscle relaxant can be classified as long-acting, intermediate- and short-acting blockers [26]. By its pharmacokinetics parameters, Az is more similar to intermediate-acting relaxants. Its excretion half-life was estimated as 15–40 min depending on administration way, while for intermediate-acting relaxants in human it varied from 17 min (atracurium [27]) to 71 min (rocuronium [28]). For Az, C_{max} was 745 ng/mL at intravenous injection (0.5 mg/kg); this value was about 1 μg/mL for vecuronium [29] and 27 μg/kg for rocuronium [23]. It should be noted that for Az we observed fairly good correlation between the time of elimination (half-life 15–40 min) and duration of muscle relaxant effect (25–55 min).

Acute toxicity of Az was determined using different administration ways: intraperitoneal (ip), intramuscular (im) and intravenous (iv) injections. The highest toxicity was observed at iv injection (LD_{50} 0.51 mg/kg). The LD_{50} of rocuronium at iv injection to rats is about 0.3 mg/kg [30], for (+)-tubocurarine in mice—0.11 mg/kg [31] and for vecuronium in mice—50 μg/kg [32]. In our experiment, intramuscular injection of rocuronium at dose of 0.13 mg/kg resulted in the death of the animal, while Az at 0.7 mg/kg was not lethal. These data indicate that the Az possesses lower acute toxicity than common peripherally acting muscle relaxants. It should be noted that the surviving animals have fully recovered in a fairly short period of time. This is a standard feature of curare-like drugs and can be considered in favor of their application in practice.

The subchronic toxicity of Az was studied at its intramuscular administration in two doses of 0.1 and 0.5 mg/kg for 14 days. Among a number of different parameters investigated, several were found to be influenced by Az administration. Thus, in males receiving Az the differences from control males receiving the carrier were observed in the following parameters: the relative heart mass was increased by 0.076%, the triglyceride level was decreased by 0.28 mmol/L and the platelet count was increased by 81 g/L. In the group of females treated with the drug at a dose of 0.1 mg/kg, the differences from the control animals in some biochemical parameters were also found—the cholesterol level was lowered by 0.73 mmol/L and calcium concentration by 0.11 mmol/L. However, these changes did not exceed the physiological norm and fall within the range of normal physiological values for male and female rats [14]. It would be incorrect to speak in this case about the toxic effects of Az, since the fluctuations of the parameters are within the range of the physiological norm. For the same reason, we cannot say that there are differences in the effects of Az on males and females. No other signs of toxicity were observed during chronic intramuscular administration of Az to female and male rats within a two-week withdrawal period. Therefore, we concluded that peptide had no significant chronic toxicity in rats at doses tested.

Like some other chemical substances, Az may induce undesirable immune reaction or allergy. Immunotoxicity is defined as adverse effects on the functioning of the immune system that result from exposure to chemical substances. The adverse effects on the immune system include reduction in antibody production, reduction in cytokine secretion, hypersensitivity and some other effects [33]. Altered immune function may lead to the increased incidence or severity of infectious diseases or cancer, since the immune system's ability to respond adequately to invading agents is suppressed. Identifying immunotoxicants is difficult because chemicals can cause a wide variety of complicated effects on immune function. That is why several methods were used in this work to estimate the immunotoxicity of azemiopsin. The precise testing of immunotoxicity and allergenicity is required to estimate the potential hazard of the putative drug. Immunotoxicity assays are important tests for new drugs being developed for application in the humans. Considering the complexity of the immune response, in vivo studies are more relevant. In this work, three different in vivo assays were used to assess Az immunotoxicity (Section 2.8). None of them revealed immunotoxic capacity of Az. No signs of allergenicity was seen in any of the in vivo tests described above; therefore, we decided first to try a single dose of 0.15 mg/kg, which is close to the anticipated therapeutic one. Allergenicity of Az was checked as its ability to induce a delayed-type hypersensitivity reaction. This test did not reveal any allergenicity signs as well, then for ethical reasons we considered additional studies inappropriate. It should be noted that neuromuscular blocking agents contribute to 50–70% of allergic reactions during anesthesia [34]. Suxamethonium appeared to be more frequently involved, while, pancuronium and cis-atracurium are associated with the lowest incidence of anaphylaxis [34]. An increased frequency of allergic reactions to rocuronium was recently noted [35]. As no allergic reaction to Az was observed in our study it may have advantages over other relaxants in this respect.

Mutagenicity, that is the induction of permanent transmissible changes in the amount or structure of the genetic material of cells or organisms, is very important parameter of drug candidate. Highly mutagenic compounds can hardly be considered for drug development, therefore mutagenicity studies are a necessary phase in preclinical evaluations. In vitro Az mutagenicity studies using mammalian CHO-k1 cells showed no mutagenic capacity. While studies of rocuronium using cultured human peripheral blood lymphocytes indicated that it was capable of causing genotoxicity via clastogenic effects at concentrations at which a significant cytotoxic effect does not occur [36]. Thus, Az is safer as compared to rocuronium.

The main application of NMRs is their use in surgery to relax the muscles during operative interventions. They are administered during anesthesia and allow to reduce the dose of anesthetics thus decreasing their adverse effects. There are some general requirements to NMR and ideally, it should have a rapid onset and short duration of action, no cardiovascular side effects, no accumulation in the body, no active metabolites, organ-dependent drug metabolism and elimination as well as an available and adequate antagonist [37,38]. Az satisfies most these requirements: it is fast acting agent with relatively short duration of action, does not accumulates in the body and has no active metabolites. All this allows considering Az for the possible application as NMR.

As it was described in introduction, other area for muscle relaxant application is the treatment of dystonia. Nowadays the main drug for dystonia treatment is extremely toxic botulinum toxin. In addition to high toxicity, there are several other side effects associated with its application. Az at non-lethal doses showed good muscle relaxing activity (Figure 3). It is deprived of some shortcomings (e.g., long action period) inherent in the botulinum toxin and may be regarded as a candidate for dystonia treatment.

The above considerations allow to conclude that Az has good drug-like properties for the application as local muscle relaxant, however further studies should be conducted to confirm it safety and applicability including investigation on humans.

4. Conclusions

In summary, we investigated the preclinical profile of Az in regard to in vitro and in vivo efficacy, acute and chronic toxicity, pharmacokinetics, allergenic capacity, immunotoxicity and mutagenic potency. Our in vitro studies confirmed the high affinity and specificity of Az for muscle type nAChR. The peptide effectively inhibited muscle nAChR but was less potent at $\alpha7$ nAChR and had a low affinity to $\alpha4\beta2$ and $\alpha3$-containing nAChR, as well as to GABAA or 5HT3 receptors. Its muscle relaxant effect was demonstrated at intramuscular injection to mice at doses of 30–300 µg/kg, the relaxant activity being higher than that of commonly used peripheral muscle relaxants. The highest acute toxicity was observed at intravenous injection of Az, at intramuscular or intraperitoneal administration it was less toxic. No toxicity signs were observed at doses inducing muscle relaxant effects. Toxicology studies in mice indicated that Az was well tolerated during chronic dosing and showed no immunotoxicity, allergenic or mutagenic activity, which differ if from most currently used muscle relaxants. Overall, the results demonstrate that azemiopsin has good drug-like properties for application as a local muscle relaxant and in its parameters is not inferior to the relaxants currently used. It should be noted that Az possessed narrow therapeutic window, which is a typical characteristic and a weak point for all muscle relaxants with a similar mechanism of action. Modifications are required that will not affect (or preferably increase) the effectiveness of the drug but reduce its toxicity and extend its narrow therapeutic window.

5. Materials and Methods

5.1. Materials

A polystyrene-poly(ethylene glycol) 2000 block-copolymer resin, modified with Knorr linker Tentagel S RAM was from Rapp Polymere GmbH (Tübingen, Germany). Fmoc-protected amino acids, 4-methyl piperidine were from Mosinter (Ningbo, China). 1-[Bis(dimethylamino)methylene]- 1H-1,2,3-triazolo[4,5-b]pyridinium 3-oxid hexafluorophosphate (HATU), 1-Hydroxy-7-azabenzotriazole (HOAt) and DL-Dithiothreitol (DTT) were from DEMO Medical (Shanghai, China). *N,N*-diisopropylethylamine (DIPEA) was from Iris Biotech GmbH (Marktredwitz, Germany), trifluoroacetic acid from Solvay S.A. (Bruxelles, Belgium), normal saline (sterile 0.9% NaCl solution), complete Freund's adjuvant and trinitrobenzenesulfonic acid (TNBS) from Sigma-Aldrich Chemie Gmbh (Munich, Germany), bovine serum albumin (BSA) from Amresko, Rocuronium Bromide from Fresenius Kabi (Bad Homburg, Germany). All other reagents and solvents of the highest purity available were purchased from local manufacturers and used without additional purification.

5.2. Animals

Specific pathogen-free (SPF) ICR mice (6–8 weeks old, weight 29–34 g) and SD rats (9–11 weeks old, weight of males 254–310 g, weight of females 188–220 g) of both sexes were obtained from the Animal House of the Branch of the Shemyakin-Ovchinnikov Institute of Bioorganic Chemistry, Russian Academy of Sciences and used for studies in vivo. Animals were housed in groups of 5–6 mice or 2 rats at 20–25 °C, 30–70% relative humidity and under a 12-h light-dark cycle (lights on at 08:00). Standard chow for rodents and filtered tap water were provided *ad libitum*. All studies involving animals were approved by the Institutional Animal Care and Use Committee (IACUC) of the Branch of the Shemyakin-Ovchinnikov Institute of Bioorganic Chemistry, Russian Academy of Sciences, the experimental protocol codes are No. 528/16, 531/16, 534/16, 560/16, 576/17, 577/17. The dates of approval are 31 March 2016, 20 April 2016, 20 May 2016, 15 September 2016, 10 February 2017, 10 February 2017, respectively. All solutions for in vivo studies were prepared fresh before the administration: the necessary amount of freeze-dried Az was dissolved in normal saline (0.9% sterile NaCl solution).

5.3. Az Synthesis

5.3.1. Solid Phase Az Synthesis

Az synthesis was performed on automatic peptide synthesizer based on Gilson automated liquid handler system according to Gilson application note 228. The peptide was synthesized utilizing a solid phase methodology with Fmoc/t-Bu protection scheme. A polystyrene-poly(ethylene glycol) 2000 block-copolymer resin Tentagel S (extent of loading 0.3 meq/g) was modified with Knorr linker followed by peptide chain assembly. A 4-fold excess of protected amino acids was used; condensation reagent was HATU/HOAt in amount equimolar to that of protected amino acids and activation reagent—2.4 equivalents of DIPEA. Coupling time for 7 C-terminal amino acid residues (up to Pro15) was 1 h and for subsequent residues the coupling time was increased to 2 h. For His9 and Pro15 residues a repetitive coupling was required to achieve their complete acylation. After chain assembly, peptidyl-polymer was subjected to a total deprotection/cleavage by the treatment with 12 mL of reagent L [39] per 1 g of dry resin for 2 h. After that, the resin was filtered, washed with trifluoroacetic acid and combined filtrate was evaporated under vacuum to ca. 30% of initial volume. The residue was diluted tenfold with dry diethyl ether; the precipitated crude peptide was filtered out, washed with ether and dried under vacuum. The crude peptide was dissolved in starting buffer and 1 g was applied on a ECOPlus column (35 × 250 mm) packed with 60 mL of SPS-Bio CM (12 µm, Purolite Corporation, Bala Cynwyd, PA, USA) resin. The column was eluted with 8 column volumes of linear concentration gradient of ammonium bicarbonate from 50 mM to 1 M (pH 8.9) in 10% isopropyl alcohol. Fractions containing peptide were collected and isopropyl alcohol was evaporated under vacuum, the remaining solution was freeze-dried. Freeze-dried peptide was dissolved in water to a final concentration of 50 mg/mL, titrated with acetic acid to a pH 3.5 and applied on a Thermo Scientific Hypersil GOLD aQ (12 µm, 250 × 50 mm) column. Elution was carried out with a linear gradient of acetonitrile in water from 10 to 35% in 30 min in the presence of 1% acetic acid at a flow rate of 150 mL/min. Main fraction was collected and freeze-dried; the obtained azemiopsin acetate had a purity greater than 97% as confirmed by UPLC-MS analysis. The total yield of pure peptide was about 20%, based on resin loading.

5.3.2. Formylation of Az

For this modification, the peptide was dissolved in formic acid at a concentration of 50 mg/mL, to the resulting solution hydrogen chloride in formic acid was added to a final concentration of 200 mM HCl. The mixture was stirred for 12 h, then formic acid was evaporated on a rotary evaporator, the residue was dissolved in water and freeze-dried. The formylation was complete, as evidenced by HPLC-MS analysis.

5.3.3. Iodination of Formylated Az

The iodination was carried out in general according to the published procedure [13]. In brief, 500 µL of diformyl azemiopsin solution in water (2.2 mM) were mixed with 100 µL of 1 M Tris-HCl buffer (pH 6.8), 30 µL of Na^{127}I solution (100 mM) and 15 µL of Na^{125}I solution with specific radioactivity of 2000 Ci/mmole. Then 27 µL of a 100 mM chloramine T solution was added to the mixture, the solution was mixed vigorously and incubated for 30 min at room temperature. The radioiodinated product was isolated by HPLC on Jupiter C18 column (10 µm, 10 × 250 mm, Phenomenex) in a linear gradient of acetonitrile in water from 10 to 25% in 15 min in the presence of 0.1% trifluoroacetic acid at a flow rate of 1 mL/min. The fraction containing diiodinated diformyl azemiopsin was concentrated on Savant SpeedVac Centrifuge Concentrator SVC100D and the solution obtained was used for deprotection.

5.3.4. Deprotection of Radioiodinated Az

To remove formyl protecting groups, the radioiodinated Az derivative was treated with 100 mM sodium hydroxide solution for 12 h. HPLC-MS analysis showed almost complete (>90%) removal of the formyl groups. After deblocking, the reaction mixture was neutralized with dilute hydrochloric acid, giving a solution of diiodoazemiopsin in an isotonic solution of sodium chloride. The concentration of Az in the resulting solution was determined spectrophotometrically from the absorbance at 280 nm and the radioactivity was measured using Wizard 1470 Automatic Gamma Counter (Perkin Elmer, Waltham, MA, USA). The specific radioactivity of the derivative was 0.15 Ci/mmole.

5.4. Toxicity Studies

5.4.1. Acute Toxicity

Az Acute Toxicity

Az acute toxicity was estimated for its intravenous (iv) and intramuscular (im) administration to male ICR mice in a stepwise procedure.

For the iv route of administration, 20 mice were randomly divided in 5 groups of 4 animals each. The animals of each group received a single dose of Az (0.3, 0.4, 0.5, 0.6, or 0.7 mg/kg, respectively) injected into the lateral tail vein at a volume of 1 mL/kg. For the im route of administration, first two experimental groups of 3 animals each were formed. Az at doses of 0.75 and 1.0 mg/kg, respectively, was injected into the quadriceps muscle of the thigh (quadriceps femoris muscle) of two mouse hind limbs at a volume of 0.5 mL/kg to each muscle. Then, five experimental groups of 5 animals each were formed. Az was injected into the quadriceps muscle of the thigh (quadriceps femoris muscle) of the two hind limbs of male ICR mice at doses of 0.8, 0.775, 0.75, 0.725 and 0.7 mg/kg. The injection volume to each limb was 1 mL/kg. Further, neurotoxic manifestations were recorded in animals at 5, 10, 20, 30, 60 and 90 min and 24 h after administration using the functional observation battery, the number of death was counted as well.

After iv and im injections of Az, the mice were observed for 24 h and signs of toxicity or lethality were recorded. For the iv and im route of administration, the median lethal dose (LD_{50}) was calculated using a probit analysis [40]. For the im route of administration, the maximum tolerated dose was determined.

Rocuronium Acute Toxicity

Rocuronium was administered intramuscularly at doses of 0.13 mg/kg ($n = 1$), 0.1 mg/kg ($n = 4$) and 0.08 mg/kg ($n = 4$) in the triceps of the forelimbs of male ICR mice (8–9 weeks old). Control animals were injected with saline ($n = 4$). The volume of administration was 0.5 mL/kg in each limb.

5.4.2. Subchronic Toxicity

A repeated dose 14-day intramuscular toxicity study was conducted on 36 SD male and 36 SD female adult rats. They were divided into 3 groups. The first control group was given normal saline, the second and the third experimental groups received Az at doses of 0.1 mg/kg and 0.5 mg/kg, respectively, daily for 14 days. The substances were injected into the quadriceps muscle of the thigh. Every day the animals were examined and any clinical signs of intoxication, body weight and food intake were recorded. The one half of the animals were euthanized on the 15th day of the study, the second half—after a 2-week cancellation period on the 29th day of the study. All animals were autopsied and their organs were inspected for any pathological signs, weighed and histologically examined. The ratio of organ-to-body weight was calculated for several organs: brain, heart, liver, spleen, thymus, kidneys, lungs, testicles and ovaries. The histological analysis was carried out for a number of organs: kidney, adrenal glands, testis, ovary, spleen, thymus, brain, heart, lung, liver, lymph node (mesenteric and mandibular), stomach, skin and muscle from the site of the administration.

A biochemical analysis of blood serum parameters was performed. A level of aspartate aminotransferase, alanine aminotransferase, glutamate dehydrogenase, alkaline phosphatase, gamma-glutamyl transferase, urea nitrogen, creatinine, total bilirubin, total protein, albumin, globulin, phosphorus, calcium, total cholesterol, triglycerides, albumin/globulin ratio was estimated. The analysis was performed using Randox GB reagent kits for each tested parameter and the automatic biochemical analyzer Sapphire-400 (Tokyo Boeki Ltd., Toyko, Japan).

A hematological analysis of animal blood included the evaluation of red and white blood cell number, a hemoglobin level, hematocrit, red cell distribution width, mean corpuscular volume (MCV), mean corpuscular hemoglobin (MCH), mean corpuscular hemoglobin concentration (MCHC), platelets number, a mean platelet volume, a mean platelet component and a cell number of neutrophils, eosinophils, basophils, lymphocytes, monocytes, large unstained cells, reticulocytes. The analysis was performed using a hematological analyzer Mythic 18 Vet (C2 DIAGNOSTICS S.A., Montpellier, France).

5.5. Pharmacokinetics

The pharmacokinetic study was performed using 163 male ICR mice. Four experimental groups with 40 animals in each were formed. Three intact mice were used as control animals. The animals of the first two experimental groups were intravenously injected with 0.25 and 0.50 mg/kg [^{125}I]-Az, respectively. The animals of the other two experimental groups were intramuscularly injected with 0.25 and 0.50 mg/kg [^{125}I]-Az, respectively. To estimate Az elimination rate, animal blood samples were taken from the orbital sinus 5, 15, 30, 60 min and 1, 2, 4 and 24 h after [^{125}I]-Az administration. The obtained blood samples were weighed. For each indicated time point, [^{125}I]-Az concentration was calculated as a mean value in blood samples of five animals. The radioactivity (cpm) in the blood samples was counted using a Wallac 1470 WIZARD® Gamma Counter (Perkin Elmer, Waltham, MA, USA). The specific radioactivity of [^{125}I]-Az was 6.26×10^7 cpm/mg. Radioactivity data (cpm) were re-calculated to the concentration of [^{125}I]-Az in the obtained blood samples. The specific pharmacokinetic parameters (AUC $(0 \to t)$, C_{max}, $T_{1/2}$) of [^{125}I]-Az were estimated.

5.6. Mutagenicity

The ability of Az to induce mutations in the hypoxanthine guanine phosphoribosyltransferase (hprt) gene of Chinese hamster ovary CHO-k1 cells was tested. The cells were purchased from the Russian collection of cell cultures (Institute of Cytology, Russian Academy of Sciences, Saint Petersburg, Russia). CHO-k1 cells were cultured in the growth medium DMEM/F12 with high glucose, glutamine and without Na_2CO_3, HEPES (Sigma-Aldrich Chemie Gmbh, Munich, Germany) supplemented with 10% FBS (BioSera, Nuaille, France), 0.1 M HEPES, 80 mg/mL gentamicin and 10 mg/mL fluconazole at 37 °C, 5% CO_2 in a CO_2 incubator. Before Az treatment CHO-k1 cells were subcultured and incubated in HAT medium (5 mM hypoxanthine, 20 mM aminopterin and 0.8 mM thymidine (Sigma-Aldrich Chemie Gmbh, Munich, Germany)) for three days, then the medium was changed and they were cultured for one day in HT medium (5 mM hypoxanthine and 0.8 mM thymidine (Sigma-Aldrich Chemie Gmbh, Munich, Germany)). After that, the cells were subcultured at a density of 40,000 cells per cm^2 in 10-cm Petri dishes. Next day they were treated for 4 h with different Az concentrations ranging from 2.8 to 2000 µg/mL in the presence or absence of a metabolic activation system (S9 mixture [15]). As positive controls, two high-mutagenic agents were used: ethylnitrosurea (6.25 and 12.5 µg/mL) and methylcholanthrene (2.5 and 5 µg/mL) in the absence and in the presence of the metabolic activation system, respectively.

After the treatment, the CHO-k1 cells were partly subcultured at a density of 150–500 cells per 55 cm^2 to determine the cytotoxicity of the different Az doses and control substances. The substance cytotoxicity was determined by a relative survival (RS) capacity of the cells, calculated as a ratio between the cloning efficiency (CE) of the cells plated immediately after the treatment and the normal cellular CE of non-treated cells (negative controls) after 7 days of culturing.

$$RS = \frac{CE \ of \ the \ treated \ cells}{CE \ of \ non - treated \ cells} \times 100$$

$$CE = \frac{Number \ of \ colonies}{Number \ of \ cells \ plated}$$

The rest of the treated CHO-k1 cells were maintained in the growth medium for 8 days to allow near-optimal phenotypic expression of any induced in hprt gene mutations. Then, to determine the frequency of the induced mutations, the cells were subcultured and maintained in the growth medium in the presence (2,000,000 cells) or in the absence (500 cells) of a selective agent (2.2 µg/mL 6-thioguanine) for 7 days. After that, the number of cell colonies in both media was counted and the frequency of mutations was calculated with the formula:

$$Mutation\ frequency = \frac{CE\ of\ mutant\ cells\ in\ a\ selective\ medium}{CE\ of\ cells\ in\ a\ non-selective\ medium}$$

5.7. Immunotoxicity

Three separate immunotoxicity tests were performed: (1) evaluation of cellular immunity in a delayed-type hypersensitivity test, (2) evaluation of animal immune response to a standard antigen, (3) evaluation of the phagocytic activity of peritoneal macrophages.

For carrying out these experiments 90 male ICR mice were used, they were divided equally into three groups (30 animals per group) for each study and in these groups they were subdivided into subgroups of 10 mice each and treated as follows: mice of the first subgroup were injected with normal saline im, the second subgroup animals were treated with Az (0.15 mg/kg im) and the animals of the third subgroup were treated with Az (0.50 mg/kg im). The drugs were injected into the quadriceps muscle of the thigh (musculus quadriceps femoris) daily for seven days (1 mL/kg).

5.7.1. Evaluation of Cellular Immunity in a Delayed-Type Hypersensitivity Test

On the last 7th day of the Az administration, the mice of the first group (30 animals) were immunized with a solution of trinitrobenzenesulfonic acid (TNBS) (200 µL, 10 mM) subcutaneously at the base of a tail. After 6 days the animals were secondarily immunized with TNBS (50 µL, 10 mM) injected in the pad of the left hind paw, the same volume of normal saline was injected into the right hind paw. 24 h after the second immunization, the animals were euthanized (CO_2 inhalation) and the weights of their experimental and control paws were determined. The reaction index was calculated using the indicated formula:

$$R_i = \frac{W_{exp} - W_{cont}}{W_{exp}} \times 100\%$$

R_i—reaction index, W_{exp}—weight of the experimental paw, W_{cont}—weight of the control paw.

5.7.2. Evaluation of Animal Immune Response to a Standard Antigen

On the last 7th day of the Az administration, the mice of the second group (30 animals) were immunized intraperitoneally with a 1:1 mixture of 200 µL of bovine serum albumin (BSA, 0.5 mg/mL) with a complete Freund's adjuvant. 10 days after the 1st immunization, the mice received the second 200 µL ip injection of the antigen (BSA, 0.5 mg/mL) in an incomplete Freund's adjuvant (1:1 ratio). After 7 days the venous mouse blood was collected from the inferior vena cava, then the blood serum was isolated and the titers of IgG antibodies to BSA were determined using a standard enzyme immunoassay.

5.7.3. Phagocytic Activity of Peritoneal Macrophages

On the next day after the seven-day course of the Az administration, the mice of the third group (30 animals) were injected intraperitoneally with 2 mL of ink particles suspension. After 10 min the mice were euthanized (CO_2 inhalation) with a subsequent isolation of the peritoneal exudate from their abdominal cavity. In the exudate (1 µL) the total number of peritoneal macrophages and the number of the ink particle-containing (phagocytic) cells among them were counted using a Gorjaev's chamber.

5.8. Allergenicity

Allergenicity of Az was studied in a delayed-type hypersensitivity test on male (20 animals) and female (20 animals) adult ICR mice. The mice were divided into four groups (two control and two experimental groups) of 10 animals (males or females) per group. Mice of the experimental groups were injected subcutaneously at the base of a tail with Az solution (0.15 mg/kg) emulsified in Freund's complete adjuvant in a ratio of 1:1 (2 mL/kg). Similarly, the mice in the control groups were administered with a suspension of normal saline in Freund's complete adjuvant. Five days later all animals received an injection of Az (0.15 mg/kg, 1.33 mL/kg) in the pad of the left hind paw. To estimate the intensity of the studied allergic reaction, 6, 12 and 24 h after the second Az injection the thickness of the left and right hind mouse paws was measured with a digital caliper.

5.9. Efficacy and Specificity Studies

5.9.1. Neuroblastoma Cell Culturing and Transient Transfection

Human neuroblastoma cells SH-SY5Y were cultured in DMEM/F12 medium (ThermoFisher Scientific, Waltham, MA, USA) supplemented with 10% fetal bovine serum (FBS) (PAA Laboratories GmbH, Pasching, Austria), 2.5 µg/mL amphotericin B and 50 µg/mL gentamicin in a CO_2 incubator at 37 °C and 5% CO_2 atmosphere. Cells were sub-cultured and plated at a density of 5000–10,000 cells per well in a 96-well black plate (Corning Inc., Corning, NY, USA). They were grown in a CO_2 incubator for 48–72 h before testing the functional activity of natively expressed nAChRs by calcium imaging.

Mouse neuroblastoma Neuro2a cells were purchased from the Russian collection of cell cultures (Institute of Cytology, Russian Academy of Sciences, Saint Petersburg, Russia). Cells were cultured in DMEM (Paneco, Moscow, Russia) supplemented with 10% FBS. They were sub-cultured the day before transfection and were plated at a density of 10,000 cells per well in a 96-well black plate. On the next day Neuro2a cells were transiently transfected with plasmids coding mouse muscle α1β1δε nAChR (pRBG4-vector) and a fluorescent calcium sensor Case12 (pCase12-cyto vector, Evrogen, Moscow, Russia) in a molar ratio of 2:1 following a lipofectamine transfection protocol (ThermoFisher Scientific, Waltham, MA, USA). Human α7 nAChR (α7 nAChR-pCEP4) was expressed accordingly with a co-expression of the human chaperone Ric-3 (Ric3-pCMV6-XL5, OriGene, Rockville, MD, USA) in a molar ratio 4:1. The transfected cells were grown at 37 °C in a CO_2 incubator for 48–72 h, before performing the calcium imaging assay.

5.9.2. Calcium Imaging

Calcium imaging procedure was performed as published earlier [11]. Briefly, after removing the growth medium, the transfected Neuro2a cells and cultivated SH-SY5Y cells were washed with a buffer containing 140 mM NaCl, 2 mM $CaCl_2$, 2.8 mM KCl, 4 mM $MgCl_2$, 20 mM HEPES, 10 mM glucose; pH 7.4. Neuro2a cells expressing muscle nAChR and the protein calcium sensor Case12 were proceeded directly, while SH-SY5Y cells expressing human α3-containing nAChRs natively were loaded with a fluorescent dye Fluo-4, AM (1.824 µM, ThermoFisher Scientific, Waltham, MA, USA) and a water-soluble probenecid (1.25 mM, ThermoFisher Scientific, Waltham, MA, USA) for 30 min at 37 °C and then were kept for 30 min at room temperature according to the manufacturer's protocol.

To detect the human α7 nAChR-mediated calcium response, transfected Neuro2a cells were incubated with its positive allosteric modulator PNU120596 (10 µM, Tocris Bioscience, Bristol, UK) for 20 min at room temperature before acetylcholine (Sigma-Aldrich Chemie Gmbh, Munich, Germany) addition. To assess mouse muscle or human α3-containing nAChRs this step was skipped. Transfected Neuro2a cells expressing muscle or α7 nAChRs and SH-SY5Y cells expressing α3-containing nAChRs natively were preincubated with Az for 15 minutes at room temperature before agonist addition.

The plates were transferred to the multimodal microplate reader Hidex Sence (Hidex, Turku, Finland) where the cells were excited by light of 485 nm wavelength and emitted fluorescence was detected at 535 ± 10 nm. Fluorescence was recorded every 2 s for three minutes following agonist addition. Responses were measured as peak intensity minus basal fluorescence level and were expressed as a percentage of the maximal response obtained to agonist. Data files were analyzed using Hidex Sence software (Hidex, Turku, Finland) and OriginPro 7.5 software (OriginLab, Northampton, MA, USA, for statistical analysis).

5.9.3. Electrophysiological Experiments

Ovary tissue from adult female *Xenopus laevis* was cut into small pieces and these pieces were digested with collagenase A (4 mg mL^{-1}, Worthington Biochemical Corp., Lakewood, NJ, USA) in Barth's solution without calcium (88.0 mM NaCl, 1.1 mM KCl, 2.4 mM NaHCO$_3$, 0.8 mM MgSO$_4$, 15.0 mM HEPES/NaOH, pH 7.6) for 1.5 ± 2 h at 20 °C. The oocytes were stored in Barth's solution with calcium (88.0 mM NaCl, 1.1 mM KCl, 2.4 mM NaHCO$_3$, 0.3 mM Ca (NO$_3$)$_2$, 0.4 mM CaCl$_2$, 0.8 mM MgSO$_4$, 15.0 mM HEPES/NaOH, pH 7.6) supplemented with 63.0 µg/mL penicillin-G sodium salt, 40.0 µg/mL streptomycin sulfate and 40.0 µg/mL gentamicin. Stage V $\pm$ VI oocytes were selected and injected with 3 ng plasmids coding the rat $\alpha4$ and $\beta2$ nAChR subunits (pcDNA3.1 vector) in a molar ratio of 1:1 using an Auto-Nanoliter Injector NanoJect-2 (Drummond Scientific Company, Broomall, PA, USA) in a total injection volume of 23 nL. After injection, oocytes were incubated at 18 °C in Barth's solution with calcium for 48–120 h. Electrophysiological recordings were made using a Turbo TEC-03X amplifier (npi electronic GmbH, Tamm, Germany) and WinWCP recording software (University of Strathclyde, Glasgow, UK). Oocytes were placed in a small recording chamber with a working volume of 50 µL and 100 µL of ligands (50 µM Az, 20 µM nicotine) solution in Barth's buffer were applied to an oocyte. Az was pre-applied to an oocyte for 5 min before its co-application with agonist nicotine. To allow receptor recovery from desensitization, the oocytes were superfused for 5–10 min with buffer (1 mL/min) between ligand applications. Electrophysiological recordings were performed at a holding potential of -60 mV.

5.9.4. In Vivo Muscle Relaxant Effect

In Vivo Muscle Relaxant Effect of Az

40 male ICR mice were divided equally into four groups: one control and three experimental groups. Animals of the control and experimental groups were treated with normal saline and with Az at doses of 0.03, 0.1 and 0.3 mg/kg, respectively. In addition 6 mice were treated with Az at dose of 0.01 mg/kg. The corresponding solutions were injected into the triceps muscles of the mouse forelimbs at a volume of 0.5 mL/kg per each limb. For all animals their basic forelimb grip strength was recorded before the substance administration with a 1027 grip strength meter (Columbus Instruments, Columbus, OH, USA). Further, their grip strength was measured 5, 10, 15, 20, 30, 60, 90 min after the Az (or normal saline) im administration.

In Vivo Muscle Relaxant Effect of Rocuronium

Rocuronium was administered intramuscularly as described for the measurement of acute toxicity (Section 5.4.1). Muscle strength was assessed prior to administration of the substance and then 1, 2, 3, 4, 5 min after administration.

5.10. Statistical Analysis

The statistical analysis of the obtained results was performed using Statistica 7.1 (TIBCO Software Inc., Palo Alto, CA, USA) and OriginPro 9.1 (Microcal, Northampton, MA, USA) software. The data of Az efficacy in vivo were analyzed using one-way repeated measures ANOVA test. To statistically evaluate the efficacy and specificity of Az in vitro, two-tailed Mann–Whitney U test was used. The data of Az subchronic toxicity, immunotoxicity and allergenicity were analyzed using Kruskall-Wallis ANOVA on ranks test. Results are expressed as mean of data $\pm$ SEM unless otherwise stated. In all tests, $p < 0.05$ was taken as significant.

Supplementary Materials: The following are available online at www.mdpi.com/2072-6651/10/1/34/s1, Figure S1: Muscle relaxant effect of Az at dose of 0.01 mg/kg.

Acknowledgments: This study was supported by the Russian Science Foundation (project No. 16-14-00215).

Author Contributions: I.V.S., M.N.Z. and A.V.L. conceived and designed the experiments; I.V.S., M.N.Z., A.V.L., I.A.I., A.I.G., I.N.K., E.A.R., M.A.S., E.A.T., V.A.R., G.A.S. and N.S.E. performed the experiments; I.V.S., M.N.Z., A.V.L., I.S.M., V.I.T. and Y.N.U. analyzed the data; I.A.I. and N.S.E contributed materials; I.V.S., A.V.L., I.A.I, A.I.G., V.I.T. and Y.N.U. wrote the paper.

Conflicts of Interest: The authors declare no conflict of interest. The founding sponsors had no role in the design of the study; in the collection, analyses, or interpretation of data; in the writing of the manuscript and in the decision to publish the results.

References

1. Utkin, Y.N.; Weise, C.; Kasheverov, I.E.; Andreeva, T.V.; Kryukova, E.V.; Zhmak, M.N.; Starkov, V.G.; Hoang, N.A.; Bertrand, D.; Ramerstorfer, J.; et al. Azemiopsin from *Azemiops feae* viper venom, a novel polypeptide ligand of nicotinic acetylcholine receptor. *J. Biol. Chem.* **2012**, *287*, 27079–27086. [CrossRef] [PubMed]

2. Unwin, N. Nicotinic acetylcholine receptor and the structural basis of neuromuscular transmission: Insights from Torpedo postsynaptic membranes. *Q. Rev. Biophys.* **2013**, *46*, 283–322. [CrossRef] [PubMed]

3. Jonsson, F.M.; Dabrowski, M.; Eriksson, L.I. Pharmacological characteristics of the inhibition of nondepolarizing neuromuscular blocking agents at human adult muscle nicotinic acetylcholine receptor. *Anesthesiology* **2009**, *110*, 1244–1252. [CrossRef]

4. Bowman, W.C. Neuromuscular block. *Br. J. Pharmacol.* **2006**, *147*, S277–S286. [CrossRef] [PubMed]

5. Habre, W.; Adamicza, A.; Lele, E.; Novák, T.; Sly, P.D.; Petak, F. The involvement of histaminic and muscarinic receptors in the bronchoconstriction induced by myorelaxant administration in sensitized rabbits. *Anesth. Analg.* **2008**, *107*, 1899–1906. [CrossRef] [PubMed]

6. Bornia, E.C.; Bando, E.; Machinski, M., Jr.; Pereira, M.W.; Alves-Do-Prado, W. Presynaptic M1, M2, and A1 receptors play roles in tetanic fade induced by pancuronium or cisatracurium. *J. Anesth.* **2009**, *23*, 513–519. [CrossRef] [PubMed]

7. Albanese, A.; Bhatia, K.; Bressman, S.B.; Delong, M.R.; Fahn, S.; Fung, V.S.; Hallett, M.; Jankovic, J.; Jinnah, H.A.; Klein, C.; et al. Phenomenology and classification of dystonia: A consensus update. *Mov. Disord.* **2013**, *28*, 863–873. [CrossRef] [PubMed]

8. Kudryavtsev, D.; Shelukhina, I.; Vulfius, C.; Makarieva, T.; Stonik, V.; Zhmak, M.; Ivanov, I.; Kasheverov, I.; Utkin, Y.; Tsetlin, V. Natural compounds interacting with nicotinic acetylcholine receptors: From low-molecular weight ones to peptides and proteins. *Toxins* **2015**, *7*, 1683–1701. [CrossRef] [PubMed]

9. Utkin, Y.N.; Zhmak, M.N.; Andreeva, T.V.; Weise, C.; Kryukova, E.V.; Tsetlin, V.I.; Kasheverov, I.E.; Starkov, V.G. Peptide Azemiopsin Selectively Interacting with Nicotinic Cholinoreceptors of the Muscle Type and Suitable for Use as Muscle Relaxant in Medicine and Cosmetology. Russian Patent R.U. 2,473,559, 26 October 2011. (In Russian)

10. Kasheverov, I.; Zhmak, M.; Chivilyov, E.; Saez-Brionez, P.; Utkin, Y.; Hucho, F.; Tsetlin, V. Benzophenone-type photoactivatable derivatives of α-neurotoxins and α-conotoxins in studies on Torpedo nicotinic acetylcholine receptor. *J. Recept. Signal Transduct. Res.* **1999**, *19*, 559–571. [CrossRef] [PubMed]

11. Shelukhina, I.; Spirova, E.; Kudryavtsev, D.; Ojomoko, L.; Werner, M.; Methfessel, C.; Hollmann, M.; Tsetlin, V. Calcium imaging with genetically encoded sensor Case12: Facile analysis of α7/α9 nAChR mutants. *PLoS ONE* **2017**, *12*, e0181936. [CrossRef] [PubMed]

12. Tsomides, T.J.; Eisen, H.N. Stoichiometric labeling of peptides by iodination on tyrosyl or histidyl residues. *Anal. Biochem.* **1993**, *210*, 129–135. [CrossRef] [PubMed]

13. Kasheverov, I.E.; Zhmak, M.N.; Khruschov, A.Y.; Tsetlin, V.I. Design of new α-conotoxins: From computer modeling to synthesis of potent cholinergic compounds. *Mar. Drugs* **2011**, *9*, 1698–1714. [CrossRef] [PubMed]

14. Hall, R.L. Clinical Pathology of Laboratory Animals. In *Animal Models in Toxicology*, 2nd ed.; Gad, S.C., Ed.; CRC Press Taylor & Francis Group: Boca Raton, FL, USA, 2007; pp. 787–830.

15. Johnson, G.E. Mammalian Cell *HPRT* Gene Mutation Assay: Test Methods. In *Genetic Toxicology. Methods in Molecular Biology (Methods and Protocols)*; Parry, J., Parry, E., Eds.; Springer: New York, NY, USA, 2012; Volume 817, ISBN 978-1-61779-420-9.

16. Matsuoka, M.; Hayashi, M.; Ishidate, M., Jr. Chromosomal aberration tests on 29 chemicals combined with S9 mix in vitro. *Mutat. Res.* **1979**, *66*, 277–290. [CrossRef]

17. Legal Portal of Eurasian Economic Union. Available online: https://docs.eaeunion.org/docs/ru-ru/01411927/cncd_21112016_81 (accessed on 5 January 2018). (In Russian)

18. Jonsson, M.; Gurley, D.; Dabrowski, M.; Larsson, O.; Johnson, E.C.; Eriksson, L.I. Distinct pharmacologic properties of neuromuscular blocking agents on human neuronal nicotinic acetylcholine receptors: A possible explanation for the train-of-four fade. *Anesthesiology* **2006**, *105*, 521–533. [CrossRef] [PubMed]

19. Montilla-García, Á.; Tejada, M.Á.; Perazzoli, G.; Entrena, J.M.; Portillo-Salido, E.; Fernández-Segura, E.; Cañizares, F.J.; Cobos, E.J. Grip strength in mice with joint inflammation: A rheumatology function test sensitive to pain and analgesia. *Neuropharmacology* **2017**, *125*, 231–242. [CrossRef] [PubMed]

20. Nevins, M.E.; Nash, S.A.; Beardsley, P.M. Quantitative grip strength assessment as a means of evaluating muscle relaxation in mice. *Psychopharmacology* **1993**, *110*, 92–96. [CrossRef] [PubMed]

21. Torii, Y.; Kiyota, N.; Sugimoto, N.; Mori, Y.; Goto, Y.; Harakawa, T.; Nakahira, S.; Kaji, R.; Kozaki, S.; Ginnaga, A. Comparison of effects of botulinum toxin subtype A1 and A2 using twitch tension assay and rat grip strength test. *Toxicon* **2011**, *57*, 93–99. [CrossRef] [PubMed]

22. Wright, P.M. Population based pharmacokinetic analysis: Why do we need it; what is it; and what has it told us about anaesthetics? *Br. J. Anaesth.* **1998**, *80*, 488–501. [CrossRef] [PubMed]

23. Yuan, G.; Zhang, R.; Wang, B.; Wei, C.; Liu, X.; Zhao, W.; Guo, R. Determination of Rocuronium in Human Plasma by High Performance Liquid Chromatography-Tandem Mass Spectrometry and its Pharmacokinetics in Patients. *J. Bioequiv. Availab.* **2012**, *4*, 7. [CrossRef]

24. Van Miert, M.M.; Eastwood, N.B.; Boyd, A.H.; Parker, C.J.; Hunter, J.M. The pharmacokinetics and pharmacodynamics of rocuronium in patients with hepatic cirrhosis. *Br. J. Clin. Pharmacol.* **1997**, *44*, 139–144. [CrossRef] [PubMed]

25. Dragne, A.; Varin, F.; Plaud, B.; Donati, F. Rocuronium pharmacokinetic-pharmacodynamic relationship under stable propofol or isoflurane anesthesia. *Can. J. Anaesth.* **2002**, *49*, 353–360. [CrossRef] [PubMed]

26. Naguib, M.; Lien, C.A. Pharmacology of muscle relaxants and their antagonists. In *Miller's Anesthesia*, 6th ed.; Churchill Livingstone: London, UK, 2005; Volume 3, pp. 493–515.

27. Vandenbrom, R.H.; Wierda, J.M.; Agoston, S. Pharmacokinetics and neuromuscular blocking effects of atracurium besylate and two of its metabolites in patients with normal and impaired renal function. *Clin. Pharmacokinet.* **1990**, *19*, 230–240. [CrossRef] [PubMed]

28. Szenohradszky, J.; Fisher, D.M.; Segredo, V.; Caldwell, J.E.; Bragg, P.; Sharma, M.L.; Gruenke, L.D.; Miller, R.D. Pharmacokinetics of rocuronium bromide (ORG 9426) in patients with normal renal function or patients undergoing cadaver renal transplantation. *Anesthesiology* **1992**, *77*, 899–904. [CrossRef] [PubMed]

29. Lebrault, C.; Berger, J.L.; D'Hollander, A.A.; Gomeni, R.; Henzel, D.; Duvaldestin, P. Pharmacokinetics and pharmacodynamics of vecuronium (ORG NC 45) in patients with cirrhosis. *Anesthesiology* **1985**, *62*, 601–605. [CrossRef] [PubMed]

30. SAGENT Pharmaceuticals. Available online: http://www.sagentpharma.com/wp-content/uploads/2015/05/Rocuronium_SDS.pdf (accessed on 5 January 2018).

31. Ginsburg, S.; Kitz, R.J.; Savarese, J.J. Neuromuscular blocking activity of a new series of quaternary N-substituted choline esters. *Br. J. Pharmacol.* **1971**, *43*, 107–126. [CrossRef] [PubMed]

32. SAGENT Pharmaceuticals. Available online: http://www.sagentpharma.com/wp-content/uploads/2017/04/Vecuronium_SDS.pdf (accessed on 5 January 2018).

33. Descotes, J. Immunotoxicology: Role in the safety assesment of drugs. *Drug Saf.* **2005**, *28*, 127–136. [CrossRef] [PubMed]

34. Peroni, D.G.; Sansotta, N.; Bernardini, R.; Crisafulli, G.; Franceschini, F.; Caffarelli, C.; Boner, A.L. Muscle relaxants allergy. *Int. J. Immunopathol. Pharmacol.* **2011**, *24*, S35–S46. [CrossRef] [PubMed]

35. Reddy, J.I.; Cooke, P.J.; van Schalkwyk, J.M.; Hannam, J.A.; Fitzharris, P.; Mitchell, S.J. Anaphylaxis is more common with rocuronium and succinylcholine than with atracurium. *Anesthesiology* **2015**, *122*, 39–45. [CrossRef] [PubMed]

36. Zan, U.; Topaktas, M.; Istifli, E.S. In vitro genotoxicity of rocuronium bromide in human peripheral lymphocytes. *Cytotechnology* **2011**, *63*, 239–245. [CrossRef] [PubMed]

37. Moore, E.W.; Hunter, J.M. The new neuromuscular blocking agents: Do they offer any advantages? *Br. J. Anaesth.* **2001**, *87*, 912–925. [CrossRef] [PubMed]

38. Kim, Y.B.; Sung, T.-Y.; Yang, H.S. Factors that affect the onset of action of non-depolarizing neuromuscular blocking agents. *Korean J. Anesthesiol.* **2017**, *70*, 500–510. [CrossRef] [PubMed]

39. Bonner, A.G.; Udell, L.M.; Creasey, W.A.; Duly, S.R.; Laursen, R.A. Solid-phase precipitation and extraction, a new separation process applied to the isolation of synthetic peptides. *J. Pept. Res.* **2001**, *57*, 48–58. [CrossRef] [PubMed]
40. Finney, D.J. *Probit Analysis*, 3rd ed.; Cambridge University Press: Cambridge, MA, USA, 1971.

© 2018 by the authors. Licensee MDPI, Basel, Switzerland. This article is an open access article distributed under the terms and conditions of the Creative Commons Attribution (CC BY) license (http://creativecommons.org/licenses/by/4.0/).

MDPI

Article

Bee Venom Suppresses the Differentiation of Preadipocytes and High Fat Diet-Induced Obesity by Inhibiting Adipogenesis

Se-Yun Cheon [1], Kyung-Sook Chung [2], Seong-Soo Roh [3] (ID), Yun-Yeop Cha [4] and Hyo-Jin An [1,*]

[1] Department of Pharmacology, College of Korean Medicine, Sang-ji University, Wonju-si, Gangwon-do 26339, Korea; chunsay1008@naver.com
[2] Catholic Precision Medicine Research Center, College of Medicine, The Catholic University of Korea, 222, Banpo-daero, Seocho-gu, Seoul 06591, Korea; adella76@hanmail.net
[3] Department of Herbology, College of Korean Medicine, Daegu Hanny University, Suseong-gu, Deagu 42158, Korea; ddede@dhu.ac.kr
[4] Department of Rehabilitation Medicine of Korean Medicine and Neuropsychiatry, College of Korean Medicine, Sang-ji University, Wonju-si, Gangwon-do 26339, Korea; omdcha@sangji.ac.kr
* Correspondence: hjan@sj.ac.kr; Tel.: +82-33-738-7503; Fax: +82-33-730-0679

Received: 8 December 2017; Accepted: 21 December 2017; Published: 24 December 2017

Abstract: Bee venom (BV) has been widely used in the treatment of certain immune-related diseases. It has been used for pain relief and in the treatment of chronic inflammatory diseases. Despite its extensive use, there is little documented evidence to demonstrate its medicinal utility against obesity. In this study, we demonstrated the inhibitory effects of BV on adipocyte differentiation in 3T3-L1 cells and on a high fat diet (HFD)-induced obesity mouse model through the inhibition of adipogenesis. BV inhibited lipid accumulation, visualized by Oil Red O staining, without cytotoxicity in the 3T3-L1 cells. Male C57BL/6 mice were fed either a HFD or a control diet for 8 weeks, and BV (0.1 mg/kg or 1 mg/kg) or saline was injected during the last 4 weeks. BV-treated mice showed a reduced body weight gain. BV was shown to inhibit adipogenesis by downregulating the expression of the transcription factors CCAAT/enhancer-binding proteins (C/EBPs) and the peroxisome proliferator-activated receptor gamma (PPARγ), using RT-qPCR and Western blotting. BV induced the phosphorylation of AMP-activated kinase (AMPK) and acetyl-CoA carboxylase (ACC) in the cell line and in obese mice. These findings demonstrate that BV mediates anti-obesity/differentiation effects by suppressing obesity-related transcription factors.

Keywords: Bee venom; PPARγ; AMPK; MAPK; adipogenesis

1. Introduction

Obesity is defined as an abnormal or excessive fat accumulation that presents a risk to health. Adipocyte hyperplasia and hypertrophy are determinant factors of obesity [1]. Adipocytes differentiate from stem cells or other precursor cells [2]. Differentiating and maturing adipocytes involve a complex program of gene expression that is important for obesity-related diseases [3]. 3T3-L1 preadipocytes have been used in studies regarding adipogenesis and differentiation. These cells differentiate in response to adipogenic inducers including insulin, dexamethasone, and 3-isobutyl-1-methylxanthine (IBMX) [4]. The differentiation sequence from preadipocytes to adipocytes comprises confluence, mitotic clonal expansion (MCE), and terminal differentiation. In the first stage, confluent cells enter a growth arrest phase [5,6]. These growth-arrested cells subsequently restart the cell cycle and increase cell numbers three- to four-fold during the MCE phase [7]. This hyperplasia during cell differentiation is related to the production of specific adipogenic transcription factors [8].

Peroxisome proliferator-activated receptors (PPAR)γ and CCAAT/enhancer-binding proteins (C/EBPs) promote the differentiation of adipocytes [9]. During early-stage differentiation of 3T3-L1 cells, the expression of C/EBPβ and C/EBPδ is increased after hormonal induction, followed by increases in the expression of C/EBPα and PPARγ [10]. C/EBPδ is important for MCE to occur during differentiation in the early stage of adipogenesis [11]. Gene expression of C/EBPβ induces the expression of PPARγ and C/EBPα [12,13]. The activation of the C/EBP family and of PPARγ regulates the expression of various adipogenic factors that promote fat accumulation.

Adenosine monophosphate–activated protein kinase (AMPK), known as a regulator of energy homeostasis, is an important target for controlling obesity [14]. In adipogenesis, AMPK activation regulates glucose and lipid metabolism by inactivating metabolic enzymes [15]. Phosphorylation by AMPK inactivates acetyl-coenzyme A carboxylase (ACC) and 3-hydroxy-3-methylglutaryl-coenzyme A reductase (HMGCR), leading to the inhibition of fatty acids and cholesterol syntheses, as well as increased fatty acid oxidation [16]. AMPK regulates the channeling of acyl-CoA towards β-oxidation and lipid biosynthesis, leading to the inhibition of glycerol-3-phosphate acyltransferase (GPAT) [17]. In addition, the phosphorylation of AMPK also inhibits the expression of adipogenic transcription factors, such as C/EBPβ, C/EBPδ, C/EBPα, and PPARγ [18].

Bee venom (BV), a complex mixture of proteins, peptides, and low molecular weight components, is an effective defense tool used for the protection of the hive by the honey bee [19]. Despite causing pain to humans who are stung, BV has been used as a traditional medicine to treat a diverse range of conditions, including tumors, skin diseases, and pain [20]. It has also been reported that the inhibition of atherosclerotic lesions via anti-inflammatory mechanisms and the suppression of benign prostatic hyperplasia in rats are additional beneficial properties [21,22]. BV is known to contain a complex mixture of active enzymes and peptides, including phospholipase A2, melittin, and apamin [23]. Melittin is a major component of BV that has been shown to improve atherosclerotic lesions and to downregulate pro-inflammatory cytokines, adhesion molecules, proatherogenic proteins, and the NF-κB signal pathway in high-fat treated atherosclerotic animal models [24]. In addition, apamin attenuated lipids, proinflammatory cytokines, adhesion molecules, fibrotic factors, and macrophage infiltration in LPS/fat-induced atherosclerotic mice [25]. BV was also reported to exhibit anti-obesity effects [26] but the mechanism has not yet been clarified. In the present study, we investigated the anti-obesity effects of BV in 3T3-L1 preadipocytes and in an HFD-induced obesity animal model.

2. Results

2.1. BV Suppressed Cell Hyperplasia and Lipid Accumulation during Differentiation of 3T3-L1 Adipocytes

Hyperplasia and lipid accumulation are known to occur in the 3T3-L1 cell line during differentiation [8,27]. 3T3-L1 cells were treated with BV (concentrations from 1.25 to 40 μg/mL) in the differentiation media (MDI) or the growth media (BS). After treatment with BV, the cell viability was determined by a 3-(4,5-Dimethylthiazol-2-yl)-2,5-diphenyl tetrazolium bromide (MTT) assay (Figure 1A). BV had no effect on the cell viability in the culture media. However, BV significantly reduced the cell viability at 2.5 μg/mL or higher concentrations during the differentiation of adipocytes.

To determine the inhibitory effect of BV on lipid accumulation in adipocytes, 3T3-L1 cells were treated with and without BV (2.5, 5, or 10 μg/mL) for 9 days. As shown in Figure 1B, differentiation-induced adipocytes drastically increased their lipid storage approximately 2.32-fold, compared with undifferentiated cells. In contrast, treatment with BV significantly reduced lipid droplet accumulation in a dose-dependent manner.

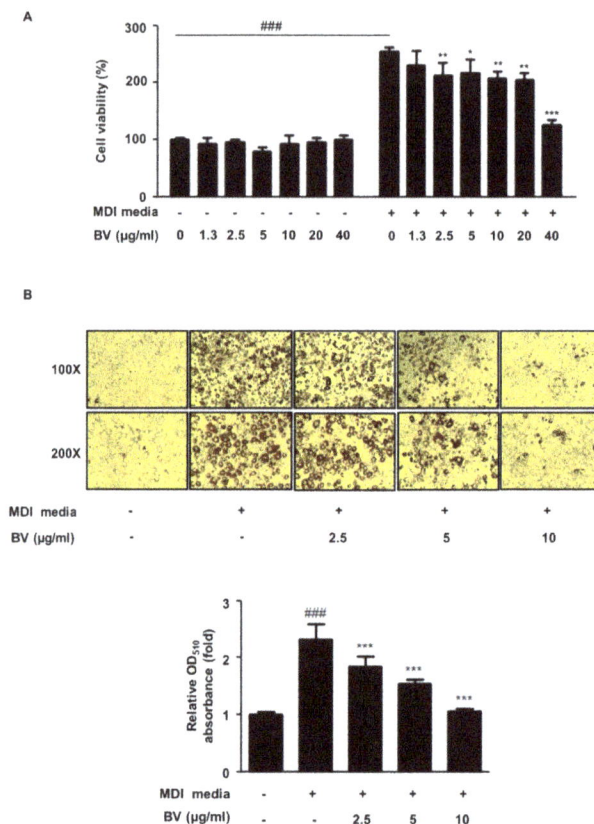

Figure 1. Bee venom (BV) suppressed lipid accumulation in 3T3-L1 preadipocytes. (**A**) Cells were cultured in the growth medium or the differentiation medium containing concentrations ranging from 1.3 to 40 μg/mL of BV for 3 days. (**B**) Preadipocytes were differentiated with and without BV (2.5, 5, and 10 μg/mL) for 9 days. Differentiated cells were stained with Oil red O and images were taken with a Leica DM IL LED microscope (100× and 200× magnifications). (**C**) Oil red O was extracted from lipid droplets using isopropanol and was measured at 510 nm. ### $p < 0.001$ vs. non-differentiation cells. * $p < 0.05$, ** $p < 0.01$, *** $p < 0.001$ vs. differentiation cells.

2.2. BV Suppressed the Expression of Adipogenic Markers during Differentiation of 3T3-L1 Adipocytes

Transcription factors, such as PPARγ and the C/EBP family, are known to play important roles during the maturation and the differentiation of adipocytes [2,28]. In this study, we investigated the anti-adipogenesis activity of BV in differentiated adipocytes using qRT-PCR and Western blot analysis. As shown in Figure 2, the mRNA expression of C/EBPα, C/EBPβ, C/EBPδ, and PPARγ was upregulated in differentiation-induced adipocytes, compared to what was seen in undifferentiated cells. However, the treatment with BV significantly downregulated the mRNA expression of the C/EBP family and decreased PPARγ expression (both mRNA and protein levels; Figure 2B).

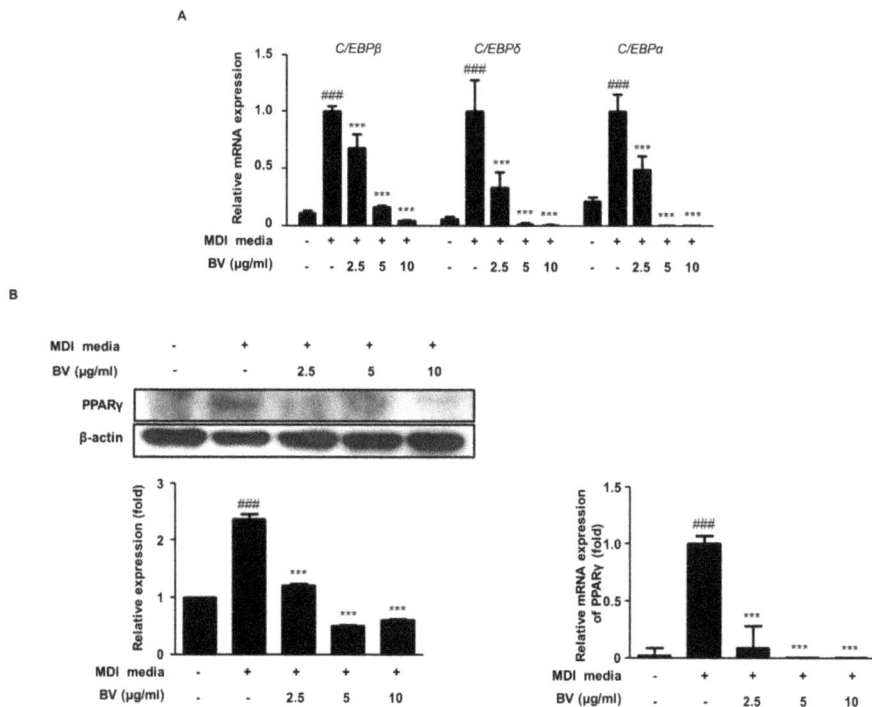

Figure 2. The effects of BV on the expression levels of lipid metabolism and adipogenesis in 3T3-L1 cells. (**A**) The mRNA levels of C/EBPβ, C/EBPδ, and C/EBPα were measured by qRT-PCR. (**B**) The protein and the mRNA expression levels of PPARγ were determined by Western blot analysis or by qRT-PCR. β-actin was used as an internal control. ### $p < 0.001$ vs. non-differentiation cells, *** $p < 0.001$ vs. differentiation cells.

2.3. BV Regulated the MAPK Pathway during Differentiation of 3T3-L1 Preadipocytes

Mitogen-activated protein kinases (MAPKs) play a crucial role in many essential cellular responses, including proliferation, apoptosis, and differentiation [29]. In particular, MAPKs are associated with regulatory effects on adipocyte differentiation [30]. To investigate the effect of BV on the MAPK pathway, we determined the protein expression of factors involved in the MAPK pathway using Western blot analysis in differentiated adipocytes. As shown in Figure 3A, the phosphorylation of ERK and JNK decreased, whereas the phosphorylation of p38 increased during differentiation, compared to that of the undifferentiated control cells. The phosphorylation of ERK and JNK was upregulated in the BV-treated adipocytes and the phosphorylation of p38 was significantly decreased.

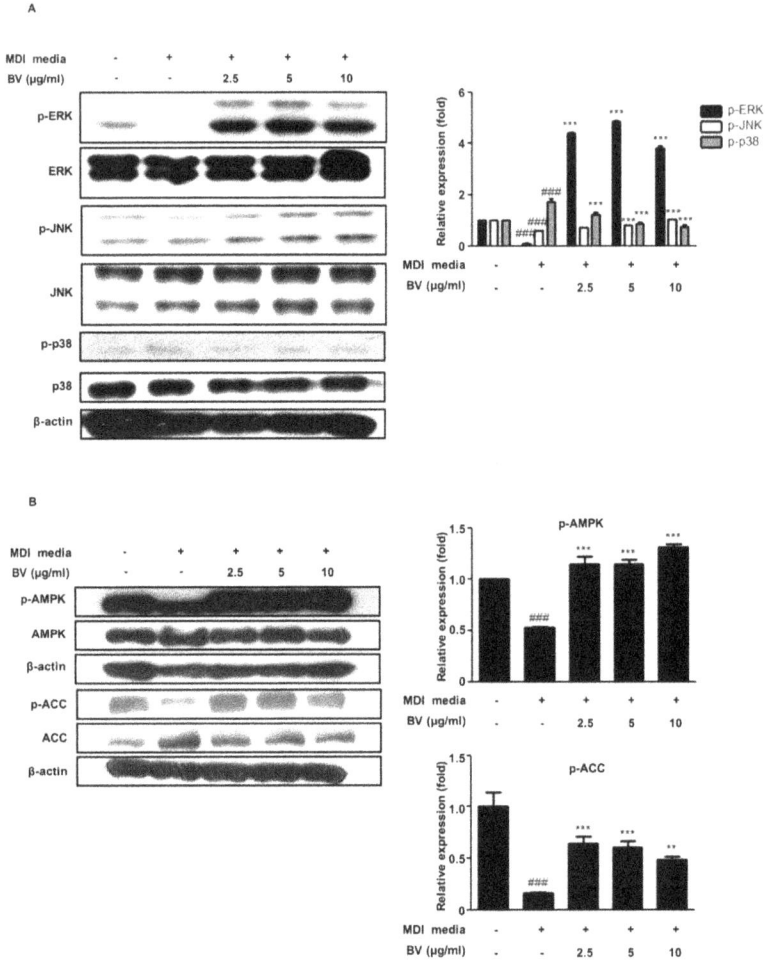

Figure 3. The effects of BV on the mitogen-activated protein kinase (MAPK) and the adenosine monophosphate-activated protein kinase (AMPK) pathways in 3T3-L1 preadipocytes. 3T3-L1 cells were treated with BV (2.5, 5, and 10 µg/mL for 3 days). The total cell lysates were analyzed by Western blotting to determine the expression of (**A**) MAPKs (**B**) AMPK and acetyl coenzyme A carboxylase (ACC). β-actin was used as an internal control. $^{\#\#\#}$ $p < 0.001$ vs. non-differentiation cells. ** $p < 0.01$, *** $p < 0.001$ vs. differentiation cells.

2.4. BV Activated the AMPK Pathway during the Differentiation of 3T3-L1 Preadipocytes

AMPK, a key enzyme of energy metabolism, regulates glucose and lipid metabolism [31]. The phosphorylation of AMPK inhibits the expression of adipogenic transcription factors, such as C/EBPs and PPARγ [18]. To investigate the effect of BV on the activation of AMPK, we determined the degree of AMPK phosphorylation using Western blotting. As shown in Figure 3B, AMPK phosphorylation was decreased in the differentiated 3T3-L1 preadipocytes, compared with the undifferentiated control cells. Meanwhile, our results revealed that treatment with BV significantly recovered the AMPK phosphorylation in differentiated adipocytes.

2.5. BV Suppressed Body Weight, Fat, and Lipid Accumulation in HFD-Induced Obese Mice

As shown in Figure 4, the total body weight of mice in the high fat diet (HFD) group, including both fat and body weight gain, were significantly increased, compared to that of the normal diet (ND) group. In contrast to the HFD group, the BV injection suppressed the total body weight, fat, and body weight gain. Obesity is characterized by the hypertrophy and the hyperplasia of adipose tissue [32], therefore we examined the inhibitory effect of BV using hematoxylin and eosin (H&E) staining. As shown in Figure 4E, H&E analysis data indicated that the HFD-fed mice showed hypertrophy of adipocytes in the epididymal adipose tissue, whereas BV treated groups showed an inhibited hypertrophy of adipocytes.

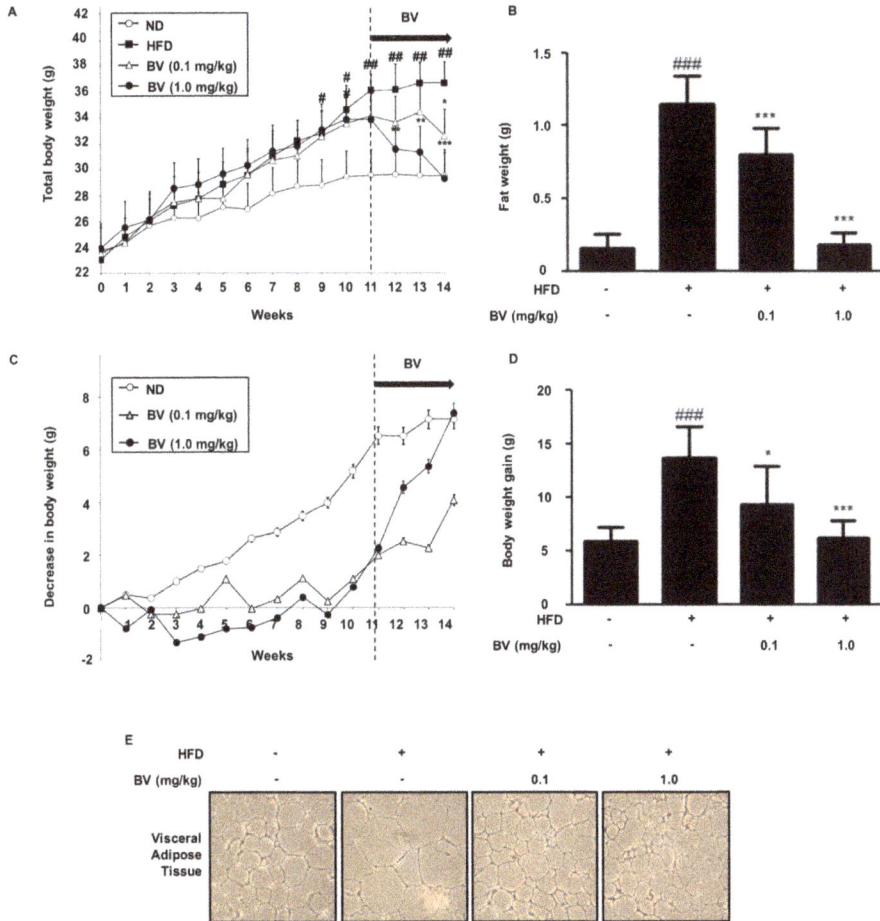

Figure 4. Effects of BV injection on (**A**) total body weight, (**B**) fat weight, (**C**) a decrease in body weight, (**D**) body weight gain, and (**E**) Epididymis adipose tissue (100× magnifications) from mice. The duration of the experimental window was 14 weeks during which mice were fed on normal diet or a high-fat diet. The representative photographic images of mice were from different treatment/feeding groups at the time of sacrifice. Values are expressed as the mean ± SEM of 10 mice per group. $^{\#} p < 0.05$, $^{\#\#} p < 0.01$, $^{\#\#\#} p < 0.001$ vs. the normal diet (ND) group. $^{*} p < 0.05$, $^{**} p < 0.01$, $^{***} p < 0.001$ vs. the high fat diet (HFD) group.

2.6. BV Suppressed Adipogenic Markers and Activated the AMPK Pathway in HFD-Induced Obese Mice

The adipogenic transcription factors, PPARγ and C/EBPα, regulate the genes for fat accumulation. Several in vivo studies have shown that PPARγ plays a crucial role in adipogenesis [9,33]. The inhibition of adipogenesis in the BV-treated groups was investigated and was found to be associated with molecular signaling, by the adipogenic markers involved in lipid metabolism and adipogenesis. We examined the protein expression of PPARγ and C/EBPα in adipose tissue, and it was significantly upregulated in the HFD group. In contrast, the expression of factors related to adipogenesis decreased in the BV-treated groups, compared to the expression in the HFD group.

The activation of AMPK to p-AMPK inhibited the activity of ACC, which resulted in an increase of carnitine palmitoyl-CoA transferase-1 [16]. ACC regulates the metabolism of fatty acids, and in particular, the ACC1 isoform of ACC regulates fatty acid synthesis [34]. In experiments with mice, the phosphorylated level of AMPK was not significantly altered in the HFD-fed group. However, the phosphorylation of ACC in mice was downregulated by HFD. BV treatment significantly increased the phosphorylated levels of AMPK and of ACC (Figure 5B,C).

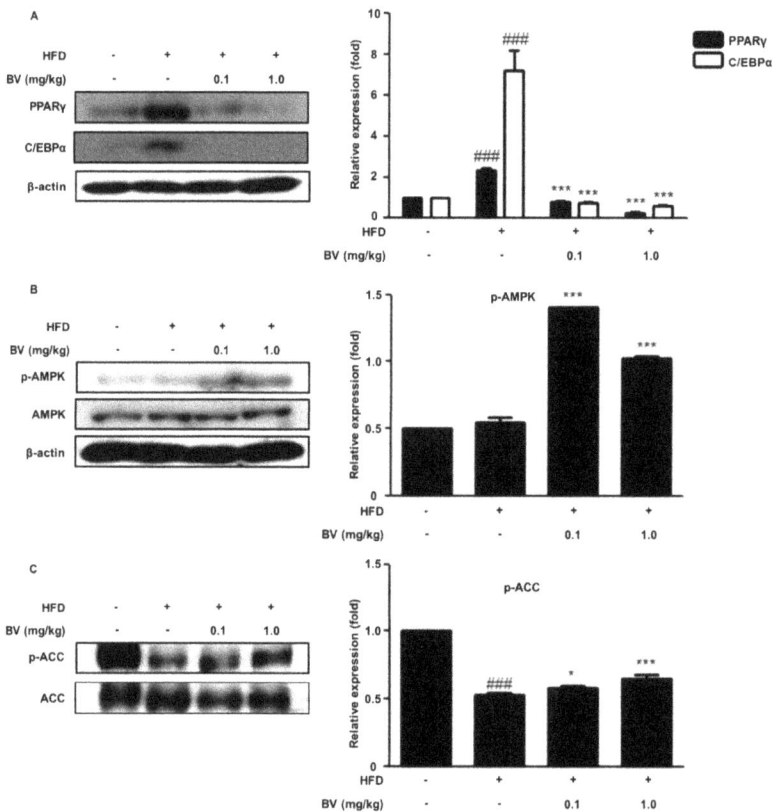

Figure 5. BV regulated adipogenic markers and the AMPK pathway in HFD-induced mice. (**A**) Important adipogenic transcription factors, PPAR γ and C/EBPα, determined using Western blot analysis. (**B**) AMPK and (**C**) ACC determined using Western blot analysis. β-actin was used as an internal control. ### $p < 0.001$ vs. ND group. * $p < 0.05$, *** $p < 0.001$ vs. HFD group.

3. Discussion

Adipocyte differentiation and lipid accumulation are both processes related to the development of obesity [35]. The 3T3-L1 cell line was derived from Swiss 3T3 mouse embryos that are used as a model of adipocyte differentiation. Adipocyte differentiation is regulated by signaling molecules from numerous pathways [4]. In murine preadipocyte models, differentiation proceeds as follows: achievement of one hundredth confluence and growth arrest, hormonal induction, re-entry into the cell cycle, post-confluent mitosis, known as MCE. MCE is an important step in the differentiation of adipocytes [8]. During MCE, the number of cells increases 3- to 4-fold [7]. Several studies have identified that the suppression of adipocyte differentiation occurs through the inhibition of MCE [36,37]. Therefore, the differentiation of preadipocytes and their proliferation are two intimately linked processes.

We investigated the inhibitory effect of BV on adipocyte proliferation by determining the effect of BV on MCE in differentiating preadipocytes, by applying the MTT assay. As shown in Figure 1, differentiating cells markedly increased cell numbers, compared to the preadipocyte with BS media (DMEM + 10% BS media). BV treatment inhibited cell proliferation in a dose-dependent manner. BV showed inhibitory effects at the lowest investigated concentration of 2.5 μg/mL. In addition, lipid accumulation was determined by Oil Red O staining (Figure 1B) on Day 8. The data indicated that BV suppressed the MCE process in MDI-induced 3T3-L1 preadipocytes. Our results demonstrate that BV significantly reduced lipid accumulation in differentiating 3T3-L1 cells.

As differentiation progresses, lipid accumulation and numerous adipogenic genes upregulate adipogenesis through the adipocyte-specific transcription factors, C/EBPs and PPARγ [38,39]. C/EBPβ and C/EBPδ, the first transcription factors in directing the differentiation process, are increased after induction of differentiation. C/EBPβ is responsive mainly to DEX and C/EBPδ is responsive mainly to IBMX. After removal of these differentiation inducers, expression of C/EBPβ and C/EBPδ are decreased. In addition, C/EBPβ and δ are known to mediate the expression of PPARγ and C/EBPα [10,40]. C/EBPδ is related to the expression of C/EBPβ in the early phase of adipogenesis [11]. In this study, BV decreased the expression of C/EBPs and PPARγ in 3T3-L1 preadipocytes (Figure 2) and in adipose tissue in the HFD-fed obese mice (Figure 5A). Our findings suggest that BV inhibits early adipogenic processes and lipid accumulation by downregulating the expression of C/EBPs and PPARγ.

In this study, we induced obesity by feeding a high-fat diet to mice for 11 weeks, followed by an injection of BV for 4 weeks. The BV injected group exhibited decreased HFD-induced body and fat weight (Figure 4). Increase of body and fat weight are closely related warning signs for health issues [41], i.e., the accumulation of epididymis adipose tissue is related to metabolic problems, such as insulin resistance, hypertension, and elevated plasma triglyceride levels [42,43]. Histological analysis revealed a greater number of hypertrophied cells in the adipose tissue of the HFD group, whereas the BV injection suppressed adipocyte size in HFD-induced adipose tissue (Figure 4). This result indicates that BV inhibits the hypertrophy of adipocytes in HFD-fed obese mice.

The activity of AMPK in adipose tissue is a useful marker for metabolic disease [44]. ACC, which is a major fatty acid synthetic enzyme, is reduced by the activation of AMPK [45]. It is well known that aminoimidazole-4-carboxamide riboside (AICAR) and metformin, known as AMPK activators, decrease the transcriptional activity of the PPARγ/retinoid X receptor (RXR) in the rat hepatoma cell line H4IIEC3, whereas compound C (6-4[4-(2-piperidin-1-yl-ethoxy)-phenyl]-3-pyridin-4-yl-pyrazolo[1,5-*a*]pyrimidine), known as an AMPK inhibitor, reversed the effects of AICAR and metformin [46]. Furthermore, it has been reported that metformin decreased the plasma levels of glucose and triglycerides by inhibiting sterol regulatory element-binding protein (SREBP)-1 activity [47]. Our present results indicated that BV enhanced the phosphorylation of AMPK and ACC during the differentiation of cultured adipocytes and in the HFD-induced obese mice (Figures 3B and 5B,C). These data suggest that BV regulates the AMPK pathway, which may be involved in the fatty acid metabolism in HFD-fed obese mice.

MAPKs are key regulators of cell growth factors, cytokines, cell proliferation, differentiation, motility, and many other cellular processes [48,49]. The MAPKs are divided into three main groups:

the ERK 1/2, the c-Jun NH2-terminal kinases (JNK 1/2/3), and the p38 MAP kinases [50]. ERK 1/2 is involved in the differentiation of adipocytes; however, continual activation inhibits the differentiation of adipocytes [30,51]. Downregulation of ERK1/2 led to a reduction in adipocyte differentiation [52]. However, some studies reported that ERK activation attenuates the differentiation of adipocytes [53,54]. JNK is known to be involved in insulin resistance [55]. However, a previous study reported that the inhibition of JNK increased lipid accumulation and the expression of PPARγ in 3T3-L1 adipocytes [56]. P38 MAPK plays a key role in adipogenesis. SB203580, an inhibitor of p38 MAP kinase, blocks adipogenesis during only the early stages of adipocyte differentiation [57]. Our results showed that the BV treatment of cells enhanced the phosphorylation of ERK and JNK (Figure 3A). However, BV did not change the p38 phosphorylation. These data suggest that BV suppressed lipid accumulation and regulated adipogenic factors by regulating MAPK signaling.

In conclusion, the present study has demonstrated that BV inhibits early adipogenic processes by downregulating the MCE stage by regulating C/EBPs, PPARγ, ERK, and AMPK signaling. Based on these findings, we conclude that BV may be a useful preventive and therapeutic agent in the treatment of obesity.

4. Materials and Methods

4.1. Chemicals and Reagents

BV, IBMX, Dexamethasone (DEX), insulin, Oil red O, and all other chemicals were purchased from Sigma Chemical Co. (St. Louis, MO, USA). Dulbecco's modified Eagles medium (DMEM), bovine serum (BS), fetal bovine serum (FBS), and penicillin-streptomycin (PS) were purchased from Life Technologies, Inc. (Grand Island, NY, USA). Antibodies against PPARγ (E-8; cat. no. sc-7273), C/EBPα (C-18; cat. no. sc-9314), and β-actin (C4; cat. no. sc-47778) were purchased from Santa Cruz biotechnology, Inc. (Santa Cruz, CA, USA). Phospho–extracellular signal-regulated kinase (p-ERK; Thr202/Tyr204; cat. no. #9101), ERK (cat. no. #9102), phospho-stress-activated protein kinase/Jun-amino-terminal kinase (p-JNK; Thr183/Tyr185; cat. no. #9251), JNK (cat. no. #9252), phospho-p38 MAPK (p-p38; Thr180/Thy182; cat. no. #9215), p38 (cat. no. #9212), p-AMPK (Thr172; cat. no. #2535), AMPK (cat. no. #2532), p-ACC (Ser79; cat. no. #3661), and ACC (cat. no. #3662) antibodies were purchased from Cell Signaling Technology, Inc. (Danvers, MA, USA) Horseradish peroxidase conjugated secondary antibodies were purchased from Jackson ImmunoResearch Laboratories, Inc. (West Grove, PA, USA). SYBR Green Master Mix was purchased from Applied Biosystems (Foster, CA, USA). *C/EBPα, C/EBPβ, C/EBPδ, PPARγ,* and glyceraldehyde-3-phosphate dehydrogenase (*GAPDH*) oligonucleotide primers were purchased from Bioneer (Daejeon, Korea).

4.2. Cell Culture and Treatment

Preadipocytes, 3T3-L1, were purchased from the Korean Cell Line Bank (Seoul, Korea) and were cultured in DMEM, supplemented with 10% BS, penicillin (100 U/mL), and 100 µg/mL streptomycin in an incubator at 37 °C with 5% CO_2. The analysis of adipocyte differentiation was carried out by culturing 3T3-L1 cells in 60 mm dishes at a density of 2×10^5 cells per mL to confluence. At full confluence, the cells were first differentiated with MDI media (0.5 mM IBMX, 1 µg/mL insulin, and 1 µM DEX in DMEM containing 10% (*v/v*) FBS and 1% PS). During this stage, we treated plates with various concentrations of BV. During the second stage, commencing on Day 3 of differentiation, the cells were treated with 1 µg/mL insulin in DMEM with 10% (*v/v*) FBS and 1% PS. In the final stage, cells were transferred to DMEM with 10% FBS and 1% PS, and the culture media was changed every 3 days. The differentiation of 3T3-L1 required 3 days in each stage.

4.3. MTT Assay

Cell viability was assessed using the 3-(4,5-Dimethylthiazol-2-yl)-2,5-diphenyl tetrazolium bromide (MTT) assay. Briefly, the 3T3-L1 preadipocyte cells were seeded into a 96-well plate at a density of 1×10^4 cells per well and were treated with various concentrations (1.25 to 40 μg/mL) of BV for 72 h at 37 °C in humidified air with 5% CO_2. After the treatment, the cells were stained by adding MTT solution (5 mg/mL) for 4 h at 37 °C. After removing the excess reagent, the insoluble formazan product was dissolved in DMSO. The cell viability was measured at 570 nm using an Epoch® microvolume spectrophotometer (BioTek Instruments Inc., Winooski, VT, USA).

4.4. Oil Red O Staining

In order to observe lipid accumulation in the 3T3-L1 adipocytes, the differentiated adipocytes were stained with Oil Red O. As described above, the differentiation was initiated by exchanging the medium and by adding BV at three concentrations. Following the differentiation, 3T3-L1 adipocytes were washed three times with phosphate-buffered saline (PBS, pH = 7.4) and were fixed with 10% formaldehyde solution in PBS for 1 h at 25 °C. After washing with distilled water three times, the cells were then stained with 3 mg/mL Oil Red O dye solution in 60% isopropanol for 2 h at room temperature. Any excess Oil Red O dye was washed away with distilled water. The images of the Oil Red O-stained adipocytes were acquired using a Leica DM IL LED microscope (Leica, Wetzlar, Germany). The intracellular lipid content was measured by extracting Oil Red O with isopropanol, and the absorbance at 520 nm was recorded using an Epoch® microvolume spectrophotometer (BioTek Instrument, Inc., Winooski, VT, USA).

4.5. Western Blot Analysis

The cells were lysed, and the tissue was homogenized in PRO-PREP™ protein extraction solution (Intron Biotechnology, Seoul, Korea) and was then incubated for 20 min at 4 °C. Debris was removed by microcentrifugation at $11,000 \times g$, followed by a quick freezing of the supernatants. The protein concentration was determined using the Bio-Rad protein assay reagent according to the manufacturer's instructions (Bio-Rad, Hercules, CA, USA). Proteins were electro-blotted onto a polyvinylidene difluoride (PVDF) membrane following their separation on an 8–12% SDS polyacrylamide gel. The membrane was incubated for 1 h with a blocking solution (5% skim milk) at room temperature, followed by incubation with a 1:1,000 dilution of primary antibodies, including, PPARγ, C/EBPα, p-ERK, ERK, p-JNK, JNK, p-p38, p38, p-AMPK, AMPK, p-ACC, ACC, and β-actin, overnight at 4 °C. The blots were washed three times with Tween 20/Tris-buffered saline (T/TBS) and were then incubated in a horseradish peroxidase-conjugated secondary antibody (dilution, 1:2500) for 2 h at room temperature. After washing them three times in T/TBS, the immuno-detection bands were reacted with the ECL solution (Ab signal, Seoul, Korea) and were recorded on X-ray film (Agfa, Belgium).

4.6. Isolation of Total RNA and Reverse Transcription Quantitative Polymerase Chain Reaction (RT-qPCR)

The cells were homogenized, and the total RNA was isolated using a Trizol reagent (Invitrogen, Carlsbad, CA, USA). cDNA was obtained using the isolated total RNA (1 μg), a d(T)16 primer, and avian myeloblastosis virus reverse transcriptase (AMV-RT). The relative gene expression was quantified using real-time PCR (Real-Time PCR System 7500, Applied Biosystems, Foster city, CA, USA) with a SYBR green PCR master mix (Applied Biosystems, Foster city, CA, USA). The forward and reverse primers were as follows: *PPARγ*, 5′-ATCGAGTGCCGAGTCTGTGG-3′ and 5′-GCAAGGCACTTCTGAAACCG-3′; *C/EBPα*, 5′-GGAACTTGAAGCACAATCGATC-3′ and 5′-TGGT TTAGCATAGACGTGCACA-3′; *C/EBPβ*, 5′-GGGGTTGTTGATGTTTTTGG-3′ and 5′-CGAAACGGA AAAGGTTCTCA-3′; *C/EBPδ*, 5′-GATCTGCACGGCCTGTTGTA-3′ and 5′-CTCCACTGCCCACCT GTCA-3′; *GAPDH*, 5′-GACGGCCGCATCTTCTTGT-3′ and 5′-CACACCGACCTTCACCATTTT-3′.

The gene Ct values of *PPARγ*, *C/EBPα*, *C/EBPβ*, and *C/EBPδ* were normalized using the Gene Express 2.0 program (Applied Biosystems, Foster city, CA, USA) to the Ct value of *GAPDH*.

4.7. Animals

C57BL/6 mice (6 weeks old, male) were purchased from Daehan Biolink Co. Ltd. (Daejeon, Korea) and were maintained under constant conditions (temperature, 22 ± 3 °C; humidity, 40–50%;

light/dark cycle 12/12 h). The mice were adapted to the feeding conditions for 1 week and then were provided free access to food and tap water for 14 weeks. The mice were randomly separated into groups of four each: ND (normal diet), HFD (high-fat diet, 30% fat) only, and the BV-treated groups (0.1 or 1.0 mg/kg i.p.; high-fat diet). Their body weight and dietary intake were recorded every week. On the last day of the 14th week, the animals were fasted overnight. Blood samples were collected for lipid profiling, and the adipose tissue was excised, rinsed, and stored at $-80\,^{\circ}$C until analysis. The Institutional Animal Care and Use Committee (IACUC) of the College of Sang-ji University of Korea approved the study protocol. The approval code is 2014-10 and the approval date is 22 July 2014.

4.8. Histological Analysis

The adipose samples were fixed in 10% formalin and were embedded in paraffin; the sections were of 8-μm thickness. The sections were stained with hematoxylin and eosin (H&E) for the histological analysis of fat droplets. Images were acquired using a Leica DM IL LED microscope (Leica, Wetzlar, Germany).

4.9. Analysis of Serum Lipid Profiles

The blood samples were collected and centrifuged at $1003 \times g$, for 15 min at room temperature to obtain serum samples, which were immediately frozen at $-80\,^{\circ}$C for further measurements. The serum concentrations of triglyceride and LDL cholesterol were determined by enzymatic methods with commercial kits (BioVision; Milpitas, CA, USA).

4.10. Statistical Analysis

The data are expressed as mean $\pm$ standard deviation (SD) of triplicate experiments. Statistical significance was determined using ANOVA and Dunnett's post hoc test, and p-values of less than 0.05 were considered statistically significant.

Acknowledgments: This research was supported by Basic Science Research Program through the National Research Foundation of Korea (NRF) and was funded by the grants from (grant number: NRF-2017R1D1A1B03034167).

Author Contributions: Hyo-Jin An conceived and designed the study. Se-Yun Cheon and Kyung-Sook Chung carried out the experiments, analyzed the research data, and wrote the manuscript. Seong-Soo Roh and Yun-Yeop Cha participated in the study design and analyzed the research data. All authors read and approved the final manuscript.

Conflicts of Interest: The authors have no conflict of interest to declare.

References

1. Marti, A.; Martinez-Gonzalez, M.A.; Martinez, J.A. Interaction between genes and lifestyle factors on obesity. *Proc. Nutr. Soc.* **2008**, *67*, 1–8. [CrossRef] [PubMed]
2. Lefterova, M.I.; Lazar, M.A. New developments in adipogenesis. *Trends Endocrinol. Metab.* **2009**, *20*, 107–114. [CrossRef] [PubMed]
3. Unger, R.H.; Clark, G.O.; Scherer, P.E.; Orci, L. Lipid homeostasis, lipotoxicity and the metabolic syndrome. *Biochim. Biophys. Acta* **2010**, *1801*, 209–214. [CrossRef] [PubMed]
4. Green, H.; Kehinde, O. An established preadipose cell line and its differentiation in culture II. Factors affecting the adipose conversion. *Cell* **1975**, *5*, 19–27. [CrossRef]
5. Fajas, L.; Fruchart, J.C.; Auwerx, J. Transcriptional control of adipogenesis. *Curr. Opin. Cell Biol.* **1998**, *10*, 165–173. [CrossRef]
6. Otto, T.C.; Lane, M.D. Adipose development: From stem cell to adipocyte. *Crit. Rev. Biochem. Mol. Biol.* **2005**, *40*, 229–242. [CrossRef] [PubMed]
7. Tang, Q.Q.; Otto, T.C.; Lane, M.D. Mitotic clonal expansion: A synchronous process required for adipogenesis. *Proc. Natl. Acad. Sci. USA* **2003**, *100*, 44–49. [CrossRef] [PubMed]
8. Rosen, E.D.; Walkey, C.J.; Puigserver, P.; Spiegelman, B.M. Transcriptional regulation of adipogenesis. *Genes Dev.* **2000**, *14*, 1293–1307. [PubMed]

9. Rosen, E.D.; Sarraf, P.; Troy, A.E.; Bradwin, G.; Moore, K.; Milstone, D.S.; Spiegelman, B.M.; Mortensen, R.M. PPARγ is required for the differentiation of adipose tissue in vivo and in vitro. *Mol. Cell* **1999**, *4*, 611–617. [CrossRef]

10. Darlington, G.J.; Ross, S.E.; MacDougald, O.A. The role of C/EBP genes in adipocyte differentiation. *J. Biol. Chem.* **1998**, *273*, 30057–30060. [CrossRef] [PubMed]

11. Hishida, T.; Nishizuka, M.; Osada, S.; Imagawa, M. The role of C/EBPdelta in the early stages of adipogenesis. *Biochimie* **2009**, *91*, 654–657. [CrossRef] [PubMed]

12. Tang, Q.Q.; Otto, T.C.; Lane, M.D. CCAAT/enhancer-binding protein β is required for mitotic clonal expansion during adipogenesis. *Proc. Natl. Acad. Sci. USA* **2003**, *100*, 850–855. [CrossRef] [PubMed]

13. Yeh, W.C.; Cao, Z.; Classon, M.; McKnight, S.L. Cascade regulation of terminal adipocyte differentiation by three members of the C/EBP family of leucine zipper proteins. *Genes Dev.* **1995**, *9*, 168–181. [CrossRef] [PubMed]

14. Erbayraktar, Z.; Yilmaz, O.; Artmann, A.T.; Cehreli, R.; Coker, C. Effects of selenium supplementation on antioxidant defense and glucose homeostasis in experimental diabetes mellitus. *Biol. Trace Elem. Res.* **2007**, *118*, 217–226. [CrossRef] [PubMed]

15. Sert, A.; Pirgon, O.; Aypar, E.; Yilmaz, H.; Odabas, D. Subclinical hypothyroidism as a risk factor for the development of cardiovascular disease in obese adolescents with nonalcoholic fatty liver disease. *Pediatr. Cardiol.* **2013**, *34*, 1166–1174. [CrossRef] [PubMed]

16. Carling, D.; Zammit, V.A.; Hardie, D.G. A common bicyclic protein kinase cascade inactivates the regulatory enzymes of fatty acid and cholesterol biosynthesis. *FEBS Lett.* **1987**, *223*, 217–222. [CrossRef]

17. Muoio, D.M.; Seefeld, K.; Witters, L.A.; Coleman, R.A. Amp-activated kinase reciprocally regulates triacylglycerol synthesis and fatty acid oxidation in liver and muscle: Evidence that sn-glycerol-3-phosphate acyltransferase is a novel target. *Biochem. J.* **1999**, *338*, 783–791. [CrossRef] [PubMed]

18. Gao, Y.; Zhou, Y.; Xu, A.; Wu, D. Effects of an amp-activated protein kinase inhibitor, compound C, on adipogenic differentiation of 3T3-L1 cells. *Biol. Pharm. Bull.* **2008**, *31*, 1716–1722. [CrossRef] [PubMed]

19. Moreau, S.J. "It stings a bit but it cleans well": Venoms of hymenoptera and their antimicrobial potential. *J. Insect Physiol.* **2013**, *59*, 186–204. [CrossRef] [PubMed]

20. Hider, R.C. Honeybee venom: A rich source of pharmacologically active peptides. *Endeavour* **1988**, *12*, 60–65. [CrossRef]

21. Lee, W.R.; Kim, S.J.; Park, J.H.; Kim, K.H.; Chang, Y.C.; Park, Y.Y.; Lee, K.G.; Han, S.M.; Yeo, J.H.; Pak, S.C.; et al. Bee venom reduces atherosclerotic lesion formation via anti-inflammatory mechanism. *Am. J. Chin. Med.* **2010**, *38*, 1077–1092. [CrossRef] [PubMed]

22. Chung, K.S.; An, H.J.; Cheon, S.Y.; Kwon, K.R.; Lee, K.H. Bee venom suppresses testosterone-induced benign prostatic hyperplasia by regulating the inflammatory response and apoptosis. *Exp. Biol. Med.* **2015**, *240*, 1656–1663. [CrossRef] [PubMed]

23. Lariviere, W.R.; Melzack, R. The bee venom test: A new tonic-pain test. *Pain* **1996**, *66*, 271–277. [CrossRef]

24. Kim, S.J.; Park, J.H.; Kim, K.H.; Lee, W.R.; Kim, K.S.; Park, K.K. Melittin inhibits atherosclerosis in LPS/high-fat treated mice through atheroprotective actions. *J. Atheroscler. Thromb.* **2011**, *18*, 1117–1126. [CrossRef] [PubMed]

25. Kim, S.J.; Park, J.H.; Kim, K.H.; Lee, W.R.; Pak, S.C.; Han, S.M.; Park, K.K. The protective effect of apamin on LPS/fat-induced atherosclerotic mice. *Evid.-Based Complement. Altern. Med.* **2012**, *2012*, 305454. [CrossRef] [PubMed]

26. Kim, M.K.; Lee, S.H.; Shin, J.Y.; Kim, K.S.; Cho, N.G.; Kwon, K.R.; Rhim, T.J. The effects of bee venom and sweet bee venom to the preadipocyte proliferation and lipolysis of adipocyte, localized fat accumulation. *J. Pharmacopunct.* **2007**, *10*, 15. [CrossRef]

27. Fruhbeck, G.; Gomez-Ambrosi, J.; Muruzabal, F.J.; Burrell, M.A. The adipocyte: A model for integration of endocrine and metabolic signaling in energy metabolism regulation. *Am. J. Physiol. Endocrinol. Metab.* **2001**, *280*, E827–E847. [CrossRef] [PubMed]

28. Tontonoz, P.; Hu, E.; Spiegelman, B.M. Stimulation of adipogenesis in fibroblasts by PPARγ2, a lipid-activated transcription factor. *Cell* **1994**, *79*, 1147–1156. [CrossRef]

29. Zhang, W.; Liu, H.T. MAPK signal pathways in the regulation of cell proliferation in mammalian cells. *Cell Res.* **2002**, *12*, 9–18. [CrossRef] [PubMed]

30. Sakaue, H.; Ogawa, W.; Nakamura, T.; Mori, T.; Nakamura, K.; Kasuga, M. Role of MAPK phosphatase-1 (mkp-1) in adipocyte differentiation. *J. Biol. Chem.* **2004**, *279*, 39951–39957. [CrossRef] [PubMed]

31. Kahn, B.B.; Alquier, T.; Carling, D.; Hardie, D.G. Amp-activated protein kinase: Ancient energy gauge provides clues to modern understanding of metabolism. *Cell Metab.* **2005**, *1*, 15–25. [CrossRef] [PubMed]

32. Bluher, M. Adipose tissue dysfunction in obesity. *Exp. Clin. Endocrinol. Diabetes* **2009**, *117*, 241–250. [CrossRef] [PubMed]

33. Barak, Y.; Nelson, M.C.; Ong, E.S.; Jones, Y.Z.; Ruiz-Lozano, P.; Chien, K.R.; Koder, A.; Evans, R.M. PPAR gamma is required for placental, cardiac, and adipose tissue development. *Mol. Cell* **1999**, *4*, 585–595. [CrossRef]

34. Kim, T.S.; Leahy, P.; Freake, H.C. Promoter usage determines tissue specific responsiveness of the rat acetyl-CoA carboxylase gene. *Biochem. Biophys. Res. Commun.* **1996**, *225*, 647–653. [CrossRef] [PubMed]

35. Rayalam, S.; Della-Fera, M.A.; Baile, C.A. Phytochemicals and regulation of the adipocyte life cycle. *J. Nutr. Biochem.* **2008**, *19*, 717–726. [CrossRef] [PubMed]

36. Kim, S.H.; Park, H.S.; Lee, M.S.; Cho, Y.J.; Kim, Y.S.; Hwang, J.T.; Sung, M.J.; Kim, M.S.; Kwon, D.Y. Vitisin A inhibits adipocyte differentiation through cell cycle arrest in 3T3-L1 cells. *Biochem. Biophys. Res. Commun.* **2008**, *372*, 108–113. [CrossRef] [PubMed]

37. Qiu, Z.; Wei, Y.; Chen, N.; Jiang, M.; Wu, J.; Liao, K. DNA synthesis and mitotic clonal expansion is not a required step for 3T3-L1 preadipocyte differentiation into adipocytes. *J. Biol. Chem.* **2001**, *276*, 11988–11995. [CrossRef] [PubMed]

38. Kwon, J.Y.; Seo, S.G.; Heo, Y.S.; Yue, S.; Cheng, J.X.; Lee, K.W.; Kim, K.H. Piceatannol, natural polyphenolic stilbene, inhibits adipogenesis via modulation of mitotic clonal expansion and insulin receptor-dependent insulin signaling in early phase of differentiation. *J. Biol. Chem.* **2012**, *287*, 11566–11578. [CrossRef] [PubMed]

39. Tang, Q.Q.; Lane, M.D. Adipogenesis: From stem cell to adipocyte. *Ann. Rev. Biochem.* **2012**, *81*, 715–736. [CrossRef] [PubMed]

40. Lane, M.D.; Tang, Q.Q.; Jiang, M.S. Role of the CCAAT enhancer binding proteins (C/EBPS) in adipocyte differentiation. *Biochem. Biophys. Res. Commun.* **1999**, *266*, 677–683. [CrossRef] [PubMed]

41. Adams, K.F.; Schatzkin, A.; Harris, T.B.; Kipnis, V.; Mouw, T.; Ballard-Barbash, R.; Hollenbeck, A.; Leitzmann, M.F. Overweight, obesity, and mortality in a large prospective cohort of persons 50 to 71 years old. *N. Engl. J. Med.* **2006**, *355*, 763–778. [CrossRef] [PubMed]

42. Despres, J.P.; Moorjani, S.; Lupien, P.J.; Tremblay, A.; Nadeau, A.; Bouchard, C. Regional distribution of body fat, plasma lipoproteins, and cardiovascular disease. *Arteriosclerosis* **1990**, *10*, 497–511. [CrossRef] [PubMed]

43. Wajchenberg, B.L. Subcutaneous and visceral adipose tissue: Their relation to the metabolic syndrome. *Endocr. Rev.* **2000**, *21*, 697–738. [CrossRef] [PubMed]

44. Zhang, B.B.; Zhou, G.; Li, C. AMPK: An emerging drug target for diabetes and the metabolic syndrome. *Cell Metab.* **2009**, *9*, 407–416. [CrossRef] [PubMed]

45. Kohjima, M.; Higuchi, N.; Kato, M.; Kotoh, K.; Yoshimoto, T.; Fujino, T.; Yada, M.; Yada, R.; Harada, N.; Enjoji, M.; et al. Srebp-1c, regulated by the insulin and AMPK signaling pathways, plays a role in nonalcoholic fatty liver disease. *Int. J. Mol. Med.* **2008**, *21*, 507–511. [CrossRef] [PubMed]

46. Sozio, M.S.; Lu, C.; Zeng, Y.; Liangpunsakul, S.; Crabb, D.W. Activated AMPK inhibits PPAR-α and PPAR-γ transcriptional activity in hepatoma cells. *Am. J. Physiol. Gastrointest. Liver Physiol.* **2011**, *301*, G739–G747. [CrossRef] [PubMed]

47. Zhou, G.; Myers, R.; Li, Y.; Chen, Y.; Shen, X.; Fenyk-Melody, J.; Wu, M.; Ventre, J.; Doebber, T.; Fujii, N.; et al. Role of amp-activated protein kinase in mechanism of metformin action. *J. Clin. Investig.* **2001**, *108*, 1167–1174. [CrossRef] [PubMed]

48. Dickinson, R.J.; Keyse, S.M. Diverse physiological functions for dual-specificity map kinase phosphatases. *J. Cell Sci.* **2006**, *119*, 4607–4615. [CrossRef] [PubMed]

49. Wagner, E.F.; Nebreda, A.R. Signal integration by JNK and p38 MAPK pathways in cancer development. *Nature Rev. Cancer* **2009**, *9*, 537–549. [CrossRef] [PubMed]

50. Chen, Z.; Gibson, T.B.; Robinson, F.; Silvestro, L.; Pearson, G.; Xu, B.; Wright, A.; Vanderbilt, C.; Cobb, M.H. Map kinases. *Chem. Rev.* **2001**, *101*, 2449–2476. [CrossRef] [PubMed]

51. Sale, E.M.; Atkinson, P.G.; Sale, G.J. Requirement of map kinase for differentiation of fibroblasts to adipocytes, for insulin activation of p90 s6 kinase and for insulin or serum stimulation of DNA synthesis. *EMBO J.* **1995**, *14*, 674–684. [PubMed]

52. Kimura, Y.; Taniguchi, M.; Baba, K. Antitumor and antimetastatic activities of 4-hydroxyderricin isolated from *Angelica keiskei* roots. *Planta Med.* **2004**, *70*, 211–219. [CrossRef] [PubMed]
53. Akihisa, T.; Motoi, T.; Seki, A.; Kikuchi, T.; Fukatsu, M.; Tokuda, H.; Suzuki, N.; Kimura, Y. Cytotoxic activities and anti-tumor-promoting effects of microbial transformation products of prenylated chalcones from *Angelica keiskei*. *Chem. Biodivers.* **2012**, *9*, 318–330. [CrossRef] [PubMed]
54. Hu, E.; Kim, J.B.; Sarraf, P.; Spiegelman, B.M. Inhibition of adipogenesis through map kinase-mediated phosphorylation of ppargamma. *Science* **1996**, *274*, 2100–2103. [CrossRef] [PubMed]
55. Hirosumi, J.; Tuncman, G.; Chang, L.; Gorgun, C.Z.; Uysal, K.T.; Maeda, K.; Karin, M.; Hotamisligil, G.S. A central role for JNK in obesity and insulin resistance. *Nature* **2002**, *420*, 333–336. [CrossRef] [PubMed]
56. Zhang, T.; Sawada, K.; Yamamoto, N.; Ashida, H. 4-hydroxyderricin and xanthoangelol from ashitaba (*Angelica keiskei*) suppress differentiation of preadiopocytes to adipocytes via AMPK and MAPK pathways. *Mol. Nutr. Food Res.* **2013**, *57*, 1729–1740. [CrossRef] [PubMed]
57. Engelman, J.A.; Lisanti, M.P.; Scherer, P.E. Specific inhibitors of p38 mitogen-activated protein kinase block 3T3-L1 adipogenesis. *J. Biol. Chem.* **1998**, *273*, 32111–32120. [CrossRef] [PubMed]

© 2017 by the authors. Licensee MDPI, Basel, Switzerland. This article is an open access article distributed under the terms and conditions of the Creative Commons Attribution (CC BY) license (http://creativecommons.org/licenses/by/4.0/).

toxins

MDPI

Article

Molecular Dynamics Simulation Reveals Specific Interaction Sites between Scorpion Toxins and K$_v$1.2 Channel: Implications for Design of Highly Selective Drugs

Shouli Yuan [1,2], Bin Gao [1] and Shunyi Zhu [1,*] ⓘ

[1] Group of Peptide Biology and Evolution, State Key Laboratory of Integrated Management of Pest Insects and Rodents, Institute of Zoology, Chinese Academy of Sciences, Beijing 100101, China; yuanshouli123@163.com (S.Y.); gaob@ioz.ac.cn (B.G.)

[2] College of Resources and Environment, University of Chinese Academy of Sciences, Beijing 100049, China

* Correspondence: zhusy@ioz.ac.cn

Academic Editors: Bryan Grieg Fry and Steve Peigneur
Received: 29 August 2017; Accepted: 19 October 2017; Published: 1 November 2017

Abstract: The K$_v$1.2 channel plays an important role in the maintenance of resting membrane potential and the regulation of the cellular excitability of neurons, whose silencing or mutations can elicit neuropathic pain or neurological diseases (e.g., epilepsy and ataxia). Scorpion venom contains a variety of peptide toxins targeting the pore region of this channel. Despite a large amount of structural and functional data currently available, their detailed interaction modes are poorly understood. In this work, we choose four K$_v$1.2-targeted scorpion toxins (Margatoxin, Agitoxin-2, OsK-1, and Mesomartoxin) to construct their complexes with K$_v$1.2 based on the experimental structure of ChTx-K$_v$1.2. Molecular dynamics simulation of these complexes lead to the identification of hydrophobic patches, hydrogen-bonds, and salt bridges as three essential forces mediating the interactions between this channel and the toxins, in which four K$_v$1.2-specific interacting amino acids (D353, Q358, V381, and T383) are identified for the first time. This discovery might help design highly selective K$_v$1.2-channel inhibitors by altering amino acids of these toxins binding to the four channel residues. Finally, our results provide new evidence in favor of an induced fit model between scorpion toxins and K$^+$ channel interactions.

Keywords: K$_v$1.2 channel; scorpion toxin; molecular dynamics simulation

---

1. Introduction

The voltage-gated K$^+$ channel K$_v$1.2, encoded by *KCNA2*, is a transmembrane protein that is composed of four identical α-subunits, with each subunit having six transmembrane segments (S1–S6) and a membrane reentering P-loop. As shown by their experimental structure [1], S1–S4 form a voltage-sensor domain (VSD) and S5–S6 constitute a pore that selectively passes K$^+$ ions [2]. K$_v$1.2 plays an important role in maintaining the resting membrane potential that enables efficient neuronal repolarization following an action potential [3]. Therefore, its loss or mutation will cause some neurogenic diseases, such as ataxia, myoclonic epilepsy, and premature death [4–7].

Given its key physiological function, K$_v$1.2 is frequently selected as a target by a diversity of venomous animals. In scorpions, at least six families of K$^+$ channel toxins (α-KTxs, β-KTxs, δ-KTxs, κ-KTxs, λ-KTxs, and ε-KTxs) are identified [2,8–12], in which α-KTxs are the largest source targeting this channel. These α-KTxs impair the K$^+$ channel functions by blockage of their pore region. Due to high sequence similarity in this region between K$_v$1.2 and its two paralogs (i.e., K$_v$1.1 and K$_v$1.3) (Figure 1A), toxins are not able to distinguish the three channel subtypes, resulting in undesired side effects when assayed in vivo. Hence, it remains a great challenge to improve the $K_v1.2$ selectivity of these natural toxins via engineering modification in the absence of detailed data about the toxin-channel interactions.

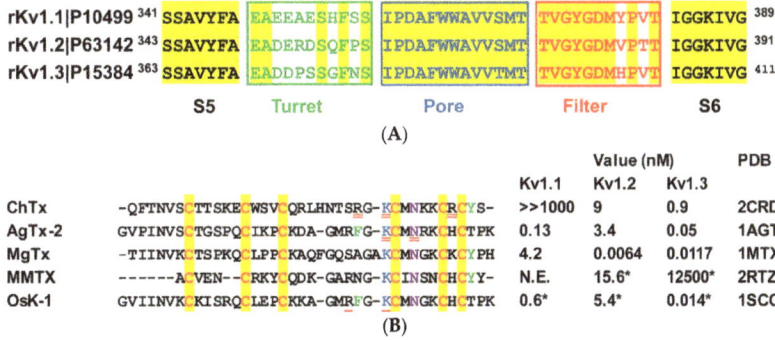

rKv1.1|P10499 341 SSAVYFA EAEEAESHFSS IPDAFWWAVVSMT TVGYGDMYPVT IGGKIVG 389
rKv1.2|P63142 343 SSAVYFA EADERDSQFPS IPDAFWWAVVSMT TVGYGDMVPTT IGGKIVG 391
rKv1.3|P15384 363 SSAVYFA EADDPSSGFNS IPDAFWWAVVTMT TVGYGDMHPVT IGGKIVG 411

| S5 | Turret | Pore | Filter | S6 |

(A)

			Value (nM)			PDB
		Kv1.1	Kv1.2	Kv1.3		
ChTx	-QFTNVSCTTSKECWSVCQRLHNTSRG-KCMNKKCRCYS-	>>1000	9	0.9		2CRD
AgTx-2	GVPINVSCTGSPQCIKPCKDA-GMRFG-KCMNRKCHCTPK	0.13	3.4	0.05		1AGT
MgTx	-TIINVKCTSPKQCLPPCKAQFGQSAGAKCMNGKCKCYPH	4.2	0.0064	0.0117		1MTX
MMTX	------ACVEN--CRKYCQDK-GARNG-KCINSNCHCYY-	N.E.	15.6*	12500*		2RTZ
OsK-1	GVIINVKCKISRQCLEPCKKA-GMRFG-KCMNGKCHCTPK	0.6*	5.4*	0.014*		1SCO

(B)

Figure 1. Scorpion toxins and K^+ channels. (**A**) The $rK_v1.1$–$rK_v1.3$ pore region sequences. Amino acids conserved across the alignment are marked in yellow, and turret, pore helix, and filter are colored in green, blue, and red, respectively; (**B**) The α-KTxs studied in this work. Their functional sites identified by mutational experiments are highlighted with a double underline. New interaction sites predicted by reported molecular dynamics (MD) simulation data are underlined once. For the affinity of each toxin, Kd (dissociation constant), Ki (equilibrium constant), or IC_{50} (half maximal inhibitory concentration) (asterisks) are shown in nanomole (nM) [8,13–17].

There is only one experimental toxin-channel complex (ChTx-$K_v1.2$) available currently [18], which hampers a detailed comparative study to draw commonality and difference among complexes. An alternative approach is to employ computative technology to solve this question. Several popular methods include homology modeling, Brownian dynamics, molecular docking, and molecular dynamics simulation [19]. Molecular dynamics simulation is a powerful tool in predicting the structures of toxin-channel complexes. Some toxin-$K_v1.2$ complexes were reported, which were constructed with molecular dynamics simulation, such as maurotoxin-$K_v1.2$ [20,21]. However, the comparison of the mechanism among these $K_v1.2$ inhibitors is lacking.

In this work, we employed molecular dynamics (MD) simulation to study the interactions of four α-KTxs (Margatoxin (abbreviated as MgTx), Agitoxin-2 (abbreviated as AgTx-2), OsK-1, and Mesomartoxin (abbreviated as MMTX) with the pore of $K_v1.2$. These toxins all bind to the channel with high affinity [8,13,14] (Figure 1B). However, as mentioned above, they are also ligands of $K_v1.1$ and $K_v1.3$, with the exception of MMTX, which lacks effect on $K_v1.1$. Our MD simulation data reveals for the first time four $K_v1.2$-specific amino acids that are involved in direct interactions with these toxins. This finding thus provides a structural basis for their $K_v1.2$ blocking activity and might help design new $K_v1.2$-targeted peptide drugs with an improved channel subtype selectivity.

2. Results

2.1. The Channel Selectivity of Four Scorpion Toxins Analyzed

All the four toxins fold into the cysteine stabilized α-helix/β-sheet (CSα/β) structure with a functional dyad comprising a conserved lysine and an aromatic amino acid in a distance of 5–7 Å (Figure 1B). These toxins reversibly block K^+ channels by interacting at the external pore of the channel protein. Of them, MgTx is a 39 amino acids peptide isolated from the venom of the scorpion *Centruroides margaritatus*, which blocks $K_v1.2$ and $K_v1.3$ at picomolar concentrations and $K_v1.1$ at nanomolar concentrations without detectable effect on other types of K^+ channels, such as $K_v1.4$–$K_v1.7$

and the insect *Shaker* K$^+$ channel [15]. K28 of MgTx is responsible for blocking K$^+$ channels [22]. AgTx-2 is a 38 amino acids peptide isolated from the venom of *Leiurus quinquestriatus hebraeus*, which reversibly inhibits K$_V$1.1 to K$_V$1.3 with Kd values of 0.13 nM, 3.4 nM, and 0.05 nM, respectively [8,16]. Its functional residues include K27 and N30 [17]. OsK-1 is a 38 amino acids peptide isolated from the venom of *Orthochirus scrobiculosus*, which blocks K$_V$1.1 to K$_V$1.3 with IC$_{50}$ values of 0.6 nM, 5.4 nM, and 0.014 nM, respectively [14]. E16K and/or K20D mutations of OsK-1 show an increased potency on K$_V$1.3 channel but do not change the effect on K$_V$1.2 [14]. MMTX is a 29 amino acids peptide isolated from *Mesobuthus martensii* that exerts a strong inhibitory effect on rK$_V$1.2 (IC$_{50}$ = 15.6 nM) and weak effect on rK$_V$1.3 (IC$_{50}$ = 12.5 μM) without affecting K$_V$1.1, even at 50 μM [13].

2.2. Modeling of Toxin-K$_v$1.2 Complexes

ChTx (also named CTX) is the most thoroughly studied scorpion K$^+$ channel toxin isolated from the venom of *Leiurus quinquestriatus hebraeus*, which inhibits K$_V$1.2 and K$_V$1.3 channels with nanomolar affinity by several crucial functional residues, such as R25, K27, and R34 [8,23]. K27 of ChTx is a key residue for blocking the pore region of the K$^+$ channel [24]. Superimposition of the four toxins to ChTx reveals root-mean-square deviations (RMSDs) of <2.5 Å in their Cα atoms (Figure 2), indicating that they are quite similar in structure. Importantly they all contain evolutionarily conserved functional motifs, K27 and N30 (numbered according to ChTx), which have been comfirmed to directly interact with the pore region of K$_V$ channels [25,26]. Supported by these observations, we assume that all these five toxins inhibit K$_V$1.2 in a similar manner. To investigate the detailed interactions between the toxins and K$_V$1.2, we constructed their complexes based on the experimental structure of ChTx-K$_V$1.2 via molecular replacement and energy minimization.

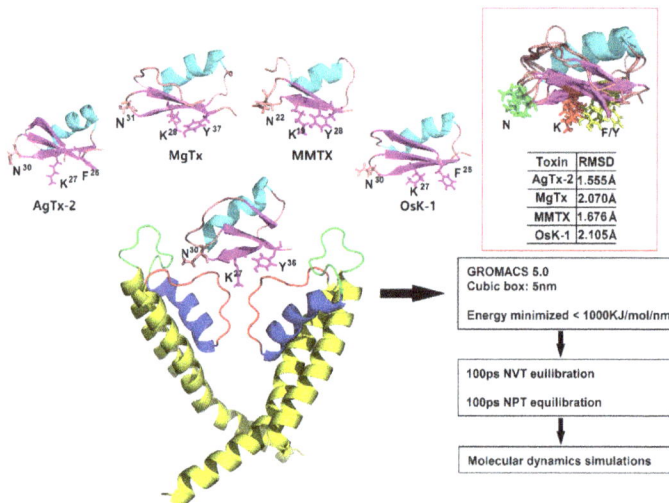

Figure 2. Construction of toxin-channel complex for MD simulation analysis. Three experimentally determined functional sites of α-KTxs are shown as sticks. Superimposed structure of these five toxins are emphasized by red box and root-mean-square deviations (RMSDs) between these four toxins and ChTx are listed below the structure.

2.3. Conformational Changes Induced by Toxin-Channel Interaction

To recognize the conformation change of toxins and K$_V$1.2 channel after combining each other, we play molecular dynamic simulations of four toxins without channel and with sole K$_V$1.2 channel. Subsequently, we compared their conformation changes.

In our molecular dynamic simulations, the equilibrated conditions of four toxin-channel complexes were established in terms of their RMSDs, residue Cα fluctuations during 40 ns time span of simulation. These systems reached equilibrium after 15 ns (Figure 3A). Simultaneously, we calculated the average Cα root-mean-square fluctuations (RMSFs) of all complexes of the $K_V1.2$ pore region. From the RMSF data, it is clear that $K_V1.2$ turret is the most flexible region besides the *N*-, *C*-terminal (Figure 3B,C). Therefore, we proposed that $K_V1.2$ channel interacts with different scorpion toxins, mainly by modulating their turret region.

The equilibrated conditions of four toxins were established in terms of their RMSDs and residue Cα wise fluctuations during 40 ns time span of simulation. MgTx and AgTx-2 reached equilibrium after 5 ns and MMTX reached equilibrium after 27 ns (Figure 4A). The system equilibrium stage of OsK-1 is from 5 ns to 35 ns (Figure 4A). We calculated the average Cα root-mean-square fluctuations (RMSFs) of their system equilibrium phase (Figure 4B). Without a doubt, the results show that these four toxins are very rigid, because α-KTxs obtain six conservative cysteines which form three intermolecular disulfide bonds. Due to their stable structure, they are developed into protein scaffolds. For example, Vita et al. designed a metal binding activity on ChTx [27]. After combining with $K_V1.2$, their structure mildly adjusted. OsK-1 and AgTx-2 have little change after combining with $K_V1.2$, which indicate that its interaction mechanism with $K_V1.2$ is similar to ChTx (Figure 4C,F). MgTx shows the increased flexibility of α-helix and γ-core region (the last two β-folds and the turn region between them) (Figure 4D). MMTX shows the increased flexibility of N-terminal (Figure 4E).

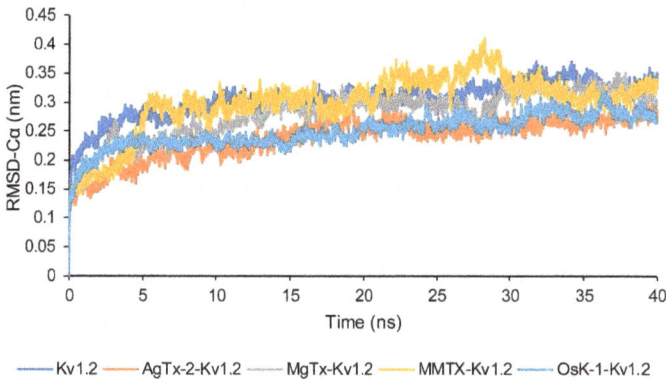

(A)

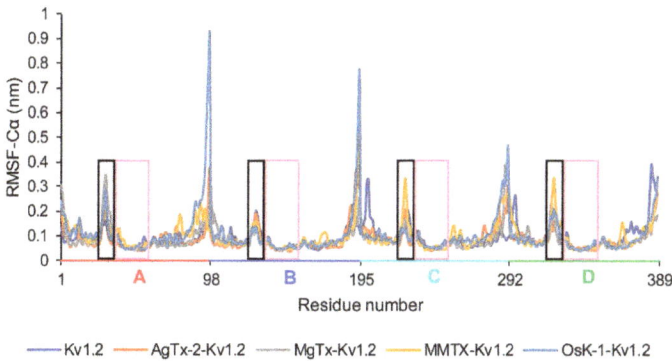

(B)

Figure 3. *Cont.*

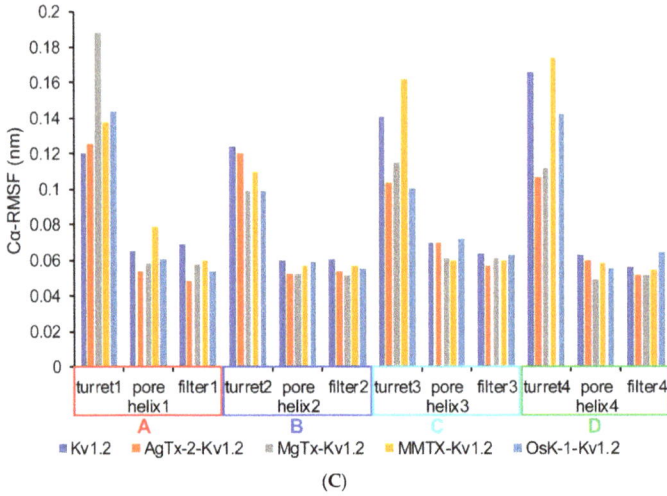

(C)

Figure 3. Structural flexibilities of four toxin-K$_V$1.2 complexes. (**A**) RMSDs of four toxin-K$_V$1.2 pore region complexes; (**B**) root-mean-square fluctuations (RMSFs) of the Cα atoms of K$_V$1.2 pore region in these four complexes from 15 ns to 40 ns. A–D indicate four different chains in K$_V$1.2. The range of A–D chains' pore region residue number is marked by red, blue, light blue, and green string, respectively. Turret regions are outlined by black rectangular boxes and pore helix and filter regions are outlined by pink rectangular boxes; (**C**) average Cα-RMSF of K$_V$1.2 pore region in sole K$_V$1.2 and these four complexes from 15 ns to 40 ns. Turret, pore helix, and filter in *x*-coordinate represent the turret, pore helix, and filter region of K$_V$1.2 channel. A–D indicate four different chains in K$_V$1.2.

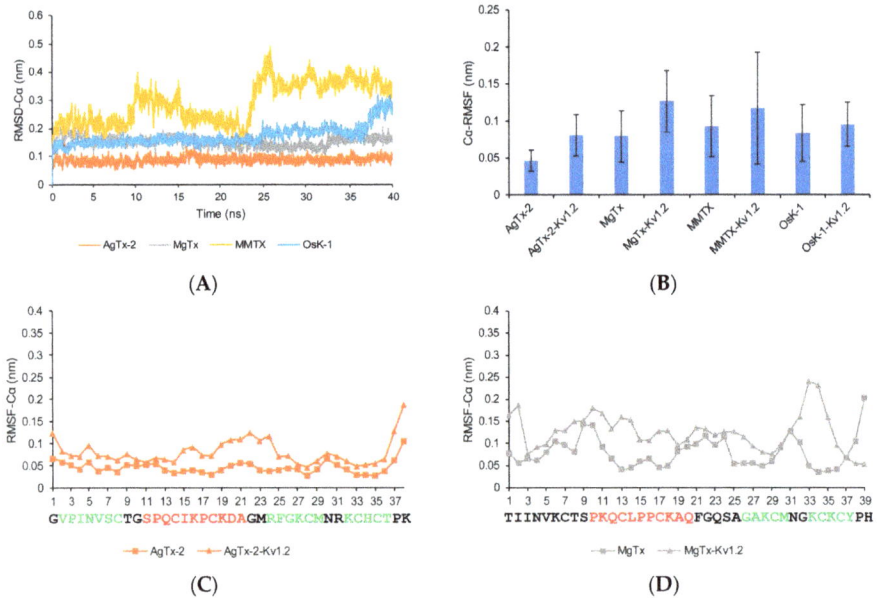

(A)

(B)

(C)

(D)

Figure 4. *Cont.*

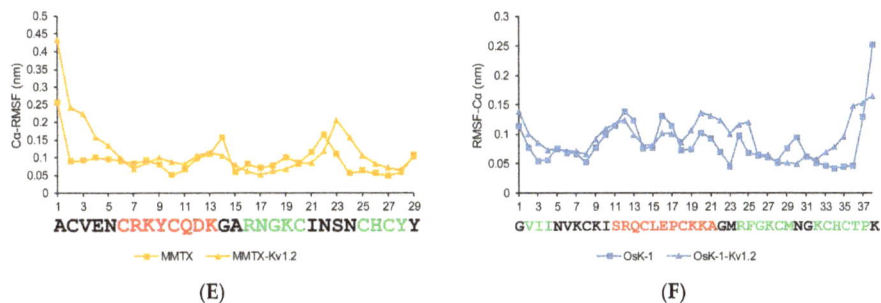

(E)

(F)

Figure 4. Structural flexibilities of four toxins. (**A**) RMSDs; (**B**) average RMSF; (**C–F**) RMSFs of the Cα atoms of four toxins. Toxins' sequences are written under x-coordinate and their α-helical and β-fold are colored in red and green, respectively.

2.4. The Interactions of α-KTxs with K_v1.2

Using LigPlot$^+$ software, we analyzed the constructed complexes, together with ChTX-K$_v$1.2 for comparison purposes. Their detailed interactions are shown in Figure 5 and Table 1. To ensure these predicted hydrogen bonds and salt bridge are reliable, we calculated these bonds' distances in 40 ns (Figure 6).

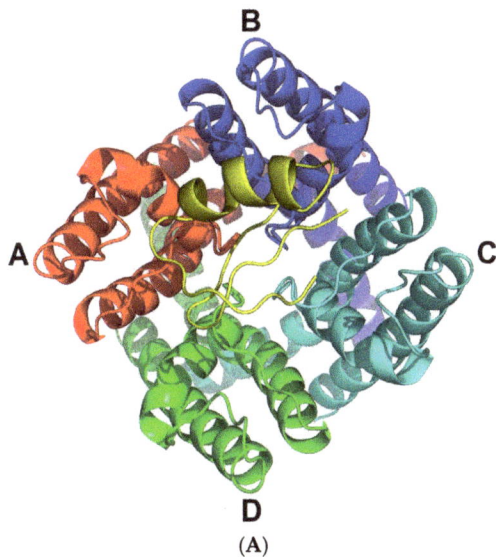

(A)

Figure 5. *Cont.*

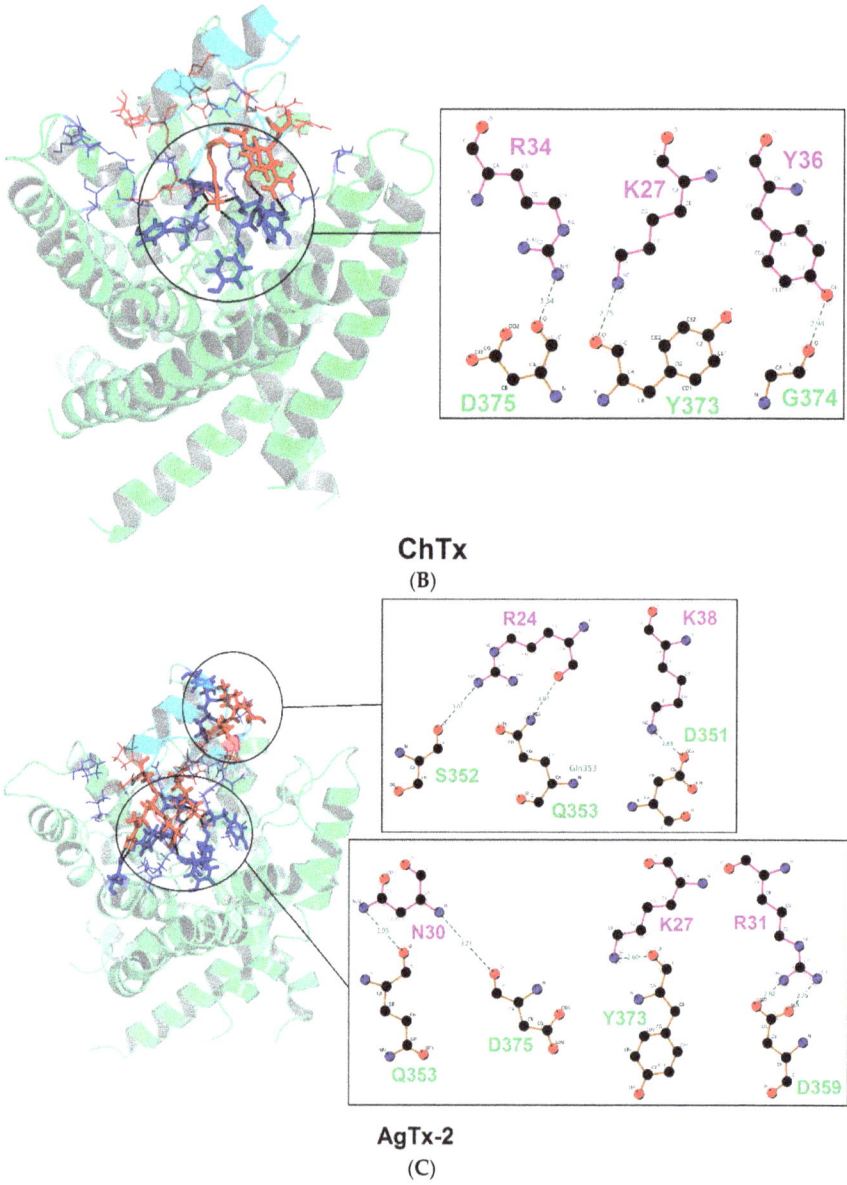

ChTx

(B)

AgTx-2

(C)

Figure 5. *Cont.*

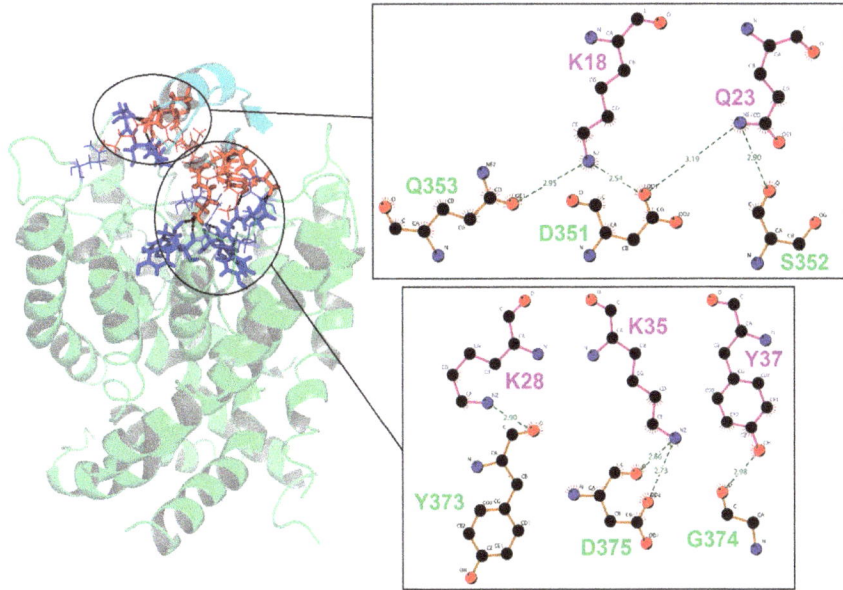

MgTx

(D)

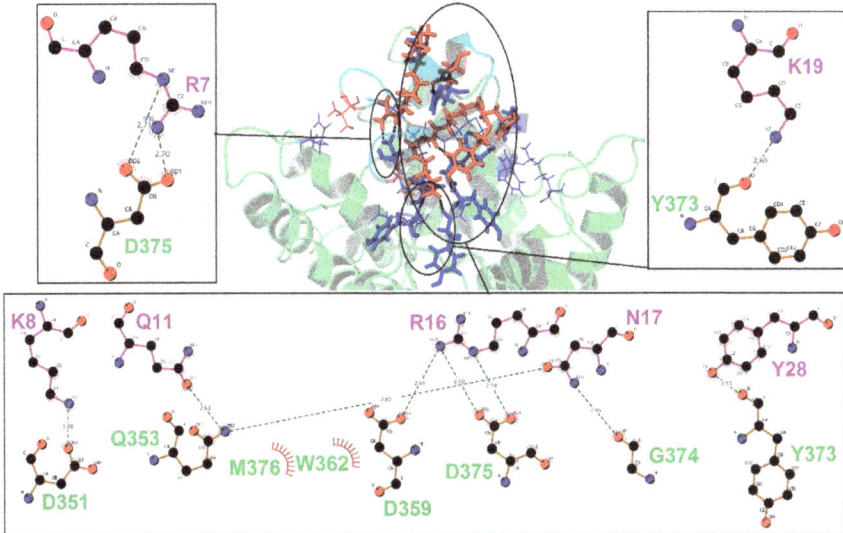

MMTX

(E)

Figure 5. *Cont.*

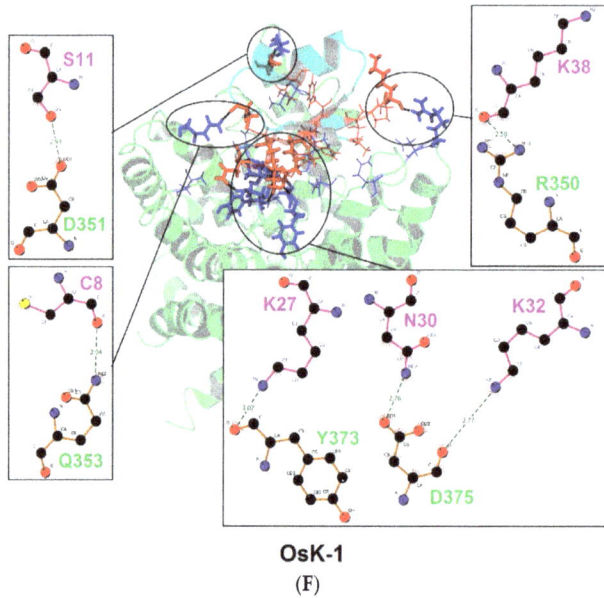

OsK-1

(F)

Figure 5. Interactions between toxins and $K_v1.2$. (**A**) A–D chains of $K_v1.2$ pore region; (**B–F**) showing the interaction residues of toxins and channel. H-bonds are shown in green and hydrophobic interactions in red. Interaction sites of toxins are highlighted in red and sites of the channel in blue. Sites involved in hydrophobic interactions are shown as lines and sticks in H-bonds or salt bridges.

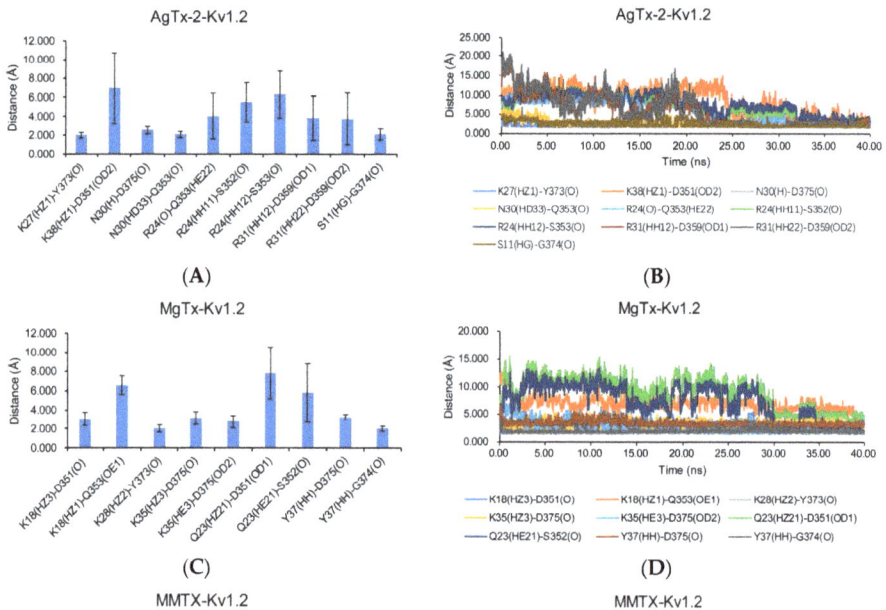

(**A**)

(**B**)

(**C**)

(**D**)

MMTX-Kv1.2

MMTX-Kv1.2

Figure 6. *Cont.*

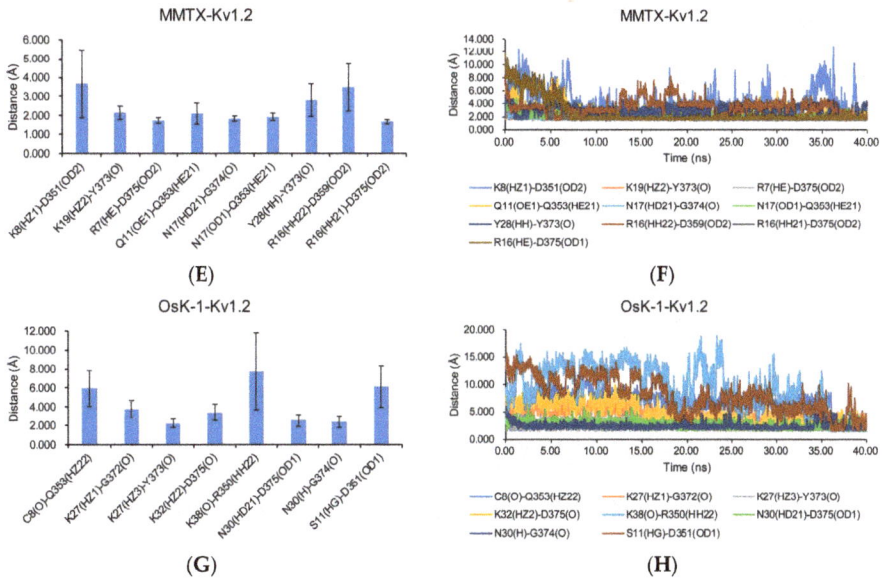

(E)

(F)

(G)

(H)

Figure 6. Hydrogen bonds and salt bridges distances of toxin-K$_V$1.2 complexes. (**A,C,E,G**) showing the average distances of hydrogen bonds and salt bridges from 15 ns to 40 ns. (**B,D,F,H**) showing the changes of hydrogen bonds and salt bridges distances between toxins and K$_V$1.2 channel in 40 ns.

Table 1. Interacting pairs between toxins and K$_V$1.2 pore region. Unconservative sites among K$_V$1.1–K$_V$1.3 pore regions are marked by asterisks (*). Chain number of channel amino acids are labeled in brackets.

Interaction Force Type	ChTx	AgTx-2	MgTx	MMTX	OsK-1
Hydrophobic contacts	T8-Q353(A) * T8-D351(A) * T9-S352(A) S10-D375(A) W14-Q353(B) * R25-M376(B) R25-Q353(B) * M29-D375(D) M29-G374(D) N30-V377(A) * N30-T379(A) * N30-D375(D) Y36-D375(B) S37-T379(C) *	G10-T379(A) * P12-Q353(A) * F25-D375(A) F25-G374(B) F25-V377(B) * F25-D375(B) M29-D375(C) M29-G374(D) M29-V377(D) *	K11-Q353(A) * S24-Q353(B) * A25-Q353(B) * G26-D375(B) M30-G374(D) M30-V377(D) * N31-Q353(D) * Y37-D375(B) H39-M376(B)	V3-Q353(A) * R7-V377(B) * R7-T379(B) * R16-K384(C) I21-G374(D) I21-D375(D) Y28-G374(C)	I3-D375(C) K7-Q353(A) * L15-Q353(B) * F25-Q353(B) * G26-D375(B) K27-G374(B) M29-G374(D) N30-T379(A) * H34-D375(C) P37-D351(C) *
H-bonds	K27-Y373(A) K27-Y373(B) K27-Y373(C) K27-Y373(D) Y36-G374(C)	S11-G374(A) R24-S352(B) R24-Q353(B) * K27-Y373(A) K27-Y373(B) K27-Y373(C) K27-Y373(D) N30-Q353(D) * N30-D375(D)	K18-Q353(B) * Q23-S352(B) K28-Y373(A) K28-Y373(B) K28-Y373(C) K28-Y373(D) Y37-G374(C)	Q11-Q353(B) * N17-Q353(B) * N17-G374(B) K19-Y373(A) K19-Y373(C) K19-Y373(C) Y28-Y373(B)	C8-Q353(A) * S11-D351(B) * K27-Y373(A) K27-G372(B) K27-Y373(C) K27-Y373(D) N30-G374(A) N30-D375(D)
Salt bridges	R34-D375(C)	R31-D359(D) K38-D351(B) *	K18-D351(B) * K35-D375(C)	R7-D375(A) K8-D351(B) * R16-D359(B) R16-D375(B)	K32-D375(D) K38-R350(C) *

Our complexes of $K_V1.2$ and toxins are consistent with the current experimental data. AgTx-2 have been reported that K27 and N30 are critical for binding affinity toward Shaker K^+ channel [17]. In our AgTx-2-$K_V1.2$ model, K27 inserts into pore region and forms H-bond with Y373, and N30 forms H-bond with Q353 and D375 (Figure 5C). Mutation experiments showed that MMTX interact with $rK_V1.2$ V379. In our MMTX-$K_V1.2$ model, R7 of MMTX interacts with V379 through hydrophobic contact (Table 1) [13]. The OsK-1-$K_V1.2$ model suggest that E16 and K20 of OsK-1 does not interact with $K_V1.2$. It accords that E16K and/or K20D mutations of OsK-1 do not change the effect on $K_V1.2$ (Table 1) [14].

As expected, these five toxins all present some commonalities in interacting with $K_V1.2$. They depend on hydrophobic contacts, hydrogen-bonds (H-bonds), and salt bridges to stay close to the channel where K27 inserts into the pore and forms H-bonds with Y373 (unless otherwise stated, all toxins and the channel are numbered according to the ChTX-$K_V1.2$ complex [18]). An aromatic amino acid (F or Y) belonging to the functional dyad interacts with D375 or Y373. With the exception of MMTX, all the toxins use M29 and N30 to contact G374 and D375 in the channel filter region through the van der Waals force. Furthermore, in our AgTx-2-$K_V1.2$ model, N30 also forms H-bond with Q353 and D375. These observations fully confirm the functional importance of K27 and N30 in this toxin previously obtained by mutational analysis [17]. The turn region preceding the α-helix of the toxins contacts the turret residue S353 and the filter D375/V377/T379 via a hydrophobic interaction force. We also observed that the basic amino acids located at the last β-strand of the toxins, except MMTX, form salt bridges with the acidic Asp of $K_V1.2$. R34 of ChTx, K35 of MgTx, and K32 of OsK-1 form salt bridges with D359 of $K_V1.2$ and R31 of AgTx-2 form salt bridges with D359 of $K_V1.2$. These toxins have conservative six cysteines that hardly participate in the interaction with $K_V1.2$, but they are related to structural stabilization [28]. Compared with other toxins, MMTX forms more hydrogen bonds and salt bridges with the channel. Especially, due to its shorter N-terminus, this toxin can enter more deeply into the channel pore to form salt bridges between K8 or R16 and the residues derived from the chain B of the channel during MD simulation. These noncovalent interactions could facilitate the formation of a more stable toxin-channel complex [29]. In our MMTX-$K_V1.2$ model, R7 of the toxin interacts with V377 of the channel through hydrophobic contact, in line with the mutational experiments that highlighted this channel residue as a target site of MMTX [13]. Taken together, our MD simulation results provide support for the toxins' binding sites mainly locating at the $K_V1.2$ turret and filter region and are thus a reasonable explanation for the lack of channel subtype selectivity in these toxins given that in these regions most sites are highly conserved among $K_V1.1$–$K_V1.3$ (Figure 7).

2.5. $K_V1.2$-Specific Amino Acids

More importantly, through analysis of the interaction between these four toxins and $K_V1.2$ channel, we observed four $K_V1.2$-specific amino acids (D355, Q358, V381, and T383 of $rK_V1.2$) that can interact with the toxins (Figures 7 and 8). In these interactions, D355 and Q358 form hydrogen bonds or salt bridges whereas V381 and T383 form hydrophobic interactions with the toxins. This observation could help answer the differential affinity of these toxins towards $K_V1.1$ to $K_V1.3$. For example, the preferred inhibition of $K_V1.2$ and $K_V1.3$ over $K_V1.1$ by Aam-KTX is explained by the variation at site 355 (corresponding to the $K_V1.2$-specific amino acid Q358). In $K_V1.1$, this site is occupied by a larger His that might hamper the toxin's entry into its pore vestibule [30]. In addition, site 381(Val) has been proposed as a main determinant of MMTX's selectivity towards $K_V1.2$ over $K_V1.1$ [13], in agreement with our MD simulation data. In the MgTx-$K_V1.2$ complex, K18 from the toxin forms an H-bond with Q358 from the channel. The corresponding amino acid at this site is a His in $K_V1.3$. Because histidine is the same kind of charge with lysine, their repulsion is adverse to the formation of an H-bond. For $K_V1.1$, a glycine occupies this position and its shorter side chain also hampers the formation of the H-bond. These observations could account for the differential in affinity of MgTx to these three channels (Figure 1A). Relative to other toxins analyzed here, MMTX possess a shorter N-terminus and is the only one without effect on $K_V1.1$. In our complex, its K8 forms salt bridges with D355, whereas sites Q11 and N17 form H-bonds with Q358. These extensive contacts likely provide a structural basis for its high activity on $K_V1.2$.

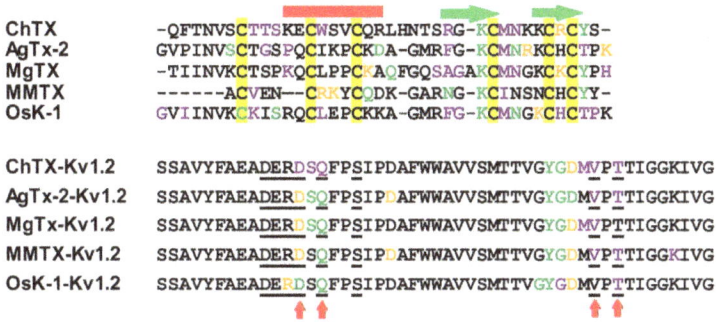

```
ChTX        -QFTNVSCTTSKECWSVCQRLHNTSRG-KCMNKKCRCYS-
AgTx-2      GVPINVSCTGSPQCIKPCKDA-GMRFG-KCMNRKCHCTPK
MgTX        -TIINVKCTSPKQCLPPCKAQFGQSAGAKCMNGKCKCYPH
MMTX        ------ACVEN--CRKYCQDK-GARNG-KCINSNCHCYY-
OsK-1       GVIINVKCKISRQCLEPCKKA-GMRFG-KCMNGKCHCTPK

ChTX-Kv1.2    SSAVYFAEADERDSQFPSIPDAFWWAVVSMTTVGYGDMVPTTIGGKIVG
AgTx-2-Kv1.2  SSAVYFAEADERDSQFPSIPDAFWWAVVSMTTVGYGDMVPTTIGGKIVG
MgTx-Kv1.2    SSAVYFAEADERDSQFPSIPDAFWWAVVSMTTVGYGDMVPTTIGGKIVG
MMTX-Kv1.2    SSAVYFAEADERDSQFPSIPDAFWWAVVSMTTVGYGDMVPTTIGGKIVG
OsK-1-Kv1.2   SSAVYFAEADERDSQFPSIPDAFWWAVVSMTTVGYGDMVPTTIGGKIVG
```

Figure 7. Interaction sites between toxins and $K_V1.2$. Sites participating in salt bridges, H-bonds, and hydrophobic effect are highlighted in golden, green, and purple, respectively. Four variable channel sites implicated in the interaction with α-KTxs are labeled with red arrows.

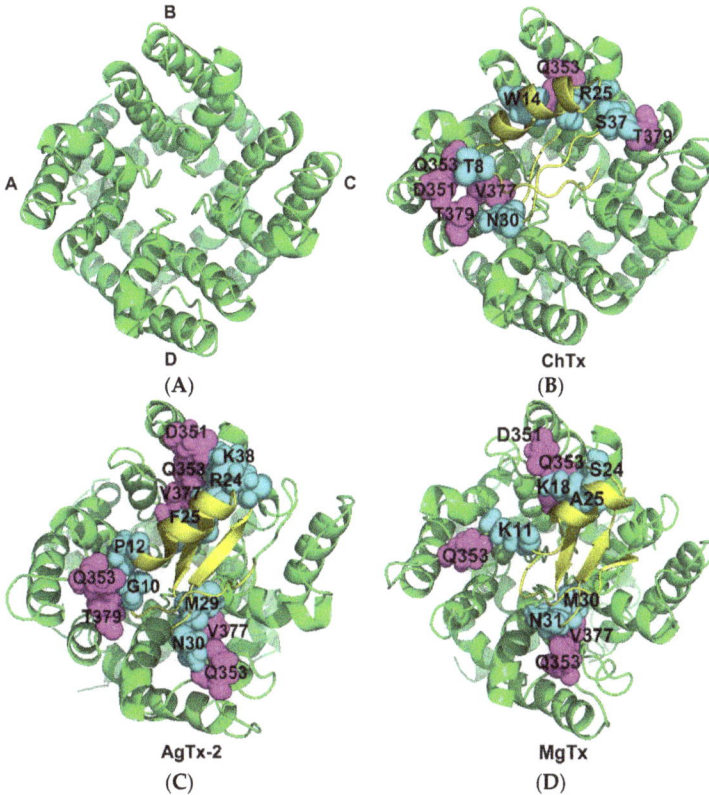

Figure 8. *Cont.*

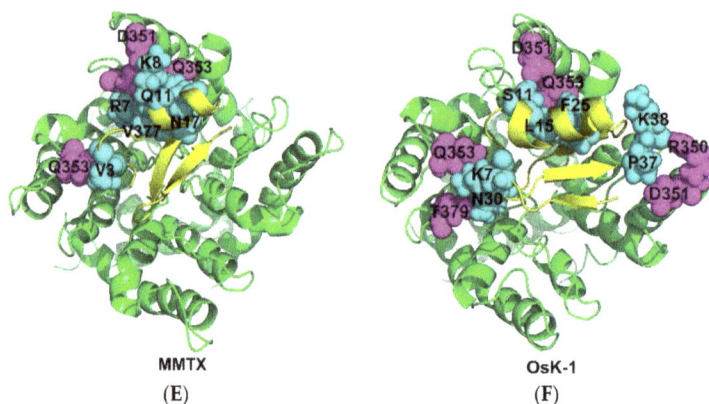

MMTX   OsK-1

(E)   (F)

Figure 8. $K_V1.2$ specific amino acids involved in the interactions with the toxins. (**A**) A–D chains of $K_V1.2$ channels pore region; (**B–F**) interaction sites of toxins are highlighted in cyan; unconservative sites in the channel are highlighted in magentas. Amino acids at the interface between a toxin and the channel are shown as spheres.

3. Discussion

It is long known that the evolutionary conservation of the pore region among $K_V1.1$ to $K_V1.3$ poses a challenge to animal toxin-based drug design. To face this challenge, three kinds of protein engineering technologies have been explored to improve their selectivity: (1) Site-directed mutation. Using this technology, several highly selective scorpion toxins against $K_V1.3$ have been obtained [31]; (2) Phage display technology. This technology led to the discovery of Mokatoxin-1, an engineering peptide that blocks $K_V1.3$ at the nanomolar level without effect on $K_V1.1$, $K_V1.2$ and $K_{Ca}1.1$ [8]; (3) MD simulation. This technology can guide molecular design based on the structural features of peptides [32].

In this work, through using MD simulation technology, we identified four $K_V1.2$-specific amino acids that are involved in the interactions with different scorpion toxins (Figure 7). These K_V specific sites might be the reason that toxins have different targets. Conservative sites of α-KTxs interact with conservative sites of K_V channels. For example, K27 of α-KTxs insert into K_V channel pore region, form H-bonds with Y373, and inhibit K^+ pass. TVGYG motif is conservative among eukaryotic K_V channel and important for binding K^+ [19]. Nevertheless, unconservative sites of α-KTxs interact with K_V specific sites which determine the selectivity of different toxins. Like scorpion Na+ channel, α-toxins has a common bipartite bioactive surface: (1) Conserved core-domain are associated with toxin potency, which interact with domain IV S1–S2 and S3–S4 of Na_V channels; (2) Variable NC-domain dictate toxin's selectivity, which interact with domain I S5–S6 of Na_V channels [33]. According to these facts, we proposed that the evolution of toxins has a common rule: conservative domain ensures their potency, which interacts with receptor and the variable domain convenient to adjusting their targets.

Furthermore, in combination with the MD simulation data presented here, we proposed a concept of "triplet-motif" for channel blockade. This motif is composed of the dyad comprising a lysine located at the first β-strand and an aromatic amino acid in a distance of 5–7 Å and an Asn in the loop linking the first and second β-strands, two residues downstream from the Lys (i.e., LysCysXaaAsn. Xaa, any amino acids). The location of these three functional sites (dyad-motif and N30) is shown on Figure 2. The inclusion of this residue into the toxins' functional motif is based on the following considerations: (1) This residue is an evolutionarily highly conserved amino acid that belongs to the scorpion toxin signature (STS), comprising Cys … CysXaaXaaXaaCys … LysCysXaaAsn … CysXaaCys; (2) Its functional significance has been highlighted in some toxins (e.g., AgTx-2, navitoxin, etc.) [17]; (3) In an NMR-based complex of KTX and KcsA-$K_V1.3$, N30 is close to D64 of KcsA-$K_V1.3$ [25]; (4) In our dynamics structures, this residue in AgTx-2, MgTx, and OsK-1 forms H-bonds or hydrophobic interactions with $K_V1.2$.

The molecular mechanism of interaction between scorpion toxins and K_V channels has always been controversial. There are currently two hypotheses: (1) The induced-fit model. Using high-resolution solid-state NMR spectroscopy, Lange et al. observed chemical shifts occurring in some residues of kaliotoxin (KTx) and KcsA-$K_V1.3$ during their interactions. For the channel, significant chemical shifts appeared in the pore helix and the selectivity filter [25]. Because of these observations, they thought that the toxin binds to the channel in an induced fit manner; (2) The lock-and-key model. This model is based on the consideration of rigidity of the K^+ channel pore region [34], evidenced by a crystal structure study that resolved the structures of $K_V1.2$–$K_V2.1$ in complex with ChTx and the channel alone. When the complex was superposed onto the $K_V1.2$–$K_V2.1$ structure, no discernible structural changes were observed in the channel [18]. Therefore, they proposed that scorpion toxins bind to K^+ channels in a lock and key manner. According to our molecular dynamics simulation result, the turret region of $K_V1.2$ is the most flexible region. Therefore, we proposed that K_V channels interact with different scorpion toxin mainly by modulating their turret region. Consider existing experimental data, different sequence and structure toxins block $K_V1.2$ though pore region inhibition and these toxins bind to more than one channel [2,35–38]. Our result and these experimental data all support induced fit model.

4. Conclusions

By MD simulation analysis combined with previous experimental data, we reveal a common mode adopted by scorpion K^+ channel toxins in binding to the channels, in which the conserved and variable toxin functional residues seem to interact with the conserved and subtype-specific channel residues, respectively. This finding provides new candidate sites in the toxins for mutations to improve their selectivity towards a specific channel subtype.

5. Materials and Methods

5.1. Atomic Coordinates and $K_v1.2$-Toxin Complexes

Atomic coordinates of ChTx-$K_V1.2$-$K_V2.1$ chimera (PDB: 4JTA), AgTx-2 (PDB: 1AGT), MgTx (PDB: 1MTX), MMTX (PDB: 2CRD), and OSK1 (PDB: 1SCO) were retrieved from the Protein Data Bank [39]. All toxin-$K_V1.2$ complex structures studied here were built by the Swiss PDB Viewer software (http://spdbv.vital-it.ch/), in which ChTx-$K_V1.2$-$K_V2.1$ chimera was used as template. These four toxins were aligned onto the ChTx-$K_V1.2$-$K_V2.1$ chimera and then deleted ChTx to build all toxin-$K_V1.2$ complexes. The channel pore region (residues 321 to 417) were used in this study.

About building the complexes of toxins and channels, some groups choose ZDOCK or HADDOCK molecular docking software to obtain the complexes [20,40,41]. Since the structure of ChTx and $K_V1.2$–2.1 chimera resolved, several groups use this resolved structure to build scorpion toxins and K^+ channel complexes [42,43]. Nekrasova et al. performed homology modeling of $K_V1.6$-toxin complexes instead of molecular docking to obtain a more reliable model [43]. Therefore, we choose ChTx and $K_V1.2$–2.1 chimera to build our toxin-channel complexes.

5.2. Molecular Dynamics Simulation

Molecular dynamics simulations were performed using Gromacs 5.0.1, in which all-atom OPLS force field was chosen [44,45]. The complex was solved with SPC water [46] and was immersed in a cubic box extending to at least 5 nm of the solvent on all sides. Also, the system was neutralized by K^+ and Cl^-. It was energy minimized by using the steepest descent algorithm for 5000 steps, and it made a maximum force of less than 1000 kJ/mol/nm. After energy minimization, the system was equilibrated in a constrained NVT (Number of Particles, Volume, Temperature) and NPT (Number of Particles, Pressure, Temperature) running for 100 ps. NVT equilibration ensured the system be brought to the temperature (300 K) which we wish to simulate, and with which we seek to establish the proper orientation about the protein. After NVT equilibration, we stabilize the pressure of the system under an NPT ensemble. Through NVT and NPT equilibration, it was well-equilibrated at 300 K and 1 bar. Bond length was constrained using the LINCS algorithm [47]. Finally, MD simulations of these complexes were carried out for 40 ns. Trajectories are saved every 10ps for analysis. For the MD simulation, the Verlet cut-off scheme and a Leap-frog integrator with a step size of 2 fs were applied. For temperature coupling, the modified Berendsen thermostat and the Parrinello-Rahman barostat for pressure coupling were used. For long-range electrostatic interaction, the Particle Mesh Ewald method was used. The method of four toxins' molecular dynamics simulation is similar to the toxin-K_v1.2 complex. The differences are that the toxins were neutralized by Na^+ and Cl^-, and they were immersed in a cubic box extending to at least 1 nm of the solvent on all sides.

The mutagenesis and simulations indicated that the scorpion toxins bind with the extracellular part of the K^+ channels and the interaction is hardly affected by the membrane and the transmembrane segment of channel [48–51]. We did not add the membrane into the simulation like the work of other study groups [20,52–59]. Also, many simulation studies on the recognition between scorpion toxins and K^+ channels without a membrane have achieved good agreements with experimental data [53,58,59]. Discarding the lipid-protein interactions has also contributed to the reduction of the computational burden and the extention of the MD simulation trajectories [19]. Certainly, a transmembrane protein system could be more reliable if we take into account the membrane around the channel.

We did not perform similar MD simulations to toxins-channels K_v1.1 and K_v1.3, because the K_v1.1–K_v1.3 channel only has eight unconservative sites (Figure 1B). And only K_v1.2–2.1 chimera structure was resolved. In this work, we aim to obtain the K_v1.2-toxin complexes and understand the detailed interaction between K_v1.2 and the toxins. This will help us to design mutations according to these complexes and obtain a K_v1.2 specific selective toxin.

5.3. Analysis of MD Simulation Results

After molecular dynamics simulation, we obtained the last MD simulation frame of the complexes. To analyse whether the protein was stable and close to the experimental structure, we measured the root-mean-square displacement (RMSD) of all the structures of $C\alpha$. LigPlot$^+$ software (http://www.ebi.ac.uk/thornton-srv/software/LigPlus/) was used to analyze the detailed interactions between the toxin and K_v1.2 [60]. LigPlot$^+$ can analyze the hydrophobic interaction, hydrogen bonds, and salt bridges between toxins and K_v1.2. To ensure the reliability of predicted hydrogen bonds and salt bridges, we calculated the distances of hydrogen bonds and salt bridges in 40 ns using GROMACS. Pymol (http://www.pymol.org/) was used to prepare all the structural images.

There is no unified standard about choosing which structure to analyse during molecular dynamics simulation. Kohl et al. chose the structure with the highest number of H-bonds between the toxin and the channel [42]. Nekrasova et al. chose 70 trajectory frames to analyse the hydrophobic interaction, the hydrogen, and the ionic bonds of the toxin [43]. Also, Yi et al. analysed the last structure of the toxin-channel structure [20,40]. Too short molecular dynamics simulation will cause proteins to have not enough time to change their conformation [61]. Therefore, we gave 40 ns to stabilize the structure and most toxin-channel dynamic studies only run several nanoseconds. Through calculating the distances of H-bonds and salt bridges, these bonds became stable with the extension of simulation time. As a result, we proposed that the last frame of the toxin-channel complex was reliable.

Acknowledgments: We thank Yingliang Wu (Wuhan University, Wuhan, China) for his valuable suggestions. This work was supported by the National Natural Science Foundation of China (31570773) to S.Z. and the State Key Laboratory of Integrated Management of Pest Insects and Rodents (Grant No. ChineseIPM1707).

Author Contributions: S.Z. conceived and designed this study; S.Y. performed the analysis; S.Y., B.G., and S.Z. jointly wrote the paper.

Conflicts of Interest: The authors declare no conflict of interest.

References

1. Chen, X.; Wang, Q.; Ni, F.; Ma, J. Structure of the full-length Shaker potassium channel Kv1.2 by normal-mode-based X-ray crystallographic refinement. *Proc. Natl. Acad. Sci. USA* **2010**, *107*, 11352–11357. [CrossRef] [PubMed]

2. Wulff, H.; Castle, N.A.; Pardo, L.A. Voltage-gated potassium channels as therapeutic targets. *Nat. Rev. Drug Discov.* **2009**, *8*, 982–1001. [CrossRef] [PubMed]

3. Cremonez, C.M.; Maiti, M.; Peigneur, S.; Cassoli, J.S.; Dutra, A.A.; Waelkens, E.; Lescrinier, E.; Herdewijn, P.; de Lima, M.E.; Pimenta, A.M.; et al. Structural and Functional Elucidation of Peptide Ts11 Shows Evidence of a Novel Subfamily of Scorpion Venom Toxins. *Toxins (Basel)* **2016**, *8*, e288. [CrossRef] [PubMed]

4. Xie, G.; Harrison, J.; Clapcote, S.J.; Huang, Y.; Zhang, J.Y.; Wang, L.Y.; Roder, J.C. A new Kv1.2 channelopathy underlying cerebellar ataxia. *J. Biol. Chem.* **2010**, *285*, 32160–32173. [CrossRef] [PubMed]

5. Brew, H.M.; Gittelman, J.X.; Silverstein, R.S.; Hanks, T.D.; Demas, V.P.; Robinson, L.C.; Robbins, C.A.; McKee-Johnson, J.; Chiu, S.Y.; Messing, A.; et al. Seizures and reduced life span in mice lacking the potassium channel subunit Kv1.2, but hypoexcitability and enlarged Kv1 currents in auditory neurons. *J. Neurophysiol.* **2007**, *98*, 1501–1525. [CrossRef] [PubMed]

6. Pena, S.D.; Coimbra, R.L. Ataxia and myoclonic epilepsy due to a heterozygous new mutation in KCNA2: Proposal for a new channelopathy. *Clin. Genet.* **2015**, *87*, e1–e3. [CrossRef] [PubMed]

7. Syrbe, S.; Hedrich, U.B.; Riesch, E.; Djémié, T.; Müller, S.; Møller, R.S.; Maher, B.; Hernandez-Hernandez, L.; Synofzik, M.; Caglayan, H.S.; et al. De-novo loss- or gain-of-function mutations in KCNA2 cause epileptic encephalopathy. *Nat. Genet.* **2015**, *47*, 393–399. [CrossRef] [PubMed]

8. Takacs, Z.; Toups, M.; Kollewe, A.; Johnson, E.; Cuello, L.G.; Driessens, G.; Biancalana, M.; Koide, A.; Ponte, C.G.; Perozo, E.; et al. A designer ligand specific for Kv1.3 channels from a scorpion neurotoxin-based library. *Proc. Natl. Acad. Sci. USA* **2009**, *106*, 22211–22216. [CrossRef] [PubMed]

9. Cerni, F.A.; Pucca, M.B.; Amorim, F.G.; Bordon, K.D.C.F.; Echterbille, J.; Quinton, L.; De Pauw, E.; Peigneur, S.; Tytgat, J.; Arantes, E.C. Isolation and characterization of Ts19 Fragment II, a new long chain potassium channel toxin from Tityus serrulatus venom. *Peptides* **2016**, *80*, 9–17. [CrossRef] [PubMed]

10. Chen, Z.Y.; Hu, Y.T.; Yang, W.S.; He, Y.W.; Feng, J.; Wang, B.; Zhao, R.M.; Ding, J.P.; Cao, Z.J.; Li, W.X.; et al. Hg1, novel peptide inhibitor specific for Kv1.3 channels from first scorpion Kunitz-type potassium channel toxin family. *J. Biol. Chem.* **2012**, *287*, 13813–13821. [CrossRef] [PubMed]

11. Moreels, L.; Peigneur, S.; Yamaguchi, Y.; Vriens, K.; Waelkens, E.; Zhu, S.; Thevissen, K.; Cammue, B.P.; Sato, K.; Tytgat, J. Expanding the pharmacological profile of κ-hefutoxin 1 and analogues: A focus on the inhibitory effect on the oncogenic channel Kv10.1. *Peptides* **2016**. [CrossRef] [PubMed]

12. Chen, Z.; Hu, Y.; Han, S.; Yin, S.; He, Y.; Wu, Y.; Cao, Z.; Li, W. ImKTx1, a new Kv1.3 channel blocker with a unique primary structure. *J. Biochem. Mol. Toxicol.* **2011**, *25*, 244–251. [CrossRef] [PubMed]

13. Wang, X.; Umetsu, Y.; Gao, B.; Ohki, S.; Zhu, S. Mesomartoxin, a new Kv1.2-selective scorpion toxin interacting with the channel selectivity filter. *Biochem. Pharmacol.* **2015**, *93*, 232–239. [CrossRef] [PubMed]

14. Mouhat, S.; Visan, V.; Ananthakrishnan, S.; Wulff, H.; Andreotti, N.; Grissmer, S.; Darbon, H.; De Waard, M.; Sabatier, J.M. K⁺ channel types targeted by synthetic OSK1, a toxin from Orthochirus scrobiculosus scorpion venom. *Biochem. J.* **2005**, *385*, 95–104. [CrossRef] [PubMed]

15. Bartok, A.; Toth, A.; Somodi, S.; Szanto, T.G.; Hajdu, P.; Panyi, G.; Varga, Z. Margatoxin is a non-selective inhibitor of human Kv1.3 K⁺ channels. *Toxicon* **2014**, *87*, 6–16. [CrossRef] [PubMed]

16. Pimentel, C.; M'Barek, S.; Visan, V.; Grissmer, S.; Sampieri, F.; Sabatier, J.M.; Darbon, H.; Fajloun, Z. Chemical synthesis and 1H-NMR 3D structure determination of AgTx2-MTX chimera, a new potential blocker for Kv1.2 channel, derived from MTX and AgTx2 scorpion toxins. *Protein Sci.* **2008**, *17*, 107–118. [CrossRef] [PubMed]

17. Ranganathan, R.; Lewis, J.H.; MacKinnon, R. Spatial localization of the K⁺ channel selectivity filter by mutant cycle-based structure analysis. *Neuron* **1996**, *16*, 131–139. [CrossRef]

18. Banerjee, A.; Lee, A.; Campbell, E.; Mackinnon, R. Structure of a pore-blocking toxin in complex with a eukaryotic voltage-dependent K⁺ channel. *eLife* **2013**, *2*, e00594. [CrossRef] [PubMed]

19. Novoseletsky, V.N.; Volyntseva, A.D.; Shaitan, K.V.; Kirpichnikov, M.P.; Feofanov, A.V. Modeling of the binding of peptide blockers to voltage-gated potassium channels: Approaches and evidence. *Acta Nat.* **2016**, *8*, 35–46.

20. Yi, H.; Qiu, S.; Cao, Z.; Wu, Y.; Li, W. Molecular basis of inhibitory peptide maurotoxin recognizing Kv1.2 channel explored by ZDOCK and molecular dynamic simulations. *Proteins* **2008**, *70*, 844–854. [CrossRef] [PubMed]

21. Chen, R.; Chung, S.H. Structural basis of the selective block of Kv1.2 by maurotoxin from computer simulations. *PLoS ONE* **2012**, *7*, e47253. [CrossRef] [PubMed]

22. Nikouee, A.; Khabiri, M.; Cwiklik, L. Scorpion toxins prefer salt solutions. *J. Mol. Model.* **2015**, *21*, 287. [CrossRef] [PubMed]

23. Park, C.S.; Miller, C. Mapping function to structure in a channel-blocking peptide: Electrostatic mutants of charybdotoxin. *Biochemistry* **1992**, *31*, 7749–7755. [CrossRef] [PubMed]

24. Park, C.S.; Miller, C. Interaction of charybdotoxin with permeant ions inside the pore of a K+ channel. *Neuron* **1992**, *9*, 307–313. [CrossRef]

25. Lange, A.; Giller, K.; Hornig, S.; Martin-Eauclaire, M.F.; Pongs, O.; Becker, S.; Baldus, M. Toxin-induced conformational changes in a potassium channel revealed by solid-state NMR. *Nature* **2006**, *440*, 959–962. [CrossRef] [PubMed]

26. Zhu, S.; Peigneur, S.; Gao, B.; Umetsu, Y.; Ohki, S.; Tytgat, J. Experimental conversion of a defensin into a neurotoxin: Implications for origin of toxic function. *Mol. Biol. Evol.* **2014**, *31*, 546–559. [CrossRef] [PubMed]

27. Vita, C.; Roumestand, C.; Toma, F.; Ménez, A. Scorpion toxins as natural scaffolds for protein engineering. *Proc. Natl. Acad. Sci. USA* **1995**, *92*, 6404–6408. [CrossRef] [PubMed]

28. Zhu, Q.; Liang, S.; Martin, L.; Gasparini, S.; Ménez, A.; Vita, C. Role of disulfide bonds in folding and activity of leiurotoxin I: Just two disulfides suffice. *Biochemistry* **2002**, *41*, 11488–11494. [CrossRef] [PubMed]

29. Lee, C.W.; Wang, H.J.; Hwang, J.W.; Tseng, C.P. Protein thermal stability enhancement by designing salt bridges: A combined computational and experimental study. *PLoS ONE* **2014**, *9*, e112751. [CrossRef] [PubMed]

30. Abbas, N.; Belghazi, M.; Abdel-Mottaleb, Y.; Tytgat, J.; Bougis, P.E.; Martin-Eauclaire, M.F. A new Kaliotoxin selective towards Kv1.3 and Kv1.2 but not Kv1.1 channels expressed in oocytes. *Biochem. Biophys. Res. Commun.* **2008**, *376*, 525–530. [CrossRef] [PubMed]

31. Ye, F.; Hu, Y.; Yu, W.; Xie, Z.; Hu, J.; Cao, Z.; Li, W.; Wu, Y. The scorpion toxin analogue BmKTX-D33H as a potential Kv1.3 channel-selective immunomodulator for autoimmune diseases. *Toxins (Basel)* **2016**, *8*, 115. [CrossRef] [PubMed]

32. Chen, R.; Robinson, A.; Gordon, D.; Chung, S.H. Modeling the binding of three toxins to the voltage-gated potassium channel (Kv1.3). *Biophys. J.* **2011**, *101*, 2652–2660. [CrossRef] [PubMed]

33. Gurevitz, M. Mapping of scorpion toxin receptor sites at voltage-gated sodium channels. *Toxicon* **2012**, *60*, 502–511. [CrossRef] [PubMed]

34. Yu, K.; Fu, W.; Liu, H.; Luo, X.; Chen, K.X.; Ding, J.; Shen, J.; Jiang, H. Computational simulations of interactions of scorpion toxins with the voltage-gated potassium ion channel. *Biophys. J.* **2004**, *86*, 3542–3555. [CrossRef] [PubMed]

35. Corzo, G.; Papp, F.; Varga, Z.; Barraza, O.; Espino-Solis, P.G.; Rodríguez de la Vega, R.C.; Gaspar, R.; Panyi, G.; Possani, L.D. A selective blocker of Kv1.2 and Kv1.3 potassium channels from the venom of the scorpion Centruroides suffusus suffusus. *Biochem. Pharmacol.* **2008**, *76*, 1142–1154. [CrossRef] [PubMed]

36. Kuzmenkov, A.I.; Peigneur, S.; Chugunov, A.O.; Tabakmakher, V.M.; Efremov, R.G.; Tytgat, J.; Grishin, E.V.; Vassilevski, A.A. C-Terminal residues in small potassium channel blockers OdK1 and OSK3 from scorpion venom fine-tune the selectivity. *Biochim. Biophys. Acta* **2017**, *1865*, 465–472. [CrossRef] [PubMed]

37. Chagot, B.; Pimentel, C.; Dai, L.; Pil, J.; Tytgat, J.; Nakajima, T.; Corzo, G.; Darbon, H.; Ferrat, G. An unusual fold for potassium channel blockers: NMR structure of three toxins from the scorpion Opisthacanthus madagascariensis. *Biochem. J.* **2005**, *388*, 263–271. [CrossRef] [PubMed]

38. Grissmer, S.; Nguyen, A.N.; Aiyar, J.; Hanson, D.C.; Mather, R.J.; Gutman, G.A.; Karmilowicz, M.J.; Auperin, D.D.; Chandy, K.G. Pharmacological characterization of five cloned voltage-gated K+ channels, types Kv1.1, 1.2, 1.3, 1.5, and 3.1, stably expressed in mammalian cell lines. *Mol. Pharmacol.* **1994**, *45*, 1227–1234. [PubMed]

39. Berman, H.M.; Westbrook, J.; Feng, Z.; Gilliland, G.; Bhat, T.N.; Weissig, H.; Shindyalov, I.N.; Bourne, P.E. The Protein Data Bank. *Nucleic Acids Res.* **2000**, *28*, 235–242. [CrossRef] [PubMed]

40. Yi, H.; Cao, Z.; Yin, S.; Dai, C.; Wu, Y.; Li, W. Interaction simulation of hERG K+ channel with its specific BeKm-1 peptide: Insight into the selectivity of molecular recognition. *J. Proteome Res.* **2007**, *6*, 611–620. [CrossRef] [PubMed]

41. Yi, H.; Qiu, S.; Wu, Y.; Li, W.; Wang, B. Differential molecular information of maurotoxin peptide recognizing IK(Ca) and Kv1.2 channels explored by computational simulation. *BMC Struct. Biol.* **2011**, *11*, 1–9. [CrossRef] [PubMed]

42. Koha, B.; Rothenberg, I.; Ali, S.A.; Alam, M.; Seebohm, G.; Kalbacher, H.; Voelter, W.; Stoll, R. Solid phase synthesis, NMR structure determination of α-KTx3.8, its in silico docking to Kv1.x potassium channels, and electrophysiological analysis provide insights into toxin-channel selectivity. *Toxicon* **2015**, *101*, 70–78.

43. Nekrasova, O.V.; Volyntseva, A.D.; Kudryashova, K.S.; Novoseletsky, V.N.; Lyapina, E.A.; Illarionova, A.V.; Yakimov, S.A.; Korolkova, Y.V.; Shaitan, K.V.; Kirpichnikov, M.P.; et al. Complexes of peptide blockers with Kv1.6 pore domain: Molecular modeling and studies with KcsA-Kv1.6 channel. *J. Neuroimmune Pharmacol.* **2017**, *12*, 260–276. [CrossRef] [PubMed]

44. Jorgensen, W.L.; Maxwell, D.S.; Tirado-Rives, J. Development and testing of the OPLS all-atom force field on conformational energetics and properties of organic liquids. *J. Am. Chem. Soc.* **1996**, *118*, 11225–11236. [CrossRef]

45. Kaminski, G.A.; Friesner, R.A.; Tirado-Rives, J.; Jorgensen, W.L. Evaluation and reparametrization of the OPLS-AA force field for proteins via comparison with accurate quantum chemical calculations on peptides. *J. Phys. Chem. B* **2001**, *105*, 6474–6487. [CrossRef]

46. Berendsen, H.J.C.; Grigera, J.R.; Straatsma, T.P. The missing term in effective pair potentials. *J. Phys. Chem.* **1987**, *91*, 6269–6271. [CrossRef]

47. Hess, B.; Bekker, H.; Berendsen, H.J.C.; Fraaije, J.G.E.M. LINCS: A linear constraint solver for molecular simulations. *J. Comput. Chem.* **1997**, *18*, 1463–1472. [CrossRef]

48. Jouirou, B.; Mouhat, S.; Andreotti, N.; de Waard, M.; Sabatier, J.M. Toxin determinants required for interaction with voltage-gated K1 channels. *Toxicon* **2004**, *43*, 909–914. [CrossRef] [PubMed]

49. Giangiacomo, K.M.; Ceralde, Y.; Mullmann, T.J. Molecular basis of alpha-KTx specificity. *Toxicon* **2004**, *43*, 877–886. [CrossRef] [PubMed]

50. Aiyar, J.; Withka, J.M.; Rizzi, J.P.; Singleton, D.H.; Andrews, G.C.; Lin, W.; Boyd, J.; Hanson, D.C.; Simon, M.; Dethlefs, B.; et al. Topology of the pore-region of a K+ channel revealed by the NMR-derived structures of scorpion toxins. *Neuron* **1995**, *15*, 1169–1181. [CrossRef]

51. Gross, A.; MacKinnon, R. Agitoxin footprinting the shaker potassium channel pore. *Neuron* **1996**, *16*, 399–406. [CrossRef]

52. Zarrabi, M.; Naderi-Manesh, H. The investigation of interactions of kappa-Hefutoxin1 with the voltage-gated potassium channels: A computational simulation. *Proteins* **2008**, *71*, 1441–1449. [CrossRef] [PubMed]

53. Wu, Y.; Cao, Z.; Yi, H.; Jiang, D.; Mao, X.; Liu, H.; Li, W. Simulation of the interaction between ScyTx and small conductance calcium-activated potassium channel by docking and MM-PBSA. *Biophys. J.* **2004**, *87*, 105–112. [CrossRef] [PubMed]

54. Qiu, S.; Yi, H.; Liu, H.; Cao, Z.; Wu, Y.; Li, W. Molecular information of charybdotoxin blockade in the large conductance calcium-activated potassium channel. *J. Chem. Inf. Model.* **2009**, *49*, 1831–1838. [CrossRef] [PubMed]

55. Han, S.; Yi, H.; Yin, S.J.; Chen, Z.Y.; Liu, H.; Cao, Z.J.; Wu, Y.L.; Li, W.X. Structural basis of a potent peptide inhibitor designed for Kv1.3 channel, a therapeutic target of autoimmune disease. *J. Biol. Chem.* **2008**, *283*, 19058–19065. [CrossRef] [PubMed]

56. Yin, S.J.; Jiang, L.; Yi, H.; Han, S.; Yang, D.W.; Liu, M.L. Different residues in channel turret determining the selectivity of ADWX-1 inhibitor peptide between Kv1.1 and Kv1.3 channels. *J. Proteome Res.* **2008**, *7*, 4890–4897. [CrossRef] [PubMed]

57. Cui, M.; Shen, J.H.; Briggs, J.M.; Luo, X.M.; Tan, X.J.; Jiang, H.L.; Chen, K.X.; Ji, R.Y. Brownian dynamics simulations of interaction between scorpion toxin Lq2 and potassium ion channel. *Biophys. J.* **2001**, *80*, 1659–1669. [CrossRef]

58. Cui, M.; Shen, J.H.; Briggs, J.M.; Fu, W.; Wu, J.L.; Zhang, Y.M.; Luo, X.M.; Chi, Z.W.; Ji, R.Y.; Jiang, H.L.; et al. Brownian dynamic simulations of the recognition of the scorpion toxin P05 with the small-conductance calcium-activated potassium channels. *J. Mol. Biol.* **2002**, *318*, 417–428. [CrossRef]

59. Eriksson, M.A.; Roux, B. Modeling the structure of agitoxin in complex with the Shaker K+ channel: A computational approach based on the experimental distance restraints extracted from thermodynamic mutant cycles. *Biophys. J.* **2002**, *83*, 2595–2609. [CrossRef]
60. Laskowski, R.A.; Swindells, M.B. LigPlot+: Multiple ligand-protein interaction diagrams for drug discovery. *J. Chem. Inf. Model.* **2011**, *51*, 2778–2786. [CrossRef] [PubMed]
61. Nussinov, R.; Tsai, C.J. Allostery without a conformational change? Revisiting the paradigm. *Curr. Opin. Struct. Biol.* **2015**, *30*, 17–24. [CrossRef] [PubMed]

© 2017 by the authors. Licensee MDPI, Basel, Switzerland. This article is an open access article distributed under the terms and conditions of the Creative Commons Attribution (CC BY) license (http://creativecommons.org/licenses/by/4.0/).

Article

Suppressive Effects of Bee Venom Acupuncture on Paclitaxel-Induced Neuropathic Pain in Rats: Mediation by Spinal α_2-Adrenergic Receptor

Jiho Choi [1,†], Changhoon Jeon [1,†], Ji Hwan Lee [2,†] (ID), Jo Ung Jang [3], Fu Shi Quan [4], Kyungjin Lee [5], Woojin Kim [1,3,*] and Sun Kwang Kim [1,2,3,*] (ID)

[1]   Department of Physiology, College of Korean Medicine, Kyung Hee University, 26 Kyungheedae-ro, Dongdamoon-gu, Seoul 02447, Korea; cyanical@hotmail.com (J.C.); cjmystars44@gmail.com (C.J.)
[2]   Department of Science in Korean Medicine, Graduate School, Kyung Hee University, 26 Kyungheedae-ro, Dongdamoon-gu, Seoul 02447, Korea; mibdna@khu.ac.kr
[3]   Department of East-West Medicine, Graduate School, Kyung Hee University, 26 Kyungheedae-ro, Dongdamoon-gu, Seoul 02447, Korea; powerfox032@naver.com
[4]   Department of Medical Zoology, School of Medicine, Kyung Hee University, 26 Kyungheedae-ro, Dongdamoon-gu, Seoul 02447, Korea; fsquan@khu.ac.kr
[5]   Department of Herbology, College of Korean Medicine, Kyung Hee University, 26 Kyungheedae-ro, Dongdamoon-gu, Seoul 02447, Korea; niceday@khu.ac.kr
*   Correspondence: wjkim@khu.ac.kr (W.K.); skkim77@khu.ac.kr (S.K.K.); Tel: +82-2-961-0334 (W.K.); +82-2-961-0491 (S.K.K.); Fax: +82-7-4194-9316 (W.K. & S.K.K.)
†   These authors contributed equally to this work.

Academic Editors: Irina Vetter and Steve Peigneur
Received: 4 July 2017; Accepted: 24 October 2017; Published: 31 October 2017

Abstract: Paclitaxel, a chemotherapy drug for solid tumors, induces peripheral painful neuropathy. Bee venom acupuncture (BVA) has been reported to have potent analgesic effects, which are known to be mediated by activation of spinal α-adrenergic receptor. Here, we investigated the effect of BVA on mechanical hyperalgesia and spinal neuronal hyperexcitation induced by paclitaxel. The role of spinal α-adrenergic receptor subtypes in the analgesic effect of BVA was also observed. Administration of paclitaxel (total 8 mg/kg, intraperitoneal) on four alternate days (days 0, 2, 4, and 6) induced significant mechanical hyperalgesic signs, measured using a von Frey filament. BVA (1 mg/kg, ST36) relieved this mechanical hyperalgesia for at least two hours, and suppressed the hyperexcitation in spinal wide dynamic range neurons evoked by press or pinch stimulation. Both melittin (0.5 mg/kg, ST36) and phospholipase A2 (0.12 mg/kg, ST36) were shown to play an important part in this analgesic effect of the BVA, as they significantly attenuated the pain. Intrathecal pretreatment with the α_2-adrenergic receptor antagonist (idazoxan, 50 µg), but not α_1-adrenergic receptor antagonist (prazosin, 30 µg), blocked the analgesic effect of BVA. These results suggest that BVA has potent suppressive effects against paclitaxel-induced neuropathic pain, which were mediated by spinal α_2-adrenergic receptor.

Keywords: bee venom acupuncture; chemotherapy-induced neuropathic pain; paclitaxel

1. Introduction

Paclitaxel is an important chemotherapeutic agent from the bark of *Taxus brevifolia* [1], which is widely used to treat various tumors [2–4]. However, despite its role against the tumors, its usage is often limited, due to the painful peripheral neuropathy occurring after its administration [5]. Symptoms commonly reported are sensory neuropathies, which are paresthesia, loss of tendon reflexes, numbness and pain in the upper and lower extremities. Although these neuropathies decrease patients' quality of life (QoL), there is still no optimal treatment method or drug to alleviate these neuropathies [5,6]. Thus, an effort to explore novel treatments is needed.

Bee venom acupuncture (BVA), a treatment method that injects diluted bee venom into acupoints, is widely used in traditional Korean medicine against various diseases, such as adhesive capsulitis [7], idiopathic Parkinson's disease [8], knee osteoarthritis [9], and musculoskeletal pain diseases [10]. Especially, BVA has been reported to have potent analgesic effect in studies conducted using various animal models of pain [11–15], and two case series also reported that BVA treatment may help to reduce the chemotherapy-induced peripheral neuropathy (CIPN), including paclitaxel-induced neuropathy [16,17]. Recently, our laboratory has demonstrated that BVA treatment could significantly alleviate mechanical and cold allodynia in a rat model of oxaliplatin-induced neuropathic pain [12,14,18,19]. Moreover, although the precise mechanism of BVA analgesic effect is unknown, we have also demonstrated that this analgesic effect was mediated by the descending noradrenergic pain modulation pathway via the activation of spinal α-adrenergic receptor, which was consistent with other previously conducted studies [11–14,20].

Thus, the aims of this study were, firstly, to examine whether the BVA has suppressive effects against paclitaxel-induced mechanical hyperalgesia and neuronal hyperexcitation in the spinal cord, and secondly, to observe the role of BVA components, such as melittin and phospholipase A2 (PLA2) in their analgesic effect, and finally, to investigate which α-adrenergic receptor subtypes mediate the analgesic effect of BVA in the spinal cord.

2. Results

2.1. Development and Maintanance of Paclitaxel-Induced Mechanical Hyperalgesia

In order to see the time-elapsed change of paclitaxel-induced mechanical hyperalgesia, we evaluated the withdrawal responses of hind paws to mechanical stimulation using a von Frey filament with 15 g bending force. In the paclitaxel group, significant increase in paw withdrawal frequency (PWF) was shown from 10 to 21 days after the first injection ($p < 0.01$, day 10 and 14; $p < 0.05$, day 21) (Figure 1). Therefore, we performed the following experiments on day 10 through to 21.

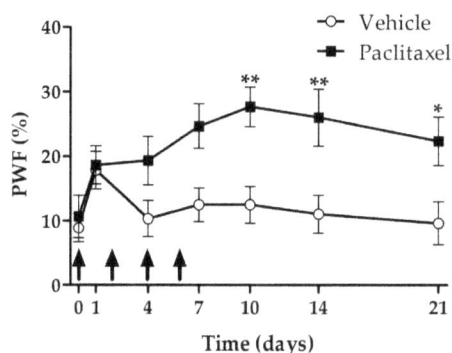

Figure 1. Time course of paclitaxel-induced mechanical hyperalgesia. Rats were divided into two groups; paclitaxel ($n = 7$), vehicle ($n = 7$). Paclitaxel (2.0 mg/kg per injection) or vehicle was injected to rats four times (arrows; days 0, 2, 4 and 6). Significant differences between two groups were observed from the day 10 to day 21. Data are presented as mean $\pm$ SEM (* $p < 0.05$, ** $p < 0.01$; two-way ANOVA followed by Bonferroni's multiple comparison test).

2.2. Effects of BVA on Paclitaxel-Induced Mechanical Hyperalgesia

Since the BVA treatment at Zusanli (ST36) acupoint, but not at Quchi (L11), showed significant anti-hyperalgesic effect (Figure 2), BVA was used at ST36 in the following experiments. Figure 3 shows the analgesic effects of BVA on paclitaxel-induced mechanical hyperalgesia with the time course. BVA treated group (paclitaxel + BVA) showed significant reduction in PWF compared to control group (paclitaxel + PBS (phosphate buffered saline)) at one and two hours after BVA (47% reduction, $p < 0.05$ and 66% reduction, $p < 0.01$, respectively). No significant difference between the two groups was shown from four hours after BVA. These results indicate that the treatment of BVA has a potent analgesic effect on paclitaxel-induced neuropathic pain, lasting at least two hours.

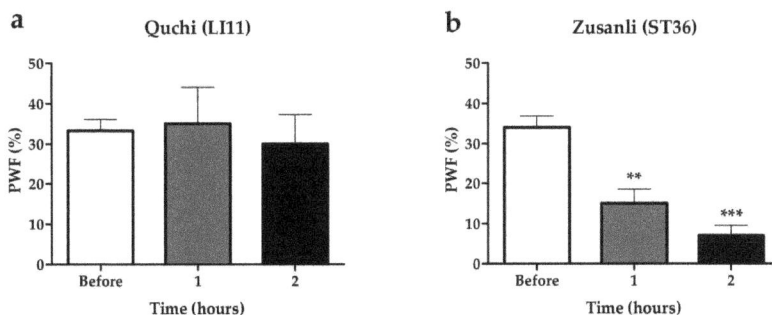

Figure 2. Effects of bee venom acupuncture (BVA) at different acupoints on paclitaxel-induced mechanical hyperalgesia. BVA (1.0 mg/kg) was used at (**a**) LI11 ($n = 6$) or (**b**) ST36 ($n = 5$) acupoints. In ST36 group, the paw withdrawal frequency (PWF) decreased significantly one or two hours after BVA, whereas no significant differences are shown in LI11 group. Data are presented as mean $\pm$ SEM (** $p < 0.01$, *** $p < 0.001$; repeated measures one-way ANOVA followed by Dunnett's post hoc test).

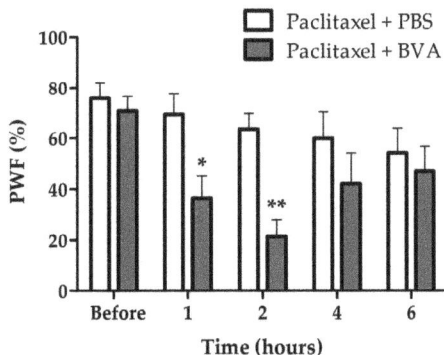

Figure 3. Time course of the analgesic effect of BVA on paclitaxel-induced mechanical hyperalgesia. Rats were dispensed arbitrarily into two groups; paclitaxel + BVA ($n = 7$), paclitaxel + phosphate buffered saline (PBS) ($n = 7$). BVA (1.0 mg/kg) and PBS were treated at ST36. Significant reduction of PWF was observed from one to two hours after BVA. Data are presented as mean $\pm$ SEM (* $p < 0.05$, ** $p < 0.01$; two-way ANOVA followed by Bonferroni's multiple comparison test).

2.3. Effects of BVA on Paclitaxel-Induced Hyperexcitation in the Spinal Wide Dynamic Range (WDR) Neurons

In order to see whether paclitaxel induces hyperexcitation in WDR neurons and BVA treatment reduces paclitaxel-induced hyperexcitation in WDR neurons, we conducted extracellular recording in vivo (Figure 4a–d). The number of spike responses of WDR neurons to mechanical stimulation (brush, press, and pinch) was significantly increased in paclitaxel group ($p < 0.05$; brush, $p < 0.001$; press and pinch, vs. vehicle, Figure 4e). In the BVA treatment group (1 mg/kg, ST36), significant reduction of paclitaxel-induced hyperexcitation in WDR neurons was observed ($p < 0.01$; press, $p < 0.001$; pinch, vs. before BVA, Figure 4f).

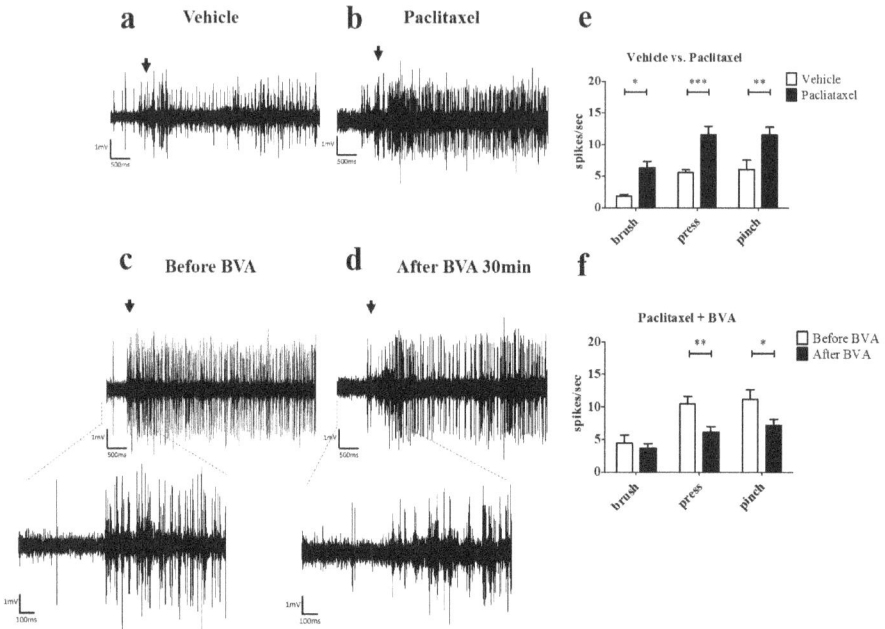

Figure 4. Paclitaxel-induced hyperexcitation in wide dynamic range (WDR) neurons and inhibition of paclitaxel-induced hyperexcitation by BVA treatment. (**a**–**d**) Representative extracellular recording raw traces of WDR neuron's responses to pressing with hard stick (arrows, during 5 s) in vehicle group (**a**), paclitaxel group (**b**), and BVA (1 mg/kg) treated group (**c**,**d**). Before BVA treatment (**c**) and 30 min after BVA treatment (**d**). (**e**,**f**) The spike response of WDR neurons to mechanical stimulation (brush, press, and pinch). Data are presented as mean ± SEM (* $p < 0.05$, ** $p < 0.01$, *** $p < 0.001$; two-way ANOVA followed by Bonferroni's multiple comparison test).

2.4. Effect of BVA, Melittin, or PLA2 on Paclitaxel-Induced Mechanical Hyperalgesia

To observe the role of different BV components in the analgesic effect of the BVA, BVA (1 mg/kg), melittin (0.5 mg/kg), or PLA2 (0.12 mg/kg) were injected at ST36. The two major protein components of the honey bee are melittin and PLA2, which occupies 50 and 12% of its dry weight, respectively [21]. Behavioral assessments were conducted one hour after the injection of BVA, melittin, or PLA2, as BVA showed its strongest analgesic effect one hour after the injection (Figure 3). This result showed that melittin had a stronger analgesic effect against paclitaxel-induced mechanical hyperalgesia than BVA or PLA2 (Figure 5).

2.5. Effects of Intrathecal α-Adrenergic Receptor Subtype Antagonists on BVA- or Melittin-Induced Anti-Hyperalgesia

To investigate which α-adrenergic receptor subtypes mediate BVA- or melittin-induced anti-hyperalgesic action, prazosin (α_1-adrenergic receptor antagonist, 30 μg, i.t.) or idazoxan (α_2-adrenergic receptor antagonist, 50 μg, i.t.) was administered 20 min before treatments. Prazosin and dimethyl sulfoxide (DMSO) showed significant decrease in PWF after BVA or melittin treatments (Figure 6a–c). This demonstrate that neither BVA nor melittin acted on spinal α_1-adrenergic receptor to reduce the hyperalgesia evoked by paclitaxel. In contrast, idazoxan, but not PBS ($p < 0.001$), blocked the BVA- or melittin-induced anti-hyperalgesic effect (Figure 6d–f). These results altogether, indicate that the spinal α_2-adrenergic receptor, but not the α_1-adrenergic receptor, mediates BVA- or melittin-induced analgesia.

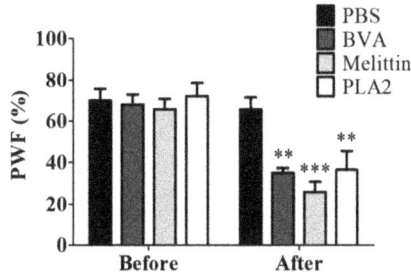

Figure 5. The analgesic effect of BVA, melittin, or PLA2 on paclitaxel-induced mechanical hyperalgesia. Rats showing signs of mechanical allodynia were dispensed arbitrarily into four groups; PBS ($n = 7$), BVA (1 mg/kg, $n = 5$), melittin (0.5 mg/kg, $n = 6$), and PLA2 (0.12 mg/kg, $n = 7$). All drugs were injected at ST36. PBS was used as control. Behavioral tests were conducted one hour after the drug administrations. Data are presented as mean ± SEM (** $p < 0.01$, *** $p < 0.001$; two-way ANOVA followed by Bonferroni's multiple comparison test).

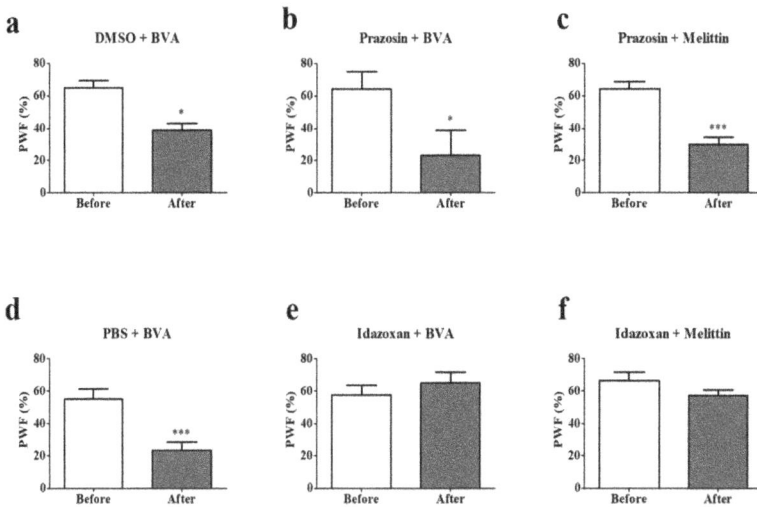

Figure 6. Effects of intrathecal adrenergic antagonists on BVA- or melittin-induced analgesic action. Rats were divided into six groups; (a) DMSO + BVA ($n = 5$), (b) prazosin + BVA ($n = 5$), (c) prazosin + melittin ($n = 7$), (d) PBS + BVA ($n = 6$), (e) idazoxan + BVA ($n = 6$), (f) idazoxan + melittin ($n = 7$). Data are presented as mean ± SEM (* $p < 0.05$, *** $p < 0.001$; paired *t*-test).

3. Discussion

Multiple injection of paclitaxel can occur peripheral neuropathy, which can limit its usage and decreases patients' QoL. Although the treatments such as gabapentin, pregabalin, and morphine have been used to alleviate the neuropathic pain, these treatments have, themselves, various side effects, such as nausea, vomiting, somnolence, dizziness, suicidal thought, and drug dependence [22–25]. Therefore, an effort to search for effective treatment options is critically needed. In traditional Korean medicine, BVA has been used to treat musculoskeletal pain and arthritis from the past [10,26]. In addition, these days, BVA has also been founded to be effective in treating patients with CIPN [16,17]. Thus, in this study, we experimented to find out whether BVA can alleviate the paclitaxel-induced neuropathy and to clarify the mechanism that lies behind it.

Our data showed that BVA treatment at ST36, not LI11, had a significant analgesic effect. It should be noted that ST36 acupoint is closer to the hind paw, where mechanical test was performed, than LI11 acupoint. It is consistent with the previous study in which BVA had more potent analgesic effect when treated closer to the tested area [14]. Then, we examined the time course of the analgesic effect of the BVA at ST36. The result showed that the analgesic effect was significant until two hours after BVA treatment. Our previous study also showed that the analgesic effect of BVA was effective until two hours after BVA treatment in oxaliplatin-induced cold allodynia [14]. Considering that moderate concentration of morphine without side effects was no longer effective in oxaliplatin-induced cold allodynia at two hours after administration [12,27], this result would be clinically significant.

The spinal wide dynamic range (WDR) neuron receives non-nociceptive and nociceptive inputs via A- and C-fibers, and descending pain modulatory systems synapse at the WDR neuron [28]. Therefore, the spinal WDR neuron is suitable for assessing the degree of pain. In addition, the hyperexcitation of spinal WDR neuron was observed previously in a rat model of paclitaxel-induced hyperalgesia [29]. In our study, electrophysiological data confirmed that hyperexcitation of WDR neurons is induced by paclitaxel. We further demonstrated that BVA treatment could significantly inhibit this paclitaxel-induced hyperexcitation in the spinal WDR cells.

In subsequent experiments, we administered BVA, melittin, or PLA2 at ST 36, to observe the role of different BV components in the analgesic effect of BVA against paclitaxel-induced mechanical hyperalgesia. Melittin is a major component of the BV, occupying 50% of its total dry weight. PLA2 occupies 12%. Our results showed that 0.5 mg/kg of melittin was more powerful than 1 mg/kg of BVA or 0.12 mg/kg of PLA2. In our previous study, we showed that intraperitoneal injection of PLA2 could significantly decrease the cold and mechanical allodynia induced by single oxaliplatin injection in mice [30]. Moreover, although not on chemotherapy induced pain model, other lab has reported that melittin injected at ST36 had a powerful analgesic effect against complete Freund's adjuvant-induced rheumatoid arthritis, showing a similar effect to BVA [31]. In this study, the analgesic effect of BVA or melittin was blocked by spinal α_2-adrenergic receptor antagonist (idazoxan), showing that BVA and melittin act on similar spinal adrenergic receptors to inhibit mechanical hyperalgesia induced by paclitaxel.

EA (electro-acupuncture) is a modified acupuncture which utilizes electrical current to treat pain. BVA is another form of acupuncture which uses chemical compounds; bee venom. The two different forms of acupuncture have similarities and differences. One of the similarities is that the endogenous analgesic systems are involved in both of their analgesic mechanisms, and the difference is that the analgesic effects of EA are mainly mediated by the opioidergic system [32], whereas those of the BVA are mostly mediated by the noradrenergic system [33]. However, despite this difference, EA and BVA were both reported to be effective in different types of allodynia assessed using thermal [34] and chemical [12] stimulations. These results show that other inhibitory systems, such as serotonergic, GABA, and/or cholinergic systems, may also play an important role, along with opioidergic and adrenergic system, in the action of EA and BVA. Furthermore, interaction of periaqueductal gray (PAG) and locus coeruleus (LC) in the brain should also play an important part in their analgesic effect, as both the EA and BVA were reported to activate PAG [35] and LC [36], which are important opioid and noradrenaline producing site in the CNS, respectively.

BVA induced analgesia was shown to be mediated by spinal α_2-adrenergic receptor [11–15], and it increased c-Fos expression in LC and A5 cell group (A5) [37,38]. Moreover, BVA reduced c-Fos expression in the spinal dorsal horn of rats with formalin or acetic acid-induced pain [13,39]. Considering that both the LC and A5 are part of the descending noradrenergic pathway [40], it is suggested that BVA suppresses conduction of afferent nociceptive signals in the spinal dorsal horn affecting descending noradrenergic pathway. Our data are consistent with previous studies showing that spinal α_2-adrenergic receptor mediates BVA-induced analgesia. Furthermore, the pain attenuating effect of melittin was also blocked by spinal α_2-adrenergic receptor antagonist (idazoxan) showing that melittin, the richest component of the BV, also acts on spinal α_2-adrenergic receptor to inhibit mechanical hyperalgesia induced by paclitaxel.

Drug combination is widely used to treat dreadful diseases, such as AIDS and cancer. The main aim of drug combination is to reduce dose and toxicity, and to delay the induction of drug resistance. Our previous study showed the combined effect of BVA and morphine on oxaliplatin-induced neuropathic pain [12]. BVA treated with morphine showed prolonged analgesic effects compared to the BVA or morphine alone. Moreover, another article showed that BVA could enhance the analgesic effect of intrathecal injection of clonidine in chronic constriction injury-induced neuropathic pain model [41]. Because such combined effect on paclitaxel-induced neuropathic pain has yet to be studied, further studies are needed to examine the combined effect of BVA with other drugs, like morphine, clonidine, SSRI, SNRI, gabapentin, and cannabinoid. Furthermore, in the future studies, it will be interesting to investigate the effect of various components of the BVA on paclitaxel-induced neuropathic pain model, as several active components exist in the BV, such as melittin [42] and PLA2 [30], which have been reported to be effective in other pain models.

4. Conclusions

In conclusion, BVA (1 mg/kg) at ST36 significantly attenuated mechanical hyperalgesia induced by paclitaxel. The significant analgesic effect lasted two hours, which was long enough compared to the effect of morphine. Suppressive action was verified by conducting extracellular recording in the spinal WDR neurons. Moreover, both melittin (0.5 mg/kg) and PLA2 (0.12 mg/kg), which are major components of the BV, significantly attenuated the paclitaxel-induced mechanical hyperalgesia. This analgesic effect of BVA or melittin was significantly blocked by intrathecal injection of idazoxan, but not by prazosin, demonstrating that the action of spinal α_2-adrenergic receptor, but not α_1-adrenergic receptor, is involved in the mechanism of analgesic effect.

5. Materials and Methods

5.1. Animals

Adult Sprague-Dawley rats (male, 180–210 g, 6 weeks old) (Daehan Biolink, Chungbuk, Korea) were housed in cages with free access to food and water, and were sustained at 23 ± 2 °C room temperature with a 12 hour light/dark cycle. Prior to any experiments, all animals were acclimated in their cages (3–4 rats per cage) for a week. All experiments using animals were ratified by the Institutional Animal Care and Use Committee of Kyung Hee University (KHUASP(SE)-16-153), and were performed on the ground of the guidelines of the International Association for the Study of Pain [43].

5.2. Administration of Paclitaxel

Paclitaxel (Wako Pure Chemical Industries, Osaka, Japan) was dissolved in cremophor EL polyethoxylated castor oil (Sigma, St. Louis, MO, USA) and 100% ethanol (Merck KGaA, Marmstadt, Germany) (1:1 solution), and 6 mg/mL stocks were made. Then, stocks were diluted by phosphate buffered saline (PBS) at a concentration of 2 mg/ml and administered at an amount of 2 mg/kg on four alternate days (days 0, 2, 4, and 6). As control, the same volume of vehicle was intraperitoneally injected. The formula of paclitaxel was slightly modified from previous studies [44,45].

5.3. Behavior Tests

Twenty to thirty minutes before the behavior test, animals were adapted to the experimental circumstances. The experimenters were blinded to paclitaxel and any other treatments. The animals were placed on a metal mesh, enclosed within a 20 (d) × 20 (w) × 14 (h) cm clear plastic cage. Mechanical hyperalgesia was assessed using von Frey filament (Stoelting Co., Wood Dale, IL, USA). The measurement method of mechanical hyperalgesia was modified from the previous studies [44–46].

On the mid-plantar area of both hind paws, the von Frey filament (bending force of 15 g) was stimulated for 10 times each, with the applications held for 5 s. The percentage of withdrawal responses to the von Frey filament application was calculated, and then expressed as an overall percentage response.

5.4. Experimental Schedule

The time schedule of this experiment is shown in Figure 7. After baseline mechanical sensitivity was measured at day 0, paclitaxel was injected intraperitoneally on four alternate days (days 0, 2, 4, and 6) (Figure 7a). Behavior tests were performed after paclitaxel administration. The time course of BVA effect was measured at 1, 2, 4, and 6 hours after administration of BVA (Figure 7b). Antagonists were treated 20 min before BVA, and then, behavior tests were conducted 1 hour later (Figure 7c).

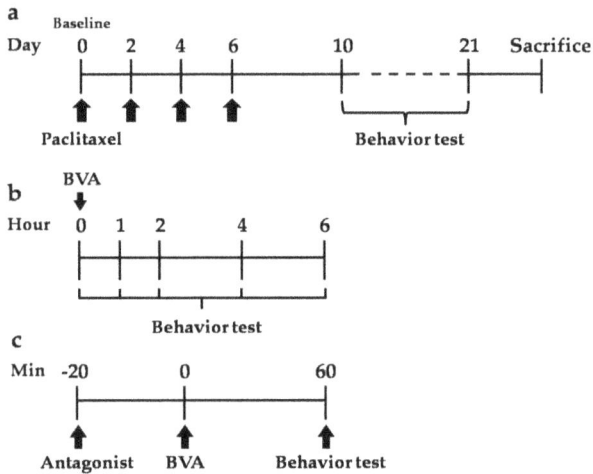

Figure 7. Time schedule of the experiment. (**a**) Paclitaxel was administered four alternate days (0, 2, 4, 6 days, i.p.); (**b**) the time course of BVA effect was conducted at 1, 2, 4, and 6 hours after administration of BVA; (**c**) antagonists were treated 20 min before administration of BVA or melittin, and behavior tests were conducted one hour after administration of BVA or melittin.

5.5. BVA, Melittin, or PLA2 Treatment

To verify the optimal acupoint for the BVA treatment, paclitaxel administered rats were divided randomly into two groups; Quchi (LI11) and Zusanli (ST36). LI11 is located at the depression medial to the extensor carpi radialis, at the lateral end of the cubital crease. ST36 is located in the anterior tibial muscle, 5 mm lateral and distal from the anterior tubercle of the tibia [47].

BV was manufactured by Jayeonsaeng TJ (Kyeonggi-Do, Korea), and its quality is strictly controlled by regular HPLC analysis (SNU National Instrumentation Center for Environmental Management, Seoul, Korea; see Supplementary Materials Figure S1). BV (1.0 mg/kg), as reported as an effective concentration without side effects from a previous study [12], dissolved in PBS was respectively injected at right side LI11 or ST36 acupoints subcutaneously, after baseline mechanical sensitivity was measured. The mechanical behavior test was performed following time course schedule (Figure 7b). Melittin (0.5 mg/kg) and PLA2 (0.12 mg/kg) were also injected at ST36. All drugs injected at acupoints were injected subcutaneously.

5.6. Extracellular Recording

Extracellular recordings were made from animals 10–21 days following administration of paclitaxel, when rats exhibited significant mechanical hyperalgesia. Extracellular recordings were carried out as previously described [48]. In brief, rats were anesthetized with urethane (Sigma, St. Louis, MO, USA; 1.5 g/kg, i.p.). The spinal cords of animals, which were fixed in a stereotaxic frame, were exposed from T13–L2 and irrigated with oxygenated (95% O_2–5% CO_2 gas) Krebs solution (in mM: 117 NaCl, 3.6 KCl, 2.5 $CaCl_2$, 1.2 $MgCl_2$, 1.2 NaH_2PO_4, 11 glucose, and 25 $NaHCO_3$) at a flow rate of 10 to 15 mL/min at 38 ± 1 °C. By their responses to brush, pressure, and pinch, WDR cells were classified. Cells were isolated in the L3–L5 segments medial to the dorsal root entry zone up to a depth of 1000 mm. Extracellular single-unit recordings were made with a low-impedance insulated tungsten microelectrode (impedance of 10 MΩ, FHC, Bowdoin, ME, USA).

For mechanical stimuli, brush, press, and pinch stimulation were applied to the lateral and ventral surfaces of the hind paw. Brush stimulus was given by brushing the receptive field five times with a camel brush. Press stimulus was given by pressing the receptive field five seconds using the blunt tip of the camel brush with a diameter of 0.5 cm and a magnitude of about 20 g. Finally, pinch stimulation was given by pinching the skin using toothed forceps (11022-14, Fine Science Tools, Heidelberg, Germany) for five seconds.

5.7. Antagonists

To investigate the mechanism of BVA, paclitaxel administered rats were divided randomly into four groups: dimethyl sulfoxide (DMSO; Sigma, St. Louis, MO, USA) + BVA, prazosin + BVA, prazosin + melittin, PBS + BVA, idazoxan + BVA, and idazoxan + melittin. α_1-Adrenergic receptor antagonist prazosin (Sigma, St. Louis, MO, USA; 30 μg) was dissolved in 20% DMSO. α_2-Adrenergic receptor antagonist idazoxan (Sigma; 50 μg) was dissolved in PBS. Under isoflurane anesthesia (Hana Pharm. Co., Kyeonggi-Do, Korea), all antagonists were treated intrathecally with a direct lumbar puncture as previously described [12,48].

5.8. Statistical Analysis

All the data are presented as mean $\pm$ SEM. Statistical analysis and graphic works were performed with Prism 5.0 (GraphPad software, La Jolla, CA, USA, 2008). Paired *t*-test, one-way ANOVA followed by Dunnett's post hoc test, and two-way ANOVA followed by Bonferroni's multiple comparison test were used for statistical analysis. In all cases, $p < 0.05$ was considered significant.

Supplementary Materials: The following are available online at www.mdpi.com/2072-6651/9/11/351/s1, Figure S1: Representative HPLC analysis of melittin and phospholipase A2 (PLA2) in BV.

Acknowledgments: This work was supported by an undergraduate research program (URP) grant from the Korea Institute of Oriental Medicine (Y16113) and an URP grant from Kyung Hee University College of Korean Medicine, and by a grant from the society of immune and pain.

Author Contributions: W.K. and S.K.K. conceived and designed the experiments; J.C., C.J. and J.H.L. performed the experiments; J.C., C.J., J.H.L., K.L. and F.S.Q analyzed the data; J.U.J. and F.S.Q contributed reagents and materials; J.C., C.J., J.H.L., J.U.J., W.K. and S.K.K. wrote the paper. All authors read and approved the final manuscript.

Conflicts of Interest: The authors declare no conflict of interest. The funding sponsors had no role in the design of the study; in the collection, analyses, or interpretation of data; in the writing of the manuscript, and in the decision to publish the results.

References

1. Wani, M.C.; Taylor, H.L.; Wall, M.E.; Coggon, P.; McPhail, A.T. Plant antitumor agents. Vi. Isolation and structure of taxol, a novel antileukemic and antitumor agent from taxus brevifolia. *J. Am. Chem. Soc.* **1971**, *93*, 2325–2327. [CrossRef] [PubMed]

2. Sparano, J.A. Taxanes for breast cancer: An evidence-based review of randomized phase ii and phase iii trials. *Clin. Breast Cancer* **2000**, *1*, 32–40. [CrossRef] [PubMed]

3. Goffin, J.; Lacchetti, C.; Ellis, P.M.; Ung, Y.C.; Evans, W.K. First-line systemic chemotherapy in the treatment of advanced non-small cell lung cancer: A systematic review. *J. Thorac. Oncol.* **2010**, *5*, 260–274. [CrossRef] [PubMed]

4. Covens, A.; Carey, M.; Bryson, P.; Verma, S.; Fung Kee Fung, M.; Johnston, M. Systematic review of first-line chemotherapy for newly diagnosed postoperative patients with stage ii, iii, or iv epithelial ovarian cancer. *Gynecol. Oncol.* **2002**, *85*, 71–80. [CrossRef] [PubMed]

5. Lee, J.J.; Swain, S.M. Peripheral neuropathy induced by microtubule-stabilizing agents. *J. Clin. Oncol.* **2006**, *24*, 1633–1642. [CrossRef] [PubMed]

6. Argyriou, A.A.; Koltzenburg, M.; Polychronopoulos, P.; Papapetropoulos, S.; Kalofonos, H.P. Peripheral nerve damage associated with administration of taxanes in patients with cancer. *Crit. Rev. Oncol. Hematol.* **2008**, *66*, 218–228. [CrossRef] [PubMed]

7. Koh, P.S.; Seo, B.K.; Cho, N.S.; Park, H.S.; Park, D.S.; Baek, Y.H. Clinical effectiveness of bee venom acupuncture and physiotherapy in the treatment of adhesive capsulitis: A randomized controlled trial. *J. Shoulder Elbow Surg.* **2013**, *22*, 1053–1062. [CrossRef] [PubMed]

8. Cho, S.Y.; Shim, S.R.; Rhee, H.Y.; Park, H.J.; Jung, W.S.; Moon, S.K.; Park, J.M.; Ko, C.N.; Cho, K.H.; Park, S.U. Effectiveness of acupuncture and bee venom acupuncture in idiopathic parkinson's disease. *Parkinsonism Relat. Disord.* **2012**, *18*, 948–952. [CrossRef] [PubMed]

9. Kwon, Y.-B.; Kim, J.-H.; Yoon, J.-H.; Lee, J.-D.; Han, H.-J.; Mar, W.-C.; Beitz, A.J.; Lee, J.-H. The analgesic efficacy of bee venom acupuncture for knee osteoarthritis: A comparative study with needle acupuncture. *J. Chin. Med.* **2001**, *29*, 187–199. [CrossRef] [PubMed]

10. Lee, M.S.; Pittler, M.H.; Shin, B.C.; Kong, J.C.; Ernst, E. Bee venom acupuncture for musculoskeletal pain: A review. *J. Pain* **2008**, *9*, 289–297. [CrossRef] [PubMed]

11. Kim, H.W.; Kwon, Y.B.; Han, H.J.; Yang, I.S.; Beitz, A.J.; Lee, J.H. Antinociceptive mechanisms associated with diluted bee venom acupuncture (apipuncture) in the rat formalin test: Involvement of descending adrenergic and serotonergic pathways. *Pharmacol. Res.* **2005**, *51*, 183–188. [CrossRef] [PubMed]

12. Kim, W.; Kim, M.; Go, D.; Min, B.-I.; Na, H.; Kim, S. Combined effects of bee venom acupuncture and morphine on oxaliplatin-induced neuropathic pain in mice. *Toxins* **2016**, *8*, 33. [CrossRef] [PubMed]

13. Kwon, Y.-B.; Kang, M.-S.; Han, H.-J.; Beitz, A.J.; Lee, J.-H. Visceral antinociception produced by bee venom stimulation of the zhongwan acupuncture point in mice: Role of α_2 adrenoceptors. *Neurosci. Lett.* **2001**, *308*, 133–137. [CrossRef]

14. Lim, B.S.; Moon, H.J.; Li, D.X.; Gil, M.; Min, J.K.; Lee, G.; Bae, H.; Kim, S.K.; Min, B.I. Effect of bee venom acupuncture on oxaliplatin-induced cold allodynia in rats. *Evid. Based Complement. Altern. Med. eCAM* **2013**, *2013*, 369324. [CrossRef] [PubMed]

15. Roh, D.H.; Kwon, Y.B.; Kim, H.W.; Ham, T.W.; Yoon, S.Y.; Kang, S.Y.; Han, H.J.; Lee, H.J.; Beitz, A.J.; Lee, J.H. Acupoint stimulation with diluted bee venom (apipuncture) alleviates thermal hyperalgesia in a rodent neuropathic pain model: Involvement of spinal alpha 2-adrenoceptors. *J. Pain* **2004**, *5*, 297–303. [CrossRef] [PubMed]

16. Park, J.W.; Jeon, J.H.; Yoon, J.; Jung, T.Y.; Kwon, K.R.; Cho, C.K.; Lee, Y.W.; Sagar, S.; Wong, R.; Yoo, H.S. Effects of sweet bee venom pharmacopuncture treatment for chemotherapy-induced peripheral neuropathy: A case series. *Integr. Cancer Ther.* **2012**, *11*, 166–171. [CrossRef] [PubMed]

17. Yoon, J.; Jeon, J.H.; Lee, Y.W.; Cho, C.K.; Kwon, K.R.; Shin, J.E.; Sagar, S.; Wong, R.; Yoo, H.S. Sweet bee venom pharmacopuncture for chemotherapy-induced peripheral neuropathy. *J. Acupunct. Meridian Stud.* **2012**, *5*, 156–165. [CrossRef] [PubMed]

18. Lee, J.H.; Li, D.X.; Yoon, H.; Go, D.; Quan, F.S.; Min, B.I.; Kim, S.K. Serotonergic mechanism of the relieving effect of bee venom acupuncture on oxaliplatin-induced neuropathic cold allodynia in rats. *BMC Complement. Altern. Med.* **2014**, *14*, 471. [CrossRef] [PubMed]

19. Yoon, H.; Kim, M.J.; Yoon, I.; Li, D.X.; Bae, H.; Kim, S.K. Nicotinic acetylcholine receptors mediate the suppressive effect of an injection of diluted bee venom into the gv3 acupoint on oxaliplatin-induced neuropathic cold allodynia in rats. *Biol. Pharm. Bull.* **2015**, *38*, 710–714. [CrossRef] [PubMed]

20. Baek, Y.H.; Huh, J.E.; Lee, J.D.; Choi, D.Y.; Park, D.S. Antinociceptive effect and the mechanism of bee venom acupuncture (apipuncture) on inflammatory pain in the rat model of collagen-induced arthritis: Mediation by alpha2-adrenoceptors. *Brain Res.* **2006**, *1073–1074*, 305–310. [CrossRef] [PubMed]

21. Eze, O.B.; Nwodo, O.F.; Ogugua, V.N. Therapeutic effect of honey bee venom. *Proteins (enzymes)* **2016**, *1*, 2.

22. Gilron, I.; Bailey, J.M.; Tu, D.; Holden, R.R.; Weaver, D.F.; Houlden, R.L. Morphine, gabapentin, or their combination for neuropathic pain. *N. Engl. J. Med.* **2005**, *352*, 1324–1334. [CrossRef] [PubMed]

23. Ormseth, M.J.; Scholz, B.A.; Boomershine, C.S. Duloxetine in the management of diabetic peripheral neuropathic pain. *Patient Prefer. Adher.* **2011**, *5*, 343–356.
24. Serpell, M.G. Gabapentin in neuropathic pain syndromes: A randomised, double-blind, placebo-controlled trial. *Pain* **2002**, *99*, 557–566. [CrossRef]
25. Vinik, A.I.; Casellini, C.M. Guidelines in the management of diabetic nerve pain: Clinical utility of pregabalin. *Diabetes Metab. Syndr. Obes. Targets Ther.* **2013**, *6*, 57–78. [CrossRef] [PubMed]
26. Son, D.J.; Lee, J.W.; Lee, Y.H.; Song, H.S.; Lee, C.K.; Hong, J.T. Therapeutic application of anti-arthritis, pain-releasing, and anti-cancer effects of bee venom and its constituent compounds. *Pharmacol. Ther.* **2007**, *115*, 246–270. [CrossRef] [PubMed]
27. Ling, B.; Coudore, F.; Decalonne, L.; Eschalier, A.; Authier, N. Comparative antiallodynic activity of morphine, pregabalin and lidocaine in a rat model of neuropathic pain produced by one oxaliplatin injection. *Neuropharmacology* **2008**, *55*, 724–728. [CrossRef] [PubMed]
28. Baron, R.; Binder, A.; Wasner, G. Neuropathic pain: Diagnosis, pathophysiological mechanisms, and treatment. *Lancet Neurol.* **2010**, *9*, 807–819. [CrossRef]
29. Cata, J.P.; Weng, H.R.; Chen, J.H.; Dougherty, P.M. Altered discharges of spinal wide dynamic range neurons and down-regulation of glutamate transporter expression in rats with paclitaxel-induced hyperalgesia. *Neuroscience* **2006**, *138*, 329–338. [CrossRef] [PubMed]
30. Li, D.; Lee, Y.; Kim, W.; Lee, K.; Bae, H.; Kim, S.K. Analgesic effects of bee venom derived phospholipase a2 in a mouse model of oxaliplatin-induced neuropathic pain. *Toxins* **2015**, *7*, 2422–2434. [CrossRef] [PubMed]
31. Li, J.; Ke, T.; He, C.; Cao, W.; Wei, M.; Zhang, L.; Zhang, J.-X.; Wang, W.; Ma, J.; Wang, Z.-R. The anti-arthritic effects of synthetic melittin on the complete freund's adjuvant-induced rheumatoid arthritis model in rats. *Am. J. Chin. Med.* **2010**, *38*, 1039–1049. [CrossRef] [PubMed]
32. Zhang, R.; Lao, L.; Ren, K.; Berman, B.M. Mechanisms of acupuncture–electroacupuncture on persistent pain. *Anesthesiology* **2014**, *120*, 482–503. [CrossRef] [PubMed]
33. Chen, J.; Lariviere, W.R. The nociceptive and anti-nociceptive effects of bee venom injection and therapy: A double-edged sword. *Prog. Neurobiol.* **2010**, *92*, 151–183. [CrossRef] [PubMed]
34. Gim, G.-T.; Lee, J.-h.; Park, E.; Sung, Y.-H.; Kim, C.-J.; Hwang, W.-w.; Chu, J.-P.; Min, B.-I. Electroacupuncture attenuates mechanical and warm allodynia through suppression of spinal glial activation in a rat model of neuropathic pain. *Brain Res. Bull.* **2011**, *86*, 403–411. [CrossRef] [PubMed]
35. Liu, W.-C.; Feldman, S.C.; Cook, D.B.; Hung, D.-L.; Xu, T.; Kalnin, A.J.; Komisaruk, B.R. Fmri study of acupuncture-induced periaqueductal gray activity in humans. *Neuroreport* **2004**, *15*, 1937–1940. [CrossRef] [PubMed]
36. Kwon, Y.-b.; Kang, M.-s.; Ahn, C.-j.; Han, H.-j.; Ahn, B.-c.; Lee, J.-h. Effect of high or low frequency electroacupuncture on the cellular actitivy of catecholaminergic neurons in the brain stem. *Acupunct. Electro Ther. Res.* **2000**, *25*, 27–36. [CrossRef]
37. Young Bae, K.; Ho Jae, H.; Alvin, J.B.; Jang Hern, L. Bee venom acupoint stimulation increases fos expression in catecholaminergic neurons in the rat brain. *Mol. Cells* **2004**, *17*, 329–333.
38. Kwon, Y.B.; Yoon, S.Y.; Kim, H.W.; Roh, D.H.; Kang, S.Y.; Ryu, Y.H.; Choi, S.M.; Han, H.J.; Lee, H.J.; Kim, K.W.; et al. Substantial role of locus coeruleus-noradrenergic activation and capsaicin-insensitive primary afferent fibers in bee venom's anti-inflammatory effect. *Neurosci. Res.* **2006**, *55*, 197–203. [CrossRef] [PubMed]
39. Kim, H.-W.; Kwon, Y.-B.; Ham, T.-W.; Roh, D.-H.; Yoon, S.-Y.; Lee, H.-J.; Han, H.-J.; Yang, I.-S.; Beitz, A.J.; Lee, J.-H. Acupoint stimulation using bee venom attenuates formalin-induced pain behavior and spinal cord fos expression in rats. *J. Vet. Med. Sci.* **2003**, *65*, 349–355. [CrossRef] [PubMed]
40. Jones, S.L. Chapter 29—descending noradrenergic influences on pain. In *Progress in Brain Research*; Barnes, C.D., Pompeiano, O., Eds.; Elsevier: Amsterdam, The Netherlands, 1991; Volume 88, pp. 381–394.
41. Yoon, S.Y.; Roh, D.H.; Kwon, Y.B.; Kim, H.W.; Seo, H.S.; Han, H.J.; Lee, H.J.; Beitz, A.J.; Lee, J.H. Acupoint stimulation with diluted bee venom (apipuncture) potentiates the analgesic effect of intrathecal clonidine in the rodent formalin test and in a neuropathic pain model. *J. Pain* **2009**, *10*, 253–263. [CrossRef] [PubMed]
42. Lin, L.; Zhu, B.-P.; Cai, L. Therapeutic effect of melittin on a rat model of chronic prostatitis induced by complete freund's adjuvant. *Biomed. Pharmacother.* **2017**, *90*, 921–927. [CrossRef] [PubMed]
43. Zimmermann, M. Ethical guidelines for investigations of experimental pain in conscious animals. *Pain* **1983**, *16*, 109–110. [CrossRef]

44. Flatters, S.J.; Bennett, G.J. Ethosuximide reverses paclitaxel- and vincristine-induced painful peripheral neuropathy. *Pain* **2004**, *109*, 150–161. [CrossRef] [PubMed]
45. Polomano, R.C.; Mannes, A.J.; Clark, U.S.; Bennett, G.J. A painful peripheral neuropathy in the rat produced by the chemotherapeutic drug, paclitaxel. *Pain* **2001**, *94*, 293–304. [CrossRef]
46. Flatters, S.J.L.; Xiao, W.-H.; Bennett, G.J. Acetyl-l-carnitine prevents and reduces paclitaxel-induced painful peripheral neuropathy. *Neurosci. Lett.* **2006**, *397*, 219–223. [CrossRef] [PubMed]
47. Yin, C.S.; Jeong, H.S.; Park, H.J.; Baik, Y.; Yoon, M.H.; Choi, C.B.; Koh, H.G. A proposed transpositional acupoint system in a mouse and rat model. *Res. Vet. Sci.* **2008**, *84*, 159–165. [CrossRef] [PubMed]
48. Choi, S.; Yamada, A.; Kim, W.; Kim, S.K.; Furue, H. Noradrenergic inhibition of spinal hyperexcitation elicited by cutaneous cold stimuli in rats with oxaliplatin-induced allodynia: Electrophysiological and behavioral assessments. *J. Physiol. Sci. JPS* **2016**. [CrossRef] [PubMed]

© 2017 by the authors. Licensee MDPI, Basel, Switzerland. This article is an open access article distributed under the terms and conditions of the Creative Commons Attribution (CC BY) license (http://creativecommons.org/licenses/by/4.0/).

toxins

MDPI

Article

Overlooked Short Toxin-Like Proteins: A Shortcut to Drug Design

Michal Linial [1],*, Nadav Rappoport [2] and Dan Ofer [1]

[1] Department of Biological Chemistry, Silberman Institute of Life Sciences, The Hebrew University of Jerusalem, Jerusalem 91904, Israel; ddofer@gmail.com

[2] Institute for Computational Health Sciences, UCSF, San Francisco, CA 94158, USA; nadav.rappoport@ucsf.edu

* Correspondence: michall@cc.huji.ac.il; Tel.: +972-(0)2-658-5889

Academic Editor: Steve Peigneur
Received: 19 September 2017; Accepted: 25 October 2017; Published: 29 October 2017

Abstract: Short stable peptides have huge potential for novel therapies and biosimilars. Cysteine-rich short proteins are characterized by multiple disulfide bridges in a compact structure. Many of these metazoan proteins are processed, folded, and secreted as soluble stable folds. These properties are shared by both marine and terrestrial animal toxins. These stable short proteins are promising sources for new drug development. We developed ClanTox (classifier of animal toxins) to identify toxin-like proteins (TOLIPs) using machine learning models trained on a large-scale proteomic database. Insects proteomes provide a rich source for protein innovations. Therefore, we seek overlooked toxin-like proteins from insects (coined iTOLIPs). Out of 4180 short (<75 amino acids) secreted proteins, 379 were predicted as iTOLIPs with high confidence, with as many as 30% of the genes marked as uncharacterized. Based on bioinformatics, structure modeling, and data-mining methods, we found that the most significant group of predicted iTOLIPs carry antimicrobial activity. Among the top predicted sequences were 120 termicin genes from termites with antifungal properties. Structural variations of insect antimicrobial peptides illustrate the similarity to a short version of the defensin fold with antifungal specificity. We also identified 9 proteins that strongly resemble ion channel inhibitors from scorpion and conus toxins. Furthermore, we assigned functional fold to numerous uncharacterized iTOLIPs. We conclude that a systematic approach for finding iTOLIPs provides a rich source of peptides for drug design and innovative therapeutic discoveries.

Keywords: neurotoxin; protein families; disulfide bonds; antimicrobial peptide; ion channel inhibitor; ClanTox; complete proteome; comparative proteomics; machine learning; insects

---

1. Introduction

Short proteins are strong candidates for peptide-based therapy and drug development [1–3]. The search for peptide-based drugs is driven by the urge to improve specificity and affinity over classical drugs [4]. At present, the search for new leads for peptide therapy is mostly restricted to known peptides that act as hormones, neuropeptides, and growth factors [5–7].

Venomous proteins are found in diverse taxonomical branches including scorpions, snakes, spiders, and marine cone snails [8]. Venomous animals have developed sophisticated array of delivery systems for defense and offense. Evolutionary studies suggest that venomous toxins often reuse common folds that are abundant in the animal phyla (e.g., lipases [9]). Sequences of short proteins that are characterized by having numerous cysteines often fold into compact, stable structural folds. The resulting different folds are often found in proteins that carry diverse functions (e.g., lectins, protease, and protease inhibitors [10]). Venomous organisms are sporadically scattered within the phylogenetic tree of life. Venomous proteins represent cases of both divergent and convergent evolution, as well as repeated use of several existing, successful and abundant folds. However, the pool of bioactive short peptides resembling animal toxins is larger than anticipated [11]. The toxins' innovation is exemplified by their high degree of sequence variation and broad specificities, with only minimal alterations in the structural scaffolds [12].

In recent years, additional bioactive peptides were identified via systematic searches in the transcriptomes and proteomes of venomous animals [13,14]. Secreted short proteins from venomous glands may include hundreds of poorly studied bioactive peptides [6]. Approximately 2000 toxins out of an estimated >70,000 bioactive peptides have been identified in the genus Conus to date [15]. Evolutionary perspective based on the huge sequence diversity among toxins provides a rich source for rational protein design [16,17].

Toxins are extremely varied in their functions and mode of action. The potency of toxins' function is associated with an extremely broad collection of ion channel inhibitors (ICIs), phospholipases, protease inhibitors, disintegrins, membrane pore inducers, and more [18]. Some animal toxins affect the most basic cellular properties [19]. Examples include the non-reversible effect of amphipathic peptides on the membrane integrity [20] from spider venom [21] to marine hydrozoan toxins [22]. These toxins may cause non-specific hemolysis [23]. However, most toxin proteins act via highly specific binding to their cognate molecular target, making them attractive for drug design. The neuronal [24] and immune systems [25] are often affected by toxin-target molecular recognition. A well-studied example for reuse of a fold that acts on numerous receptors of the cholinergic system was described by Gibbons et al. [26]. The three-finger proteins (TFP) fold is found in numerous mammalian proteins acting in the innate immune system [27], and was also identified as Elapidae α-neurotoxins [28,29]. Two striking examples of human toxin-like proteins are Lynx1 [30] and SLURP-1 [31]. These are human proteins that possess similarity to snake α-neurotoxins, and modulate nicotinic acetylcholine receptors (nAChR), as does the snake α-neurotoxins. The identification of SLURP-1 as a neuromodulator has contributed to the understanding of the genetic effect of the Mal de Meleda disease, a skin disease that results from over activation of TNF-alpha [31].

Many short bioactive molecules are ion channels blockers (ICIs) and toxins with antimicrobial activity [32]. ICIs constitute the most widely studied group of toxins. A large group of ICIs whose evolution has been studied are the K$^+$ ICIs [33]. It is estimated that more than 10 different structural folds and 40 structural families represent this extremely diverse (structurally and evolutionarily) group [34]. In spite of that, two amino residues are critical for all K$^+$ ICIs' function: Lys and a Tyr/Phe, known as the functional dyad [35]. Surprisingly, even though these residues appear in very different positions along the sequences of K$^+$ ICIs, the solved structures show they are similarly aligned in space relatively to each other [36]. The same principle of sequence plasticity and structural rigidity apply for ICIs that affect other channels (e.g., [37–40]). Different ICIs targeting the same channel can vary in both sequence and structural folds [41].

The evolutionary mechanisms underlying the extreme diversity of toxins have been investigated [42]. Direct approaches for assessing the rapid mutation rate of a variety of toxins sharing the same fold have been reported (e.g., for phospholipases A2 [43]). TFP topology is also a strong example of the accelerated evolution and functional diversification reported for many snake toxins [44]. 3D complexes of short toxins and their cognate channels provide the best lead for the design of toxin-based pharmaceutical agents (e.g., [45]). A number of short toxins are already being used in the clinic for pain management [46], antiviral and antibacterial applications [47].

A common ICI design principle is conserved spacing, and the number of cysteines that form a stable scaffold in a few disulfide bridges [11]. In many cases, the core elements of the fold remain untouched by the preservation of at least two cysteine bridges, while the surfaces of the toxins undergo a natural dynamic adaptive evolution process. The extreme stability of the cysteine knot motif in peptide toxins makes these folds attractive for molecular engineering and drug design [48].

Based on the observation that many short animal toxins are rich in cysteines [49,50], we focused on a subset of short proteins (<75 amino acids) that can be used for discoveries towards peptide therapy [51]. The goal of our study is to present a systematic approach for identifying insects' toxin-like proteins TOLIPs (iTOLIPs). We analyzed a large number of published proteomes [52]. A rich catalogue of short bioactive proteins will have the potential to benefit the pharma and medical communities that seek new leads for drugs [53].

Insects represent one of the most diversified metazoan phyla. Many insect species evolved in unique ecological niches (e.g., parasitoid wasp) [54], and exhibit complex social behavior with rapidly evolving genomes [55,56]. In this study, we show that despite limited sequence similarity between short sequences, many toxin-like candidate sequences have been revealed via a machine learning predictor (ClanTox [57]). ClanTox was trained only on features extracted from ion channels inhibitors (ICI) from venomous proteins, for identifying TOLIPs. Using a rigorous bioinformatics and structural modeling scheme, we assigned a potential functional relevance for numerous iTOLIPs. We present dozens of new candidates for peptide-based therapy and discuss their potential for drug design.

2. Results and Discussion

2.1. Thousands of Toxin-Like Secreted Short Proteins in Insects

UniProtKB is the largest existing proteomic database (about 90 million sequences, August 2017) and is the main source of new templates for drug development. In recent years many new genomes have been sequenced including >30 insects. Despite a tsunami of genome sequences, only a few model organisms (e.g., *Drosophila melanogaster*) have high quality, manually annotated proteomes. While DNA sequencing quality has improved dramatically, current gene finding methodologies are still geared towards finding transcripts based on length (usually >100 amino acids, AA). Functional inference of genes' function from a transcribed genome remains an unsolved challenge [58]. Short proteins often have missing or faulty annotations (e.g., [59]).

We focused our discovery platform on short proteins. For the rest of the analyses we considered two thresholds on the proteins' length: (i) proteins of length <100 AA (Figure 1); (ii) a subset of shorter proteins, length <75 AA, that are attractive for drug development.

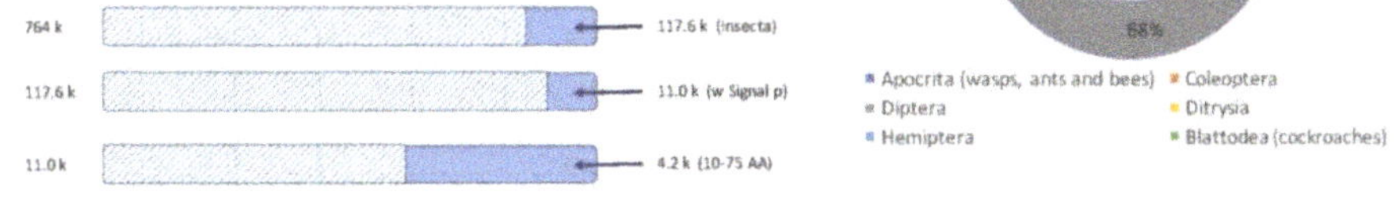

Figure 1. Selection of short secreted proteins from insect proteomes. (**A**) A sequence of filtration steps for protein sequences from UniProtKB is shown (top to bottom). Each step shows the number of proteins (left) and the resulting protein (right). The dashed bar marks the fraction of the data that is excluded from the following step. Sequences marked as "fragments" by UniProtKB were excluded. The final set used in this study includes proteins from Insecta with a "signal peptide" sequence annotation keyword, a restricted length of 10–100 AA and a further selection for proteins length of 10–75 AA. (**B**) A partition of the main orders of insects and their representation from the set of about 11,000 proteins.

We started with all proteins shorter than 100 AA (after removing all fragmented proteins), restricted to the insects' taxon, which resulted in ~117,600 proteins. Of these, 11,000 proteins were predicted to be secreted, and thus function in the extracellular space (Figure 1A).

Analyzing the ~11,000 protein'-origins show that the proteomes of major orders of insects are biased towards the previously sequenced genomes (Figure 1B). Diptera, which includes mosquitos and flies, dominates the collection (68%). The rest of the candidate short proteins belong to Hymenoptera (mostly bees, wasp, and ants, 10%), Ditrysia (including moth, bumblebee, and butterfly, 9%) and a smaller amount of Hemiptera (e.g., aphids), Coleoptera (mostly beetles) and Blattodea (mostly termites).

While most insects are not venomous [19], some bees, ants, and wasps developed mechanisms to release their venomous proteins and toxic peptides. Many of the short proteins are uncharacterized (see discussion in [56]). Moreover, annotations of genes from fast evolving organisms are often missing. Due to these fast evolutionary innovation in many insects, we anticipate a rich repertoire of overlooked bioactive peptides [60] and iTOLIPs [61].

We used ClanTox [57] to investigate the abundance of iTOLIPs among the 11,000 short, secreted proteins (<100 AA). To this end, we divided the protein according to the major orders of insects, and further investigated the ClanTox predictions, according to the confidence level of the predictor (marked as P1–P3, see Methods). We have previously shown that many valid TOLIPs are identified at all confidence levels, including the least confident one (P1, see Methods, [57]). ClanTox was trained only on ICIs from venomous animals for seeking TOLIPs from all organisms. While it was trained on a limited function, predictions are associated with a much broader spectrum of functions that specify known toxins and proteins with no known homologues in venoms [11].

Figure 2 shows the results from ClanTox prediction with iTOLIPs cover the two largest orders of insects, the Diptera (Figure 2A) and Hymenoptera (Figure 2B). A bias in the prediction towards model organisms is evident. The iTOLIPs from Drosophilae (fruit fly) accounts for 44% of the predicted sequences. Still, >1000 sequences are detected in less studied organisms, such as the Tsetse fly, Aedes, blowfly, and more (Figure 2A). The fraction of iTOLIPs among the cysteine rich short proteins from Hymenoptera (wasp, bees, and ants) is 24%. The high number of iTOLIPs from ant proteomes is a reflection of the many recently sequenced ant genomes (Figure 2C) [56]. Note that the number of predictions from *Nasonia vitripennis* (Parasitic wasp) is disproportionally high. Of 145 *Nasonia vitripennis*'-short proteins, 57 (39%) were predicted as iTOLIPs (Figure 2C).

From a therapeutic perspective, often, the shorter the protein, the easier it is to produce it synthetically, and to introduce it to laboratory and clinical trials. We restricted the search to 4181 sequences are shorter than 75 AA (Figure 1A).

Figure S1 shows the distribution of the 4181 sequences according to ClanTox's prediction confidence (N, P1–P3, see Methods). Note that most proteins (76%) are predicted as negative, and do not comply with the definition of iTOLIPs (Clantox's label N stands for—"not a toxin-like"). The high confidence predictions (P3, top prediction for Toxin-like) include 379 proteins (9%, Figure S1). The rest of the analyses will focus on these high confidence-predicted iTOLIPs (P3).

Table 1 shows the partition of the top predicted iTOLIPs among the major orders of insects. The most outstanding observation is the abundance of iTOLIPs in termites (52%), and the low discovery of top prediction iTOLIPs among Ditrysia (5%). A list of 379 predicted sequences is available (Table S1).

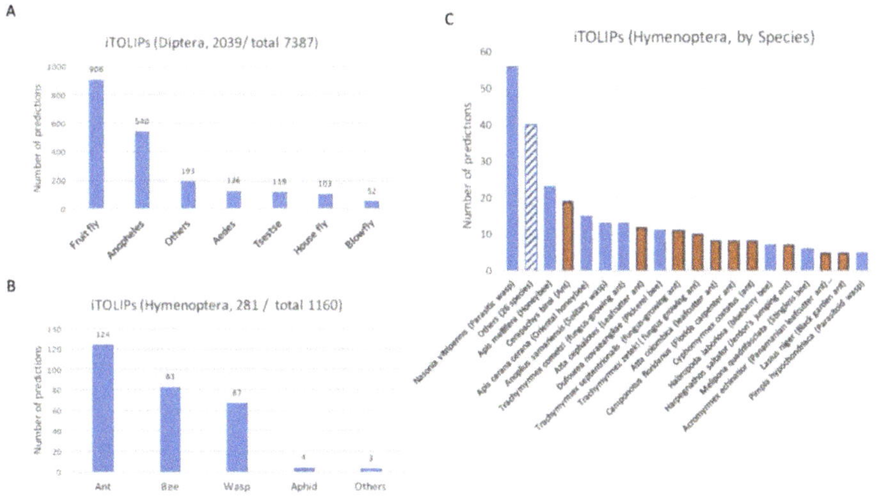

Figure 2. Partition of ClanTox prediction for mini-proteins of toxin-like proteins from insects (iTOLIPs). The fraction of iTOLIPs that was identified as iTOLIPs by ClanTox is shown for the orders Diptera (**A**), and Hymenoptera (**B**). Only major genus representatives are shown. The total numbers indicate the number of sequences that were introduced to ClanTox. (**C**) A detailed partition of the species that are associated with iTOLIPs. Only species having ≥5 proteins are listed. The dashed bar is an aggregation of iTOLIPs from 26 different species. Orange bar are different ant species, and blue bars are other representatives of Hymenoptera.

Table 1. iTOLIPs top predictions by major insects' order.

Insects	Number of Short Proteins	Number of Top Predictions	% Top Predictions from Total	Representative Family
Blattoidea	238	124	52.1	Termite
Hymenoptera (wasps, ants and bees)	460	35	7.6	Honeybee
Ditrysia	403	20	5	Butterfly
Polyphaga	139	12	8.6	Beetle
Hemiptera	230	16	7	Aphid
Pulicidae	17	2	11.8	Flea
Acrididae	9	2	22.2	Grasshopper
Pseudagrion	2	0	0	Damselfly
Psocodea	24	0	0	Lice
All insects	**4196**	**379**	**9**	

2.2. Most iTOLIP Mini-Proteins Resemble Antibacterial and Antifungal Peptides

Antimicrobial peptides (AMPs) are very abundant among insects [62]. At present, >150 insect AMPs have been identified [63]. A total of 121 peptides out of 379 iTOLIPs are from the Blattodea order, and named by UniProtKB as "termicin". Among the top predicted iTOLIPs, these proteins comprise the largest group. Termicins are restricted to the order Blattodea (termites and cockroaches). These are a collection of secreted AMP mini-proteins (25–40 AA), sharing a moderate sequence similarity. A termicin-like peptide (25 AA) from the cockroach *Eupolyphaga sinensis* exhibits anti-fungal activity, and a weak activity against bacteria [63]. We hypothesize that other sequences among the al iTOLIPs resemble antimicrobial proteins and potentially act as such.

Structurally, termicin is characterized by three disulfide bridges forming a rigid fold. The tertiary structure of termicin contains an α-helical segment and a two-stranded antiparallel β-sheet (called cysteine-stabilized α-helix/β-sheet, CSαβ, Figure 3A). The structural motif of CSαβ is similar to that of short insect defensins. The cysteine positions and pairing suggest that despite a minimal sequence similarity with insect defensins, the structure is shared by all defensins [64]. Expending the analysis of ClanTox top predictions suggests that the AMP and defensin-like fold could be subjected for a design approach aiming to improve the peptide specificity in the current post-antibiotic era (Figure 3A).

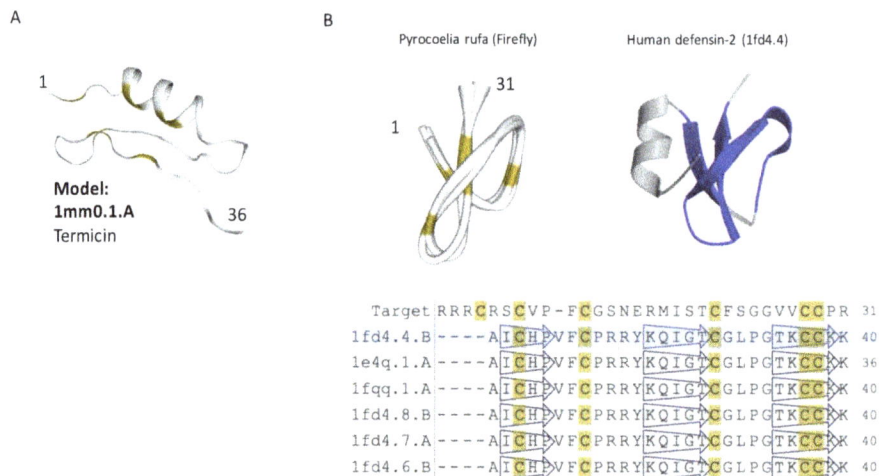

Figure 3. Structural model of iTOLIPs with antifungal activity. (**A**) The tertiary structure of D2D008_9NEOP from *Macrotermes barneyi* is shown. The structure is a representative of 120 related sequences of 35–36 AA identified as iTOLIPs. The model shows the α-helix stabilized next to two-stranded antiparallel β-sheet (called CSαβ). (**B**) A structural model for the mature Q95UJ8 protein (25–55 AA) from firefly (*Pyrocoelia rufa*) is shown. The best model for this sequence is the human defensin-2 protein (PDB:1fd4.4) (right). The light green shades indicate the overlap between the two proteins. Representatives for the structural model and their multiple sequence alignments are shown. The positions of the β-sheets are shown by the hollow arrows. Yellow color marks the position of the cysteines.

The insect defensin protein is a shorter version of the human defensin-2 (Figure 3B). Furthermore, the human defensin's N-terminal helix is completely missing in the firefly protein. It is plausible that functionality as an AMP comes from the core folded structure of (31 AA) of the firefly version of the defensin, and therefore, the N'-terminal helix is redundant (Figure 3B, light green shade). Structural variations of insect antimicrobial peptides illustrate the resemblance to a short version of the defensin fold. The diversity of AMP peptides in view of scorpion toxins had been extensively studied [65,66]. Defensins were also found among sponge, platypus, and scorpion toxins [67]. The assumption is that short specific structural motifs are used as templates by animal toxins [68]. Note that many additional versions of insect defensin genes are longer than 75 AA, and thus will not be further discussed [69,70].

The other major shared function among the top predicted iTOLIPs (Table S1) is the antifungal activity associated with the many Drosomicin genes, including two large sets of DRO and DRS genes [71]. Drosomycins (DRS) are inducible antifungal peptides, and were isolated from the hemolymph of immune-challenged Drosophilae. A similar antifungal specificity applies for DRO1–DRO6 cassette, which responds to injury and microbial infection [72]. The DRS scaffold is a typical cysteine-stabilized α-helical and β-sheet (CSαβ) that specifies many of the known defensins (Figure 4). The hallmark of DRS gene is its extra-stability, which is gained by clamping the N'- and C'-termini by an additional disulfide bond. This solution for extreme stability was also found in the spider toxin ω-hexatoxin-Hv1a. This innovation in protein stability is beneficial for a protein design approach for a biochemical stable scaffold [48].

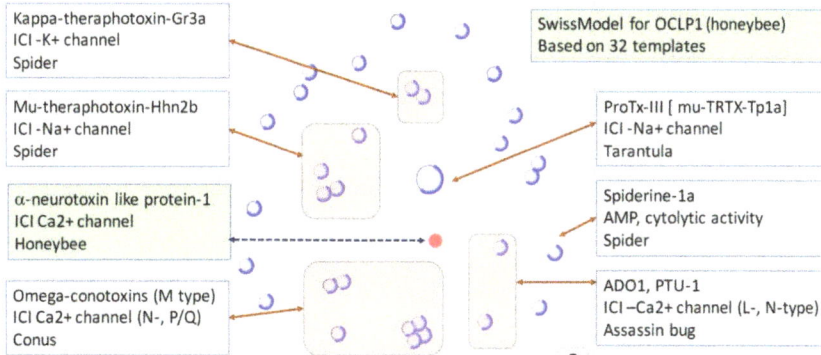

Figure 4. Omega conotoxin –like protein 1 (OCLP1) and its similarity to structurally solved proteins used as templates. The protein OCTP1 (red circle) is shown in view of a sequence similarity from the best SwissModel for H9KQJ7 (AA 26–74) from *Apis mellifera* (Honeybee). Each blue circle is one of the 32 template proteins. The functions of the listed proteins and the relevant organism are listed. ICI, ion channel inhibitor.

Short versions of the AMP peptide, with three disulfide bonds resembling defensin were identified in marine sponges [73] and jellyfish [74]. In jellyfish, a similarity to defensin is extended also to the K$^+$ ICIs of sea anemones. Multiple functionalities had been experimentally validated for the short CSαβ scaffold of DRS, and the truncated scorpion toxin. Both peptides are effective as ion channel modulators (on *D. melanogaster* voltage-gated sodium channel) and exhibit anti-fungal activity [75].

2.3. iTOLIPs as Ion Channel Inhibitors

We analyzed proteins whose structural similarity to toxins have been identified. Table 2 lists nine instances in which a toxin related function is revealed. All 9 proteins exhibit channel blocker similarity to various channels [76]. Interestingly, two sequences from the *Apis mellifera* (Honeybee) and *Aphidius ervi* (Aphid parasite) show a clear homology to ω-conotoxin MVIIC and GVIA, a potent conus peptide that effectively blocks Ca^{2+} channels. The OCLP1 was initially identified using ClanTox, and its function as ICI had been validated [11].

Table 2. Toxin-like mini-proteins from insects.

UniProtKB	AA (Mature) [a]	Protein Name	Species	PDB	% Seq. Sim	Description
H9KQJ7	74 (54)	ω-conotoxin-like protein 1	*A. mellifera*	2n86.1	44.1	Spiderine-1a
A0A084WJA1	71 (46)	K-channel toxin α-KTx 18.3	*A. sinensis*	2b68.1	24.1	defensin
J7HBU2	70 (47)	Salivary toxin-like peptide	*N. intermedia*	5t4r.1	51.5	Mu-theraphotoxin-Pn3a
J7HIK0	70 (47)	Salivary toxin-like peptide	*N. intermedia*	5t4r.1	51.5	Mu-theraphotoxin-Pn3a
J7HBS6	70 (46)	Salivary toxin-like peptide	*N. intermedia*	5t4r.1	51.5	Mu-theraphotoxin-Pn3a
J7HBT1	75 (50)	Salivary toxin-like peptide	*N. intermedia*	1d1h.1	46.7	Hanatoxin Type 1
A0A034WXR3	60 (36)	Venom toxin-like peptide	*A. ervi*	1q3j.1	33.3	ALO3
A0A034WY34	61 (37)	Venom toxin-like peptide	*A. ervi*	2lqa.1	43.8	Asteropsin A
A0A034WWW1	51 (37)	Venom toxin-like peptide	*A. ervi*	1omn.1	48.0	ω-Conotoxin MVIIC

[a] Full length of the protein, and the length of the mature protein (in parentheses). Mature protein is a cleaved product after removal of the N′-terminal signal sequence. Seq. sim, sequence similarity.

We retested the OCLP1 structural model in view of the doubling of proteins with 3D-structures in the last decade. The most likely structural model for OCLP1 benefited from structural relatedness (Figure 4). The similarity in the cysteine distribution locations along the sequence, and the cysteines that contribute to the disulfide bridges applies for ω-conotoxin MVIIC (1cnn.1, 1omn.1), Ptu-1 (1i26.1), Toxin Ado1 (1lmr. 1), SVIB (1mvj.1), ω-conotoxin GVIA (1omc.1, 1tr6.1, 1ttl.1, 2cco.1), Robustoxin (1qdp.1), Hainantoxin-3 (2jtb.1), Spiderine-1a (2n86.1), and more. Importantly, the OCLP1 model indicates a comparable sequence similarity to a large number of ICIs. The related sequences exhibiting ICI function blocks Na^+, K^+, and all major types of Ca^{+2} channels (L-, N-, and P/Q-types, Figure 4). As such, these sequences are attractive templates for drug development seeking feature determinants that dictate a detailed specificity. Actually, the specificity is not restricted to the selective ion but to the exact version of the ion channel. For example, the protein μ-theraphotoxin-Pn3a that was isolated from venom of the tarantula *Pamphobeteus nigricolor,* is a potent inhibitor of Nav1.7, a subtype of the sodium ion channel (Nav). Its specificity for the other Nav subtypes is lower by 2–3 order of magnitudes [77].

A detailed report for the five top templates that are used for construction of a structural model for each of the 9 proteins (Table 2) is available (Table S2).

2.4. Uncharacterized iTOLIPs Reveal New Cysteine-Rich Patterns

Among the identified mini-proteins are 110 sequences that are annotated as "uncharacterized" (and genes named by their genomic index). About 65% of them are from Diptera (55 from Drosophilae, and 16 from Anopheles). Inspecting the spacing and number of the cysteines among the "uncharacterized" mini-proteins shows numerous recurring patterns (Figure 5).

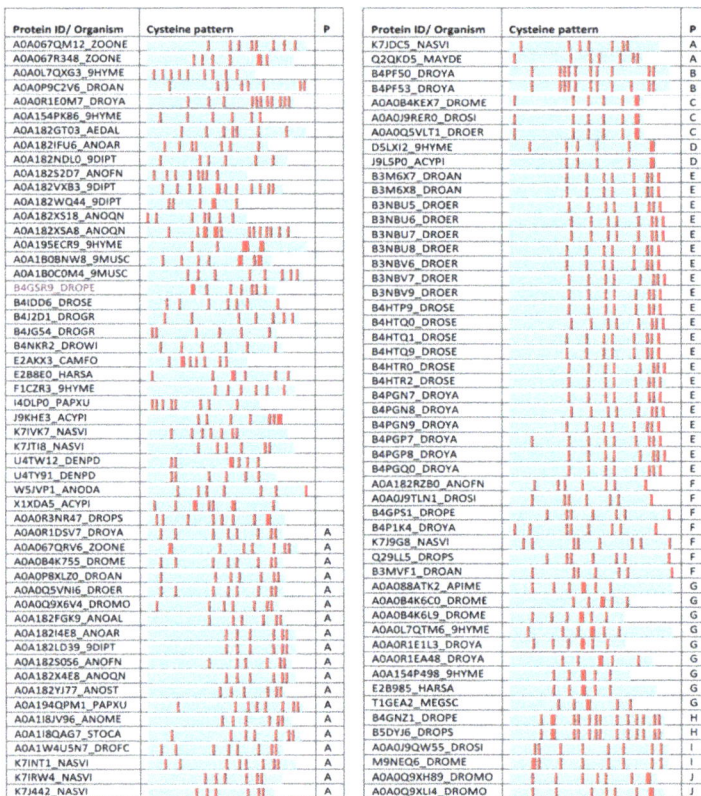

Protein ID/ Organism	Cysteine pattern	P		Protein ID/ Organism	Cysteine pattern	P
A0A067QM12_ZOONE				K7JDC5_NASVI		A
A0A067R348_ZOONE				Q2QKD5_MAYDE		A
A0A0L7QXG3_9HYME				B4PF50_DROYA		B
A0A0P9C2V6_DROAN				B4PF53_DROYA		B
A0A0R1E0M7_DROYA				A0A0B4KEX7_DROME		C
A0A154PK86_9HYME				A0A0J9RER0_DROSI		C
A0A182GT03_AEDAL				A0A0Q5VLT1_DROER		C
A0A182IFU6_ANOAR				D5LXI2_9HYME		D
A0A182NDL0_9DIPT				J9LSP0_ACYPI		D
A0A182S2D7_ANOFN				B3M6X7_DROAN		E
A0A182VXB3_9DIPT				B3M6X8_DROAN		E
A0A182WQ44_9DIPT				B3NBU5_DROER		E
A0A182XS18_ANOQN				B3NBU6_DROER		E
A0A182XSA8_ANOQN				B3NBU7_DROER		E
A0A195ECR9_9HYME				B3NBU8_DROER		E
A0A1B0BNW8_9MUSC				B3NBV6_DROER		E
A0A1B0C0M4_9MUSC				B3NBV7_DROER		E
B4GSR9_DROPE				B3NBV9_DROER		E
B4ID06_DROSE				B4HTP9_DROSE		E
B4J2D1_DROGR				B4HTQ0_DROSE		E
B4JG54_DROGR				B4HTQ1_DROSE		E
B4NKR2_DROWI				B4HTQ9_DROSE		E
E2AKX3_CAMFO				B4HTR0_DROSE		E
E2B8E0_HARSA				B4HTR2_DROSE		E
F1CZR3_9HYME				B4PGN7_DROYA		E
I4DLP0_PAPXU				B4PGN8_DROYA		E
J9KHE3_ACYPI				B4PGN9_DROYA		E
K7IVK7_NASVI				B4PGP7_DROYA		E
K7JTI8_NASVI				B4PGP8_DROYA		E
U4TW12_DENPD				B4PGQ0_DROYA		E
U4TY91_DENPD				A0A182RZB0_ANOFN		F
W5JVP1_ANODA				A0A0J9TLN1_DROSI		F
X1XDA5_ACYPI				B4GPS1_DROPE		F
A0A0R3NR47_DROPS		A		B4P1K4_DROYA		F
A0A0R1DSV7_DROYA		A		K7J9G8_NASVI		F
A0A067QRV6_ZOONE		A		Q29LL5_DROPS		F
A0A0B4K755_DROME		A		B3MVF1_DROAN		F
A0A0P8XLZ0_DROAN		A		A0A088ATK2_APIME		G
A0A0Q5VNI6_DROER		A		A0A0B4K6C0_DROME		G
A0A0Q9X6V4_DROMO		A		A0A0B4K6L9_DROME		G
A0A182FGK9_ANOAL		A		A0A0L7QTM6_9HYME		G
A0A182I4E8_ANOAR		A		A0A0R1E1L3_DROYA		G
A0A182LD39_9DIPT		A		A0A0R1EA48_DROYA		G
A0A182S0S6_ANOFN		A		A0A154P498_9HYME		G
A0A182X4E8_ANOQN		A		E2B9B5_HARSA		G
A0A182YJ77_ANOST		A		T1GEA2_MEGSC		G
A0A194QPM1_PAPXU		A		B4GNZ1_DROPE		H
A0A1I8JV96_ANOME		A		B5DYJ6_DROPS		H
A0A1I8QAG7_STOCA		A		A0A0J9QW55_DROSI		I
A0A1W4U5N7_DROFC		A		M9NEQ6_DROME		I
K7INT1_NASVI		A		A0A0Q9XH89_DROMO		J
K7IRW4_NASVI		A		A0A0Q9XLI4_DROMO		J
K7J442_NASVI		A				

Figure 5. Uncharacterized iTOLIPs and a graphical representations of the mini-proteins. The cysteine residues are marked by red bars. The proteins are grouped according to the recurrent pattern of cysteines based on their number and location along the protein sequence (P, pattern).

A recurring pattern is illustrated by the B3M6X8_DROAN (*Drosophila ananassae*). This pattern is identified in *Drosophila erecta* and *Drosophila yabuba*, and appears in 20 proteins (with small variations, Figure 5, Patten E). Using structural modeling, we found that the strongest sequence similarity is to PDB: 1myn.1 (Drosomycin). Yet, another set of toxins such as the α-like toxin Lqh3 and BmαTX47 toxins from old and new world scorpions [78] seems to share a structural fold (Figure 6A). All these neurotoxins are specific to different Nav subtypes [79]. The stiff structure is visible mainly through the α-helix and the antiparallel β-sheets (Figure 6A). However, the substantial variations in the loops indicate the potential site for specificity of AMP, and the K^+ and Na^+ ion channel blocking. The overlap of B3M6X relative to 7 protein representatives that contributed to the model is shown along their multiple sequence alignment (Figure 6A, bottom).

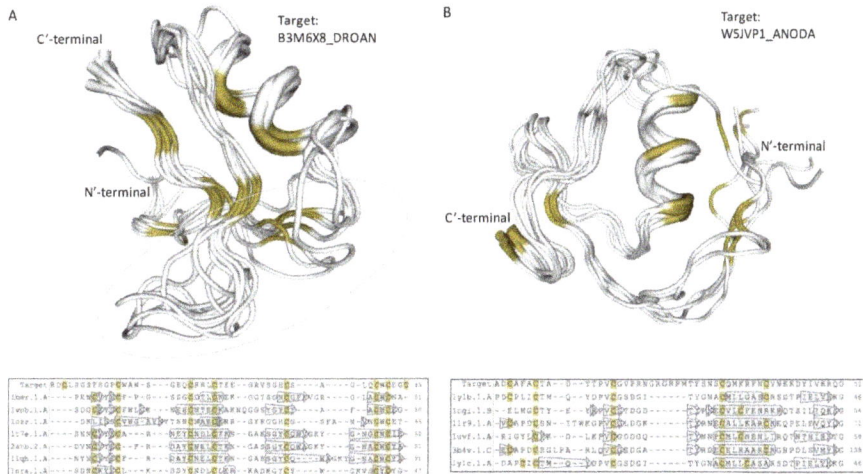

Figure 6. **Structural model of iTOLIPs uncharacterized proteins.** (**A**) Structural model of the protein B3M6X8_DROAN is shown. The structural model is a defensin fold. The overlap of 7 patterns is shown along with their multiple sequence alignments. The light green marks the area in which the sequences vary the most among the representative proteins. (**B**) Structural model of the protein W5JVP1_ANODA is shown. The structural model is of the Kazal protease inhibitor fold. The overlap of 6 patterns is shown along with their multiple sequence alignments. The positions of the α-helix and β-sheets are shown by the hollow frame and arrows, respectively. Yellow color marks the position of the cysteines.

A systematic search for a model for the uncharacterized proteins showed that for A0A182S0S6_ANOFN (*Anopheles funestus*, Figure 5, Patten A), the best model is similar to gamma 1-P thionins from barley and wheat endosperm (PDB: 1gps). These proteins are common motifs among toxic arthropod proteins and defensins. Still, the most likely defensin that was associated with *Anopheles funestus* protein is from a plant origin (PDB: 5nce.1).

Modeling the structure of the uncharacterized W5JVP1_ANODA (Figure 5, Pattern F) revealed a strong and highly conserved structure similar to a "non-classical" Kazal-type inhibitor (Figure 6B). All six structure representatives are aligned, and support its function as protease inhibitor. Kazal protease inhibitor fold was identified from some snakes, sea anemone, and skin of tree frogs. However, most proteinase inhibitor from toxins are associated with Kunitz fold that display a broader taxonomical coverage and a robust protease inhibition [80]. Other proteins predicted by structural modeling to have the Kazal protease inhibitor fold include A0A182RZB0_ANOFN, A0A0J9TLN1_DROSI, Q29LL5_DROPS, K7J9G8_NASVI, B3MVF1_DROAN, and B4GPS1_DROPE (Figure 5, Patten F).

Testing other uncharacterized proteins from the list (Figure 5) resulted in poor or no supportive models. Note that some cysteine-based patterns appear with multiple examples in the list. For example, B4PF50_DROYA and B4PF53_DROYA share the same pattern in terms of their cysteine number and spacing (Figure 5, Pattern B). Additional proteins are associated with structurally new shapes that could not be modeled to reach a satisfactory level (e.g., A0A0P9C2V6_DROAN). These findings suggest that the uncharacterized proteins provide a rich, yet unexplored scaffold for future drug design.

3. Materials and Methods

3.1. Protein Databases

We used datasets from UniProtKB Release Aug_2017 [81] including 90 million protein sequences, combining the SwissProt and TrEMBL datasets [82]. We used the current data from RCSB protein data bank [83] with the collection of about 124,000 proteins' structural information.

3.2. Bioinformatics Analysis Tools

SignalP 4.0 was used to predict signal peptides [84]. This self-standing predictive tool is also provided as an annotation in UniProtKB [KW-0732]. The average length of the signal sequence in mammals is about 25 AA. We consider a protein length of 75 AA to account for a mature protein of about 50 AA. EBI's ClustalW and alignment viewer tools were used. Swiss-Model [82] was applied with default parameters for building a model according to the templates from the RCSB database. In the automated mode, both BLAST and HHblits (profile -profile search) are used. HHpred and HHblits [85] provide sensitive structural prediction by HMM -HMM- comparison. The HHblits builds HMM from a query sequence and compares it with a library of HMMs representing all known structures from PDB [83]. All structural predictions obtained from Swiss-Model, and HHblits were compared for testing the quality of the results.

Template quality is estimated along the process of the model building, for maximization of the quality and coverage of the model. In some cases, more than one model is presented to reflect the structural diversity. The quality of the models is estimated using calculated statistical parameters of the model (GMQE and QMEAN). These values are determined with respect to experimental parameters of proteins with a similar length ([82]). Only sufficiently supported quality models are presented. The visualization tool used are embedded in Swiss-Model. A sequence similarity map shows the proteins that were used as templates, and contributed to the final model from a set of non-redundant structurally solved proteins.

3.3. ClanTox Prediction and Scoring

ClanTox (classifier of animal toxins) is a machine learning classifier ensemble for ranking protein sequences according to their toxin-like properties. ClanTox provides characterization for these mostly uncharacterized proteins. ClanTox uses about 600 features, including the stability and the spacing of the cysteine residues [57]. However, features are not restricted to cysteine-related features. ClanTox was trained on few hundreds of ICIs from a broad range of animal toxins. ClanTox's method represents each sequence as a vector of numerical sequence-derived features. The test set performance of ClanTox in cross-validation is very high, with a mean area under the curve (AUC) of >0.99 [86].

The sequences from the selected subset of insect proteomes downloaded from UniProtKB were used as input for ClanTox. The classifier outputs four labels: N for negative prediction, and P1–P3, reflecting three levels of positive predictions for toxin-like proteins (TOLIPs). The most significant predictions (labeled P3) accounts for proteins with a mean score >0.2, as well as having a coefficient of variation (CV) <0.5. The negative predictions (N, predicted as non-toxin) account for all sequences with a mean score <−0.2. The confidence of the prediction indirectly considers the robustness of the prediction. Formally, P3 are predictions with a mean score >0.2 or mean score >2*SD; P2 are predictions with mean score >0.2 or mean score between SD and 2*SD; and P1 are predictions with mean score >−0.2 or mean score <SD [57].

4. Conclusions

From the evolutionary perspective, toxins that possess similar functions (e.g., ICIs) may appear in unrelated venomous species, which is in accord with an accelerated evolution and innovation among toxins. Detecting endogenous toxin-like proteins from insects (iTOLIPs) confirmed that much of the innovation associated with bioactive peptides and mini-proteins links to defense against microbes, mainly fungi, and modulating of ion channels. Potentially, these functions are not mutually exclusive, and short proteins may carry more than one function. The rich collection identified in insects is instrumental in searching for particular AA that can enhance specificity towards specific fungi, or bacterium in the case of AMPs. In this study, we discussed a collection of top predictions from ClanTox. Note that hundreds of additional iTOLIPs are reported at somewhat lower predicted confidence. We conclude that the overlooked iTOLIPs characterized by structural stability and enhanced specificity are attractive templates for drug design.

Supplementary Materials: The following are available online at www.mdpi.com/2072-6651/9/11/350/s1, Table S1: Top prediction of ClanTox (P3) for insect < 75 AA with 379 iTOLIPs. Table S2: Top 5 templates selected by Swiss-Model for constructing the structural model of nine mature iTOLIP mini-protein. Figure S1. Scoring of ClanTox predictions for insects' secreted mini-proteins. Distribution of ClanTox predictions of 4180 insects' secreted proteins shorter than 75 AA. The top scoring iTOLIPs are marked by P3 (dark red), the intermediate confidence is P2 and P1 is the least confident predictions. The gray marks the bulk of the sequences (76%) with negative prediction (i.e., not a TOLIPs). All together there are 379 proteins that are scored as P3 (Table S1).

Author Contributions: M.L. N.R and D.O. analyzed the data, and wrote the paper. N.R. is part of developing team of the ClanTox webtool (www.clantox.cs.huji.ac.il), which was used throughout this study.

Conflicts of Interest: The authors declare no conflict of interest.

Abbreviations

AMP	antimicrobial peptides
CSαβ	cysteine-stabilized α-helical and β-sheet
ClanTox	classifier of animal toxins
CRISP	cysteine rich short proteins
ICI	ion channel inhibitor
DRS	Drosomycin
nAChR	nicotinic acetylcholine receptors
OCLP	omega conotoxin-like protein
TFP	three-finger proteins
iTOLIP	insect toxin-like proteins

References

1. Adermann, K.; John, H.; Standker, L.; Forssmann, W.G. Exploiting natural peptide diversity: Novel research tools and drug leads. *Curr. Opin. Biotechnol.* **2004**, *15*, 599–606. [CrossRef] [PubMed]
2. Alonso, D.; Khalil, Z.; Satkunanthan, N.; Livett, B.G. Drugs from the sea: Conotoxins as drug leads for neuropathic pain and other neurological conditions. *Mini Rev. Med. Chem.* **2003**, *3*, 785–787. [CrossRef] [PubMed]
3. King, G.F. Venoms as a platform for human drugs: Translating toxins into therapeutics. *Expert Opin. Biol. Ther.* **2011**, *11*, 1469–1484. [CrossRef] [PubMed]
4. Proksch, P.; Edrada, R.; Ebel, R. Drugs from the seas-current status and microbiological implications. *Appl. Microbiol. Biotechnol.* **2002**, *59*, 125–134. [PubMed]
5. Bock, J.E.; Gavenonis, J.; Kritzer, J.A. Getting in shape: Controlling peptide bioactivity and bioavailability using conformational constraints. *ACS Chem. Biol.* **2013**, *8*, 488–499. [CrossRef] [PubMed]

6. Vetter, I.; Davis, J.L.; Rash, L.D.; Anangi, R.; Mobli, M.; Alewood, P.F.; Lewis, R.J.; King, G.F. Venomics: A new paradigm for natural products-based drug discovery. *Amino Acids* **2011**, *40*, 15–28. [CrossRef] [PubMed]

7. Bulaj, G. Integrating the discovery pipeline for novel compounds targeting ion channels. *Curr. Opin. Chem. Biol.* **2008**, *12*, 441–447. [CrossRef] [PubMed]

8. Harvey, A.L. Toxins and drug discovery. *Toxicon* **2014**, *92*, 193–200. [CrossRef] [PubMed]

9. Fry, B.G.; Roelants, K.; Champagne, D.E.; Scheib, H.; Tyndall, J.D.; King, G.F.; Nevalainen, T.J.; Norman, J.A.; Lewis, R.J.; Norton, R.S.; et al. The toxicogenomic multiverse: Convergent recruitment of proteins into animal venoms. *Annu. Rev. Genom. Hum. Genet.* **2009**, *10*, 483–511. [CrossRef] [PubMed]

10. Wong, E.S.; Belov, K. Venom evolution through gene duplications. *Gene* **2012**, *496*, 1–7. [CrossRef] [PubMed]

11. Kaplan, N.; Morpurgo, N.; Linial, M. Novel families of toxin-like peptides in insects and mammals: A computational approach. *J. Mol. Biol.* **2007**, *369*, 553–566. [CrossRef] [PubMed]

12. Fry, B.G.; Wuster, W.; Kini, R.M.; Brusic, V.; Khan, A.; Venkataraman, D.; Rooney, A.P. Molecular evolution and phylogeny of elapid snake venom three-finger toxins. *J. Mol. Evol.* **2003**, *57*, 110–129. [CrossRef] [PubMed]

13. Craik, D.J.; Fairlie, D.P.; Liras, S.; Price, D. The future of peptide-based drugs. *Chem. Biol. Drug Des.* **2013**, *81*, 136–147. [CrossRef] [PubMed]

14. Han, T.S.; Teichert, R.W.; Olivera, B.M.; Bulaj, G. Conus venoms—A rich source of peptide-based therapeutics. *Curr. Pharm. Des.* **2008**, *14*, 2462–2479. [CrossRef] [PubMed]

15. Lavergne, V.; Harliwong, I.; Jones, A.; Miller, D.; Taft, R.J.; Alewood, P.F. Optimized deep-targeted proteotranscriptomic profiling reveals unexplored conus toxin diversity and novel cysteine frameworks. *Proc. Natl. Acad. Sci. USA* **2015**, *112*, E3782–E3791. [CrossRef] [PubMed]

16. Drabeck, D.H.; Dean, A.M.; Jansa, S.A. Why the honey badger don't care: Convergent evolution of venom-targeted nicotinic acetylcholine receptors in mammals that survive venomous snake bites. *Toxicon* **2015**, *99*, 68–72. [CrossRef] [PubMed]

17. Zambelli, V.; Pasqualoto, K.; Picolo, G.; Chudzinski-Tavassi, A.; Cury, Y. Harnessing the knowledge of animal toxins to generate drugs. *Pharmacol. Res.* **2016**, *112*, 30–36. [CrossRef] [PubMed]

18. Fry, B.G. From genome to "venome" Molecular origin and evolution of the snake venom proteome inferred from phylogenetic analysis of toxin sequences and related body proteins. *Genome Res.* **2005**, *15*, 403–420. [CrossRef] [PubMed]

19. Casewell, N.R.; Wuster, W.; Vonk, F.J.; Harrison, R.A.; Fry, B.G. Complex cocktails: The evolutionary novelty of venoms. *Trends Ecol. Evol.* **2013**, *28*, 219–229. [CrossRef] [PubMed]

20. Sitprija, V.; Sitprija, S. Renal effects and injury induced by animal toxins. *Toxicon* **2012**, *60*, 943–953. [CrossRef] [PubMed]

21. Corzo, G.; Villegas, E.; Gomez-Lagunas, F.; Possani, L.D.; Belokoneva, O.S.; Nakajima, T. Oxyopinins, large amphipathic peptides isolated from the venom of the wolf spider oxyopes kitabensis with cytolytic properties and positive insecticidal cooperativity with spider neurotoxins. *J. Biol. Chem.* **2002**, *277*, 23627–23637. [CrossRef] [PubMed]

22. Edwards, L.P.; Whitter, E.; Hessinger, D.A. Apparent membrane pore-formation by portuguese man-of-war (physalia physalis) venom in intact cultured cells. *Toxicon* **2002**, *40*, 1299–1305. [CrossRef]

23. Slotta, K.H.; Gonzalez, J.; Roth, S. The direct and indirect hemolytic factors from animal venoms. In *RUSSELL Animal Toxins*; Elsevier: Amsterdam, The Netherlands, 2016; pp. 369–377.

24. Estrada, G.; Villegas, E.; Corzo, G. Spider venoms: A rich source of acylpolyamines and peptides as new leads for cns drugs. *Nat. Prod. Rep.* **2007**, *24*, 145–161. [CrossRef] [PubMed]

25. Petricevich, V.L. Scorpion venom and the inflammatory response. *Mediat. Inflamm.* **2010**, *2010*, 903295. [CrossRef] [PubMed]

26. Gibbons, A.; Dean, B. The cholinergic system: An emerging drug target for schizophrenia. *Curr. Pharm. Des.* **2016**, *22*, 2124–2133. [CrossRef] [PubMed]

27. Tirosh, Y.; Ofer, D.; Eliyahu, T.; Linial, M. Short toxin-like proteins attack the defense line of innate immunity. *Toxins* **2013**, *5*, 1314–1331. [CrossRef] [PubMed]

28. Tsetlin, V.I. Three-finger snake neurotoxins and ly6 proteins targeting nicotinic acetylcholine receptors: Pharmacological tools and endogenous modulators. *Trends Pharmacol. Sci.* **2015**, *36*, 109–123. [CrossRef] [PubMed]

29. Kini, R.M. Evolution of three-finger toxins—A versatile mini protein scaffold. *Acta Chim. Slovenica* **2011**, *58*, 693–701.
30. Ibanez-Tallon, I.; Miwa, J.M.; Wang, H.L.; Adams, N.C.; Crabtree, G.W.; Sine, S.M.; Heintz, N. Novel modulation of neuronal nicotinic acetylcholine receptors by association with the endogenous prototoxin lynx1. *Neuron* **2002**, *33*, 893–903. [CrossRef]
31. Chimienti, F.; Hogg, R.C.; Plantard, L.; Lehmann, C.; Brakch, N.; Fischer, J.; Huber, M.; Bertrand, D.; Hohl, D. Identification of slurp-1 as an epidermal neuromodulator explains the clinical phenotype of mal de meleda. *Hum. Mol. Genet.* **2003**, *12*, 3017–3024. [CrossRef] [PubMed]
32. Kalia, J.; Milescu, M.; Salvatierra, J.; Wagner, J.; Klint, J.K.; King, G.F.; Olivera, B.M.; Bosmans, F. From foe to friend: Using animal toxins to investigate ion channel function. *J. Mol. Biol.* **2015**, *427*, 158–175. [CrossRef] [PubMed]
33. Mouhat, S.; Andreotti, N.; Jouirou, B.; Sabatier, J.-M. Animal toxins acting on voltage-gated potassium channels. *Curr. Pharm. Des.* **2008**, *14*, 2503–2518. [CrossRef] [PubMed]
34. Norton, R.S. Structure and function of peptide and protein toxins from marine organisms. *J. Toxicol. Toxin Rev.* **1998**, *17*, 99–130. [CrossRef]
35. Terlau, H.; Olivera, B.M. Conus venoms: A rich source of novel ion channel-targeted peptides. *Physiol. Rev.* **2004**, *84*, 41–68. [CrossRef] [PubMed]
36. Quintero-Hernández, V.; Jiménez-Vargas, J.; Gurrola, G.; Valdivia, H.; Possani, L. Scorpion venom components that affect ion-channels function. *Toxicon* **2013**, *76*, 328–342. [CrossRef] [PubMed]
37. Bohlen, C.J.; Chesler, A.T.; Sharif-Naeini, R.; Medzihradszky, K.F.; Zhou, S.; King, D.; Sánchez, E.E.; Burlingame, A.L.; Basbaum, A.I.; Julius, D. A heteromeric texas coral snake toxin targets acid-sensing ion channels to produce pain. *Nature* **2011**, *479*, 410. [CrossRef] [PubMed]
38. Guo, M.; Teng, M.; Niu, L.; Liu, Q.; Huang, Q.; Hao, Q. Crystal structure of the cysteine-rich secretory protein stecrisp reveals that the cysteine-rich domain has a K$^+$ channel inhibitor-like fold. *J. Biol. Chem.* **2005**, *280*, 12405–12412. [CrossRef] [PubMed]
39. Gibbs, G.M.; Orta, G.; Reddy, T.; Koppers, A.J.; Martínez-López, P.; de la Vega-Beltràn, J.L.; Lo, J.C.; Veldhuis, N.; Jamsai, D.; McIntyre, P. Cysteine-rich secretory protein 4 is an inhibitor of transient receptor potential m8 with a role in establishing sperm function. *Proc. Natl. Acad. Sci. USA* **2011**, *108*, 7034–7039. [CrossRef] [PubMed]
40. Diochot, S.; Salinas, M.; Baron, A.; Escoubas, P.; Lazdunski, M. Peptides inhibitors of acid-sensing ion channels. *Toxicon* **2007**, *49*, 271–284. [CrossRef] [PubMed]
41. Mouhat, S.; Jouirou, B.; Mosbah, A.; De Waard, M.; Sabatier, J.-M. Diversity of folds in animal toxins acting on ion channels. *Biochem. J.* **2004**, *378*, 717–726. [CrossRef] [PubMed]
42. Ohno, M.; Menez, R.; Ogawa, T.; Danse, J.M.; Shimohigashi, Y.; Fromen, C.; Ducancel, F.; Zinn-Justin, S.; Le Du, M.H.; Boulain, J.C.; et al. Molecular evolution of snake toxins: Is the functional diversity of snake toxins associated with a mechanism of accelerated evolution? *Prog. Nucl. Acid Res. Mol. Biol.* **1998**, *59*, 307–364.
43. Chang, L.-S. Genetic diversity in snake venom three-finger proteins and phospholipase a2 enzymes. *Toxin Rev.* **2007**, *26*, 143–167. [CrossRef]
44. Casewell, N.R.; Wagstaff, S.C.; Harrison, R.A.; Renjifo, C.; Wüster, W. Domain loss facilitates accelerated evolution and neofunctionalization of duplicate snake venom metalloproteinase toxin genes. *Mol. Biol. Evol.* **2011**, *28*, 2637–2649. [CrossRef] [PubMed]
45. Banerjee, A.; Lee, A.; Campbell, E.; MacKinnon, R. Structure of a pore-blocking toxin in complex with a eukaryotic voltage-dependent K$^+$ channel. *Elife* **2013**, *2*, e00594. [CrossRef] [PubMed]
46. Strix, G. A toxin against pain. *Sci. Am.* **2005**, *292*, 88–93. [CrossRef]
47. Góngora-Benítez, M.; Tulla-Puche, J.; Albericio, F. Multifaceted roles of disulfide bonds. Peptides as therapeutics. *Chem. Rev.* **2013**, *114*, 901–926. [CrossRef] [PubMed]
48. Herzig, V.; King, G.F. The cystine knot is responsible for the exceptional stability of the insecticidal spider toxin ω-hexatoxin-hv1a. *Toxins* **2015**, *7*, 4366–4380. [CrossRef] [PubMed]
49. Kuzmenkov, A.I.; Fedorova, I.M.; Vassilevski, A.A.; Grishin, E.V. Cysteine-rich toxins from lachesana tarabaevi spider venom with amphiphilic c-terminal segments. *Biochim. Biophys. Acta* **2013**, *1828*, 724–731. [CrossRef] [PubMed]

50. Lavergne, V.; Alewood, P.F.; Mobli, M.; King, G.F. The structural universe of disulfide-rich venom peptides. In *Venoms to Drugs: Venoms as a Source for the Development of Human Therapeutics*; Royal Society of Chemistry: London, UK, 2015.

51. Avrutina, O. Synthetic cystine-knot miniproteins—Valuable scaffolds for polypeptide engineering. *Adv. Exp. Med. Biol.* **2016**, *917*, 121–144. [PubMed]

52. Rappoport, N.; Karsenty, S.; Stern, A.; Linial, N.; Linial, M. Protonet 6.0: Organizing 10 million protein sequences in a compact hierarchical family tree. *Nucl. Acids Res.* **2012**, *40*, D313–D320. [CrossRef] [PubMed]

53. Ofer, D.; Rappoport, N.; Linial, M. The little known universe of short proteins in insects: A machine learning approach. In *Short Views on Insect Genomics and Proteomics*; Springer: Berlin, Germany, 2015; pp. 177–202.

54. Werren, J.H.; Richards, S.; Desjardins, C.A.; Niehuis, O.; Gadau, J.; Colbourne, J.K.; Group, N.G.W. Functional and evolutionary insights from the genomes of three parasitoid nasonia species. *Science* **2010**, *327*, 343–348. [CrossRef] [PubMed]

55. Nygaard, S.; Zhang, G.; Schiøtt, M.; Li, C.; Wurm, Y.; Hu, H.; Zhou, J.; Ji, L.; Qiu, F.; Rasmussen, M. The genome of the leaf-cutting ant acromyrmex echinatior suggests key adaptations to advanced social life and fungus farming. *Genome Res.* **2011**, *21*, 1339–1348. [CrossRef] [PubMed]

56. Rappoport, N.; Linial, M. Trends in genome dynamics among major orders of insects revealed through variations in protein families. *BMC Genom.* **2015**, *16*, 583. [CrossRef] [PubMed]

57. Naamati, G.; Askenazi, M.; Linial, M. Clantox: A classifier of short animal toxins. *Nucleic Acids Res.* **2009**, *37*, W363–W368. [CrossRef] [PubMed]

58. Radivojac, P.; Clark, W.T.; Oron, T.R.; Schnoes, A.M.; Wittkop, T.; Sokolov, A.; Graim, K.; Funk, C.; Verspoor, K.; Ben-Hur, A. A large-scale evaluation of computational protein function prediction. *Nat. Methods* **2013**, *10*, 221–227. [CrossRef] [PubMed]

59. Kaplan, N.; Linial, M. Automatic detection of false annotations via binary property clustering. *BMC Bioinform.* **2005**, *6*, 46. [CrossRef] [PubMed]

60. Ofer, D.; Linial, M. Neuropid: A predictor for identifying neuropeptide precursors from metazoan proteomes. *Bioinformatics* **2013**, *30*, 931–940. [CrossRef] [PubMed]

61. Tirosh, Y.; Linial, I.; Askenazi, M.; Linial, M. Short toxin-like proteins abound in cnidaria genomes. *Toxins* **2012**, *4*, 1367–1384. [CrossRef] [PubMed]

62. Tassanakajon, A.; Somboonwiwat, K.; Amparyup, P. Sequence diversity and evolution of antimicrobial peptides in invertebrates. *Dev. Comp. Immunol.* **2015**, *48*, 324–341. [CrossRef] [PubMed]

63. Liu, Z.; Yuan, K.; Zhang, R.; Ren, X.; Liu, X.; Zhao, S.; Wang, D. Cloning and purification of the first termicin-like peptide from the cockroach eupolyphaga sinensis. *J. Venom. Anim. Toxins Incl. Trop. Dis.* **2016**, *22*, 5. [CrossRef] [PubMed]

64. Fjell, C.D.; Hiss, J.A.; Hancock, R.E.; Schneider, G. Designing antimicrobial peptides: Form follows function. *Nat. Rev. Drug Discov.* **2011**, *11*, 37–51. [CrossRef] [PubMed]

65. Froy, O.; Gurevitz, M. Arthropod defensins illuminate the divergence of scorpion neurotoxins. *J. Pept. Sci.* **2004**, *10*, 714–718. [CrossRef] [PubMed]

66. Froy, O.; Gurevitz, M. New insight on scorpion divergence inferred from comparative analysis of toxin structure, pharmacology and distribution. *Toxicon* **2003**, *42*, 549–555. [CrossRef]

67. Bun Ng, T.; Chi Fai Cheung, R.; Ho Wong, J.; Juan Ye, X. Antimicrobial activity of defensins and defensin-like peptides with special emphasis on those from fungi and invertebrate animals. *Curr. Protein Pept. Sci.* **2013**, *14*, 515–531.

68. Whittington, C.M.; Papenfuss, A.T.; Bansal, P.; Torres, A.M.; Wong, E.S.; Deakin, J.E.; Graves, T.; Alsop, A.; Schatzkamer, K.; Kremitzki, C. Defensins and the convergent evolution of platypus and reptile venom genes. *Genome Res.* **2008**, *18*, 986–994. [CrossRef] [PubMed]

69. Varkey, J.; Singh, S.; Nagaraj, R. Antibacterial activity of linear peptides spanning the carboxy-terminal beta-sheet domain of arthropod defensins. *Peptides* **2006**, *27*, 2614–2623. [CrossRef] [PubMed]

70. Zhu, S.; Li, W.; Jiang, D.; Zeng, X. Evidence for the existence of insect defensin-like peptide in scorpion venom. *IUBMB Life* **2000**, *50*, 57–61. [PubMed]

71. Gao, B.; Zhu, S. The drosomycin multigene family: Three-disulfide variants from drosophila takahashii possess antibacterial activity. *Sci. Rep.* **2016**, *6*, 32175. [CrossRef] [PubMed]

72. Deng, X.J.; Yang, W.Y.; Huang, Y.D.; Cao, Y.; Wen, S.Y.; Xia, Q.Y.; Xu, P. Gene expression divergence and evolutionary analysis of the drosomycin gene family in drosophila melanogaster. *J. Biomed. Biotechnol.* **2009**, *2009*, 315423. [CrossRef] [PubMed]

73. Li, H.; Su, M.; Hamann, M.T.; Bowling, J.J.; Kim, H.S.; Jung, J.H. Solution structure of a sponge-derived cystine knot peptide and its notable stability. *J. Nat. Prod.* **2014**, *77*, 304–310. [CrossRef] [PubMed]

74. Ovchinnikova, T.V.; Balandin, S.V.; Aleshina, G.M.; Tagaev, A.A.; Leonova, Y.F.; Krasnodembsky, E.D.; Men'shenin, A.V.; Kokryakov, V.N. Aurelin, a novel antimicrobial peptide from jellyfish aurelia aurita with structural features of defensins and channel-blocking toxins. *Biochem. Biophys. Res. Commun.* **2006**, *348*, 514–523. [CrossRef] [PubMed]

75. Cohen, L.; Moran, Y.; Sharon, A.; Segal, D.; Gordon, D.; Gurevitz, M. Drosomycin, an innate immunity peptide of drosophila melanogaster, interacts with the fly voltage-gated sodium channel. *J. Biol. Chem.* **2009**, *284*, 23558–23563. [CrossRef] [PubMed]

76. Stehling, E.G.; Sforca, M.L.; Zanchin, N.I.; Oyama, S., Jr.; Pignatelli, A.; Belluzzi, O.; Polverini, E.; Corsini, R.; Spisni, A.; Pertinhez, T.A. Looking over toxin-k(+) channel interactions. Clues from the structural and functional characterization of alpha-ktx toxin tc32, a kv1.3 channel blocker. *Biochemistry* **2012**, *51*, 1885–1894. [CrossRef] [PubMed]

77. Deuis, J.R.; Dekan, Z.; Wingerd, J.S.; Smith, J.J.; Munasinghe, N.R.; Bhola, R.F.; Imlach, W.L.; Herzig, V.; Armstrong, D.A.; Rosengren, K.J.; et al. Pharmacological characterisation of the highly nav1.7 selective spider venom peptide pn3a. *Sci. Rep.* **2017**, *7*, 40883. [CrossRef] [PubMed]

78. Jablonsky, M.J.; Jackson, P.L.; Krishna, N.R. Solution structure of an insect-specific neurotoxin from the new world scorpion centruroides sculpturatus ewing. *Biochemistry* **2001**, *40*, 8273–8282. [CrossRef] [PubMed]

79. Krimm, I.; Gilles, N.; Sautiere, P.; Stankiewicz, M.; Pelhate, M.; Gordon, D.; Lancelin, J.M. Nmr structures and activity of a novel alpha-like toxin from the scorpion leiurus quinquestriatus hebraeus. *J. Mol. Biol.* **1999**, *285*, 1749–1763. [CrossRef] [PubMed]

80. Mourao, C.B.; Schwartz, E.F. Protease inhibitors from marine venomous animals and their counterparts in terrestrial venomous animals. *Mar. Drugs* **2013**, *11*, 2069–2112. [CrossRef] [PubMed]

81. Boutet, E.; Lieberherr, D.; Tognolli, M.; Schneider, M.; Bansal, P.; Bridge, A.J.; Poux, S.; Bougueleret, L.; Xenarios, I. Uniprotkb/swiss-prot, the manually annotated section of the uniprot knowledgebase: How to use the entry view. In *Plant Bioinformatics: Methods and Protocols*; Springer: Berlin, Germany, 2016; pp. 23–54.

82. Bienert, S.; Waterhouse, A.; de Beer, T.A.; Tauriello, G.; Studer, G.; Bordoli, L.; Schwede, T. The swiss-model repository-new features and functionality. *Nucleic Acids Res.* **2017**, *45*, D313–D319. [CrossRef] [PubMed]

83. Rose, P.W.; Prlić, A.; Altunkaya, A.; Bi, C.; Bradley, A.R.; Christie, C.H.; Costanzo, L.D.; Duarte, J.M.; Dutta, S.; Feng, Z. The rcsb protein data bank: Integrative view of protein, gene and 3d structural information. *Nucleic Acids Res.* **2016**, *45*, D271–D281. [PubMed]

84. Petersen, T.N.; Brunak, S.; von Heijne, G.; Nielsen, H. Signalp 4.0: Discriminating signal peptides from transmembrane regions. *Nat. Methods* **2011**, *8*, 785–786. [CrossRef] [PubMed]

85. Remmert, M.; Biegert, A.; Hauser, A.; Söding, J. Hhblits: Lightning-fast iterative protein sequence searching by hmm-hmm alignment. *Nat. Methods* **2012**, *9*, 173–175. [CrossRef] [PubMed]

86. Naamati, G.; Askenazi, M.; Linial, M. A predictor for toxin-like proteins exposes cell modulator candidates within viral genomes. *Bioinformatics* **2010**, *26*, i482–i488. [CrossRef] [PubMed]

© 2017 by the authors. Licensee MDPI, Basel, Switzerland. This article is an open access article distributed under the terms and conditions of the Creative Commons Attribution (CC BY) license (http://creativecommons.org/licenses/by/4.0/).

Article

The Effects of Melittin and Apamin on Airborne Fungi-Induced Chemical Mediator and Extracellular Matrix Production from Nasal Polyp Fibroblasts

Seung-Heon Shin [1],*, Mi-Kyung Ye [1], Sung-Yong Choi [1] and Kwan-Kyu Park [2]

[1] Department of Otolaryngology-Head and Neck Surgery, School of Medicine, Catholic University of Daegu, Daegu 42472, Korea; miky@cu.ac.kr (M.-K.Y.); iroom-ent@naver.com (S.-Y.C.)

[2] Department of Pathology, School of Medicine, Catholic University of Daegu, Daegu 42472, Korea; kkpark@cu.ac.kr

* Correspondence: hsseungl@cu.ac.kr; Tel.: +82-53-650-4530; Fax: +82-53-650-4533

Academic Editor: Steve Peigneur
Received: 20 September 2017; Accepted: 25 October 2017; Published: 27 October 2017

Abstract: Melittin and apamin are the main components of bee venom and they have been known to have anti-inflammatory and anti-fibrotic properties. The aim of this study was to evaluate the effect of melittin and apamin on airborne fungi-induced chemical mediator and extracellular matrix (ECM) production in nasal fibroblasts. Primary nasal fibroblasts were isolated from nasal polyps, which were collected during endoscopic sinus surgery. Nasal fibroblasts were treated with *Alternaria* and *Aspergillus*. The effects of melittin and apamin on the production of interleukin (IL)-6 and IL-8 were determined with enzyme linked immunosorbent assay. ECM mRNA and protein expressions were determined with the use of quantitative RT-PCR and Western blot. *Alternaria*-induced IL-6 and IL-8 production was significantly inhibited by apamin. However, melittin did not influence the production of IL-6 and IL-8 from nasal fibroblasts. Melittin or apamin significantly inhibited collagen type I, TIMP-1, and MMP-9 mRNA expression and protein production from nasal fibroblasts. Melittin and apamin inhibited *Alternaria*-induced phosphorylation of Smad 2/3 and p38 MAPK. Melittin and apamin can inhibit the fungi-induced production of chemical mediators and ECM from nasal fibroblasts. These results suggest the possible role of melittin and apamin in the treatment of fungi induced airway inflammatory diseases.

Keywords: melittin; apamin; *Alternaria*; *Aspergillus*; nasal fibroblast; chemical mediator; extracellular matrix

1. Introduction

Nasal polyps are swellings and there is damage to the mucosal epithelium with inflammatory cell infiltration. Although the pathogenesis of nasal polyps is not fully understood, mucosal epithelial damage, extracellular matrix (ECM) accumulation, and increased local inflammatory mediators are the characteristic pathophysiologic findings of nasal polyps [1]. Fibroblasts are the main structural components of the nasal mucosa and they participate in the local immune response through the production of biological mediators which are involved in the recruitment of inflammatory cells and the cellular source of ECM proteins [2].

Fungi are ubiquitous saprophytes in nature and airway mucosa is exposed by inhalation of fungal spores. A fine immunologic balance to maintain a stable host–fungi relationship is important in order to maintain the physiologic condition, and disruption of the host immune response causes pathologic conditions of the airway. Fungi are commonly found in the nasal mucosa and relatively few species are implicated in airway inflammatory disease. *Alternaria* and *Aspergillus* are known common pathogens found in nasal secretion and they induce the production of chemical mediators from nasal epithelial cells and fibroblasts [3,4]. TLR2, TLR4, and TLR5 seem to be important pattern recognition receptors for fungi [3,5].

Bee venom (BV) has been used to treat several inflammatory diseases, such as rheumatoid arthritis, tendonitis, and Parkinson's disease [6,7]. BV is a complex mixture with peptides, enzymes, and biogenic amines with various pharmaceutical properties. Melittin and apamin are the main components of BV peptide. Melittin comprises about 40–50% of the dried BV and has antibacterial, antiviral, anti-inflammatory, and anti-fibrotic properties [8–10]. Apamin also has anti-inflammatory and anti-fibrotic properties in various cells [11].

Melittin and apamin have anti-fibrotic activities that suppress the pro-fibrotic gene and protein expression through the inhibition of TGF-βRII-Smad, ERK1/2, and JNK phosphorylation in rat kidney fibroblasts [10]. Melittin and apamin have immunomodulatory activities, and in this study, we evaluated the effect of melittin and apamin on airborne fungi induced chemical mediator and ECM production in nasal fibroblasts.

2. Results

2.1. The Cytotoxicity of BV, Melittin, and Apamin

MTT assay was used to determine the cytotoxicity of these three agents. The cells were treated with various concentrations of BV (0.1 to 5 μg/mL), melittin (0.1 to 5 μg/mL), and apamin (0.1 to 10 μg/mL) for 24 h. The viability of fibroblasts was significantly suppressed by BV at a concentration of 5 μg/mL ($55.8 \pm 6.8\%$), melittin at a concentration of 3 μg/mL ($82.6 \pm 8.3\%$), and apamin at a concentration of 10 μg/mL ($55.8 \pm 11.7\%$) (Figure 1). Based on these results, we used up to 3 μg/mL of BV, 1 μg/mL of melittin, and 5 μg/mL of apamin for further experiments.

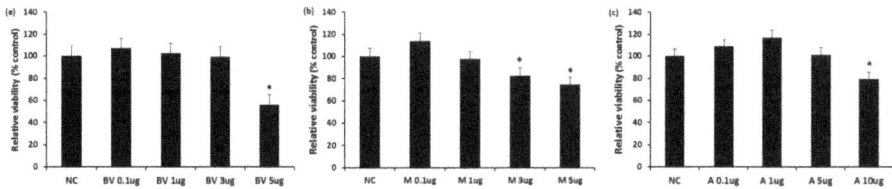

Figure 1. Effect of bee venom, melittin, and apamin on the proliferation of nasal fibroblasts. Nasal fibroblasts were treated with various concentrations of various concentrations of (**a**) bee venom; (**b**) melittina; and (**c**) apamin for 24 h. Values are expressed as the mean ± SE of four independent experiments. 5 μg/mL of BV, 3 μg/mL of melittin, and 10 μg/mL of apamin inhibited the proliferation of nasal fibroblasts. NC: negative control; BV: bee venom; M: melittin; A: apamin; μg: μg/mL; * $p < 0.05$.

2.2. The Effect of BV, Melittin, and Apamin on the Production of Chemical Mediators

When the fibroblasts were stimulated with *Alternaria*, IL-6 (6476.1 ± 352.4 pg/mL at a concentration of 50 μg/mL, 3368.1 ± 247.5 pg/mL at a concentration of 25 μg/mL) and IL-8 (7969.4 ± 690.2 pg/mL at a concentration of 50 μg/mL, 2399.0 ± 175.5 pg/mL at a concentration of 25 μg/mL) production was significantly increased compared with that in the non-stimulated group (IL-6; 1812.1 ± 93.6 pg/mL, IL-8; 1010.6 ± 132.4 pg/mL). However, *Aspergillus* did not significantly enhance the production of IL-6 and IL-8 from nasal fibroblasts. IL-6 and IL-8 production induced by *Alternaria* was significantly inhibited by BV and apamin in a dose dependent manner. However, melittin did not influence the production of IL-6 and IL-8 from nasal fibroblasts (Figure 2).

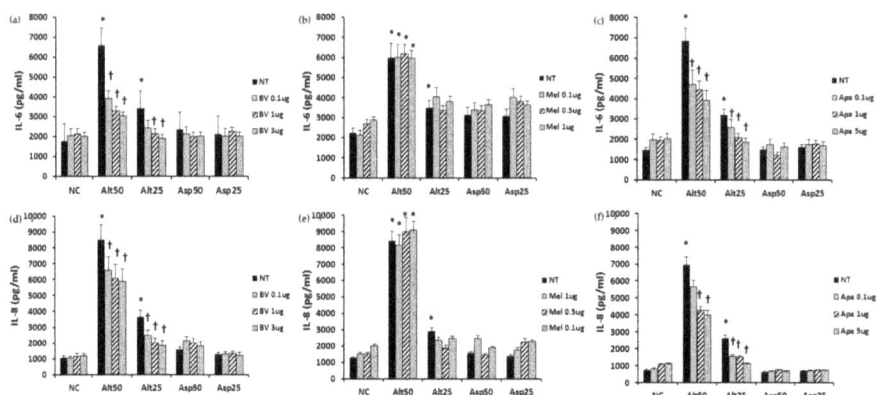

Figure 2. Effect of bee venom, melittin, and apamin on the production of IL-6 and IL-8 from nasal fibroblasts. Nasal fibroblasts were treated with *Alternaria* and *Aspergillus* for 24 h with or without various concentrations of these three agents. (**a**,**d**) *Alternaria* enhanced IL-6 and IL-8 production from nasal fibroblasts and the production of IL-6 and IL-8 was significantly inhibited by bee venom and (**c**,**f**) apamin; (**b**,**e**) Melittin did not influence the production of IL-6 and IL-8 from nasal fibroblasts. Alt 50: *Alternaria* 50 μg/mL; Asp 50: *Aspergillus* 50 μg/mL; NC: negative control; NT: non-treated; BV: bee venom; Mel: melittin; Apa: apamin, μg: μg/mL; * $p < 0.05$ compared with negative control; † $p < 0.05$ compared with the non-treated group.

2.3. The Effect of Melittin and Apamin on the Expression of ECM

Collagen type I mRNA and protein expression was significantly increased with 50 μg/mL of *Aspergillus*. *Aspergillus* induced collagen type I mRNA and protein expression was significantly suppressed when treated with melittin. Apamin suppressed both *Alternaria* and *Aspergillus* induced collagen type I mRNA and protein expression (Figure 3). When the nasal polyp fibroblasts were stimulated with *Alternaria* TIMP-1 mRNA and protein expressions were significantly increased. However, TIMP-1 mRNA and protein expressions were not significantly increased by stimulation with *Aspergillus*. TIMP-1 mRNA expression induced by *Alternaria* was significantly inhibited by apamin in a dose dependent manner, and TIMP-1 protein expression was also significantly inhibited by melittin and apamin in a dose dependent manner (Figure 4). *Alternaria* induced MMP-9 mRNA expression, but *Aspergillus* did not induce MMP-9 mRNA expression. *Alternaria* induced MMP-9 mRNA expression was significantly inhibited by meittin and apamin. Although fungi did not influence the production of MMP-9 protein, melittin and apamin tended to inhibit the production of MMP-9 from nasal polyp fibroblasts (Figure 5).

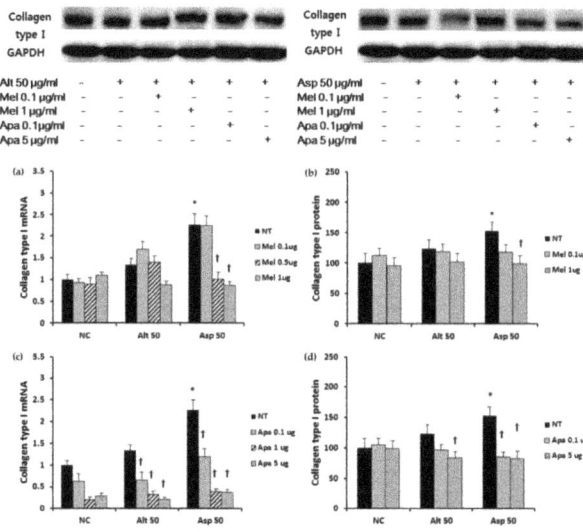

Figure 3. Effect of melittin and apamin on the expression of collagen type I mRNA and protein in nasal fibroblasts. *Aspergillus* induced collagen type I mRNA and protein expressions were significantly suppressed by (**a**,**b**) melittin and (**c**,**d**) apamin. Apamin also significantly suppressed collagen type I mRNA expression in negative control and the *Alternaria* stimulated group. Alt 50: *Alternaria* 50 μg/mL; Asp 50: *Aspergillus* 50 μg/mL; NC: negative control; NT: non-treated; Mel: melittin; Apa: apamin; μg: μg/mL; *: $p < 0.05$ compared with negative control; † $p < 0.05$ compared with the non-treated group.

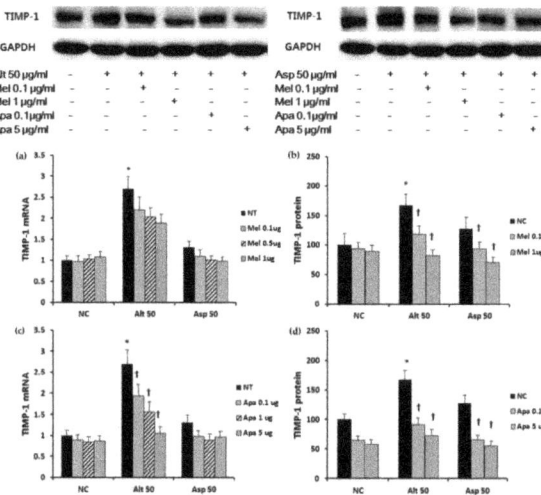

Figure 4. Effect of melittin and apamin on the expression of TIMP-1 mRNA and protein in nasal fibroblasts. *Alternaria* induced TIMP-1 protein production was significantly inhibited by (**b**) melittin and (**d**) apamin; (**c**) Apamin also significantly suppressed *Alternaria* induced TIMP-1 mRNA expression; (**a**) Melittin did no inhibit *Alternaria* induced TIMP-1 mRNA expression. Alt 50: *Alternaria* 50 μg/mL; Asp 50: *Aspergillus* 50 μg/mL; NC: negative control; NT: non-treated; Mel: melittin; Apa: apamin, μg: μg/mL; * $p < 0.05$ compared with negative control; † $p < 0.05$ compared with the non-treated group.

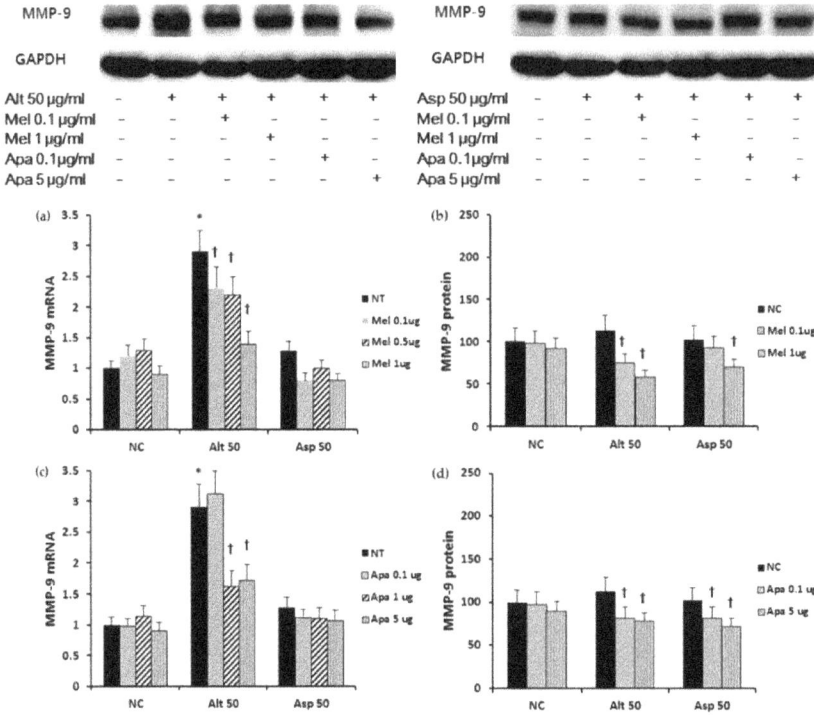

Figure 5. Effect of melittin and apamin on the expression of MMP-9 mRNA and protein in nasal fibroblasts. *Alternaria* induced MMP-9 mRNA expression was significantly inhibited by (**a**) melittin and (**c**) apamin. (**b**,**d**) Melittin and apamin tended to inhibit the production of MMP-9 protein. Alt 50: *Alternaria* 50 µg/mL; Asp 50: *Aspergillus* 50 µg/mL; NC: negative control; NT: non-treated; Mel: melittin; Apa: apamin; µg: µg/mL; * $p < 0.05$ compared with negative control; † $p < 0.05$ compared with the non-treated group.

2.4. Effect of Melittin and Apamin on Phosphorylation of Smad 2/3 and p38 MAPK

To determine the inhibitory mechanism of ECM production, we identified the effect of melittin and apamin on phosphorylation of Smad2/3 and p38 MAPK. Duration of treatment was 30 min for p-p38 and 60 min for pSmad 2/3. The densitometric quantification results showed that *Alternaria* potently induced the activation of Smad 2/3 and p38 MAPK. When the fibroblasts were treated with 1 µg/mL of melittin, Smad 2/3 (approximately 30.9%), and p38 MAPK (approximately 37.4%) expressions were significantly suppressed. Apamin also significantly suppressed Smad 2/3 (approximately 27.4% and 36/4% at 1 and 5 µg/mL of apamin) and p38 MAPK (approximately 29.2% and 34.7% at 1 and 5 µg/mL of apamin) expressions (Figure 6).

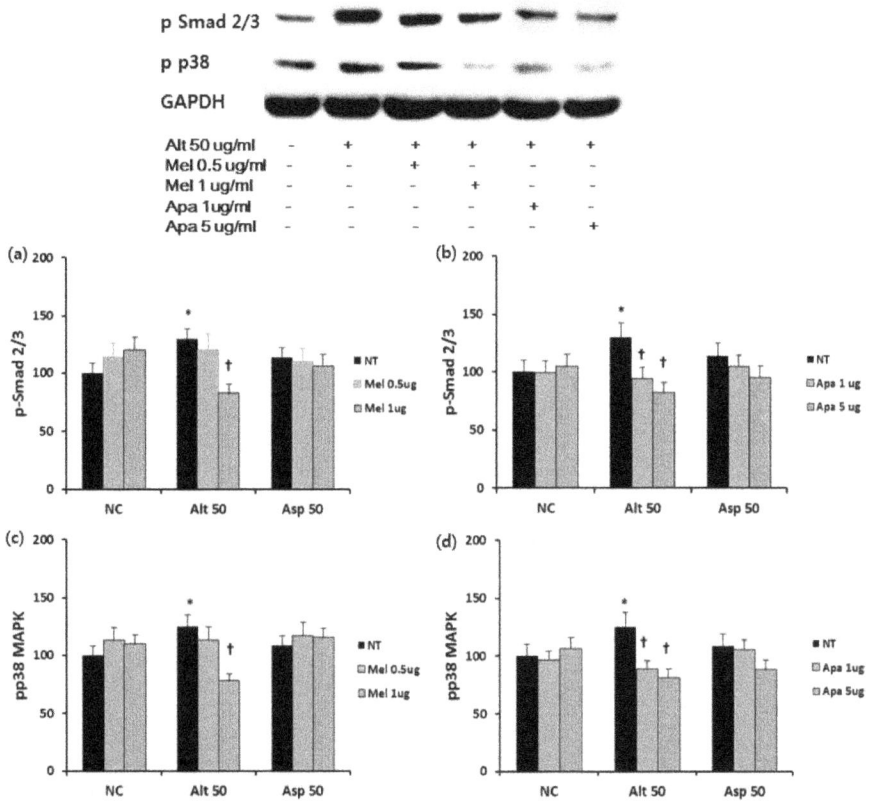

Figure 6. Effect of melittin and apamin on phosphorylation of Smad 2/3 and p38 MAPK expression in nasal fibroblasts. Phosphorylation of Smad 2/3 and p38 MAPK was measured using Western blotting and density analysis. *Alternaria* induced Smad 2/3 and p38 MAPK phosphorylation was significantly inhibited by (**a**,**c**) melittin and (**b**,**d**) apamin. Alt 50: *Alternaria* 50 μg/mL; Asp 50: *Aspergillus* 50 μg/mL; NC: negative control; NT: non-treated; Mel: melittin; Apa: apamin; μg: μg/mL; * $p < 0.05$ compared with negative control; † $p < 0.05$ compared with the non-treated group.

3. Discussion

BV has been used as a traditional medicine with satisfactory results for the treatment of some inflammatory, cancer, and immune related diseases [6,7,12]. Melittin and apamin are the most well-known components of BV. Melittin is the main component of BV with hyaluronidase and phospholipase A2 [13]. Melittin is the active component of apitoxin with antimicrobial, anti-inflammatory, and anti-atherosclerotic properties [8,9]. Apamin is a peptide neurotoxin with a crucial role in repetitive activities in neurons, inducing alpha-adrenergic, cholinergic, purinergic, and neurotensin-induced relaxation [7,14]. In this study, we tried to determine the effect of melittin and apamin on fungi induced chemical mediator and extraceullar matrix production from nasal fibroblasts.

Fungi have been associated with upper and lower airway inflammatory diseases and *Alternaria* and *Aspergillus* are commonly found in the nasal secretion and the respiratory tract [15]. Fungi can induce chemical mediator production through the interaction with toll-like receptors (TLRs) [3]. IL-6 and IL-8 productions were increased with *Alternaria* and production of these chemical mediators was significantly inhibited when the fibroblasts were treated with apamin. BV also significantly inhibited the production of IL-6 and IL-8. BV encompasses a mixture of many types of compounds, proteins, peptides, and enzymes. However, BV did not strongly inhibit chemical mediator production compared to apamin in nasal fibroblasts. IL-6 production was more strongly inhibited by BV (50.1% at 1 μg/mL of BV vs. 42.6% at 1 μg/mL of apamin), and IL-8 production was more strongly inhibited by apamin (28.8% at 1 μg/mL of BV vs. 38.7% at 1 μg/mL of apamin). Although we cannot explain the exact cause of this discrepancy, some components of BV may enhance the anti-inflammtory effect and the other components may suppress the anti-inflammatory effect of apamin. Anti-inflammatory properties of BV for treating skin disease, neurodegenerative disease, and joint diseases have been commonly studied, and these anti-inflammatory effects are associated with decreased expression of TLRs, chemical mediators, nitric oxides, and phosphorylation of inflammatory transcription factors [7]. According to the previous study, *Alternaria* induces the production of chemical mediators through TLR 2 and TLR5 [16]. Although we do not know the action mechanism of apamin and melittin, they may not influence the expression of TLR2 or 5 or the concentration of melittin used in this study may not enough to suppress the production of chemical mediators from nasal fibroblasts.

The damage to mucosal epithelium, ECM accumulation, and inflammatory cell infiltration are important pathologic findings of nasal polyps [2]. Fibroblasts are the cellular source of ECM and they are involved in the development of nasal polyps. The MMP-9 level was elevated in nasal polyps and the TIMP-1 level was elevated in chronic rhinosinusitis [17]. We performed a kinetic study with *Alternaria* and *Aspergillus* for 8, 24, and 48 h. Collagen type I, TIMP-1, and MMP-9 mRNA expression levels were the highest at 24 h stimulation. Expression levels of these ECM proteins were highest at 6 h after stimulation with fungi. However, fungi did not influence the fibronectin mRNA and protein expression in nasal fibroblasts [18]. Therefore, we evaluated collagen type I, TIMP-1, and MMP-9 mRNA expression in nasal fibroblasts at 24 h and their protein expression at 6 h after stimulation with fungi. In this study, the expression pattern of ECM was different depending on the type of fungi used for stimulation. These differences may be associated with the unique molecular pattern, the peptides, or the immune triggering components of fungi [19]. Also, different fungi may interact with different pattern recognition receptors, such as TLRs, protease activated receptors, or G-protein-coupled receptors. TLR2, TLR4, and TLR9 are the main TLRs involved in sensing the fungal components [20]. *Alternaria* and *Aspergillus* enhance the production of chemical mediators through TLR4 in nasal epithelial cells [21]. Kao et al. suggested that TLR4 triggering activates the MAPK signaling pathway, which cross-talks with the Smad2 cascade and promotes the production of ECM [22]. When the nasal fibroblasts were stimulated with fungi, collagens type I, TIMP-1, and MMP-9 mRNA and/or protein expressions were increased. Also, this study showed that *Alternaria* can induce Smad 2/3 hyperphosphorylation in a TGF-β independent manner and p38 MAPK hyperphosphorylation. Our results suggest that *Alternaria* can directly activate the Smad 2/3 cascade or indirectly induce phosphorylation of Smad 2/3 through the TLR4/MAPK signaling pathway.

Melittin and apamin have been known to inhibit ECM production and tissue fibrosis from kidney and liver [10,11,23]. In this study, the inhibition of ECM expression in nasal fibroblasts by melittin and apamin differed depending on the types of ECM mRNA and protein. Although melittin and apamin show anti-fibrotic properties, they have different pharmacological characteristics, chemical structures, and they may control different signaling pathways. Melittin and apamin showed different patterns in suppressing the pro-fibrotic activity in TGF-β treated fibroblasts. Melittin attenuates fibrogenesis by inhibiting NF-κB and AP-1 dependent collagen type I and MMP-9 expression [10,24]. Apamin attenuates fibrogenesis by inhibiting phosphorylated Smad 2/3 and Smad dependent ECM deposition [25]. Melittin and apamin inhibited *Alternaria* induced phosphorylation of Smad 2/3 and MAPK in nasal fibroblasts. Although we do not know whether melittin and apamin directly suppress phosphorylation of Smad 2/3, melittin and apamin can directly or indirectly inhibit the *Alternaria* induced Smad 2/3 cascade.

The principal finding of this study is the anti-inflammatory and anti-fibrotic effects of melittin and apamin. Fungi can induce production of chemical mediators and ECM deposition in nasal fibroblasts. The production of these chemical mediators and ECM production were inhibited by melittin and apamin. In particular, melittin and apamin inhibited ECM production through direct suppression of the Smad cascade or indirect inhibition of Smad 2/3 phosphorylation through the MAPK signaling pathway. These results suggest a novel pharmacological rationale for the treatment of fungi induced airway inflammatory diseases. Because BV contains melittin and apamin, BV may have strong anti-inflammatory and anti-fibrotic effects, and could be a good candidate as a therapeutic agent for airway inflammatory diseases.

4. Materials and Methods

4.1. Isolation of Primary Nasal Polyp Fibroblasts

Primary nasal fibroblasts were isolated from 11 patients (7 men and 4 women; 43.5 ± 8.2 years) with chronic rhinosinusitis with nasal polyps during endoscopic sinus surgery. Subjects were excluded if they had an active inflammation, allergy, or aspirin hypersensitivity, had received antibiotics, antihistamine, steroids, or other medications for at least four weeks preceding the surgery. Allergy status was defined using the skin prick test. The study was approved by the Institutional Review Board of Daegu Catholic University Medical Center. A duly completed written informed consent form that outlined the objectives of the research and experiments was obtained from each patient.

The tissues were cut into 0.3 to 0.5 mm fragments and washed with phosphate buffered saline. These tissues were suspended and cultured in Dulbecco's Modified Eagle's Medium F-12 (DMEM/F-12) (Gibco, Grand Island, NY, USA) that contained 10% fetal bovine serum, penicillin at 100 U/mL, streptomycin at 100 µg/mL, and amphotericin B at 1.5 µg/mL at 37 °C and 5% CO_2. The second to third passages of fibroblasts were used for this experiment.

4.2. The Cytotoxic Effect of BV, Melittin, and Apamin on Nasal Polyp Fibroblasts

The cytotoxic effect of BV (melittin comprise approximately 50% and apamin comprise 3% of dried BV) (Chung Jin Biotech Co., Ansan, Korea) [26], melittin (Enzo Life Sciences AG, Lausen, Switzerland), and apamin (Sigma-Aldrich, St. Louis, Mo, USA) was evaluated using a CellTiter-96® aqueous cell proliferation assay kit (Promega, Madison, WI, USA). On a 96-well microstate plate, NP fibroblasts were cultured in the presence of 0.1, 1, 3, and 5 µg/mL of BV, 0.1, 1, 3, and 5 µg/mL of melittin, and 0.1, 1, 5, and 10 µg/mL of apamin for 24 h at 37 °C in a 5% CO_2. The reduced tetrazolium compound produces a colored formazan product due to the mitochondrial activity in the cell. The amount of formazan is directly proportional to the number of viable cells. Color intensities were assessed with a fluorescence microplate reader at 490 nm.

4.3. The Effect of Bee Venom, Melittin, and Apamin on IL-6 and IL-8 Production from Nasal Polyp Fibroblasts

The fibroblasts were incubated with endotoxin removed *Alternaria alternate* and *Aspergillus fumigatus* at 50 and 25 µg/mL, respectively (Greer Lab, Lenoir, NC, USA). After 24 h of stimulation, the cell culture supernatants and cells were harvested and stored at -70 °C until they were assayed. To determine the effect of melittin and apamin on the production of Interleukin (IL)-6 and IL-8, fibroblasts were incubated with or without various concentrations of melittin and apamin. IL-6 and IL-8 were quantified by using commercially available ELISA kits (R&D system, Minneapolis, MN, USA).

4.4. Real Time Reverse Transcription–Polymerase Chain Reaction (RT-PCR) for ECM mRNA from Nasal Polyp Fibroblasts

Fibroblasts were exposed to fungi with or without melittin and apamin for 24 h. The total RNA was extracted from fibroblasts with Trizol reagent (Invitrogen, Carlsbad, CA, USA) according to the manufacturer's instructions. Total RNA, 1 µg, was reverse transcribed using SuPrimeScript RT Premix (Genetbio Inc., Daejeon, Korea) and Quantitative PCR was then carried out on a mini opticon system (Bio-Rad Lab., Hercules, CA, USA) according to the manufacturer's protocol. The forward and reverse primers were as follows: β-actin, 5-ACAGGAAGTCCCTTGCCATC-3 and 5-AGGGAGACCAAAAGCCTTCA-3; α-SMA, 5-ATAGAACATGGCATCATCACCAAC-3, and 5-GGGCAACACGAAGCTCATTGTA-3; fibronectin, 5-GCCAGATGATGAGCTGCAC-3, and 5-GAGCAAATGGCACCGAGATA-3; tissue inhibitors of matrix metalloproteinase-1 (TIMP-1) 5-CCTTATACCAGCGTTATGAGATCAA-3 and 5-AGTGATGTGCAAGAGTCCATCC-3; and matrix metalloproteinase-9 (MMP-9), 5-ATTTCTGCCAGGACCGCTTCTACT-3, and 5-CAGTTTGTATCCGGCAAACTGGCT-3. The cDNA was amplified with initial denaturation at 95 °C for 10 min, followed PCR for 40 cycles of 95 °C for 5 s, 58 °C for 30 s, and finally one cycle of melting curve following cooling at 60 °C for 60 s. To confirm the amplification specificity, the PCR products from each primer pair were subjected to a melting curve analysis. Analysis of relative gene expression was performed by evaluating q-RT-PCR data by the 2(-DDCt) method. The gene expression levels were determined by normalization relative to β-actin expression.

4.5. Western Blot Analysis of Nasal Polyp Fibroblasts

Nasal fibroblast lysates were subjected to sodium dodecyl sulfate polyacrylamide gel electrophoresis and transferred onto NC membranes (Bio-Rad, Berkeley, CA, USA). Membranes were blocked with 5% skim milk solution and they were incubated with antibodies against MMP-9 (Cell signaling, Beverly, MA, USA), collagen type I, fibronectin, TIMP-1, phosphorylated Smad (pSmad) 2/3, p38 mitogen-activated protein kinase (MAPK), and GAPDH (Santa Cruz Biotechnology, Santa Cruz, CA, USA). After incubation for 1 h, the membranes were washed and then treated with peroxidase-conjugated anti-rabbit immunoglobulin G (Santa Cruz Biotechnology). Bands were visualized using horseradish peroxidase conjugated secondary antibodies and an ECL system (Pierce, Rockford, IL, USA). The band densities were measured using the multi Gauge v.2.02 software (Fujifilm, Tokyo, Japan). The band intensities were expressed as a percentage of treated cells versus untreated cells.

4.6. Statistical Analysis

The experimental data are presented as mean $\pm$ SE. The statistical significance of the differences between control and experimental data was analyzed using paired or unpaired Student's *t*-test and one-way analysis of variance followed by Tukey's test (SPSS ver. 21.0, SPSS Inc., Chicago, IL, USA). *p* value < 0.05 was considered to indicate a statistically significant difference. All results were obtained from at least four independent individuals and every experiment was performed in duplication.

Acknowledgments: This work was carried out with the support of "Cooperative Research Program for Agriculture Science & Technology Development (Project No. PJ01132501)" Rural Development Administration, Republic of Korea. This research was supported by the Basic Science Research Program through the National Research Foundation of Korea (NRF) funded by the Ministry of Education, Science and Technology (2010-0023163).

Author Contributions: Seung-Heon Shin conceived and designed the experiments; Mi-Kyung Ye and Sung-Yong Choi performed the experiments; Seung-Heon Shin and Kwan-Kyu Park analyzed the data; Mi-Kyung Ye contributed reagents/materials/analysis tools; Seung-Heon Shin and Sung-Yong Choi wrote the paper.

Conflicts of Interest: The authors declare no conflict of interest.

References

1. Cho, J.S.; Moon, Y.M.; Park, I.H.; Um, J.Y.; Moon, J.H.; Park, S.J.; Lee, S.H.; Kang, H.J.; Lee, H.M. Epigenetic regulation of myofibroblast differentiation and extracellular matrix production in nasal polyp-derived fibroblasts. *Clin. Exp. Allergy J. Br. Soc. Allergy Clin. Immunol.* **2012**, *42*, 872–882. [CrossRef] [PubMed]
2. Nakagawa, T.; Yamane, H.; Nakai, Y.; Shigeta, T.; Takashima, T.; Takeda, Z. Comparative assessment of cell proliferation and accumulation of extracellular matrix in nasal polyps. *Acta Oto-Laryngol. Suppl.* **1998**, *538*, 205–208. [CrossRef]

3.   Shin, S.H.; Kim, Y.H.; Jin, H.S.; Kang, S.H. *Alternaria* induces production of thymic stromal lymphopoietin in nasal fibroblasts through toll-like receptor 2. *Allergy Asthma Immunol. Res.* **2016**, *8*, 63–68. [CrossRef] [PubMed]

4.   Gao, F.S.; Cao, T.M.; Gao, Y.Y.; Liu, M.J.; Liu, Y.Q.; Wang, Z. Effects of chronic exposure to *Aspergillus fumigatus* on epidermal growth factor receptor expression in the airway epithelial cells of asthmatic rats. *Exp. Lung Res.* **2014**, *40*, 298–307. [CrossRef] [PubMed]

5.   Roeder, A.; Kirschning, C.J.; Rupec, R.A.; Schaller, M.; Korting, H.C. Toll-like receptors and innate antifungal responses. *Trends Microbiol.* **2004**, *12*, 44–49. [CrossRef] [PubMed]

6.   Kim, J.I.; Yang, E.J.; Lee, M.S.; Kim, Y.S.; Huh, Y.; Cho, I.H.; Kang, S.; Koh, H.K. Bee venom reduces neuroinflammation in the mptp-induced model of parkinson's disease. *Int. J. Neurosci.* **2011**, *121*, 209–217. [CrossRef] [PubMed]

7.   Moreno, M.; Giralt, E. Three valuable peptides from bee and wasp venoms for therapeutic and biotechnological use: Melittin, apamin and mastoparan. *Toxins* **2015**, *7*, 1126–1150. [CrossRef] [PubMed]

8.   Choi, J.H.; Jang, A.Y.; Lin, S.; Lim, S.; Kim, D.; Park, K.; Han, S.M.; Yeo, J.H.; Seo, H.S. Melittin, a honeybee venomderived antimicrobial peptide, may target methicillinresistant staphylococcus aureus. *Mol. Med. Rep.* **2015**, *12*, 6483–6490. [CrossRef] [PubMed]

9.   Lee, G.; Bae, H. Anti-inflammatory applications of melittin, a major component of bee venom: Detailed mechanism of action and adverse effects. *Molecules* **2016**, *21*, 616. [CrossRef] [PubMed]

10. Park, S.H.; Cho, H.J.; Jeong, Y.J.; Shin, J.M.; Kang, J.H.; Park, K.K.; Choe, J.Y.; Park, Y.Y.; Bae, Y.S.; Han, S.M.; et al. Melittin inhibits tgf-beta-induced pro-fibrotic gene expression through the suppression of the tgfbetarii-smad, erk1/2 and jnk-mediated signaling pathway. *Am. J. Chin. Med.* **2014**, *42*, 1139–1152. [CrossRef] [PubMed]

11. Kim, J.Y.; An, H.J.; Kim, W.H.; Park, Y.Y.; Park, K.D.; Park, K.K. Apamin suppresses biliary fibrosis and activation of hepatic stellate cells. *Int. J. Mol. Med.* **2017**, *39*, 1188–1194. [CrossRef] [PubMed]

12. Huh, J.E.; Baek, Y.H.; Lee, M.H.; Choi, D.Y.; Park, D.S.; Lee, J.D. Bee venom inhibits tumor angiogenesis and metastasis by inhibiting tyrosine phosphorylation of vegfr-2 in llc-tumor-bearing mice. *Cancer Lett.* **2010**, *292*, 98–110. [CrossRef] [PubMed]

13. Bilo, B.M.; Rueff, F.; Mosbech, H.; Bonifazi, F.; Oude-Elberink, J.N. Diagnosis of hymenoptera venom allergy. *Allergy* **2005**, *60*, 1339–1349. [CrossRef] [PubMed]

14. Lazdunski, M.; Fosset, M.; Hughes, M.; Mourre, C.; Romey, G.; Schmid-Antomarchi, H. The apamin-sensitive Ca^{2+}-dependent K^+ channel molecular properties, differentiation and endogenous ligands in mammalian brain. *Biochem. Soc. Symp.* **1985**, *50*, 31–42. [PubMed]

15. Shin, S.H.; Ye, M.K.; Lee, Y.H. Fungus culture of the nasal secretion of chronic rhinosinusitis patients: Seasonal variations in Daegu, Korea. *Am. J. Rhinol.* **2007**, *21*, 556–559. [CrossRef] [PubMed]

16. Shin, S.H.; Ye, M.K.; Kim, Y.H.; Kim, J.K. Role of TLRs in the production of chemical mediators in nasal polyp fibroblasts by fungi. *Auris Nasus Larynx* **2016**, *43*, 166–170. [CrossRef] [PubMed]

17. Watelet, J.B.; Bachert, C.; Claeys, C.; Van Cauwenberge, P. Matrix metalloproteinases mmp-7, mmp-9 and their tissue inhibitor timp-1: Expression in chronic sinusitis vs. nasal polyposis. *Allergy* **2004**, *59*, 54–60. [CrossRef] [PubMed]

18. Shin, S.H.; Ye, M.K.; Choi, S.Y.; Kim, Y.H. Effect of eosinophils activated with *Alternaria* on the production of extracellular matrix from nasal fibroblasts. *Ann. Allergy Asthma Immunol. Off. Publ. Am. Coll. Allergy Asthma Immunol.* **2016**, *116*, 559–564. [CrossRef] [PubMed]

19. Matsuwaki, Y.; Wada, K.; White, T.A.; Benson, L.M.; Charlesworth, M.C.; Checkel, J.L.; Inoue, Y.; Hotta, K.; Ponikau, J.U.; Lawrence, C.B.; et al. Recognition of fungal protease activities induces cellular activation and eosinophil-derived neurotoxin release in human eosinophils. *J. Immunol. (1950)* **2009**, *183*, 6708–6716. [CrossRef] [PubMed]

20. Chen, K.; Huang, J.; Gong, W.; Iribarren, P.; Dunlop, N.M.; Wang, J.M. Toll-like receptors in inflammation, infection and cancer. *Int. Immunopharmacol.* **2007**, *7*, 1271–1285. [CrossRef] [PubMed]

21. Shin, S.H.; Lee, Y.H. Airborne fungi induce nasal polyp epithelial cell activation and toll-like receptor expression. *Int. Arch. Allergy Immunol.* **2010**, *153*, 46–52. [CrossRef] [PubMed]

22. Kao, Y.H.; Chen, P.H.; Wu, T.Y.; Lin, Y.C.; Tsai, M.S.; Lee, P.H.; Tai, T.S.; Chang, H.R.; Sun, C.K. Lipopolysaccharides induce smad2 phosphorylation through pi3k/akt and mapk cascades in hsc-t6 hepatic stellate cells. *Life Sci.* **2017**, *184*, 37–46. [CrossRef] [PubMed]

23. An, H.J.; Kim, K.H.; Lee, W.R.; Kim, J.Y.; Lee, S.J.; Pak, S.C.; Han, S.M.; Park, K.K. Anti-fibrotic effect of natural toxin bee venom on animal model of unilateral ureteral obstruction. *Toxins* **2015**, *7*, 1917–1928. [CrossRef] [PubMed]

24. Park, J.H.; Jeong, Y.J.; Park, K.K.; Cho, H.J.; Chung, I.K.; Min, K.S.; Kim, M.; Lee, K.G.; Yeo, J.H.; Park, K.K.; et al. Melittin suppresses pma-induced tumor cell invasion by inhibiting nf-kappab and ap-1-dependent mmp-9 expression. *Mol. Cells* **2010**, *29*, 209–215. [CrossRef] [PubMed]

25. Lee, W.R.; Kim, K.H.; An, H.J.; Kim, J.Y.; Lee, S.J.; Han, S.M.; Pak, S.C.; Park, K.K. Apamin inhibits hepatic fibrosis through suppression of transforming growth factor beta1-induced hepatocyte epithelial-mesenchymal transition. *Biochem. Biophys. Res. Commun.* **2014**, *450*, 195–201. [CrossRef] [PubMed]

26. Lee, W.R.; Pak, S.C.; Park, K.K. The protective effect of Bee venom on Fibrosis causing inflammatory diseases. *Toxins* **2015**, *16*, 4758–4772. [CrossRef] [PubMed]

© 2017 by the authors. Licensee MDPI, Basel, Switzerland. This article is an open access article distributed under the terms and conditions of the Creative Commons Attribution (CC BY) license (http://creativecommons.org/licenses/by/4.0/).

toxins

MDPI

Article

Anticoagulant Activity of Low-Molecular Weight Compounds from *Heterometrus laoticus* Scorpion Venom

Thien Vu Tran [1,2], Anh Ngoc Hoang [1], Trang Thuy Thi Nguyen [3], Trung Van Phung [4], Khoa Cuu Nguyen [1], Alexey V. Osipov [5], Igor A. Ivanov [5], Victor I. Tsetlin [5] and Yuri N. Utkin [5,*] ⓘ

[1]   Institute of Applied Materials Science, Vietnam Academy of Science and Technology, Ho Chi Minh City 700000, Vietnam; vuthien82@yahoo.com (T.V.T.); hnanh52@yahoo.com (A.N.H.); nckhoavnn@yahoo.com (K.C.N.)
[2]   Vietnam Academy of Science and Technology, Graduate University of Science and Technology, Ho Chi Minh City 700000, Vietnam
[3]   Faculty of Pharmacy, Nguyen Tat Thanh University, Ho Chi Minh City 700000, Vietnam; thuytrangd07@yahoo.com
[4]   Istitute of Chemical Technology, Vietnam Academy of Science and Technology, Ho Chi Minh City 700000, Vietnam; trung_cnhh@yahoo.com
[5]   Shemyakin-Ovchinnikov Institute of Bioorganic Chemistry, Russian Academy of Sciences, Moscow 117997, Russia; osipov-av@ya.ru (A.V.O.); chai.mail0@gmail.com (I.A.I.); victortsetlin3f@gmail.com (V.I.T.)
*   Correspondence: yutkin@yandex.ru or utkin@mx.ibch.ru; Tel.: +7-495-336-6522

Academic Editor: Steve Peigneur
Received: 9 September 2017; Accepted: 21 October 2017; Published: 26 October 2017

Abstract: Scorpion venoms are complex polypeptide mixtures, the ion channel blockers and antimicrobial peptides being the best studied components. The coagulopathic properties of scorpion venoms are poorly studied and the data about substances exhibiting these properties are very limited. During research on the *Heterometrus laoticus* scorpion venom, we have isolated low-molecular compounds with anticoagulant activity. Determination of their structure has shown that one of them is adenosine, and two others are dipeptides LeuTrp and IleTrp. The anticoagulant properties of adenosine, an inhibitor of platelet aggregation, are well known, but its presence in scorpion venom is shown for the first time. The dipeptides did not influence the coagulation time in standard plasma coagulation tests. However, similarly to adenosine, both peptides strongly prolonged the bleeding time from mouse tail and in vitro clot formation in whole blood. The dipeptides inhibited the secondary phase in platelet aggregation induced by ADP, and IleTrp decreased an initial rate of platelet aggregation induced by collagen. This suggests that their anticoagulant effects may be realized through the deterioration of platelet function. The ability of short peptides from venom to slow down blood coagulation and their presence in scorpion venom are established for the first time. Further studies are needed to elucidate the precise molecular mechanism of dipeptide anticoagulant activity.

Keywords: venom; scorpion; blood coagulation; adenosine; peptide

---

1. Introduction

Scorpions (order Scorpiones) are distributed mainly in the hot areas and in the warmer regions of the temperate zone. Scorpion venoms are complex mixtures of compounds represented mainly by peptides and proteins. They manifest mostly neurotoxic effects and instantly paralyze small prey. A sting of large tropical scorpions can be fatal to humans, the main symptom being nervous system damage.

Forest scorpion *Heterometrus laoticus* (family Scorpionidae) occupies the Indochinese peninsula and can be often found in South-West Vietnam [1]. Among the symptoms of *H. laoticus* envenomation are local pain, inflammation, edema, swelling and redness of the stung area, lasting from a few hours to a few days [2]; no human fatalities have been reported so far. *H. laoticus* venom showed both anti-nociceptive and anti-inflammatory activity at subcutaneous injection [3]. A few toxins were isolated from this venom and characterized. The toxin heteroscorpine-1 [4] inhibited growth of bacteria and showed high homology to polypeptide toxins from scorpine family. Toxin HelaTx1 manifesting the moderate activity against Kv1.1 and Kv1.6 channels belongs to new κ-KTx5 subfamily of potassium channel blockers [5]. One more toxin, hetlaxin, of the scorpion alpha-toxin family possesses high affinity to Kv1.3 potassium channel [3]. The data about coagulopathic properties of this venom are absent.

However, some scorpion venoms cause blood-clotting disorders, but the number of coagulopathic compounds studied to date is quite small. It was reported that venoms of scorpions *Buthotus judaicus*, *Heterometrus spinnifer*, *Parabuthus transvaalicus*, *Androctonus australis*, *Scorpio maurus palmatus*, *Leiurus quinquestriatus habraeus* and *Pandinus imperator* caused an increase of clotting time. In particular, the venoms of *P. imperator* and *P. transvaalicus* species increased the clotting time by 2.5 and 2.3 times, respectively, while other venoms prolonged the time 0.8–2 times [6]. The crude venom of *Buthus tamulus* scorpion caused coagulopathy and might also induce disseminated intravascular coagulopathy (DIC syndrome). The intravenous injection of this scorpion venom at a sublethal dose to dogs and rabbits resulted in a change of the blood coagulation [7].

The investigation of in vitro effects of the venoms from scorpions *Palamneus gravimanus* and *Leiurus quinquestriatus* upon the coagulation of human plasma have shown that the crude venom of *P. gravimanus* has both procoagulant and anti-coagulant properties [8] and the crude venom of *L. quinquestriatus* is very weak anti-coagulant, which shortens the recalcified plasma clotting time by only 5–20% [8]. No fibrinolytic activity was found. Further experiments with fractions of *P. gravimanus* venom, partially purified by DEAE-Sephadex column chromatography, suggest that the procoagulant activity promotes Factor X activation while the anticoagulant fraction interferes with the action of thrombin upon fibrinogen.

It has been shown that a high concentration of *Tityus discrepans* venom in the human blood plasma increases the severity of envenomation symptoms by modifying activated partial thromboplastin time (APTT) and prothrombin time (PT), increasing cytokine level and amylase concentration as well as by inducing hyperglycemia [9]. This scorpion venom was separated into six fractions by gel filtration on a "Protein-Pack 125" column [10]. The investigations of effects on APTT, PT and direct clotting activity, using fresh human plasma and purified fibrinogen as substrates, for crude venom and its fractions showed that the venom and fraction F1 shortened APTT; venom, fraction F6 and fraction F2 prolonged PT. No thrombin-like activity was found with this venom on human plasma or purified fibrinogen [10].

Several fibrin(ogen)olytic enzymes were partially purified from *T. discrepans* venom by different types of liquid chromatography [11]. Two fractions had fibrinolytic, fibrinogenolytic (Aα-chains degradation) and tissue plasminogen activator (t-PA)-like activities; one was only fibrinogenolytic (fast degradation of fibrinogen Aα-chains and slower degradation of Bβ-chains). The fibrino(geno)lytic activity in these fractions was abolished by metalloprotease inhibitors. The other two fractions contained fibrinogenolytic (Aα-chains degradation) and fibronectinolytic activities. Serine protease inhibitors abolished activities in these fractions. None of the fractions degraded fibrinogen γ-chains. Fibrinogen degradation by active fractions was associated with an anticoagulant effect.

Furthermore, two anticoagulant phospholipases A2 (PLA2) were isolated: the imperatoxin (IpTxi)—from the *P. imperator* [12] and the phaiodactylipin—from *Anuroctonus pahiodactylus* [13]. Imperatoxin is a heterodimeric protein with a molecular weight of 14,314 Da. Its molecule consists of a large subunit (104 amino acid residues) that exhibits phospholipase activity, and the small subunit

(27 amino acid residues) covalently linked by disulfide bridge to the large subunit. The native heterodimer exhibits hydrolytic activity against phospholipids, although it was originally described as an inhibitor of ryanodine receptor in sarcoplasmic and cardiac reticulums [12]. Phaiodactylipin is a glycosylated dimeric PLA2 with a molecular weight of 19,172 Da. The protein consists of two subunits: the large one comprises 108 amino acid residues and the small subunit—18 residues; the protein structure is stabilized by five disulfide bonds. Phaiodactylipin exerts a hemolytic effect on human erythrocytes and prolongs the blood coagulation time. It should be noted that IpTxi is more effective than pachyodactylipin as anticoagulant; if its concentration in the blood exceeds 10 µg/ml, the clotting time is 30 min [13].

Scorpion venom, along with the above mentioned high molecular weight proteins and PLA2s, also contains polypeptides of smaller molecular mass that affect coagulation. The toxin isolated from the venom of the Chinese scorpion *Buthus martensii Karsch* and called Scorpion Venom Active Polypeptide (SVAP), induced platelet aggregation in rabbit blood in vivo and in vitro. SVAP also caused thrombus formation and a change in levels of thromboxane B2 and 6-keto-prostaglandin F1a in blood plasma. SVAP is the most abundant component of this venom [14]. Recently, a peptide called discreplasminin was isolated from the scorpion *T. discrepans* venom [15]. Discreplasminin had a pI of 8.0 and a molecular weight of about 6 kDa. It strongly inhibited plasmin activity and was suggested to have an anti-fibrinolytic mechanism, similar to that of aprotinin, and to interact with the active site of plasmin.

Previously, we have shown that the fractions obtained by gel filtration of *Heterometrus laoticus* scorpion venom affect the processes of blood coagulation [16] and the structures of low-molecular anticoagulants were established [17]. In the present work, we determined their anticoagulant activity.

2. Results

2.1. Isolation of Active Compounds

Earlier, we fractionated the *Heterometrus laoticus* scorpion venom by gel-filtration on Sephadex G-50 column and found that some fractions influenced blood coagulation in vitro and bleeding in mice in vivo [16]. The low-molecular weight fraction V was further separated by reversed phase HPLC (Figure 1) and fractions obtained were tested for effects on blood coagulation and bleeding [18].

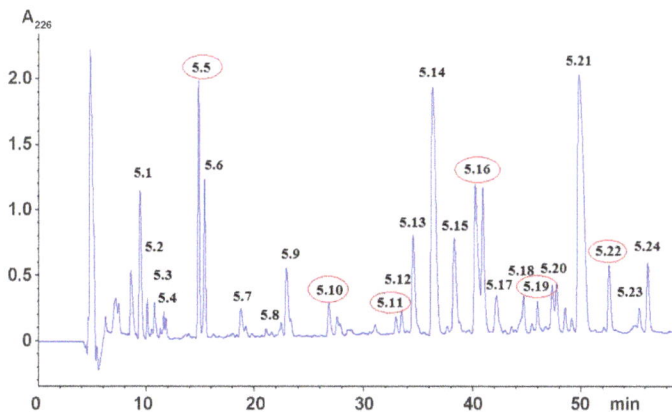

Figure 1. Separation of a low molecular weight fraction by reversed phase HPLC on Eclipse XDB C18 column (9.4 × 250 mm, 5 µm); the gradient of acetonitrile in 0.1% trifluoroacetic acid from 0% to 35% in 70 min. Flow rate 2 mL/min. The presence of polypeptide in the fractions was detected by UV absorbance at 226 nm. Fractions increasing coagulation and bleeding time are indicated by red ellipses.

From 24 fractions obtained, five fractions (5.5, 5.10, 5.11, 5.16, 5.19, and 5.22) significantly increased blood coagulation time in vitro and bleeding time in vivo [18]. The most active fractions, 5.5 and 5.22, were further purified by two more cycles of reversed phase chromatography. Figure 2 shows the first round of purification. The active fractions indicated by asterisks in Figure 2 were re-chromatographed under the same conditions and used for structure determination. The most abundant fraction, 5.21 (Figure 1), was also purified under conditions used for purification of fraction 5.22 and the structure of purified compound was analyzed as well. Analysis of purified substances indicated that fraction 5.5 corresponds to adenosine, fraction 5.21—to dipeptide LeuTrp and fraction 5.22—to dipeptide IleTrp [17]. The large quantities of dipeptides were prepared by peptide synthesis and their anticoagulant activity was analyzed in detail.

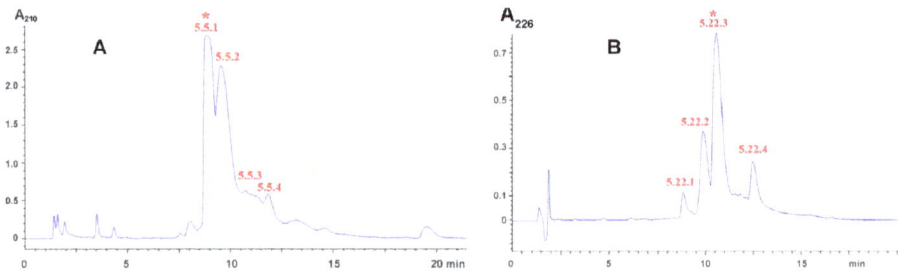

Figure 2. Isolation of active compounds by reversed phase chromatography on Analytical Eclipse XDB C18 column (4.6 × 250 mm, 5 μm) (**A**) Separation of fraction 5.5 in the gradient of acetonitrile in 0.1% trifluoroacetic acid from 0 to 10% in 40 min. Flow rate 1 mL/min; (**B**) Separation of fraction 5.22 in the gradient of acetonitrile in 0.1% trifluoroacetic acid from 15 to 30% in 30 min. Flow rate 1 mL/min.

2.2. Studies of Anticoagulant Activity

2.2.1. Influence on Blood Coagulation In Vitro

To check in vitro anticoagulant activity of synthetic dipeptides in human plasma, we used the standard coagulometric tests including the determination of the activated partial thromboplastin time (APTT), the prothrombin test (PTT), and the thrombin test (TT). In all these tests, we did not find any anticoagulant activity of the dipeptides applied at concentrations up to 100 μM. For dipeptide IleTrp, higher concentrations were used in two tests. It was inactive in TT assay up to 620 μM and in PTT— up to 1.6 mM.

The anticoagulant activity of both dipeptides on the whole mice blood was determined by the Burker's method [19], and when tested on the whole blood, the synthetic dipeptides showed significant increases in clotting time [17]. To study anticoagulant activity, the synthetic dipeptides were injected into the lateral vein of the mouse tail at a dose of 7.8 nmoles/g, which corresponds to maximal calculated concentration of 110 μM in the peripheral blood.

The clotting time was determined every 20–30 min during two hours after peptide injection. It was found that the dipeptides at dose of 7.8 nmoles/g significantly prolonged the coagulation time (Table 1). The table also includes data for adenosine and low molecular fraction obtained after the gel-filtration of the crude venom and used for isolation of dipeptides. Adenosine is a well-known inhibitor of platelet aggregation and its injection results in the increase of coagulation time. The statistically significant differences were observed for all tested samples at 20 min after injection, and at one hour after injection only adenosine and IleTrp significantly prolonged clotting time. Although the increase in coagulation time was observed for LeuTrp, it was not statistically significant. Since the most active in this test was dipeptide IleTrp, which significantly increased clotting time even 90 min after injection, its activity was studied in more detail using shorter times after injection (Figure 3): IleTrp strongly prolonged the coagulation time during observation period of two hours. The highest effect was observed immediately after injection and significant differences were registered during the whole observation time.

Table 1. Influence of low molecular weight compounds on the mice whole blood coagulation time.

Compound	Time after Injection, min				
	20	30	60	90	120
	Clotting Time, s				
Control	307.7 ± 17.4	290.67 ± 9.58	286.00 ± 6.31	256.3 ± 20.4	229.7 ± 13.2
Fraction 5	422.3 ± 8.4 [1]	391.7 ± 48.1 [1]	387.5 ± 35.0	360.2 ± 6.5	358.8 ± 26.6 [1]
Adenosin	442.5 ± 20.6 [2]	426.2 ± 25.6 [2]	366.2 ± 25.0 [2]	428.3 ± 51.1	296.3 ± 37.4
LeuTrp	401.5 ± 31.2	340.8 ± 29.1	300.3 ± 2.1	460.5 ± 41.7	313.3 ± 24.9 [1]
IleTrp	556.5 ± 87.2 [2]	426.2 ± 3.7 [1]	388.0 ± 46.3 [1]	367.0 ± 25.5 [1]	261.8 ± 16.4

[1] $p < 0.05$ compared to control; [2] $p < 0.01$ compared to control.

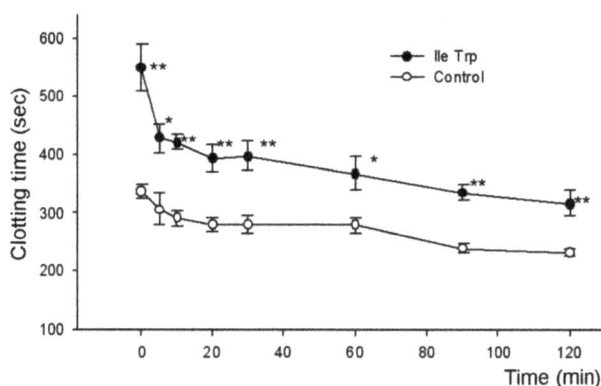

Figure 3. Influence of dipeptide IleTrp on the whole blood coagulation time. The abscissa indicates the time after dipeptide injection. * $p < 0.05$ compared to control; ** $p < 0.01$ compared to control.

2.2.2. Influence on Bleeding Time In Vivo

To determine in vivo anticoagulant activity, solutions of synthetic dipeptides and adenosine were injected in the mice as described above and tail bleeding times were evaluated. As shown in Table 2, tail bleeding times were significantly prolonged by all compounds tested. Similarly to the whole blood clotting test, all samples significantly prolonged bleeding during the first 20 min after injection and only adenosine and IleTrp were active during the first hour. In this test, again the most active was dipeptide IleTrp, which was more active than adenosine 60 min after injection. Although LeuTrp showed a higher effect during the first 20 min, the action of IleTrp was more prolonged. The activity of this dipeptide was studied in more detail using shorter times after injection (Figure 4).

It was found that IleTrp strongly prolonged the bleeding time during the first 10 min. The highest effect was observed immediately after injection and a statistically significant difference between experimental and control mice was maintained up to 90 min.

Table 2. Influence of low molecular weight compounds on the bleeding time in mice.

Compound	Time after Injection, min				
	20	30	60	90	120
	Bleeding Time, s				
Control	79.5 ± 13.7	43.33 ± 1.94	45.83 ± 3.95	40.67 ± 5.02	49.67 ± 7.85
Fraction 5	386.2 ± 88.7 [1]	187.0 ± 64.6 [1]	86 ± 2.38	119.3 ± 29.2 [1]	183 ± 80.7
Adenosin	248.2 ± 66.7 [1]	314 ± 58.6 [1]	146.7 ± 46.0 [1]	65 ± 14.5	40.2 ± 10.3
LeuTrp	314.5 ± 85.2 [1]	84.8 ± 16.7	81.2 ± 15.4	61.8 ± 14.8	68.8 ± 16.4
IleTrp	233.0 ± 30.6 [2]	179.0 ± 41.4 [1]	218.7 ± 78.5 [2]	151.5 ± 57.4	83.8 ± 13.7

[1] $p < 0.05$ compared to control; [2] $p < 0.01$ compared to control.

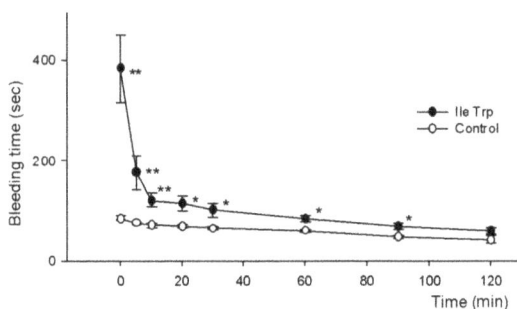

Figure 4. Influence of dipeptide IleTrp on the bleeding time. The abscissa indicates the time after dipeptide injection. * $p < 0.05$ compared to control; ** $p < 0.01$ compared to control.

We have observed no dipeptide effects on plasma coagulation, but have seen strong influence on the whole blood clotting and in vivo bleeding. Basing on these data, we suggested that dipeptides might inhibit platelet aggregation. To check this suggestion, the influence of dipeptides on platelet aggregation was studied.

2.2.3. Influence of Dipeptides on Platelet Aggregation In Vitro

The effect of dipeptides on platelet aggregation was studied using human platelet rich plasma, which was prepared immediately before use from the blood of healthy donors. The increase in light transmittance upon platelet aggregation was registered and the substance inducing the aggregation by different mechanisms were used. The peptide influence on platelet aggregation induced by addition of ADP, collagen, ristocetin and thrombin was investigated. We have observed practically no effects of IleTrp on aggregation induced by thrombin and ristocetin at peptide concentration up to about 600 µM. The IleTrp also produced no effect, when aggregation was induced by ADP at low concentration and only first aggregation phase was evident. However, at higher ADP concentration inducing two phase aggregation, IleTrp suppressed the secondary phase (Figure 5A), the LeuTrp being much weaker in this assay. When the collagen was used as aggregation inducer, a decrease in initial rate of aggregation was observed in the presence of IleTrp (Figure 5B). The LeuTrp was inactive in this test.

Toxins **2017**, *9*, 343

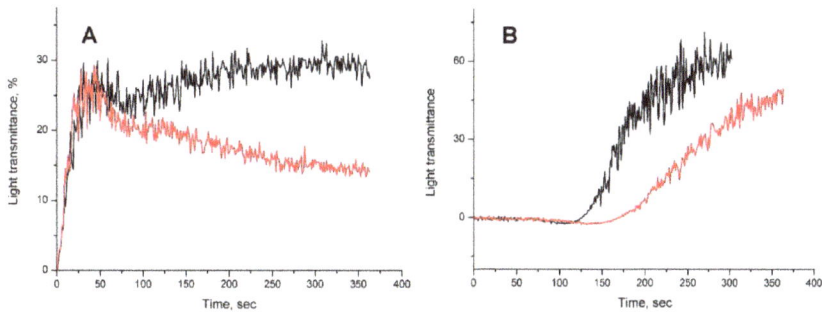

Figure 5. Influence of dipeptide IleTrp on the platelet aggregation. (**A**) ADP induced aggregation. At time zero, ADP was added to platelet rich plasma and light transmittance was registered. The black curve is control (water); the red curve was registered in the presence of IleTrp. (**B**) Collagen induced aggregation. At time zero, collagen was added to platelet rich plasma and light transmittance was measured. The black curve is control (water); the red curve was registered in the presence of IleTrp. Each curve is the mean of two independent measurements.

3. Discussion

From the venom of scorpion *H. laoticus*, we isolated low molecular weight compounds possessing anticoagulant activity. Determination of their structure showed that one compound was well known anticoagulant adenosine, while the other two were dipeptides LeuTrp and IleTrp [17]. The anticoagulant activity of the isolated substances was studied using standard tests for plasma coagulation (APTT, PTT and TT), whole blood clotting test in vitro and bleeding assay in vivo. In standard tests on plasma, no anticoagulant activity of the dipeptides was observed at concentrations up to 100 μM. The dipeptide IleTrp was inactive in TT at concentration up to 620 μM and in PTT—up to 1.6 mM. However, all three isolated compounds substantially prolonged clotting time of whole blood (Table 1). It was found that at all tested times adenosine at a dose of 2.48 mg/kg (9.3 nmole/g) showed a clotting time greater than in the control group. At 20, 30 and 60 min these differences were statistically significant, while at 90 and 120 min, although adenosine increased clotting time compared to the control group, these differences did not yet have statistical significance. Similarly, the dipeptide IleTrp at a dose of 7.8 nmoles/g significantly prolonged clotting time during all observation time of two hours (Figure 3). The dipeptide LeuTrp at the same dose showed a statistically significant increase in clotting time only at 120 min and was the least active among substances tested. In bleeding in vivo assay all three compounds also prolonged bleeding time (Table 2, Figure 4). The increase in time was very similar to that observed for blood clotting. It should be noted that directly after injection, the IleTrp effect on bleeding was much greater than on clotting (a 4.6-fold increase versus 1.6-fold one).

It should be noted that no systemic bleeding or coagulopathies were reported for *H. laoticus* envenomation [2], therefore the amount of the venom received by the human victims after scorpion sting in not sufficient to affect hemotsasis strongly. Although it is difficult to estimate the amount of venom injected in the prey, the concentrations of dipeptides used in this study are certainly exceeding those achieved in natural scorpion victims. To obtain reliable results, for individual compounds we used the doses of 2.48 mg/kg which corresponds to 1/5 of LD_{50} (12.4 mg/kg) for the crude venom. Certainly the content of each studied compound in *H. laoticus* venom is much less than 20%, however even their low amounts may be sufficient to induce coagulopathies in small animals.

The absence of anticoagulant activity of dipeptides in the standard coagulometric tests on blood plasma might indicate that both intrinsic and extrinsic coagulation pathways were not influenced by these substances, suggesting that the possible target of peptide intervention may be platelet aggregation.

Several platelet receptors involved in platelet activation are known; these are P2Y1 and P2Y12 receptors, thromboxane A2 receptor, PAR-1 and PAR-4 receptors, as well as collagen GPVI and GPIbα receptors [20]. Inhibition of platelet activation is achieved by blocking these receptors. On the other hand, activation of A_{2A} adenosine receptors by adenosine also results in inhibition of platelet aggregation [21]. Thus, sub-micromolar concentrations of adenosine have an antiaggregatory effect on whole human blood [22]. Adenosine has an extremely short lifetime in blood plasma [23]; however, a high dose of 9.3 nmole/g used in our experiments allowed for the detection of the effect. This dose corresponds to the concentration of 133 μM in peripheral blood. At this dose, the observed effects were quite long, the statistically significant increase in both time of coagulation and bleeding was observed within an hour after adenosine administration (Tables 1 and 2).

Interestingly, the dipeptide IleTrp at slightly lower dose revealed stronger effects; it showed statistically significant elongation of coagulation time during 90 min after administration, with the tendency to increase in time being seen even after two hours (Table 1). Increased bleeding time was observed for two hours after administration, while during the second hour the differences were not statistically significant (Table 2). The observed effect is the first indication that dipeptide can cause an increase in the blood clotting time in vivo.

It should be noted that the tryptophan-containing dipeptides, including IleTrp and LeuTrp, which have been found in food protein hydrolysates, are inhibitors of angiotensin converting enzyme (ACE) involved in the regulation of blood pressure [24]. The peptide IleTrp is a selective and competitive inhibitor of the C-terminal domain of the enzyme possessing a selectivity coefficient of 40 compared to the N-terminal domain [24]. At the same time, increasing evidence suggests that ACE inhibitors (ACE-I) exert antithrombotic effects [25]. Thus it was shown that ACE-I (captopril and lisinopril) enhanced the antiplatelet response to aspirin at concentrations of 15 μM [26]. It was also found that ACE-I exerted an antithrombotic effect in experimental thrombosis in rats [27]. The arterial and venous thrombus weights were reduced after the rats' treatment with some ACE-I. The same treatments resulted in significant inhibition of the collagen induced platelet aggregation in the whole blood [27]. Pretreatment with ACE-Is resulted in a significant reduction of platelet adhesion to fibrillar collagen. These data are in good agreement with our results on inhibition of collagen induced platelet aggregation by IleTrp (Figure 5).

There is also evidence that the dipeptide IleTrp under trademark BNC210 (Bionomics Limited, Thebarton, Australia) is in the second stage of clinical trials for the treatment of post-traumatic stress disorder [28]. By the mechanism of action, BNC210 is a negative allosteric modulator of nicotinic acetylcholine receptors of the alpha7 type [29]. Given the fact that the aggregation of platelets is substantially inhibited [30] by α-bungarotoxin and methyllycaconitine which are selective antagonists of alpha7 nicotinic cholinergic receptors, one can suggest that the observed anticoagulant effect the dipeptide IleTrp might be mediated by its interaction with this receptor.

Some other molecular mechanisms could be involved in anticoagulant effects of dipeptides. For example, ACE-I, to which IleTrp belongs, have generally been shown to improve the fibrinolytic balance by reducing plasma plasminogen activator inhibitor type 1 (PAI-1) level [31]. The PAI-1 binds to tissue plasminogen activator (t-PA) forming an inactive complex and preventing fibrin breakdown, thus prolonging preservation of thrombus. The reduced PAI-1 level should result in the faster thrombus retraction by t-PA.

The traditional concept of the hemostatic system regulation by a coagulation factor cascade along with platelet activation in recent years has been supplemented by new evidence that the immune system may strongly affect blood coagulation. Under certain conditions, leukocytes can express tissue factor and release proinflammatory and procoagulant molecules [32]. These molecules can influence platelet activation and adhesion as well as activation of the intrinsic and extrinsic coagulation pathways. There is also evidence about multiple interactions between the hemostatic system and innate immunity, and the coagulation and complement cascades. Thus, complement factor 3 (C3) deficiency causes prolonged bleeding, reduced thrombus incidence, thrombus size, fibrin and platelet deposition as well as diminished platelet activation in vitro [33]. Although there are no data about influence of dipeptides on immune system, one cannot exclude their possible action on blood coagulation through effects on immune system.

Interestingly, in in vitro experiments on human platelets, the dipeptides inhibited the secondary phase in aggregation induced by ADP, and IleTrp caused a decrease in initial rate of aggregation induced by collagen (Figure 5). As several mechanisms are involved in the secondary phase in aggregation induced by ADP [34], it is difficult to say which of them is affected by dipeptides. This IleTrp effect on collagen induced aggregation is in good agreement with the literature data. It was shown earlier that in human blood, adenosine A2 receptor agonist CGS 21680 attenuated both in vitro aggregation induced by collagen and flow cytometric markers of platelet activation-aggregation [35]. Moreover, platelet responsiveness to adenosine A2 receptor stimulation was species-dependent: adenosine A2 receptor stimulation inhibited platelet activation by collagen in human, but not canine models. Based on these data, one can suggest that dipeptide effects may be realized though their interaction with adenosine receptors and are stronger in mice than in humans due to species dependence.

However, the above consideration suggests the necessity of further studies to elucidate the exact molecular mechanism of anticoagulant effects of dipeptides.

4. Conclusions

From the *H. laoticus* scorpion venom, for the first time we isolated adenosine and two dipeptides LeuTrp and IleTrp possessing anticoagulant activity. The dipeptides did not influence the coagulation time in standard plasma coagulation tests, but, similarly to adenosine, strongly prolonged the bleeding time from mouse tail and in vitro clot formation. The dipeptides inhibited the secondary phase of aggregation induced by ADP, and IleTrp decreased an initial rate of collagen induced platelet aggregation in vitro, which may suggest it interaction with adenosine A_{2A} receptors. One can assume that anticoagulant effects of dipeptides may be realized through the deterioration of platelet function. The ability of short venom peptides to slow down blood coagulation was established for the first time. Further studies are needed to elucidate the precise molecular mechanism of this action and potentially apply it to clinical practice.

5. Materials and Methods

5.1. Materials

The kits for APTT, PTT, and TT tests as well as thrombin, ADP, ristocetin and collagen were obtained from NPO Renam (Moscow, Russia). Adenosine was from Merck KGaA (Darmstadt, Germany).

Scorpions and Scorpion Venom

The scorpions *H. laoticus* were collected in the An Giang province of Vietnam and bred at laboratory of Institute of Applied Materials Science VAST, Ho Chi Minh City. They were fed with crickets and locusts. The scorpions were milked by electrical stimulation, and the venom obtained by this ways was dried over anhydrous $CaCl_2$ and kept at $-20\,^{\circ}C$ until use.

5.2. Venom Fractionation and Isolation of Low Molecular Weight Compounds

Crude *H. laoticus* venom was first fractionated by gel-filtration Sephadex G50 as described [3]. Five main fractions were obtained. The fraction 5 was further separated by reversed phase HPLC on Eclipse XDB C18 column (9.4 × 250 mm, 5 μm); the gradient of acetonitrile in 0.1% trifluoroacetic acid from 0% to 35% in 70 min. Flow rate 2 mL/min. The presence of polypeptide in the fractions was detected by UV absorbance at 226 nm (Figure 1). The active compounds were further purified by reversed phase chromatography on Analytical Eclipse XDB C18 column (4.6 × 250 mm, 5 μm)

Fraction 5.5 was separated in the gradient of acetonitrile in 0.1% trifluoroacetic acid from 0 to 10% in 40 min; fraction 5.21 and 5.22—in the gradient of acetonitrile in 0.1% trifluoroacetic acid from 15 to 30% in 30 min. Flow rate 1 mL/min. The molecular masses of obtained substances were determined by mass-spectrometry on mass-spectrometer LCQ DECA XP+ (Thermo Finnigan, Somerset, NJ, USA).

5.3. Characterization of Low Molecular Weight Compounds

The structures of the compounds isolated from fractions 5.5, 5.21 and 5.22 were determined as described [17]. In brief, the molecular mass of the compound from fraction 5.5 was 267.8 Da close to the mass of adenosine (267.2 Da). Both substances co-eluted as a single peak from reversed phase HPLC column and had similar fragmentation mass-spectra. Thus, fraction 5.5 contained adenosine. The compounds from fractions 5.21 and 5.22 had an identical mass of 317.1 Da. The structure of the compound from fraction 5.21 was determined by proton nuclear magnetic resonance. Tandem mass spectrometry analysis of fraction 5.22 showed Leu/IleTrp structure for dipeptide present. As LeuTrp was found in fraction 5.21, fraction 5.22 contained IleTrp dipeptide. The determined structures of dipeptides were confirmed by their chemical synthesis. The peptide synthesis was carried out as described [36].

5.4. Mice

Male Swiss albino mice were obtained from the Pasteur Institute of Ho Chi Minh City (Vietnam). The mice were kept at least 2 days prior to the test at Faculty of Pharmacy, Nguyen Tat Thanh University, Ho Chi Minh City. All the appropriate actions were taken to minimize discomfort to mice. World Health Organization's International Guiding Principles for Biomedical Research Involving Animals were followed during experiments on animals.

5.5. Determination of Anticoagulant Activity

To study anticoagulant activity, solutions of synthetic dipeptides and adenosine in 0.9% NaCl were injected into the lateral vein of the mouse tail at a dose of 2.48 mg/kg (injection volume 0.1 mL/10 g of body mouse weight). This dose corresponded to 1/5 of LD50 (12.4 mg/kg) for *H. laoticus* venom at intravenous injection. For dipeptides, 2.48 mg/kg is equal to 7.8 nmoles/g. The average circulating blood volume for mice is 72 mL/kg [37]. The average weight of the mouse used was 20 ± 2 g, the molecular weight of dipeptides—317 Da, and the average blood volume is 1.4 mL; the amount of compounds injected yielded a maximum calculated concentration of 110 µM in the peripheral blood. The mice of the control group received only 0.9% NaCl solution. Each experimental and control group included 6 mice.

5.5.1. Determination of Blood Coagulation Time

The blood coagulation time was determined by modified Burker method [19]. In brief, a drop of blood from the mouse tail was placed on the glass. Every 30 s one tried to tear it away from the glass with the help of an injection needle. The moment when the formed fibrin threads could detach the blood clot from the glass corresponded to the end of the coagulation.

5.5.2. Determination of In Vivo Bleeding Time

Tail bleeding times were measured using the method described by Liu et al. [38]. The distal 5 mm of tail was amputated and the tail (diameter of about 1.5 mm) was immersed in 37 °C solution of 0.9% NaCl. Time to visible cessation of bleeding was recorded.

5.5.3. Platelet Aggregation Measurements

The preparation of platelet rich plasma (PRP) and platelet aggregation measurements using the platelet aggregation analyzer AR2110 (Solar, Minsk, Belarus) were performed essentially as described [39]. In brief, the solution (25 mM) of IleTrp in water (20 µL) was added to 450 µL of PRP and the light transmission was recorded for approximately 30 s. Subsequently, 50 µL of solution containing aggregation inducer was added, and the recording was continued. Thrombin (3 units), ADP at final concentration of 6 µM, ristocetin (1 mg/mL) and collagen at final concentration of 0.02% were used to induce aggregation.

5.5.4. Plasma Coagulation Tests

The plasma coagulation tests were performed using APG2-02 hemostasis analyzer (EMCO, Moscow, Russia) according to the manufacturer protocols.

5.6. Statistical Analysis

Significance of differences between experimental and control groups was analyzed by Kruskal-Wallis method and then by Mann-Whitney method using Minitab 15.0 (Minitab Inc., State College, PA, USA) program. The differences were considered significant for p values < 0.05. All results are presented as the mean $\pm$ SEM (standard error of the mean).

Animal experiments described in this study were approved by the Scientific Council of the Faculty of Pharmacy, Nguyen Tat Thanh University (Protocol No. 1). Protocol was signed by the Chairman of the Council Vice Principal Prof. Nguyen Van Thanh and the Council Secretary Dr. Vo Thi Ngoc My. Date of approval was 19 January 2017.

Acknowledgments: This study was supported by the Russian Foundation for Basic Research (project No. 14-04-93007 Viet_a). Experiments were partially carried out using the equipment provided by the core facility of the Shemyakin-Ovchinnikov Institute of Bioorganic Chemistry (CKP IBCH, supported by Russian Ministry of Education and Science, grant RFMEFI62117X0018).

Author Contributions: A.N.H., T.V.P., K.C.N. and Y.N.U. conceived and designed the experiments; T.V.T., T.T.T.N. and A.V.O. performed the experiments; A.N.H., T.V.P., K.C.N. and A.V.O. analyzed the data; I.A.I. contributed reagents and materials; A.N.H., T.V.T., V.I.T. and Y.N.U. wrote the paper.

Conflicts of Interest: The authors declare no conflict of interest. The founding sponsors had no role in the design of the study; in the collection, analyses, or interpretation of data; in the writing of the manuscript, and in the decision to publish the results.

References

1. Couzijn, H.W.C. Revision of the genus *Heterometrus* Hemprich & Ehrenberg (Scorpionidae, Arachnoidea). *Zoologische Verhandelingen* **1981**, *184*, 1–196.
2. Uawonggul, N.; Chaveerach, A.; Thammasirirak, S.; Arkaravichien, T.; Chuachan, C.; Daduang, S. Screening of plants acting against *Heterometrus laoticus* scorpion venom activity on fibroblast cell lysis. *J. Ethnopharmacol.* **2006**, *103*, 201–207. [CrossRef] [PubMed]
3. Hoang, A.N.; Vo, H.D.; Vo, N.P.; Kudryashova, K.S.; Nekrasova, O.V.; Feofanov, A.V.; Kirpichnikov, M.P.; Andreeva, T.V.; Serebryakova, M.V.; Tsetlin, V.I.; et al. Vietnamese Heterometrus laoticus scorpion venom: evidence for analgesic and anti-inflammatory activity and isolation of new polypeptide toxin acting on Kv1.3 potassium channel. *Toxicon* **2014**, *77*, 40–48. [CrossRef] [PubMed]
4. Uawonggul, N.; Thammasirirak, S.; Chaveerach, A.; Arkaravichien, T.; Bunyatratchata, W.; Ruangjirachuporn, W.; Jearranaiprepame, P.; Nakamura, T.; Matsuda, M.; Kobayashi, M.; et al. Purification and characterization of Heteroscorpine-1 (HS-1) toxin from *Heterometrus laoticus* scorpion venom. *Toxicon* **2007**, *49*, 19–29. [CrossRef] [PubMed]
5. Vandendriessche, T.; Kopljar, I.; Jenkins, D.P.; Diego-Garcia, E.; Abdel-Mottaleb, Y.; Vermassen, E.; Clynen, E.; Schoofs, L.; Wulff, H.; Snyders, D.; et al. Purification, molecular cloning and functional characterization of HelaTx1 (Heterometrus laoticus): the first member of a new κ-KTX subfamily. *Biochem. Pharmacol.* **2012**, *83*, 1307–1317. [CrossRef] [PubMed]
6. Tan, N.H.; Ponnudurai, G. Comparative study of the enzymatic, hemorrhagic, procoagulant and anticoagulant activities of some animal venoms. *Comp. Biochem. Physiol. C* **1992**, *103*, 299–302. [CrossRef] [PubMed]

7. Gajalakshmi, B.S. Coagulation studies following scorpion venom injection in animals. *Indian J. Med. Res.* **1982**, *76*, 337. [PubMed]
8. Hamilton, P.J.; Ogston, D.; Douglas, A.S. Coagulant activity of the scorpion venoms *Palamneus gravimanus* and *Leiurus quinquestriatus*. *Toxicon* **1974**, *12*, 291–296. [CrossRef]
9. D'Suze, G.; Moncada, S.; Gonzalez, C.; Sevcik, C.; Aguilar, V.; Alagon, A. Relationship between plasmatic levels of various cytokines, tumour necrosis factor, enzymes, glucose and venom concentration following *Tityus* scorpion sting. *Toxicon* **2003**, *41*, 367–375. [CrossRef]
10. Brazón, J.; Guerrero, B.; Arocha-Piñango, C.L.; Sevcik, C.; D'Suze, G. Effect of *Tityus discrepans* scorpion venom on global coagulation test. Preliminary studies. *Investig. Clin.* **2008**, *49*, 49–58.
11. Brazón, J.; Guerrero, B.; D'Suze, G.; Sevcik, C.; Arocha-Piñango, C.L. Fibrin(ogen)olytic enzymes in scorpion (*Tityus discrepans*) venom. *Comp. Biochem. Physiol. B* **2014**, *168*, 62–69. [CrossRef] [PubMed]
12. Zamudio, F.Z.; Conde, R.; Arévalo, C.; Becerril, B.; Martin, B.M.; Valdivia, H.H.; Possani, L.D. The mechanism of inhibition of ryanodine receptor channels by imperatoxin I, a heterodimeric protein from the scorpion *Pandinus imperator*. *J. Biol. Chem.* **1997**, *272*, 11886–11894. [CrossRef] [PubMed]
13. Valdez-Cruz, N.A.; Batista, C.V.; Possani, L.D. Phaiodactylipin, a glycosylated heterodimeric phospholipase A from the venom of the scorpion *Anuroctonus phaiodactylus*. *Eur. J. Biochem.* **2004**, *271*, 1453–1464. [CrossRef] [PubMed]
14. Song, Y.M.; Tang, X.X.; Chen, X.G.; Gao, B.B.; Gao, E.; Bai, L.; Lv, X.R. Effects of scorpion venom bioactive polypolypeptides on platelet aggregation and thrombosis and plasma 6-keto-PG F1α and TXB2 in rabbits and rats. *Toxicon* **2005**, *46*, 230–235. [CrossRef] [PubMed]
15. Brazón, J.; D'Suze, G.; D'Errico, M.L.; Arocha-Piñango, C.L.; Guerrero, B. Discreplasminin, a plasmin inhibitor isolated from *Tityus discrepans* scorpion venom. *Arch. Toxicol.* **2009**, *83*, 669–678. [CrossRef] [PubMed]
16. Hoang, N.A.; Vo, D.M.H.; Nikitin, I.; Utkin, Y. Isolation and preliminary study of short toxins from scorpion *Heterometrus laoticus*. *Tap Chi Hoa Hoc (J. Chem.)* **2011**, *49*, 118–122.
17. Tran, T.V.; Hoang, A.N.; Nguyen, T.T.T.; Phung, T.V.; Nguyen, K.C.; Osipov, A.V.; Dubovskii, P.V.; Ivanov, I.A.; Arseniev, A.S.; Tsetlin, V.I.; et al. Low-molecular compounds with anticoagulant activity from scorpion *Heterometrus laoticus* venom. *Dokl. Biochem. Biophys.* **2017**, *476*. in press. [CrossRef]
18. Hoang, A.N.; Hoang, T.M.Q.; Nguyen, T.T.T.; Nguyen, T.T.T.; Pham, Y.N.D.; Vo, H.D.M. Isolation and characterization of anticoagulant components from scorpion venom *Heterometrus laoticus*. *Tap Chi Hoa Hoc (J. Chem.)* **2013**, *51*, 520–524.
19. Barker, L.F. The Clinical Diagnosis of Internal Diseases. In *Monographic Medicine, D*; Appleton and Company: New York, NY, USA; London, UK, 1917; Volume III, p. 131.
20. Li, Z.; Delaney, M.K.; O'Brien, K.A.; Du, X. Signaling during platelet adhesion and activation. *Arterioscler. Thromb. Vasc. Biol.* **2010**, *30*, 2341–2349. [CrossRef] [PubMed]
21. Johnston-Cox, H.A.; Ravid, K. Adenosine and blood platelets. *Purinergic Signal.* **2011**, *7*, 357–365. [CrossRef] [PubMed]
22. Söderbäck, U.; Sollevi, A.; Wallen, N.H.; Larsson, P.T.; Hjemdahl, P. Anti-aggregatory effects of physiological concentrations of adenosine in human whole blood as assessed by filtragometry. *Clin. Sci. (Lond.)* **1991**, *81*, 691–694. [CrossRef] [PubMed]
23. Klabunde, R.E. Dipyridamole inhibition of adenosine metabolism in human blood. *Eur. J. Pharmacol.* **1983**, *93*, 21–26. [CrossRef]
24. Lunow, D.; Kaiser, S.; Rückriemen, J.; Pohl, C.; Henle, T. Tryptophan-containing dipeptides are C-domain selective inhibitors of angiotensin converting enzyme. *Food Chem.* **2015**, *166*, 596–602. [CrossRef] [PubMed]
25. Skowasch, D.; Viktor, A.; Schneider-Schmitt, M.; Lüderitz, B.; Nickenig, G.; Bauriedel, G. Differential antiplatelet effects of angiotensin converting enzyme inhibitors: comparison of ex vivo platelet aggregation in cardiovascular patients with ramipril, captopril and enalapril. *Clin. Res. Cardiol.* **2006**, *95*, 212–216. [CrossRef] [PubMed]
26. Al-Azzam, S.I.; Alzoubi, K.H.; Khabour, O.F.; Quttina, M.; Zayadeen, R. Evaluation of the effect of angiotensin converting enzyme inhibitors and angiotensin receptors blockers on aspirin antiplatelet effect. *Int. J. Clin. Pharmacol. Ther.* **2016**, *54*, 96–101. [CrossRef] [PubMed]

27. Wojewódzka-Zelezniakowicz, M.; Chabielska, E.; Mogielnicki, A.; Kramkowski, K.; Karp, A.; Opadczuk, A.; Domaniewski, T.; Malinowska-Zaprzałka, M.; Buczko, W. Antithrombotic effect of tissue and plasma type angiotensin converting enzyme inhibitors in experimental thrombosis in rats. *J. Physiol. Pharmacol.* **2006**, *57*, 231–245. [PubMed]

28. Phase II Study of BNC210 in PTSD. Available online: https://clinicaltrials.gov/ct2/show/study/NCT02933606 (accessed on 13 October 2017).

29. O'Connor, S.; Thebault, V.; Danjou, P.; Mikkelsen, J.D.; Doolin, E.; Simpson, J.; Tadie, E. A multiple ascending dose study with evidence for target engagement of BNC210; a negative allosteric modulator of alpha7 nAChR in development for anxiety. *Eur. Neuropsychopharmacol.* **2016**, *26*, S609. [CrossRef]

30. Schedel, A.; Thornton, S.; Schloss, P.; Klüter, H.; Bugert, P. Human platelets express functional alpha7-nicotinic acetylcholine receptors. *Arterioscler. Thromb. Vasc. Biol.* **2011**, *31*, 928–934. [CrossRef] [PubMed]

31. Fogari, R.; Zoppi, A. Antihypertensive drugs and fibrinolytic function. *Am. J. Hypertens.* **2006**, *19*, 1293–1299. [CrossRef] [PubMed]

32. Swystun, L.L.; Liaw, P.C. The role of leukocytes in thrombosis. *Blood* **2016**, *128*, 753–762. [CrossRef] [PubMed]

33. Subramaniam, S.; Jurk, K.; Hobohm, L.; Jäckel, S.; Saffarzadeh, M.; Schwierczek, K.; Wenzel, P.; Langer, F.; Reinhardt, C.; Ruf, W. Distinct contributions of complement factors to platelet activation and fibrin formation in venous thrombus development. *Blood* **2017**, *129*, 2291–2302. [CrossRef] [PubMed]

34. Cattaneo, M.; Gachet, C.; Cazenave, J.-P.; Packham, M.A. Adenosine diphosphate (ADP) does not induce thromboxane A2 generation in human platelets. *Blood* **2002**, *99*, 3868–3869. [CrossRef] [PubMed]

35. Linden, M.D.; Barnard, M.R.; Frelinger, A.L.; Michelson, A.D.; Przyklenk, K. Effect of adenosine A2 receptor stimulation on platelet activation-aggregation: Differences between canine and human models. *Thromb. Res.* **2008**, *121*, 689–698. [CrossRef] [PubMed]

36. Utkin, Y.N.; Weise, C.; Kasheverov, I.E.; Andreeva, T.V.; Kryukova, E.V.; Zhmak, M.N.; Starkov, V.G.; Hoang, N.A.; Bertrand, D.; Ramerstorfer, J.; et al. Azemiopsin from Azemiops feae viper venom, a novel polypeptide ligand of nicotinic acetylcholine receptor. *J. Biol. Chem.* **2012**, *287*, 27079–27086. [CrossRef] [PubMed]

37. Diehl, K.H.; Hull, R.; Morton, D.; Pfister, R.; Rabemampianina, Y.; Smith, D. A good practice guide to the administration of substances and removal of blood, including routes and volumes. *J. Appl. Toxicol.* **2001**, *21*, 15–23. [CrossRef] [PubMed]

38. Liu, Y.; Jennings, N.L.; Dart, A.M.; Du, X.J. Standardizing a simpler, more sensitive and accurate tail bleeding assay in mice. *World J. Exp. Med.* **2012**, *2*, 30–36. [CrossRef] [PubMed]

39. Osipov, A.V.; Filkin, S.Y.; Makarova, Y.V.; Tsetlin, V.I.; Utkin, Y.N. A new type of thrombin inhibitor, noncytotoxic phospholipase A2, from the Naja haje cobra venom. *Toxicon* **2010**, *55*, 186–194. [CrossRef] [PubMed]

© 2017 by the authors. Licensee MDPI, Basel, Switzerland. This article is an open access article distributed under the terms and conditions of the Creative Commons Attribution (CC BY) license (http://creativecommons.org/licenses/by/4.0/).

toxins

MDPI

Article

Can Inhibitors of Snake Venom Phospholipases A$_2$ Lead to New Insights into Anti-Inflammatory Therapy in Humans? A Theoretical Study

Thaís A. Sales [1], Silvana Marcussi [1], Elaine F. F. da Cunha [1], Kamil Kuca [2,3,*] and Teodorico C. Ramalho [1,3]

[1] Department of Chemistry, Federal University of Lavras, P.O. Box 3037, 37200-000 Lavras, MG, Brazil; thaissales194@hotmail.com (T.A.S.); marcussi@dqi.ufla.br (S.M.); Elaine_cunha@dqi.ufla.br (E.F.F.d.C.); teo@dqi.ufla.br (T.C.R.)
[2] Biomedical Research Center, University Hospital Hradec Kralove, 500 05 Hradec Kralove, Czech Republic
[3] Center for Basic and Applied Research, Faculty of Informatics and Management, University of Hradec Kralove, Rokitanskeho 62, 500 03 Hradec Kralove, Czech Republic
* Correspondence: kamil.kuca@fnhk.cz; Tel.: +420-495-833-447

Academic Editor: Steve Peigneur
Received: 24 September 2017; Accepted: 21 October 2017; Published: 25 October 2017

Abstract: Human phospholipase A$_2$ (hPLA$_2$) of the IIA group (HGIIA) catalyzes the hydrolysis of membrane phospholipids, producing arachidonic acid and originating potent inflammatory mediators. Therefore, molecules that can inhibit this enzyme are a source of potential anti-inflammatory drugs, with different action mechanisms of known anti-inflammatory agents. For the study and development of new anti-inflammatory drugs with this action mechanism, snake venom PLA$_2$ (svPLA$_2$) can be employed, since the svPLA$_2$ has high similarity with the human PLA$_2$ HGIIA. Despite the high similarity between these secretory PLA$_2$s, it is still not clear if these toxins can really be employed as an experimental model to predict the interactions that occur with the human PLA$_2$ HGIIA and its inhibitors. Thus, the present study aims to compare and evaluate, by means of theoretical calculations, docking and molecular dynamics simulations, as well as experimental studies, the interactions of human PLA$_2$ HGIIA and two svPLA$_2$s, *Bothrops* toxin II and Crotoxin B (BthTX-II and CB, respectively). Our theoretical findings corroborate experimental data and point out that the human PLA$_2$ HGIIA and svPLA$_2$ BthTX-II lead to similar interactions with the studied compounds. From our results, the svPLA$_2$ BthTX-II can be used as an experimental model for the development of anti-inflammatory drugs for therapy in humans.

Keywords: experimental model; svPLA$_2$; vanillic acid

1. Introduction

The inflammatory process involves a complex cascade of biochemical and cellular events, and it is an innate reaction of the organism that occurs in tissue in response to any cell injury from any dangerous agent: physical, chemical or biological. One of the stages of the inflammatory process is the breakdown of membrane phospholipids by phospholipases A$_2$ (PLA$_2$), which generates fatty acids, such as arachidonic acid (AA) and lysophospholipids. Oxidation of AA generates inflammatory mediators, such as prostaglandins, thromboxanes and leukotrienes, through the action of cyclooxygenase (COX) and lipoxygenase (LOX) enzymes. In addition to AA, the breakdown of membrane phospholipids generates lysophospholipids, a precursor of platelet-activating factor (PAF), another potent inflammatory mediator [1,2].

For the treatment of these inflammatory conditions, the Non-Steroidal Anti-Inflammatory Drugs (NSAIDs) are the most commonly employed drugs [3]. Their wide use throughout the world is due to the large number of diseases involving inflammatory disorders, the spread of rheumatic diseases, and an increase in the life expectancy of the population. Despite their widespread utilization, the prolonged use of this class of anti-inflammatory drugs causes several side effects, such as gastrointestinal toxicity and hepatotoxicity, among other diseases [4–6]. For this reason, there is great interest in the development of new compounds that can act as anti-inflammatory drugs, but with fewer side effects.

Despite their structural differences, all NSAIDs have a similar action mechanism, and are inhibitors of COX enzymes [7]. Recent studies have reported that the anti-inflammatory action of NSAIDs occurs by inhibition of the COX-2 isoform. However, the other products of the inflammatory cascade are also involved in inflammatory conditions, and the inhibition of the COX pathway may accentuate the LOX activity, and consequently increases leukotrienes production, the other product of arachidonic acid degradation [8–10]. In this way, the inhibition of the PLA_2, which can act at the top of the cascade, is a promising alternative, since at the same time that it decreases the COX pathway, it also regulates the production of leukotrienes and the PAF-AH. Despite having great importance, only a few theoretical studies have been devoted to this topic and currently secreted PLA_2 enzymes have not been explored as a molecular target by medicinal chemistry [11].

Among the various classes of existing PLA_2, the human secreted PLA_2 of the IIA group (HGIIA) belongs to the group of PLA_2 that is the most associated with diseases and consequently are the target enzyme for inhibition [11]. Since human enzymes are difficult to obtain, some experimental models are generally employed for their study [12–16]. In this line, some enzymes that would possibly serve as an experimental model are the snake venoms PLA_2 ($svPLA_2$). The secreted PLA_2 from snake venoms are distributed in Subgroups I and II of the secreted PLA_2 group. Of these, the crotalid and bothropic PLA_2 are part of the II group, which is the same group as the HGIIA, which is a human PLA_2. Crotoxin B (CB) (PDB ID 2QOG, UniProtKB AC: P24027) is the basic part of the Crotoxin (Cro), and its toxic part. Crotoxin was the first animal toxin to be purified and crystallized and is the main protein present in venom of the *Crotalus durissus terrificus* (South American rattlesnake) [17]. The *Bothrops* toxin II (BthTX-II) is another basic PLA_2 isolated from *Bothrops jararacussu* venom. This myotoxic toxin is also known for its edematogenic and hemolytic effects and for its ability to induce platelet aggregation and secretion [18].

The use of $svPLA_2$ for understanding the activity and action mechanisms of the human PLA_2 HGIIA has been proposed, as there is a high similarity between $svPLA_2$ and HGIIA PLA_2 [19]. Moreover, the use of snake venom toxins could also be justified, because they are rich in Group I and II secreted PLA_2, especially Elapidae and Viperidae families [20]. It should be kept in mind that despite this high similarity, it is not clear if these enzymes can perform similar interactions and if the $svPLA_2$ can really be employed as an experimental model to describe the HGIIA interactions, since some works reveal the contrary [21]. Thus, the objective of this work was to evaluate, experimentally, the phospholipasic activity of vanillic acid (VA) on $svPLA_2$ enzymes, such as BthTX-II and CB. In addition, we compare, theoretically, the interactions of these enzymes with the interaction of the same compound with HGIIA. Finally, two molecules rationally modified from the VA molecule were proposed to improve their interaction of the VA with HGIIA and to develop new potential anti-inflammatory drugs.

2. Results

2.1. Experimental Assays

Figure 1 contains the percent inhibition of both $svPLA_2$ by VA in relation to the different concentrations of this molecule. It is possible to observe that $svPLA_2$ BthTX-II presented a higher activity and value of inhibition percentage by the VA. In relation to the halo of activity (Figure S1 of the Support Material), the sample with the highest proportion of VA presents the lowest activity halo for both enzymes, which indicates that VA decreases the activity of both $svPLA_2$. In addition, the inhibition percentage values also indicate that vanillic acid is able to inhibit *sv*PLA$_2$. Figure 1 shows that the highest proportion of VA tested was responsible for the highest percentage of inhibition for both enzymes, equivalent to 23.7% for the BthTX-II and 20% for CB.

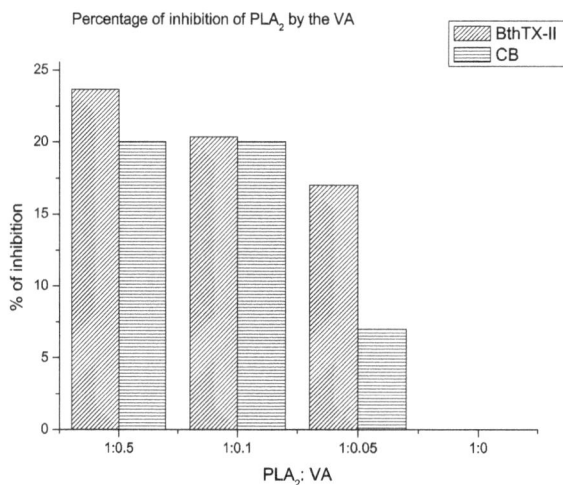

Figure 1. Percent inhibition of phospholipase A$_2$ activity caused by vanillic acid (VA), for the phospholipases A$_2$ isolated from snake venom *Bothrops* toxin II (BthTX-II) and Crotoxin B (CB).

2.2. Alignments of Amino Acid Sequences of svPLA$_2$ and Human PLA$_2$

Two alignments of primary amino acid sequences of BthTX-II (PDBID 3JR8), CB (PDBID 2QOG), and HGIIA (PDBID 3U8D) were performed and can be seen in Figures S2 and S3 of the support material. One is for calculating the percentages of identity and similarity between the enzymes, while the other is focusing on charge distributions and hydrophobicity of the three PLA$_2$. The results of the first alignment showed 64.0% identity (84.4% similar) for alignment BthTX-II vs. CB (3-143:4-144); 53.1% identity (72.0% similar) for HGIIA vs. BthTX-II (3-145:3-143); and 55.2% identity (81.1% similar) for the alignment of the HGIIA vs. CB (3-145:4-144). For the second alignment, it is possible to observe that, although there are a few differences in some residues, the enzymes present groups (hydrophobic, negatively or positively charged residues) that behave similarly in the same positions, in most cases.

2.3. Theoretical Calculations

2.3.1. Molecular Docking Calculations

To validate the methodology, re-docking was performed on the human enzyme, under the same conditions, to compare the structure obtained in the theoretical calculation with the ligand structure of the U8D present in the crystal. The root-mean-square deviation (RMSD) obtained in re-docking was zero for all structures, which means that the structures obtained presented little alteration in relation to the average structure, which is satisfactory. The overlap of the obtained poses with the U8D active ligand are shown in Figure 2. As can be seen, there was no significant variation in the structures theoretically obtained with the active ligand structure of the 3U8D complex. Therefore, this result can validate our docking study.

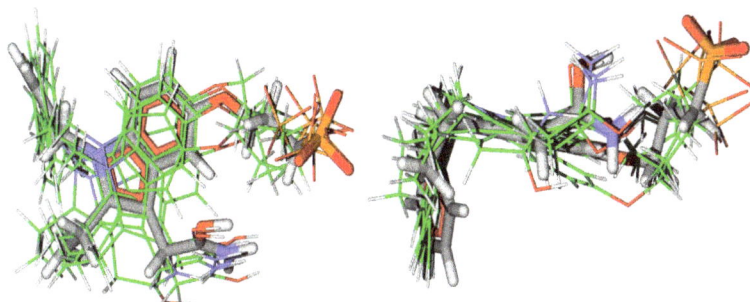

Figure 2. Superposition of the obtained poses with the active ligand U8D, obtained by re-docking calculation.

Afterwards, the docking analysis of the vanillic acid with all enzymes was performed, and the results are reported in Table 1. As can be seen in Table 1, the interaction energies and the score function obtained for the enzyme HGIIA and BthTX-II were very close. This result does not apply to the third phospholipase CB. Both $svPLA_2$ have high similarity to the human enzyme HGIIA, however, only CB has four subunits composed of two equal dimmers, as seen in Figure 3. Each of these two dimmers is similar to the other PLA_2 studied. BthTX-II and HGIIA enzymes have only two subunits in their active conformation, and this small structural difference can make the BthTX-II enzyme a little more appropriate to help describe the interactions that occur in the human enzyme.

Table 1. Values obtained for the Score Binding Functions, Interaction Energy, and Hydrogen bonds for docking calculation of human phospholipase A_2 ($hPLA_2$) of the IIA group PLA_2 (HGIIA), BthTX-II, and CB.

Enzyme	MolDock Score	Rerank Score	Interaction	HBond
	$hPLA_2$			
HGIIA	−69.38	−58.93	−75.35	−2.21
	$svPLA_2$			
BthTX-II	−71.22	−57.02	−79.45	0.00
CB	−37.87	−35.17	−44.91	−0.02

In relation to the hydrogen bond energies, the $svPLA_2$ had the lowest values, different from the HGIIA that have approximately −2.21 KJ mol^{-1}. Despite of this difference in energies, just one hydrogen bond between the Histidine 47 residue of HGIIA and VA occurs, which can be seen in Figure 4. The bond length is 2.601 Å, and bonds longer than 2.5 Å are not very stable [22].

Through the superposition of the VA and the active ligand U8D, in Figure S4 of the support material, it is possible to deduce that VA leads to similar hydrophobic interactions. This means interactions between U8D and the hydrophobic residues of HGIIA, since the aromatic rings of VA are localized very close to the aromatic ring of the U8D. In addition, the oxygen atoms of the VA carboxyl group are also near the oxygen groups of the U8D molecule. Figure S5 shows the vanillic acid molecule inside the cavity of the HGIIA enzyme. As can be seen, VA occupies only a part of the cavity.

If new radical groups are rationally added in a vanillic acid molecule to take up all the cavity space, it is possible that the interaction of these compounds increases. Based on this idea, and considering the composition of the residues that are in the active site, two VA modified molecules were supposed, and docking analysis of their energies was performed. The proposed modifications are shown in Figure 5.

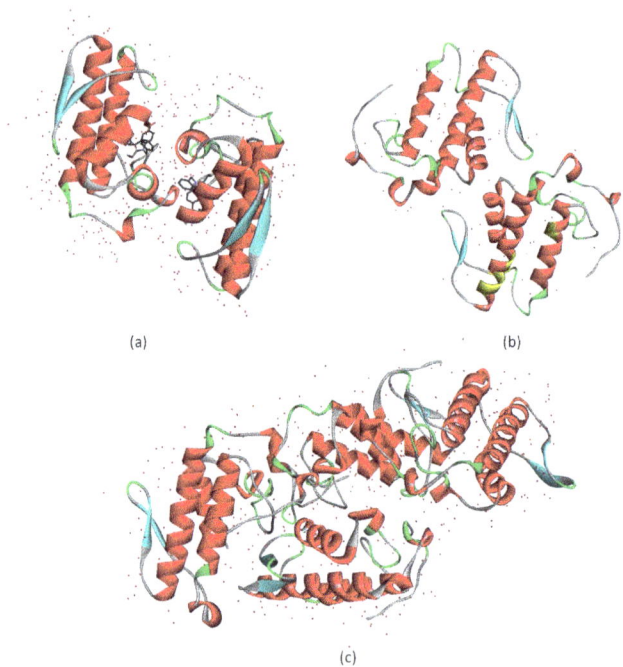

Figure 3. Three dimensional structures of secretory phospholipases A$_2$: (**a**) represents the structures of HGIIA, with 3U8D PDB code; (**b**) represents the BthTX-II structure, with 3JR8 PBD code; and (**c**) is the 3D structure of CB, PDB code 2QOG.

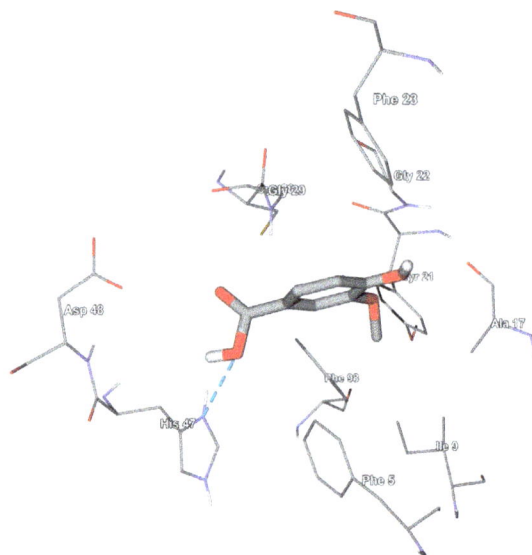

Figure 4. Hydrogen bond made between a vanillic acid molecule and the His 47 residue of PLA$_2$ HGIIA, whose length is 2601 Å.

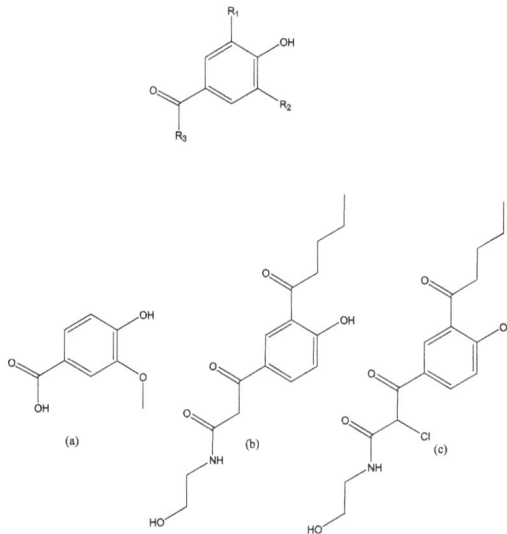

Figure 5. Modifications rationally proposed to improve the inhibitory activity of vanillic acid. On the top of the figure is the general structure; the molecule represented in (**a**) is the vanillic acid (VA), (**b**) is the first modification, named analogue I, and (**c**) is the second modification, named analogue II.

The docking calculation of the modified VA structures (analogues I and II) also were performed, and the results are displayed in Table 2. As can be seen, all results are better than the unmodified VA molecule (Table 1), which means that the modifications are satisfactory. The hydrogen bond energies have improved and are more similar between BthTX-II and HGIIA than the unmodified VA molecule. Moreover, the analogues followed the same interaction pattern, having more affinity for BthTX-II, followed by HGIIA and CB, the energies of the first two being very similar. The interaction energy between analogue I and the enzymes BthTX-II, HGIIA, and CB are -113.82, -107.12, and -71.93 KJ mol^{-1}, respectively. For the interaction of these enzymes and analogue II, the interaction energies were -126.35, -115.38, and -55.64 KJ mol^{-1}, respectively. These data also suggest that the BthTX-II serves as an experimental model to evaluate inhibitions in human secretory phospholipases of the IIA group.

Table 2. Values obtained for the Score, Interaction Energy, and Hydrogen Bond energies of the two analogues tested by docking calculation with the PLA$_2$ HGIIA, BthTX-II, and CB.

Compound	Enzyme	MolDock Score	Rerank Score	Interaction	HBond
Analogue I	HGIIA	−101.52	−73.99	−107.12	−4.99
Analogue I	BthTX-II	−111.32	−69.64	−113.82	−4.61
Analogue I	CB	−72.26	24.57	−71.94	−5.14
Analogue II	HGIIA	−117.02	−26.35	−115.38	−4.63
Analogue II	BthTX-II	−123.33	−101.81	−126.35	−7.91
Analogue II	CB	−59.03	31.78	−55.64	−9.65

2.3.2. Molecular Dynamics Simulation

After the molecular docking study of VA in both PLA$_2$, the structures obtained from the enzymes and the poses were analyzed by molecular dynamics. The root-mean square deviation (RMSD) and the number of hydrogen bonds were obtained for both systems. The first plot (Figure S6) shows the RMSD for each enzyme/inhibitor complex (HGIIA/VA; BthTX-II/VA, and CB/VA). In all systems, both VA and enzymes were stabilized, which indicates that all systems reached equilibrium. The *sv*PLA$_2$ structures have more fluctuations over time, especially the CB/VA system.

For the HGIIA/VA system, which was the most stable, the equilibrium occurred as early as in the first picoseconds of simulation, and its maximum value was approximately 0.5 nm for the VA and 0.4 nm for the protein, both low values. This indicates that the VA ligand stabilized within the active site of the enzyme and that its interactions with HGIIA are favorable, proving its inhibitory potential. For the BthTX-II/VA system, the equilibrium was reached later for the ligand after 1000 ps, but it also occurred and was relatively maintained over time. Its maximum value was less than 1.2 nm while the RMSD of the BthTX-II enzyme did not reach 0.7 nm, which means that the permanence of the VA in the active site of the PLA$_2$ BthTX-II is also favorable. The CB/VA complex provided the largest variation in position over time, but despite this, it also stabilized. The ligand varied widely in the active site of the CB enzyme, reaching a maximum RMSD of 2.5 nm for the ligand and 1 nm for the protein. Similar to the behavior adopted in the docking calculations, the BthTX-II enzyme was that which behaved more like the human enzyme HGIIA. This also suggests that BthTX-II is capable of aiding in the description of the experimental behavior of the human enzyme and that the CB PLA$_2$ does not provide information of interactions between the VA ligand and the human PLA$_2$.

Comparing the RMSD between the enzymes, the HGIIA human PLA$_2$ (*h*PLA$_2$) was the most stable during the simulation, and BthTX-II was relatively stable. At the same time, the structure of CB *sv*PLA$_2$ had many more fluctuations, as already mentioned. Turning now to the inhibitor in the three enzymes (Figure 6), it can be seen that the VA conformations in the enzymes HGIIA and BthTX-II have similar behaviors, unlike the ligand in the CB active site.

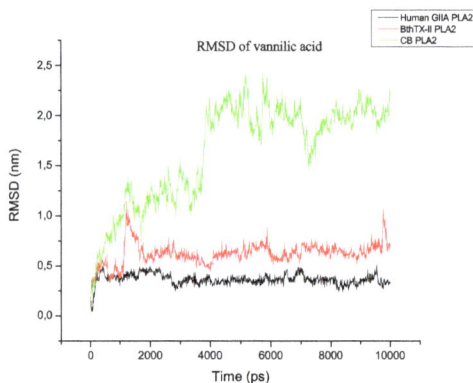

Figure 6. Comparison of root-mean square deviation (RMSD) of VA in each active site.

In relation to the hydrogen bonds carried out over time for the three studied PLA$_2$ (Figure S7 of the support material), it is possible to observe that the *h*PLA$_2$ HGIIA performed seven bonds during the molecular dynamics (MD) simulation, with approximately four being maintained most of the time. The BthTX-II *sv*PLA$_2$, similarly to HGIIA, performed six hydrogen bonds in the MD simulation, four of which are more frequent over time. Regarding CB *sv*PLA$_2$, unlike the other two phospholipases, CB PLA$_2$ showed up to eight hydrogen bonds, but these were less stable, since they appear only at a few moments throughout the time.

3. Discussion

3.1. Can svPLA$_2$/Inhibitors Describe the hPLA$_2$/Inhibitors Interactions?

From the experimental data, it is possible to observe that the BthTX-II enzyme has more affinity with the VA molecule. This pattern was maintained in the docking and MD simulations, which indicate that the theoretical studies carried out are coherent with the experiment and suggest that HGIIA performs similar interactions between BthTX-II and VA. According to the literature, a degree identity over 35% is satisfactory [23]. Despite their similarity, it is important to comment that Kim and collaborators (2017) [21] found that the *sv*PLA$_2$ purified from the venom of *Daboia russelli pulchella* (VRV-PL-VIII) is not appropriate as a model for describing the interactions between the human PLA$_2$ and its inhibitors. As we can see in this work, the *sv*PLA$_2$ CB, despite the high similarity with HGIIA, does not provide information about the interactions that occur between the HGIIA and VA, while the BthTX-II has a behavior similar to the human enzyme. This feature suggests that the structural similarity is a very important factor to consider, but is not the only factor. The other factor which plays an important role is the behavior of the enzyme in solution [21,24]. According to Kim and collaborators (2017) [21], the *sv*PLA$_2$ does not provide any useful foundation for a prediction of the binding mode to specific ligands in a HGIIA complex. The authors conclude this based on the fact that the *sv*PLA$_2$ enzymes have different behavior in solution, and because of this feature can interact with different chains (A or B) in a different mode. They found that the ligand FLSIK, in the HGIIA:FLSIK complex, does not interact with both chains, and as such, the chain B is not necessary for the inhibition activity, since the ligand interacts only with chain A. In other words, the authors found that the HGIIA acts as a monomer in solution. For the *sv*PLA$_2$ that the authors chose (PLA$_2$ purified from the venom of *Daboia russelli pulchella* (VRV-PL-VIIIA, svPLA$_2$, UniProt accession code P59071, with 49% identity to HGIIA), the behavior in solution is different, and because of this, despite the high similarity, this *sv*PLA$_2$ does not provide information of HGIIA interactions, as it acts as a monomer and *sv*PLA$_2$ act as a dimmer.

Similarly, for the authors, the simulations with HGIIA in the present work show that the ligand interacts with a single chain of the enzyme, which can be seen in Figure 7. The images represent the frames at the beginning, middle, and end of the simulation for the HGIIA/VA complex. As can be seen, the VA molecule is maintained in a single chain of the molecule at the three times. However, different from the conclusions of Kim and collaborators (2017) [21], in this work, we found that the *sv*PLA$_2$ BthTX-II can provide a useful foundation for a prediction of the HGIIA binding mode. This fact is justified because the BthTX-II behavior in solution is similar to the HGIIA (Figure 7), different from the *sv*PLA$_2$ CB tested in this work and the *sv*PLA$_2$ tested by Kim and collaborators [21]. In Figure 7, the VA molecule also remains in the only chain of the enzyme most of time. As mentioned above, the CB PLA$_2$, despite its high primary sequence similarity with HGIIA, acts as a tetramer, different from the other two tested PLA$_2$. In addition, the PLA$_2$ tested by Kim et al. (2017) [21], besides having less similarity to HGIIA, does not act as a monomer in solution.

Thus, for the similar interactions between HGIIA and BthTX-II, the similar behavior in solution, and for the high structural similarity of these compounds, it is possible that, experimentally, the vanillic acid acts in HGIIA in the same manner, with inhibition percentage values close to those of the BthTX-II results. Despite the differences in hydrogen bond energies in the docking calculations, the time dependent simulations show that the number of hydrogen bonds of BthTX-II and HGIIA are similar, and are maintained most of time, which also contributes to their similarity in interactions, contributing to the fact that the BthTX-II can be used as an experimental model for HGIIA.

Moreover, the aromatic ring of the VA is in the same position as the active ligand of the 3U8D complex, suggesting that the same hydrophobic interaction can occur. With the structures obtained in the MD simulation, it was possible to create a pharmacophoric map of the HGIIA middle structures, which is approximately correspondent to the BthTX-II interactions. The maps are shown in Figure 8. The enzymes have similar hydrophobic interactions, and these interactions can explain the similar interaction energy obtained in molecular docking. In the map, it is possible to observe that the VA molecule performs π-π stacking interactions with phenylalanine residues and a hydrogen bond with glycine residues in both enzymes. The results obtained in this work are in agreement with the results obtained by Dileep et al. (2015) [25], who analyzed the effect of some phenolics on secretory PLA$_2$ of the swine pancreas. The authors reported that vanillic acid interacts with this phospholipase by performing an H bond and by hydrophobic interactions with residues Phe 5, Leu 2, Phe 22, and Leu 31. Therefore, one secretory PLA$_2$ that has more availability and is more easily obtained, which is BthTX-II, can be used as an experimental model for the study of mechanisms and the development of new inhibitors for the HGIIA PLA$_2$ that are so important in regulations of the arachidonic acid pathway.

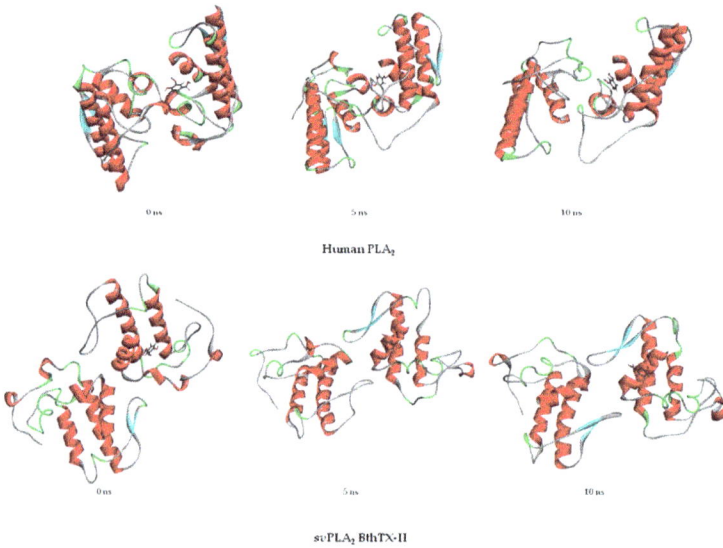

Figure 7. Comparison between the structure of the complex HGIIA/VA at the beginning (0 ns), middle (5 ns), and end (10 ns) of the molecular dynamics simulation, and comparison of the structures of the complex BthTX-II/VA at the beginning (0 ns), middle (5 ns), and end (10 ns) of the molecular dynamics simulation.

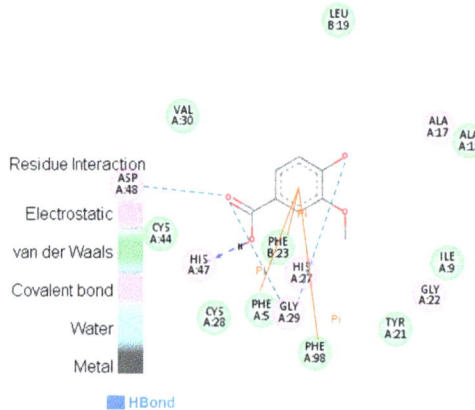

Figure 8. Pharmacophoric map of the interactions between HGIIA and vanillic acid (VA).

3.2. Searching Molecular Interactions of Vanillic Acid Analogs

With the modification of the VA molecule, the interactions increase significantly. As presented in Table 2, the interaction energies increase for both modifications with all PlA$_2$. Moreover, the hydrogen bond energies for HGIIA and BthTX-II were very similar. Through the modifications of the VA molecule, the majority of the active site was occupied with radicals that interact with specific residues. This modification brings new hydrophobic interactions and hydrogen bonds, as can be seen in Figure 9, the pharmacophoric map. In addition, the chlorine atom in analogue II performs electrostatic interactions with HGIIA. With this, it is possible to conclude that vanillic acid can act as a base molecule for the rational development of new secreted PLA$_2$ inhibitors. With better interaction, these new inhibitors can be more effective and selective for these enzymes, which enables the use of these molecules as possible anti-inflammatory drugs, with a different action mechanism from that of the current commercially available drugs.

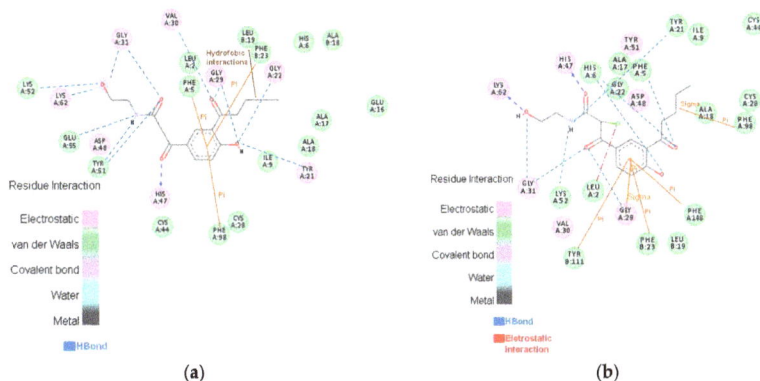

(a)   (b)

Figure 9. Interactions between the analogues I and II and HGIIA enzyme: (a) presents the interactions of the analogue I with HGIIA; (b) shows the interactions of the analogue II with the HGIIA enzyme.

4. Conclusions

In this work, a comparison of the HGIIA and *sv*PLA$_2$ interactions was performed in order to clarify the discussion about the use of *sv*PLA$_2$ as a model for analysis of human PLA$_2$ interactions. In addition, two modified molecules from vanillic acid were theoretically proposed for increasing the inhibition of the VA molecule as well as its inhibitory effect. It is possible to conclude that the enzyme BthTX-II can provide useful information about the interactions of the potential inhibitors with HGIIA *h*PLA$_2$. The other *sv*PLA$_2$ tested, the Crotoxin B, or CB, does not present the same results, and so this enzyme cannot be used as an experimental model for HGIIA. It is also concluded that the primary sequence similarity is not the only factor to be considered, and the behavior of the enzyme in solution is an important factor for the comparison of interactions between the structurally similar enzymes.

This work is of great use, because we report a proof-of-principle study that snake venom toxins, more specifically *sv*PLA$_2$, can be used as tools for studies in human PLA$_2$, taking care in choosing the correct *sv*PLA$_2$. Furthermore, it serves as evidence that both structural similarity and enzyme solution behavior are important to describe similarities in interactions of two or more enzymes. Lastly, vanillic acid has potential to inhibit secreted PLA$_2$, and can be a base molecule for the development of molecules that can interact more strongly and can be more selective. The two rationally modified molecules developed from VA show better interaction energies than VA, which means that the developed molecules are more potent inhibitors than VA, and can be potential-use candidates for new anti-inflammatory drugs.

5. Materials and Methods

5.1. Experimental Assays

For the experimental analysis, the model of secretory *PLA*$_2$ employed was the *svPLA*$_2$ isolated from the species *Crotalus durissus terrificus* (CB) and *Bothrops jararacussu* (BthTX-II). The inhibition of phospholipase activity for vanillic acid was assessed using solid medium as described by Gutiérrez et al., 1988 [26], replacing agarose with agar and without the addition of erythrocytes. The substrate used was egg yolk. The egg yolk is a source of phospholipids, mainly phosphatidylcholine and phosphatidylethanolamine, thus forming an affordable and low-cost source for the detection of phospholipase activity [27]. The medium was prepared with 1% bacteriological agar, pH 7.2, and egg yolk diluted in phosphate-buffered saline (PBS) (1:3, vv^{-1}). Also, 0.01 mol L^{-1} of CaCl$_2$ and 0.005% of sodium azide was also added in the medium. After the gel solidified in plates, the treatments were applied in wells of approximately 0.5 cm of diameter. The two PLA$_2$ isolated from snake venoms (BthTX-II and CB) were used to induce the breakdown of phospholipids. Each PLA$_2$ and vanillic acid were diluted in CaCl$_2$ solution and previously incubated in a water bath at 37 °C for 30 min, at the following ratios: 1:1, 1:0.5, 1:0.1, and 1:0.05 (PLA$_2$/vanillic acid, *w/w*). The potential of vanillic acid in inhibiting PLA$_2$ was evaluated after 18 h of incubation of the plates in a cell culture chamber at that same temperature. Controls containing only PLA$_2$ were also evaluated. The formation of a clear halo around the well in the gel characterized the phospholipase activity, which was measured according to the halo diameter. The results were expressed as percentages of activity, and inhibition and the controls containing only venom were considered as having 100% phospholipase activity.

5.2. Alignments of Amino Acid Sequences

In order to verify the similarity of these enzymes with the human secretory PLA$_2$ HGIIA, alignments were made using the LALIGN [28], a dynamic programming algorithm that determines similar regions of two protein sequences and other biomolecules. Additionally, the alignment of the UniProt [29] was employed to verify the presence of positive, negative, and hydrophobic residues. For the alignments, the primary sequences of these secretory PLA$_2$ were downloaded from Expasy [30] in the categories of proteomics on the topic of protein sequences and identification, using the UniProtKB database [31]. In order to compare the interactions that occur between the ligands and all secretory PLA$_2$, the same theoretical calculations were performed for both secretory PLA$_2$.

5.3. Simulation Methods

5.3.1. Docking Energies Calculations

To calculate the partial charges of ligands, the three-dimensional structures were previously created through the program PC Spartan® (version Pro, Wavefunction, Inc., Ivine, CA, USA) [32], and the calculation was performed by the semi-empirical method AM1. After this, the ligands were docked inside the HGIIA (PDB code 3U8D, that have a resolution of 1.8 Å and are complexed with Ca^{2+}, Cl$^-$, and 3-{[3-(2-amino-2-oxoethyl)-1-benzyl-2-ethyl-1H-indol-5-yl]oxy}propyl)phosphonic acid (PDB code U8D)) using the software Molegro Virtual Docker (MVD®, version 2011.4.3.0, Qiagen Bioinformatics, Redwood City, CA, USA) [33]. The binding site was restricted into a sphere with a radius of 11 Å, and the residues within a radius of 8 Å were considered flexible. Fifty runs were performed, with 50 poses obtained for the analysis of the ligand-protein interactions and of the overlaps with the U8D inside of the human PLA$_2$. The best conformation was selected, based on the best overlap and the interaction energy. For the analysis of the *svPLA*$_2$, the binding site, identified by the His47, was restricted into a sphere with 7 and 5 Å for the BthTX-II and CB, respectively, according to the size of the cavity. Since these enzyme structures do not have ligands, the best energy of interaction was taken into account. The selected conformations of all were used for the further MD simulation steps.

5.3.2. Molecular Dynamics Simulations

Initial ligand configurations were produced using the Gaussian 09 Program [34] to construct the structures, and the Automated Topology Builder (ATB) server [35,36] to generate the topology and structure files. For the simulations, the force field used was GROMOS 96 54a7 [37], GROMACS program [38] (Version 5.1.2, Royal Institute of Technology and Uppsala University, Uppsala, Sweden). The enzyme/inhibitor complexes (HGIIA/VA, BthTX-II/VA and CB/VA) were constructed using the mentioned force field, in a volume simulation box of 645.57, 742.71, and 938.66 nm for each complex, respectively. For the energy minimization, the steepest descent algorithm was used, minimizing when the maximum force was <10.0 kJ/mol. After the minimization step, the complexes were submitted to molecular dynamics analysis for a time interval of 10 ns, and 1000 conformations were obtained for each complex. The equations of motion were integrated using the leapfrog scheme. The results were analyzed through the VMD® program (version 1.9.2, University of Illinois at Urbana-Champaign, Champaign, IL, USA) [39] and Discovery Studio® 3.5 (Accelrys, San Diego, CA, USA). The total energy, RMSD, and hydrogen bond graphs were generated to analyze the results through the Origin® program (Version 3.5.0, Accelrys Software Inc., San Diego, CA, USA) [40–43].

Supplementary Materials: The following are available online at www.mdpi.com/2072-6651/9/11/341/s1, Figure S1: Halo of inhibition, in centimeters, formed by the inhibition of svPLA2svPLA2 isolated from BthTX-II and CB venom, by vanillic acid, Figure S2: Alignments of Human PLA2 HGIIA (3U8D) aminoacid sequences with the phospholipases A2 BthTX-II (3JR8) and PLA2 CB (2QOG), Figure S3: The amino acid sequence comparison between HGIIA, BthTX-II and PLA2 CB focusing on similar distribution of charged amino acid and hydrophobicity, Figure S4; Overlap of the active ligand of the 3U8D complex, of the enzyme HGIIA, with the vanillic acid obtained by the molecular docking, Figure S5: Volume of the cavity of the enzyme HGIIA with the molecule of vanillic acid anchored, Figure S6: Root Mean square deviation (RMSD) for the HGIIA/VA, BthTX-II/VA and CB/VA complexes, Figure S7: Hydrogen bonds carried out between vanillic acid and PLA2 enzymes.

Acknowledgments: This work was funded by Long term development plans (UHK and FNHK).

Author Contributions: T.A.S. and S.M. designed the experiments; T.A.S. and E.F.F.d.C. performed the experiments. T.A.S. and T.C.R. analyzed the data. S.M., T.A.S., T.C.R., and K.K. wrote the paper.

Conflicts of Interest: The authors declare no conflict of interest.

References

1. Joshi, V. Dimethyl ester of bilirubin exhibits anti-inflammatory activity through inhibition of secretory phospholipase A2, lipoxygenase and cyclooxygenase. *Arch. Biochem. Biophys.* **2016**, *598*, 28–39. [CrossRef] [PubMed]
2. Silva, P. *Farmacologia*, 8th ed.; Guanabara Koogan: Rio de Janeiro, Brazil, 2010; 1024p.
3. Muri, E.M.; Sposito, M.M.M.; Metsavaht, L. Nonsteroidal antiinflammatory drugs and their local pharmacology. *Acta Fisiatr.* **2009**, *16*, 186–190.
4. Anelli, M.G.; Scioscia, C.; Grattagliano, I.; Lapadula, G. Old and new antirheumatic drugs and the risk of hepatotoxicity. *Ther. Drug Monit.* **2012**, *34*, 622–628. [CrossRef] [PubMed]
5. Rafaniello, C.; Ferrajolo, C.; Sullo, M.G.; Sessa, M.; Sportiello, L.; Balzano, A.; Manguso, F.; Aiezza, M.L.; Rossi, F.; Scarpignato, C.; et al. Risk of gastrointestinal complications associated to NSAIDs, low-dose aspirin and their combinations: Results of a pharmacovigilance reporting system. *Pharm. Res.* **2016**, *104*, 108–114. [CrossRef] [PubMed]
6. Yousefpour, A.; Iranagh, S.A.; Nademi, Y.; Modarress, H. Molecular dynamics simulation of nonsteroidal antiinflammatory drugs, naproxen and relafen, in a lipid bilayer membrane. *Int. J. Quant. Chem.* **2013**, *113*, 1919–1930. [CrossRef]
7. Cronstein, B.N.; Weissmann, G. Targets for antiinflammatory drugs. *Ann. Rev. Pharmacol. Toxicol.* **1995**, *35*, 449–462. [CrossRef] [PubMed]
8. Gaddipati, R.S.; Raikundalia, G.K.; Mathai, M.L. Dual and selective lipid inhibitors of cyclooxygenases and lipoxygenase: A molecular docking study. *Med. Chem. Res.* **2014**, *23*, 3389–3402. [CrossRef]
9. Peters-Golden, M.D.M.; Henderson, M.D.W.R., Jr. Leukotrienes. *N. Engl. J. Med.* **2007**, *357*, 1841–1854. [CrossRef] [PubMed]

10. Pyasi, K.; Tufvesson, E.; Moitra, S. Evaluating the role of leukotriene-modifying drugs in asthma management: Are their benefits 'losing in translation'? *Pulm. Pharmacol. Ther.* **2016**, *41*, 52–59. [CrossRef] [PubMed]

11. Quach, N.D.; Arnold, R.D.; Cummings, B.S. Secretory phospholipase A2 enzymes as pharmacological targets for treatment of disease. *Biochem. Pharmacol.* **2014**, *90*, 338–348. [CrossRef] [PubMed]

12. Tomankova, V.; Anzenbacher, P.; Anzenbacherova, E. Effects of obesity on liver cytochromes P450 in various animal models. *Biomed. Pap. Olomouc* **2017**, *161*, 144–151. [CrossRef] [PubMed]

13. Lerch, M.M.; Adler, G. Experimental animal-models of acute-pancreatitis. *Int. J. Pancreatol.* **1994**, *15*, 159–170. [PubMed]

14. Liu, Y.; Zeng, B.H.; Shang, H.T.; Cen, Y.Y.; Wei, H. Bama Miniature Pigs (Sus scrofa domestica) as a Model for Drug Evaluation for Humans: Comparison of In Vitro Metabolism and In Vivo Pharmacokinetics of Lovastatin. *Comp. Med.* **2008**, *58*, 580–587. [PubMed]

15. Prueksaritanont, T. Use of In Vivo Animal Models to Assess Drug-Drug Interactions. Enzyme- and Transporter-Based Drug-Drug Interactions: Progress and Future Challenges. *Pharm. Res.* **2010**, *27*, 283–297.

16. Siltari, A.; Kivimäki, A.S.; Ehlers, P.I.; Korpela, R.; Vapaatalo, H. Effects of Milk Casein Derived Tripeptides on Endothelial Enzymes In Vitro; a Study with Synthetic Tripeptides. *Arzneimittelforschung* **2012**, *62*, 477–481. [CrossRef] [PubMed]

17. Marchi-Salvador, D.P.; Corrêa, L.C.; Magro, A.J.; Oliveira, C.Z.; Soares, A.M.; Fontes, M.R. Insights into the role of oligomeric state on the biological activities of crotoxin: Crystal structure of a tetrameric phospholipase A2 formed by two isoforms of crotoxin B from *Crotalus durissus terrificus* venom. *Proteins* **2008**, *72*, 883–891. [CrossRef] [PubMed]

18. Dos Santos, J.I.; Cintra-Francischinelli, M.; Borges, R.J.; Fernandes, C.A.; Pizzo, P.; Cintra, A.C.; Braz, A.S.; Soares, A.M.; Fontes, M.R. Structural, functional, and bioinformatics studies reveal a new snake venom homologue phospholipase A$_2$ class. *Proteins* **2011**, *79*, 61–78. [CrossRef] [PubMed]

19. Teixeira, C.F.; Landucci, E.C.; Antunes, E.; Chacur, M.; Cury, Y. Inflammatory effects of snake venom myotoxic phospholipases A2. *Toxicon* **2003**, *42*, 947–962. [CrossRef] [PubMed]

20. Marcussi, S.; Sant'Ana, C.D.; Oliveira, C.Z.; Rueda, A.Q.; Menaldo, D.L.; Beleboni, R.O.; Stabeli, R.G.; Giglio, J.R.; Fontes, M.R.; Soares, A.M. Snake venom phospholipase A2 inhibitors: Medicinal chemistry and therapeutic potential. *Curr. Top. Med. Chem.* **2007**, *7*, 743–756. [CrossRef] [PubMed]

21. Kim, R.R.; Malde, A.K.; Nematollahi, A.; Scott, K.F.; Church, W.B. Molecular dynamics simulations reveal structural insights into inhibitor binding modes and functionality in human Group IIA phospholipase A2. *Proteins* **2017**, *85*, 827–842. [CrossRef] [PubMed]

22. Batsanov, S.S. Van der Waals Radii of Elements. *Inorg. Mater.* **2001**, *37*, 1031–1046.

23. Da Cunha, E.E.; Barbosa, E.F.; Oliveira, A.A.; Ramalho, T.C. Molecular modeling of Mycobacterium tuberculosis DNA gyrase and its molecular docking study with gatifloxacin inhibitors. *J. Biomol. Struct. Dyn.* **2010**, *27*, 619–625. [CrossRef] [PubMed]

24. Golçalves, M.A.; Santos, L.S.; Prata, D.M.; Fernando, P.C.; Cunha, E.F.F.; Ramalho, T.C. Optimal wavelet signal compression as an efficient alternative to investigate molecular dynamics simulations: Aplication to thermal and solvent effects of MRI probes. *Theor. Chem. Acc.* **2017**, *136*, 15. [CrossRef]

25. Dileep, K.V.; Remya, C.; Cerezo, J.; Fassihi, A.; Pérez-Sánchez, H.; Sadasivan, C. Comparative studies on the inhibitory activities of selected benzoic acid derivatives against secretory phospholipase A2, a key enzyme involved in the inflammatory pathway. *Mol. Biosyst.* **2015**, *11*, 1973–1979. [CrossRef] [PubMed]

26. Gutiérrez, J.M.; Avila, C.; Rojas, E.; Cerdas, L. An alternative in vitro method for testing the potency of the polyvalent antivenom produced in Costa Rica. *Toxicon* **1998**, *26*, 411–413. [CrossRef]

27. Price, M.F.; Wilkinson, I.D.; Gentry, L.O. Plate method for detection of phospholipase activity in *Candida albicans*. *Sabouraudia* **1982**, *20*, 7–14. [CrossRef] [PubMed]

28. Huang, X.; Miller, W.; Huang, X.; Miller, W. A time-efficient linear-space local similarity algorithm. *Adv. Appl. Math.* **1991**, *12*, 337–357. [CrossRef]

29. Consortium, U. UniProt: A hub for protein information. *Nucleic Acids Res.* **2015**, *43*, D204–D212. [CrossRef] [PubMed]

30. Expasy. Available online: http://au.expasy.org/ (accessed on 23 October 2017).

31. UniProtKB Database. Available online: http://www.uniprot.org/ (accessed on 23 October 2017).

32. Hehre, W.J.; Deppmeier, B.J.; Klunzinger, P.E. *Pcspartanpro*; Wavefunction, Inc.: Irvine, CA, USA, 1999.

33. Thomsen, R.; Christensen, M.H. MolDock: A new technique for high-accuracy Molecular docking. *J. Med. Chem.* **2006**, *49*, 3315–3321. [CrossRef] [PubMed]
34. Frisch, M.J. *Gaussian 09*; Gaussian, Inc.: Wallingford, CT, USA, 2009.
35. Canzar, S.; El-Kebir, M.; Pool, R.; Elbassioni, K.; Mark, A.E.; Geerke, D.P.; Stougie, L.; Klau, G.W. Charge group partitioning in biomolecular simulation. *J. Comput. Biol.* **2013**, *20*, 188–198. [CrossRef] [PubMed]
36. Automated Topology Builder (ATB) Server. Available online: http://compbio.biosci.uq.edu.au/atb/ (accessed on 23 October 2017).
37. Van Gunsteren, W.F.; Billeter, S.R.; Eising, A.A.; Hunenberger, P.H.; Krüger, P.; Mark, A.E.; Scott, W.R.P.; Tironi, I.G. *Biomolecular Simulation: The GROMOS96 Manual and User Guide*; VDF Hochschulverlag AG an der ETH Zürich: Zurich, Switzerland, 1996.
38. Abraham, M.J.; Murtola, T.; Schulz, R.; Pall, S.; Smith, J.C.; Hess, B.; Lindahl, E. GROMACS: High performance molecular simulations through multi-level parallelism from laptops to supercomputers. *SoftwareX* **2015**, *1–2*, 19–25. [CrossRef]
39. Caddigan, E.J. *VMD User's Guide*; University of Illinois; Beckman Institute: Urbana, IL, USA, 2004.
40. Edwards, P.M. Origin 7.0: Scientific graphing and data analysis software. *J. Chem. Inf. Comput. Sci. Wash.* **2002**, *42*, 1270–1271. [CrossRef]
41. Mancini, D.T.; Matos, K.S.; da Cunha, E.F.; Assis, T.M.; Guimarães, A.P.; França, T.C.; Ramalho, T.C. Molecular modeling studies on nucleoside hydrolase from the biological warfare agent Brucella suis. *J. Biomol. Struct. Dyn.* **2012**, *30*, 125–136. [CrossRef] [PubMed]
42. Martins, T.L.C.; Ramalho, T.C.; Figueroa-Villar, J.D.; Flores, A.F.C.; Pereira, C.M.P. Theoretical and experimental C-13 and N-15 NMR investigation of guanylhydrazones in solution. *Magn. Reson. Chem.* **2013**, *41*, 983–988. [CrossRef]
43. De Castro, A.A.; Prandi, I.G.; Kamil, K.; Ramalho, T.C. Organophosphorus degrading enzymes: Molecular basis and perspectives for enzymatic bioremediation of agrochemicals. *Ciencia e Agrotecnologia* **2017**, *41*, 471–482.

© 2017 by the authors. Licensee MDPI, Basel, Switzerland. This article is an open access article distributed under the terms and conditions of the Creative Commons Attribution (CC BY) license (http://creativecommons.org/licenses/by/4.0/).

toxins

MDPI

Article

Anti-*Salmonella* Activity Modulation of Mastoparan V1—A Wasp Venom Toxin—Using Protease Inhibitors, and Its Efficient Production via an *Escherichia coli* Secretion System

Yeon Jo Ha [1,†], Sam Woong Kim [1,†], Chae Won Lee [2], Chang-Hwan Bae [2], Joo-Hong Yeo [2], Il-Suk Kim [1], Sang Wan Gal [1], Jin Hur [3], Ho-Kyoung Jung [4], Min-Ju Kim [5] and Woo Young Bang [2,*]

[1]   Swine Science and Technology Center, Gyeongnam National University of Science and Technology, Gyeongnam 52725, Korea; chakfhd@daum.net (Y.J.H.); swkim@gntech.ac.kr (S.W.K.); iskim@gntech.ac.kr (I.-S.K.); sangal@gntech.ac.kr (S.W.G.)
[2]   National Institute of Biological Resources (NIBR), Environmental Research Complex, Incheon 22689, Korea; chaewon326@korea.kr (C.W.L.); bae0072@korea.kr (C.-H.B.); y1208@korea.kr (J.-H.Y.)
[3]   Veterinary Public Health, College of Veterinary Medicine and Bio-Safety Research Institute, Chonbuk National University, Iksan 54596, Korea; hurjin@jbnu.ac.kr
[4]   Komipharm International Co. Ltd., Gyeonggi 15094, Korea; pignvet@gmail.com
[5]   Department of Alternative Medicine, Kyonggi University, Gyeonggi 16227, Korea; only1only1@naver.com
*   Correspondence: wybang@korea.kr; Tel.: +82-32-590-7203
†   These authors contributed equally to this work.

Academic Editor: Steve Peigneur
Received: 15 September 2017; Accepted: 11 October 2017; Published: 13 October 2017

Abstract: A previous study highlighted that mastoparan V1 (MP-V1), a mastoparan from the venom of the social wasp *Vespula vulgaris*, is a potent antimicrobial peptide against *Salmonella* infection, which causes enteric diseases. However, there exist some limits for its practical application due to the loss of its activity in an increased bacterial density and the difficulty of its efficient production. In this study, we first modulated successfully the antimicrobial activity of synthetic MP-V1 against an increased *Salmonella* population using protease inhibitors, and developed an *Escherichia coli* secretion system efficiently producing active MP-V1. The protease inhibitors used, except pepstatin A, significantly increased the antimicrobial activity of the synthetic MP-V1 at minimum inhibitory concentrations (determined against 10^6 cfu/mL of population) against an increased population (10^8 cfu/mL) of three different *Salmonella* serotypes, Gallinarum, Typhimurium and Enteritidis. Meanwhile, the *E. coli* strain harboring *OmpA SS::MP-V1* was identified to successfully secrete active MP-V1 into cell-free supernatant, whose antimicrobial activity disappeared in the increased population (10^8 cfu/mL) of *Salmonella* Typhimurium recovered by adding a protease inhibitor cocktail. Therefore, it has been concluded that our challenge using the *E. coli* secretion system with the protease inhibitors is an attractive strategy for practical application of peptide toxins, such as MP-V1.

Keywords: AMP; bacterial secretion system; inoculum effect; mastoparan; MP-V1; protease inhibitor; *Salmonella*; wasp venom toxin

---

1. Introduction

Salmonella infection is a major public health concern causing a primary enteric pathogenic disease in both humans and animals [1–3]. For example, *Salmonella* serotypes, such as Typhi and Gallinarum, cause typhoid fever—an acute illness—in human and domestic poultry species, respectively, and nontyphoidal *Salmonella* serotypes, including Typhimurium and Enteritidis, are the most common cause of foodborne infections [4–6]. Therefore, various antibiotics have been widely used for prevention and treatment of the infection, but this has caused the emergence and rapid dissemination of antibiotic-resistant bacteria, leading to serious problems with global human deaths due to antibiotic-resistant infections [7]. For this reason, recent studies have highlighted the discovery of novel and potent antimicrobial agents, including alterative drugs based on antimicrobial peptides (AMPs) [3,8–12].

The most promising candidates for AMPs have been discovered extensively in the venom of animals such as scorpions, snakes, spiders, ants, wasps, bees, centipedes, and so on. [13]. For example, peptide toxins, such as androtonin, parbutoporin, opistoporins, TstH and vpAmp 1.0, from scorpion venom, showed potent antimicrobial activity against Gram-positive and Gram-negative bacteria or fungi [14–17]. Cardiotoxin and crotamine from snake venom also exhibited potent antibacterial or antifungal activity [18,19]. Particularly, wasp and spider venoms offer a vast source of AMPs due to their diversity around the globe, with more than 20,000 and almost 40,000 species, respectively; mastoparans are representative AMPs from wasp venoms, and toxins including lycotoxins, lactarcins, oxyopinins and lycosin-II were identified in spider venoms [13].

Even though the toxins originated from venoms have been identified extensively as potent AMPs, there exist some limits for their practical application. For example, the AMPs can be subject to proteolytic degradation by proteases produced from an increased bacterial population [20–22], which may limit their pharmaceutical, nutraceutical and cosmeceutical uses. In addition, there are limits for their large-scale production because the chemical synthesis of large amounts of AMPs is unavailable in low unit cost and the over-collection of crude venom extracts for purification of AMPs can cause ecosystem destruction [23,24]. To the best of our knowledge, this report is the first that addresses these issues.

Recently, we reported that the mastoparan V1 (MP-V1), a de novo type of mastoparan from venom of the social wasp *Vespula vulgaris*, has superior anti-*Salmonella* activity compared with other typical mastoparans [25]. In this study, we also successfully modulated its antimicrobial activity against an increased *Salmonella* population through the use of protease inhibitors to overcome the proteolysis. In addition, we first made a cell-free supernatant including the MP-V1 with potent antimicrobial activity using an *Escherichia coli* secretion system. Therefore, our study supplies important information to set new strategies to modulate the antimicrobial activity of venom toxins and to produce them effectively for their practical application.

2. Results

2.1. Antimicrobial Activity of Synthetic MP-V1 against the Three Salmonella Serotypes

Antimicrobial activity of the synthetic MP-V1 used in the previous study [25] was examined with 25 to 250 μg/mL concentrations against 10^6 cfu/mL of three different *Salmonella* serotypes, Gallinarum—the typhoidal serotype—and Typhimurium and Enteritidis, the nontyphoidal serotypes (Table 1), as shown in Figure 1A. The minimum inhibitory concentrations (MICs) were determined as 106.95, 56.86 and 123 μg/mL against the three serotypes, Gallinarum, Typhimurium and Enteritidis, respectively. Subsequently, antimicrobial activity of the MP-V1 was examined with the MICs against 10^3 to 10^8 cfu/mL of *Salmonella* population (Figure 1B). MP-V1 at the MICs, determined by 10^6 cfu/mL, significantly inhibited the bacterial growth against the 10^3 to 10^7 cfu/mL of the three different serotypes (Figure 1B). However, it showed no antimicrobial activities against the 10^8 cfu/mL of population in all three serotypes (Figure 1B). Further challenges to recover its antimicrobial activities against the 10^8 cfu/mL of population were performed as reported in the following section.

Table 1. Bacterial strains and plasmids used for this study.

Strains or Plasmids	Genotypes or Phenotypes	Sources
Bacterial strains		
E. coli		
Top10	F-*mcr*A Δ(*mrr-hsd*RMS-*mcr*BC) F80*lacZ* ΔM15 Δ*lac*X74 *rec*A1 *ara*Δ139 Δ(*ara-leu*)7697 *gal*U *gal*K *rps*L (Strr) *end*A1 *nup*G	Invitrogen
Salmonella		
HJL331	*Salmonella* Typhimurium HJL331, Wild type, Sm^R (isolated from swine)	Chonbok National University, Korea
HJL462	*Salmonella* Gallinarum HJL462, Wild type, Na^R (isolated from chicken)	Chonbok National University, Korea
HJL390	*Salmonella* Enteritidis HJL390, Wild type, Cm^R (isolated from swine)	Chonbok National University, Korea
Plasmids		
T-vector	Cloning vector; pUC*ori* Amp^R	Promega
pMMP319	A T-vector derivative harboring *OmpA SS*	This study
pMMP320	A T-vector derivative harboring *OmpA SS::MP-V1*	This study

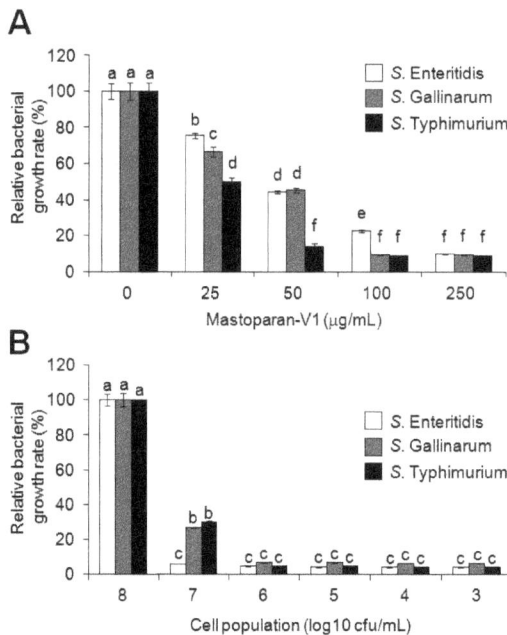

Figure 1. Antimicrobial activity of mastoparan V1 (MP-V1) against three *Salmonella* serotypes. (**A**) Determination of minimum inhibitory concentration (MIC) at against the three *Salmonella* serotypes, *Salmonella* Enteritidis, *Salmonella* Gallinarum and *Salmonella* Typhimurium, shown in Table 1. MIC of the synthetic MP-V1 was determined by using 25 to 250 μg/mL doses against 10^6 CFU/mL of the *Salmonella* serotypes. (**B**) Examination of antimicrobial activity of the MP-V1 according to the *Salmonella* population density. Antimicrobial activity of the MP-V1 was examined with the MICs, determined by 10^6 cfu/mL, against 10^3 to 10^8 cfu/mL of *Salmonella* population. Data are means ± standard error (SE) (n = 3). Different letters indicate significant differences by the one-way ANOVA/Duncan ($p < 0.05$).

2.2. Anti-Salmonella Activity Modulation of the Synthetic MP-V1 Using Various Protease Inhibitors

Previous studies have highlighted that bacteria have an intrinsic AMP resistance mechanism through proteolysis using their proteases [21,22]. Thus, we here investigated the effect of a protease inhibitor cocktail (Sigma-Aldrich, Milwaukee, WI, USA) on antimicrobial activities of MP-V1 with MICs against 10^8 cfu/mL of the three different *Salmonella* serotypes. The protease inhibitor cocktail exhibited a dose-dependent effect on the increase of the antimicrobial activity against 10^8 cfu/mL of the all three serotypes (Figure 2). Next, each of the inhibitors were independently assessed as to whether they also have an effect on the increase of antimicrobial activity because the protease inhibitor cocktail consists of various inhibitors, such as 23 mM 4-(2-aminoethyl)benzenesulfonyl fluoride (AEBSF), 2 mM bestatin, 0.3 mM pepstatin A, 0.3 mM E-64 and 100 mM ethylenediaminetetraacetic acid (EDTA) (Figure 2). Except for pepstatin A, all inhibitors used showed a significant dose-dependent effect on the increase of antimicrobial activity against 10^8 cfu/mL of all three serotypes (Figure 2). Among them, AEBSF in particular exhibited the most superior effect on the increase of antimicrobial activity against all three serotypes (Figure 2). Furthermore, EDTA, whose unit price is the lowest among the inhibitors, was shown to effectively increase the antimicrobial activity against the increased population of all three serotypes (Figure 3). These results indicate that MP-V1 is subjected to the proteolysis by bacterial proteases in the increased *Salmonella* population, and thus, protease inhibitors can be used as effective tools to modulate its antimicrobial activity.

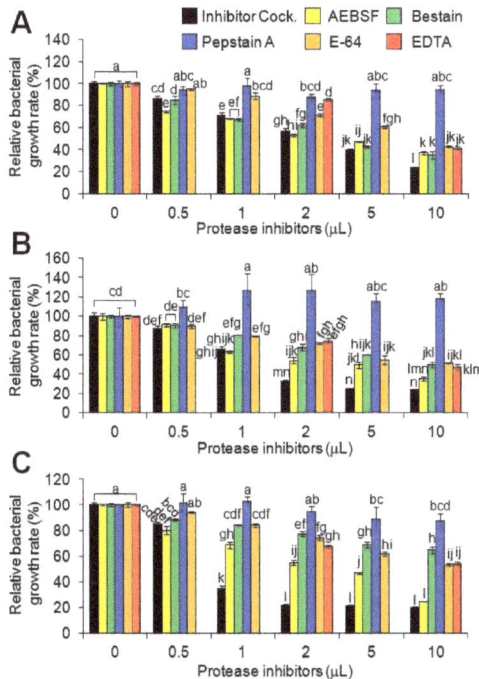

Figure 2. Effect of protease inhibitors on antimicrobial activities of MP-V1 in the increased *Salmonella* population density. The effect of a protease inhibitor cocktail (Sigma-Aldrich, Milwaukee, WI, USA) and its components (23 mM 4-(2-aminoethyl)benzenesulfonyl fluoride (AEBSF), 2 mM bestatin, 0.3 mM pepstatin A, 0.3 mM E-64 and 100 mM EDTA) on antimicrobial activities was examined using 0.5 to 10 µL doses against 10^8 cfu/mL of *Salmonella* Enteritidis (**A**); *Salmonella* Gallinarum (**B**); and *Salmonella* Typhimurium (**C**). The MP-V1 was used at the MICs determined in 10^6 cfu/mL. Data are means ± SE ($n = 3$). Different letters indicate significant differences by the one-way ANOVA/Duncan ($p < 0.05$).

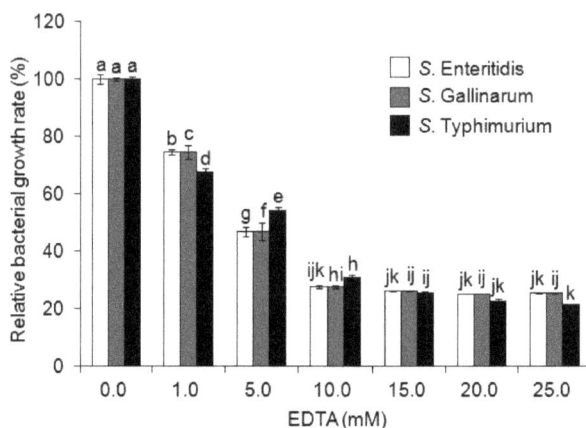

Figure 3. Effect of EDTA on antimicrobial activities of MP-V1 in the increased *Salmonella* population density. The effect of EDTA on antimicrobial activities was examined using 1 to 25 mM doses against 10^8 cfu/mL of *Salmonella* Enteritidis, *Salmonella* Gallinarum and *Salmonella* Typhimurium. The MP-V1 was used at the MICs determined in 10^6 cfu/mL. Data are means ± SE (*n* = 3). Different letters indicate significant differences by the one-way ANOVA/Duncan (*p* < 0.05).

2.3. Construction of the E. coli Secretion System to Efficiently Produce Active MP-V1

The second aim for this study is to efficiently obtain the MP-V1 with antimicrobial activity at a low unit price for a practical application. Here, we constructed an *E. coli* secretion system to efficiently express active MP-V1 peptides, subsequently secreted into the cell-free supernatant by the Sec-dependent type II secretion system, consisting of Sec and GSPs (general secretory proteins) [26–28]. The OmpA signal sequence (OmpA SS) was used as a signal peptide for the secretion of MP-V1 through the type II secretion system. In short, the nucleotide sequence fused with the *OmpA SS* and the *MP-V1* sequence were prepared by an artificial gene synthesis, and finally were cloned as the pMMP320 plasmid (Figure 4A).

To identify the secretion of MP-V1 into the cell-free supernatant, the cell-free supernatant from the *OmpA SS::MP-V1* strain, an *E. coli* cell harboring *pMMP320*, was subjected to the examination of antimicrobial activity, which was performed with 10 to 100 μL doses against 10^6 cfu/mL of *Salmonella* Typhimurium (Figure 4B). The *OmpA SS::MP-V1* strain dose-dependently inhibited the growth of *Salmonella* Typhimurium, while a negative control from the *OmpA SS* strain had no effect on the *Salmonella* growth (Figure 4B). The scanning-electron micrographs also revealed that the cell-free supernatant from the *OmpA SS::MP-V1* strain effectively caused cellular lysis through the damage of the *Salmonella* membrane via pore formation, whereas a negative control from the *OmpA SS* strain did not (Figure 4C). Thus, these results clearly prove that the cell-free supernatant, produced by the *E. coli* system, contains the active MP-V1, forming pores into the *Salmonella* membrane.

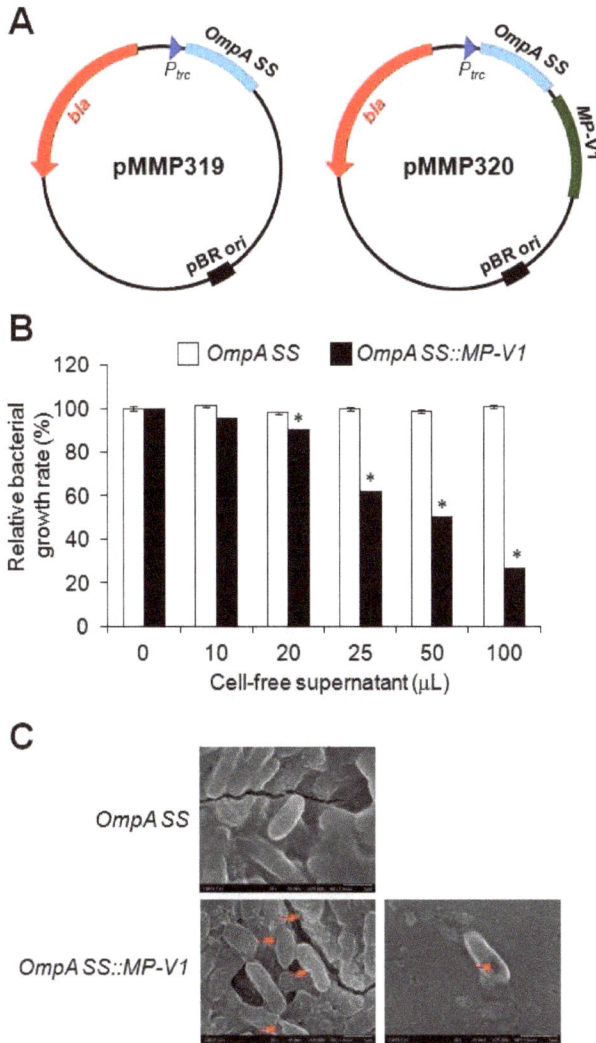

Figure 4. Examination of antimicrobial activity with the *OmpA SS::MP-V1* secretion system. (**A**) Construction of an *E. coli* secretion system for the production of MP-V1. The *MP-V1* sequence was designed to be directly fused with the *OmpA SS*, connecting to the translational start-site derived from the *Ptrc* promoter. The *OmpA SS*, fused directly to the one from the *Ptrc* promoter without the *MP-V1*, was designed to be used as a negative control. The designed artificial genes were cloned into the T-vector, finally resulting in the pMMP319 (left) and pMMP320 (right), respectively. (**B**) Comparison of antimicrobial activity of the *OmpA SS::MP-V1* secretion system with that of the *OmpA SS* one. Antimicrobial activity of the cell-free supernatants from the *OmpA SS::MP-V1* strain, an *E. coli* cell harboring pMMP320, and the *OmpA SS* strain, an *E. coli* cell harboring pMMP319, was examined with 10 to 100 μL doses against 10^6 cfu/mL of *Salmonella* Typhimurium. Data are means ± SE (*n* = 3). Asterisks indicate significant effect of the *OmpA SS::MP-V1* strain as compared to the *OmpA SS* one by the two-way ANOVA/Duncan ($p < 0.05$). (**C**) Scanning-electron micrographs of *Salmonella* Typhimurium treated with the cell-free supernatants from the *OmpA SS::MP-V1* strain and the *OmpA SS* one. The red arrows indicate the pores forming into the *Salmonella* membrane.

2.4. Anti-Salmonella Activity Modulation of the Cell-Free Supernatant Using Protease Inhibitors

When antimicrobial activity of 100 µL of the cell-free supernatant was examined against 10^3 to 10^8 cfu/mL populations of *Salmonella* Typhimurium, there were no antimicrobial activities against the 10^7 and 10^8 cfu/mL populations, while inhibitory effects on the bacterial growth rate against the 10^3 to 10^6 CFU/mL populations were observed (Figure 5A). Accordingly, as in Figure 2, the effect of the protease inhibitor cocktail on antimicrobial activities of 100 µL of the cell-free supernatant against 10^8 cfu/mL of *Salmonella* Typhimurium was investigated to increase its antimicrobial activities against the increased population. As shown in Figure 5B, the protease inhibitor cocktail exhibited a dose-dependent effect on the increase of antimicrobial activity against the 10^8 cfu/mL population. EDTA also effectively increased antimicrobial activity against the increased population of *Salmonella* Typhimurium (Figure 5C). Taken together, our results represent that the *E. coli* secretion system producing the active MP-V1 can be considered together with protease inhibitors as a successful strategy for its practical application.

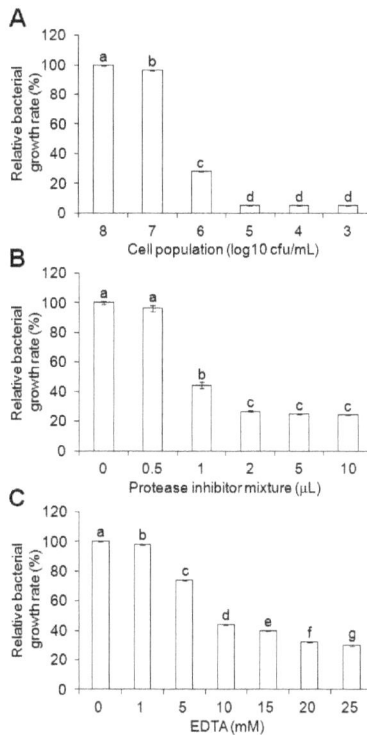

Figure 5. Effect of protease inhibitors on antimicrobial activities of the cell-free supernatant including MP-V1 in increased *Salmonella* population density. (**A**) Examination of antimicrobial activity of the cell-free supernatant including MP-V1 according to the *Salmonella* population density. The antimicrobial activity of 100 µL cell-free supernatant from the *OmpA SS::MP-V1* strain was examined against 10^3 to 10^8 cfu/mL of *Salmonella* Typhimurium. (**B,C**) The effect of the protease inhibitor cocktail and EDTA on antimicrobial activities of the cell-free supernatant including MP-V1 in increased *Salmonella* population density. The effect of a protease inhibitor cocktail (**B**) and EDTA (**C**) on antimicrobial activities of 100 µL cell-free supernatant from the *OmpA SS::MP-V1* strain was examined against 10^8 cfu/mL of *Salmonella* Typhimurium using 0.5 to 10 µL doses and 1 to 25 mM doses, respectively. Data are means ± SE (*n* = 3). Different letters indicate significant differences by the one-way ANOVA/Duncan (*p* < 0.05).

3. Discussion

3.1. Protease Inhibitors Can Modulate the Anti-Salmonella Activity of MP-V1 through Avoidance of the Inoculum Effect

Antimicrobial agents often decrease in their activity with increasing density of the starting bacterial population, and this phenomenon is known as the inoculum effect (IE) [29]. We identified that MP-V1 showed no anti-*Salmonella* activities at its MICs, determined in 10^6 cfu/mL of population, against the increased population (10^8 cfu/mL) of three different *Salmonella* serotypes, suggesting that due to the IE, its MICs might increase significantly when the number of bacteria inoculated was increased to 10^8 cfu/mL. The IE is known to be generally attributed to enzymatic degradation of the antimicrobial agents, despite the recent studies reporting other potential mechanisms, such as heat-shock-mediated growth instability, intercellular signaling between resistant and sensitive cells, and so on [29–31]. In addition, it has been reported that bacteria have an intrinsic AMP resistance mechanism through proteolysis using their proteases [21,22]. Thus, we investigated the effect of a protease inhibitor cocktail (Sigma-Aldrich, Milwaukee, WI, USA) on antimicrobial activities of MP-V1 at the MICs when the *Salmonella* inoculum density was increased to 10^8 cfu/mL. The inhibitor cocktail showed a significantly dose-dependent effect on the increase of its antimicrobial activity against 10^8 cfu/mL of all three serotypes (Figure 2), indicating that they contributed to lowering threshold levels of MP-V1 at the increased population density. Furthermore, when each of the inhibitors were examined, as in the inhibitor cocktail, all inhibitors used, except pepstatin A, showed significantly dose-dependent effects on the increase of antimicrobial activity against 10^8 cfu/mL of all three serotypes (Figure 2). The inhibitor cocktail (Sigma-Aldrich, Milwaukee, WI, USA) used in this study is optimized commercially for only bacterial uses and is a mixture of inhibitors including 23 mM AEBSF, 2 mM bestatin, 0.3 mM pepstatin A, 0.3 mM E-64 and 100 mM EDTA, which inhibit serine proteases, aminopeptidases, aspartic acid proteases, cysteine peptidases and metalloproteases, respectively [32]. Accordingly, this indicates that the IE of MP-V1 against the increased *Salmonella* population density is at least partly caused by bacterial proteases, such as serine proteases, aminopeptidases, cysteine peptidases, or metalloproteases, except for aspartic acid proteases inhibited by pepstatin A. Therefore, these results suggest that protease inhibitors can be used as effective tools for modulating anti-*Salmonella* activity of toxins, such as MP-V1 in the increased bacterial population density, through avoidance of the IE. In addition, the anti-*Salmonella* activity modulation in the range of 10^8 cfu/mL has a very important meaning for industrial applications, because that population density represents the general number of bacteria that grow in a culture medium.

3.2. Efficient Production of Active MP-V1 Using the OmpA SS-Mediated E. coli Secretion System

Obtaining efficiently potent AMPs, such as venom toxins, at low unit cost is a bottleneck in their practical application because their chemical synthesis is very expensive and the over-collection of crude venom extracts for their purification may result in ecosystem destruction [23,24]. Accordingly, instead of the above conventional methods, overexpression of an AMP in bacteria using recombinant technologies has been considered as an attractive strategy for its efficient production [24]. However, their bacterial toxicity also limits the use of the bacterial expression system for their efficient production and, thus, very few AMPs have been produced successfully in bacterial expression systems [24]. For example, AMPs such as moricin and cecropin from silkworms, defensin from humans and OG2 from frogs have been produced in bacterial expression systems using carrier proteins such as maltose-binding protein (MBP), glutathione *S*-transferase (GST), or thioredoxin (Trx) for the soluble expression of AMPs [33–36]. Furthermore, a recent *E. coli* expression system using green fluorescent protein (GFP) as a scaffold showed the efficient production of AMPs, such as protegrin-1 and PMAP-36 from pig, buforin-2 from toad and bactridin-1 from scorpion venom, in high yields [24]. However, the above bacterial expression systems still have a disadvantage in needing an additional step, such as chemical or enzymatic digestion of the MBP, the GST, the Trx or the GFP that is fused to AMPs, which may increase the production cost [24,33–36].

For efficient production of MP-V1, we constructed an *E. coli* secretion system expressing OmpA SS fusion peptides (Figure 4A). The OmpA SS was used as a signal peptide for delivery of the MP-V1 into the cell-free supernatant and in general, it exerts its role for the secretion of target peptides through the Sec-dependent type II secretion system consisting of Sec and GSPs (Figure 6). In detail, the SecA protein, existing in the cytosol, recognizes the OmpA SS, a signal sequence of a translated target protein, such as the *OmpA SS::MP-V1*, and then guides it to the plasma membrane (Figure 6) [37–40]. In the cytosol, random folding of the target protein can be prevented by SecB, a chaperone protein (Figure 6) [41]. Subsequently, the OmpA SS fusion protein is translocated to the periplasm through the SecYEG complex in the inner membrane [42], the OmpA SS signal peptide is digested by the LepB, a peptidase, and then the digested protein, such as MP-V1, is released freely into the periplasm (Figure 6) [43]. Finally, the digested protein is secreted to the extracellular space through the type II secretion system consisting of the GSPs (Figure 6) [26–28]. Through the *E. coli* secretion system, we obtained a cell-free supernatant from the *OmpA SS::MP-V1* strain, which dose-dependently inhibited the growth of *Salmonella* Typhimurium (Figure 4B) and formed pores into the *Salmonella* membrane (Figure 4C), while the one from the *OmpA SS* strain, a negative control, did not (Figure 4B,C). Thus, these results strongly support the hypothesis that the cell-free supernatant from the *OmpA SS::MP-V1* strain contains the active MP-V1, suggesting that our *E. coli* system can be a simple and efficient strategy, as opposed to previous methods that need an additional step, such as chemical or enzymatic digestion of the fusion tags such as the MBP, the GST, the Trx or the GFP.

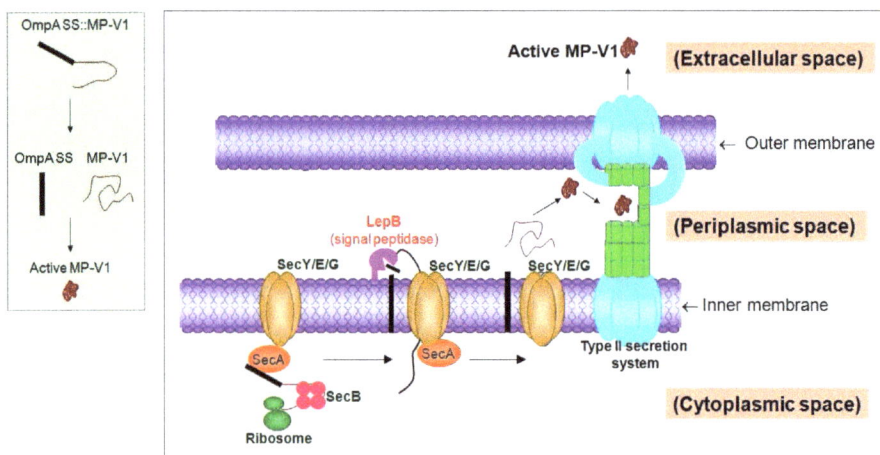

Figure 6. Secretion of MP-V1 to the extracellular space through the Sec-dependent type II secretion system. In the cytosol, the OmpA SS, a signal sequence of the *OmpA SS::MP-V1*, a translated target protein, is recognized by the SecA protein, which guides it to the plasma membrane, and the random folding of the target protein can be prevented by SecB, a chaperone protein. Subsequently, the OmpA SS fusion protein is translocated to the periplasm through the SecYEG complex in the inner membrane, the OmpA SS signal peptide is digested by the LepB, a peptidase, and then the digested MP-V1 is released freely into the periplasm. Finally, the digested peptide is secreted as the active MP-V1 to the extracellular space through the type II secretion system consisting of the general secretory proteins (GSPs). The left green box summarizes the procedure from the *OmpA SS::MP-V1* protein to the active MP-V1.

4. Conclusions

We successfully modulated the antimicrobial activity of MP-V1 in an increased *Salmonella* population density by avoiding the IE through the use of protease inhibitors, and also showed that the *OmpA SS*-mediated *E. coli* secretion system is an efficient method to produce active MP-V1 in a cell-free supernatant and can be used together with protease inhibitors in an increased *Salmonella* population density (Figure 5). Altogether, these results suggest that our *E. coli* secretion system combined with protease inhibitors may be an attractive strategy for the practical application and production of AMPs such as venom toxins.

5. Materials and Methods

5.1. Materials

The three *Salmonella* serotypes, *Salmonella* Typhimurium, *Salmonella* Enteritidis and *Salmonella* Gallinarum (Table 1), were obtained from Dr. Jin Hur (Chonbuk National University, Iksan, Korea). The synthetic MP-V1 used in the previous study [25] was also used in this study. The protease inhibitor cocktail for bacterial use and the protease inhibitors, including AEBSF, bestatin, pepstatin A, E-64 and EDTA, were purchased from Sigma-Aldrich (Milwaukee, WI, USA). An *E. coli* strain and plasmids, used for the construction of the *E. coli* secretion system, are listed in Table 1, and Top10 (an *E. coli* competent cell) and the T-vector were purchased from Invitrogen (Carlsbad, CA, USA) and Promega (Madison, WI, USA), respectively.

5.2. Analysis of Minimal Inhibitory Concentration (MIC)

MIC assays of synthetic MP-V1 against *Salmonella* Typhimurium, *Salmonella* Enteritidis and *Salmonella* Gallinarum were performed by the microtiter plate method. The synthetic MP-V1 was dissolved to an appropriate concentration (10 mg/mL) and then applied for the assay of antimicrobial activity.

The precultured strains were used for the MIC assay via adjusting to 10^6 cfu/mL or an appropriate concentration. Generally, the concentrations of synthetic MP-V1 applied for the MIC assay were 0, 25, 50, 100 and 250 µg/mL. After adding the reaction solution into a microtiter plate, the antimicrobial activity was observed for 16 h at 37 °C. The determination of MIC was performed by measurement at a wavelength of 600 nm (Multiscan GO, Thermo Scientific Co. Ltd., Rochester, NY, USA).

5.3. Examination of Antimicrobial Activity Depending on Protease Inhibitor

The starting stock solutions of each protease inhibitor used were prepared according to the information about the components of the protease inhibitor cocktail (Sigma-Aldrich, Milwaukee, WI, USA), comprising 23 mM AEBSF, 2 mM bestatin, 0.3 mM pepstatin A, 0.3 mM E-64 and 100 mM EDTA. The determination of MIC was performed at 10^8 cfu/mL and each protease inhibitor concentration as indicated in Figures 2 and 5B. The procedure for the determination of the antimicrobial activity was carried out in the same manner as described above.

5.4. Plasmid Construction for the Secretion of MP-V1 and Transformation into a General Host Strain

Nucleotide sequences of *OmpA SS* and *MP-V1* were collected via the National Center for Biotechnology Information (NCBI) and a previous study [25], respectively. The collected *OmpA SS* nucleotide sequence was designed to connect directly to the translational start site derived from the *Ptrc* promoter, and the *MP-V1* was also designed to directly connect *OmpA SS*. The *OmpA SS* was fused directly to the *Ptrc* promoter without *MP-V1*, and was designed to be used as a negative control. The designed artificial genes were prepared by an artificial-gene synthesis (Bioneer Corp., Daejon, Korea) and then cloned into T-vector (Promega, Madison, WI, USA), finally resulting in the pMMP319 and pMMP320 (Figure 4 and Table 1).

General DNA manipulations were conducted as described by Sambrook et al. [44]. Plasmids were introduced into Top10 (Invitrogen, Carlsbad, CA, USA), *E. coli* competent cells, by heat-shock with RbCl$_2$ treatment. Nucleotide sequencing was conducted by using an ABI 3730XI automatic sequencer (Applied Biosystems, Foster City, CA, USA). The *E. coli* strain and plasmids used for this study are listed in the Table 1.

5.5. Antimicrobial Activity Analysis of Cell-Free Supernatant from the E. coli Secretion System

The pMMP319 containing the *OmpA SS* and the pMMP320 containing the *OmpA SS::MP-V1* were transformed into Top10 cells, which were aerobically precultured at 37 °C until optical density at a wavelength of 600 nm is 0.5. The cultured broth was centrifuged for 20 min at 3000 rpm, the supernatant was recovered, and then the solution was filtered by a 0.2 μm syringe filter. The determinations of MICs were performed for each cell population and protease inhibitor, as indicated in Figures 4 and 6. The procedure for the determination of the antimicrobial activity was carried out in the same manner as described above.

5.6. Scanning-Electron Microscope (SEM) Analysis

Salmonella cells treated by the cell-free supernatants from the *E. coli* secretion system were fixed with a volume fraction of 2.5% glutaraldehyde (Sigma-Aldrich, Milwaukee, WI, USA) for 24 h at 4 °C. The samples were rinsed with sterile PBS buffer thrice, and then dehydrated with 30%, 50%, 70%, 80%, 90% and 100% (*v/v*) graded ethanol, successively, with 15 min incubation at each concentration. The samples were dried at room temperature and sprayed with a gold coating before the SEM observation.

5.7. Statistical Analysis

The one- or two-way analysis of variance (ANOVA) was followed by Duncan test using IBM SPSS software (IBM Corp., Armonk, NY, USA). Results are expressed as means ± standard errors (SEs) of at least three independent experiments. Different letters and asterisks indicate significant differences ($p < 0.05$).

Acknowledgments: This work was mainly supported by a grant from the National Institute of Biological Resources (NIBR), funded by the Ministry of Environment (MOE) of the Republic of Korea (NIBR201711104). S.W.K. and I.-S.K. were supported by Priority Research Centers Program through the National Research Foundation of Korea (NRF) funded by the Ministry of Education (2009-0093813).

Author Contributions: W.Y.B., I-S.K. and S.W.K. conceived and designed the experiments; Y.J.H., S.W.K., C.W.L. and M.-J.K. performed the experiments; J.H., S.W.G. and H.-K.J. analyzed the data; J.H., C.-H.B., and J.-H.Y. contributed reagents/materials/analysis tools; and W.Y.B. wrote the paper.

Conflicts of Interest: The authors declare no conflict of interest.

References

1. Dougan, G.; John, V.; Palmer, S.; Mastroeni, P. Immunity to salmonellosis. *Immunol. Rev.* **2011**, *240*, 196–210. [CrossRef] [PubMed]
2. Maiti, S.; Patro, S.; Purohit, S.; Jain, S.; Senapati, S.; Dey, N. Effective control of *Salmonella* infections by employing combinations of recombinant antimicrobial human beta-defensins hBD-1 and hBD-2. *Antimicrob. Agents Chemother.* **2014**, *58*, 6896–6903. [CrossRef] [PubMed]
3. Yeom, J.-H.; Lee, B.; Kim, D.; Lee, J.-K.; Kim, S.; Bae, J.; Park, Y.; Lee, K. Gold nanoparticle-DNA aptamer conjugate-assisted delivery of antimicrobial peptide effectively eliminates intracellular *Salmonella* enterica serovar Typhimurium. *Biomaterials* **2016**, *104*, 43–51. [CrossRef] [PubMed]
4. Kaiser, P.; Rothwell, L.; Galyov, E.E.; Barrow, P.A.; Burnside, J.; Wigley, P. Differential cytokine expression in avian cells in response to invasion by *Salmonella* typhimurium, *Salmonella* enteritidis and *Salmonella* gallinarum. *Microbiology* **2000**, *146*, 3217–3226. [CrossRef] [PubMed]

5. McClelland, M.; Sanderson, K.E.; Spieth, J.; Clifton, S.W. Complete genome sequence of *Salmonella* enterica serovar Typhimurium LT2. *Nature* **2001**, *413*, 852. [CrossRef] [PubMed]
6. Parkhill, J.; Dougan, G.; James, K.D.; Thomson, N.R. Complete genome sequence of a multiple drug resistant *Salmonella* enterica serovar Typhi CT18. *Nature* **2001**, *413*, 848. [CrossRef] [PubMed]
7. Watkins, R.R.; Bonomo, R.A. Overview: Global and local impact of antibiotic resistance. *Infect. Dis. Clin.* **2016**, *30*, 313–322. [CrossRef] [PubMed]
8. Hancock, R.E.W.; Sahl, H.-G. Antimicrobial and host-defense peptides as new anti-infective therapeutic strategies. *Nat. Biotechnol.* **2006**, *24*, 1551. [CrossRef] [PubMed]
9. Steckbeck, J.D.; Deslouches, B.; Montelaro, R.C. Antimicrobial peptides: New drugs for bad bugs? *Expert Opin. Biol. Ther.* **2014**, *14*, 11–14. [CrossRef] [PubMed]
10. Tsai, W.-C.; Zhuang, Z.-J.; Lin, C.-Y.; Chen, W.-J. Novel antimicrobial peptides with promising activity against multidrug resistant *Salmonella* enterica serovar Choleraesuis and its stress response mechanism. *J. Appl. Microbiol.* **2016**, *121*, 952–965. [CrossRef] [PubMed]
11. Xia, X.; Zhang, L.; Wang, Y. The antimicrobial peptide cathelicidin-BF could be a potential therapeutic for *Salmonella* typhimurium infection. *Microbiol. Res.* **2015**, *171*, 45–51. [CrossRef] [PubMed]
12. Wang, S.; Zeng, X.; Yang, Q.; Qiao, S. Antimicrobial peptides as potential alternatives to antibiotics in food animal industry. *Int. J. Mol. Sci.* **2016**, *17*, 603. [CrossRef] [PubMed]
13. Wang, K.; Li, Y.; Xia, Y.; Liu, C. Research on Peptide Toxins with Antimicrobial Activities. *Ann. Pharmacol. Pharm.* **2016**, *1*, 1006.
14. Ramirez-Carreto, S.; Jimenez-Vargas, J.M.; Rivas-Santiago, B.; Corzo, G.; Possani, L.D.; Becerril, B.; Ortiz, E. Peptides from the scorpion Vaejovis punctatus with broad antimicrobial activity. *Peptides* **2015**, *73*, 51–59. [CrossRef] [PubMed]
15. Machado, R.J.A.; Estrela, A.B.; Nascimento, A.K.L.; Melo, M.M.A.; Torres-Rego, M.; Lima, E.O.; Rocha, H.A.O.; Carvalho, E.; Silva-Junior, A.A.; Fernandes-Pedrosa, M.F. Characterization of TistH, a multifunctional peptide from the scorpion Tityus stigmurus: Structure, cytotoxicity and antimicrobial activity. *Toxicon* **2016**, *119*, 362–370. [CrossRef] [PubMed]
16. Powers, J.-P.S.; Hancock, R.E.W. The relationship between peptide structure and antibacterial activity. *Peptides* **2003**, *24*, 1681–1691. [CrossRef] [PubMed]
17. Hetru, C.; Letellier, L.; Ziv, O.; Hoffmann, J.A.; Yechiel, S. Androctonin, a hydrophilic disulphide-bridged non-haemolytic anti-microbial peptide: A plausible mode of action. *Biochem. J.* **2000**, *345*, 653–664. [CrossRef] [PubMed]
18. Oguiura, N.; Boni-Mitake, M.; Affonso, R.; Zhang, G. In Vitro antibacterial and hemolytic activities of crotamine, a small basic myotoxin from rattlesnake Crotalus durissus. *J. Antibiot.* **2011**, *64*, 327–331. [CrossRef] [PubMed]
19. Chen, L.-W.; Kao, P.-H.; Fu, Y.-S.; Lin, S.-R.; Chang, L.-S. Membrane-damaging activity of Taiwan cobra cardiotoxin 3 is responsible for its bactericidal activity. *Toxicon* **2011**, *58*, 46–53. [CrossRef] [PubMed]
20. Guina, T.; Eugene, C.Y.; Wang, H.; Hackett, M.; Miller, S.I. A PhoP-regulated outer membrane protease of *Salmonella* enterica serovar Typhimurium promotes resistance to alpha-helical antimicrobial peptides. *J. Bacteriol.* **2000**, *182*, 4077–4086. [CrossRef] [PubMed]
21. Nawrocki, K.L.; Crispell, E.K.; McBride, S.M. Antimicrobial peptide resistance mechanisms of gram-positive bacteria. *Antibiotics* **2014**, *3*, 461–492. [CrossRef] [PubMed]
22. Band, V.I.; Weiss, D.S. Mechanisms of antimicrobial peptide resistance in Gram-negative bacteria. *Antibiotics* **2014**, *4*, 18–41. [CrossRef] [PubMed]
23. Bray, B.L. Large-scale manufacture of peptide therapeutics by chemical synthesis. *Nat. Rev. Drug Discov.* **2003**, *2*, 587–593. [CrossRef] [PubMed]
24. Soundrarajan, N.; Cho, H.-S.; Ahn, B.; Choi, M.; Choi, H.; Cha, S.-Y.; Kim, J.-H.; Park, C.-K.; Seo, K.; Park, C. Green fluorescent protein as a scaffold for high efficiency production of functional bacteriotoxic proteins in *Escherichia coli*. *Sci. Rep.* **2016**, *6*, 20661. [CrossRef] [PubMed]
25. Kim, Y.; Son, M.; Noh, E.-Y.; Kim, S.; Kim, C.; Yeo, J.-H.; Park, C.; Lee, K.W.; Bang, W.Y. MP-V1 from the Venom of Social Wasp *Vespula vulgaris* Is a de Novo Type of Mastoparan that Displays Superior Antimicrobial Activities. *Molecules* **2016**, *21*, 512. [CrossRef] [PubMed]
26. Desvaux, M.; Parham, N.J.; Scott-Tucker, A.; Henderson, I.R. The general secretory pathway: A general misnomer? *Trends Microbiol.* **2004**, *12*, 306–309. [CrossRef] [PubMed]

27. Filloux, A. The underlying mechanisms of type II protein secretion. *Biochim. Biophys. Acta Mol. Cell Res.* **2004**, *1694*, 163–179. [CrossRef] [PubMed]

28. Johnson, T.L.; Abendroth, J.; Hol, W.G.; Sandkvist, M. Type II secretion: From structure to function. *FEMS Microbiol. Lett.* **2006**, *255*, 175–186. [CrossRef] [PubMed]

29. Karslake, J.; Maltas, J.; Brumm, P.; Wood, K.B. Population Density Modulates Drug Inhibition and Gives Rise to Potential Bistability of Treatment Outcomes for Bacterial Infections. *PLoS Comput. Biol.* **2016**, *12*, e1005098. [CrossRef] [PubMed]

30. Tan, C.; Smith, R.P.; Srimani, J.K.; Riccione, K.A.; Prasada, S.; Kuehn, M.; You, L. The inoculum effect and band-pass bacterial response to periodic antibiotic treatment. *Mol. Syst. Biol.* **2012**, *8*, 617. [CrossRef] [PubMed]

31. Lee, H.H.; Molla, M.N.; Cantor, C.R.; Collins, J.J. Bacterial charity work leads to population-wide resistance. *Nature* **2010**, *467*, 82–85. [CrossRef] [PubMed]

32. Hooper, N.M. Proteases: A primer. *Essays Biochem.* **2002**, *38*, 1–8. [CrossRef] [PubMed]

33. Hara, S.; Yamakawa, M. Production in *Escherichia coli* of moricin, a novel type antibacterial peptide from the silkworm, Bombyx mori. *Biochem. Biophys. Res. Commun.* **1996**, *220*, 664–669. [CrossRef] [PubMed]

34. Piers, K.L.; Brown, M.H.; Hancock, R.E.W. Recombinant DNA procedures for producing small antimicrobial cationic peptides in bacteria. *Gene* **1993**, *134*, 7–13. [CrossRef]

35. Xie, Y.-G.; Luan, C.; Zhang, H.-W.; Han, F.-F.; Feng, J.; Choi, Y.-J.; Groleau, D.; Wang, Y.-Z. Effects of thioredoxin: SUMO and intein on soluble fusion expression of an antimicrobial peptide OG2 in *Escherichia coli*. *Protein Pept. Lett.* **2013**, *20*, 54–60. [CrossRef] [PubMed]

36. Xia, L.; Zhang, F.; Liu, Z.; Ma, J.I.; Yang, J. Expression and characterization of cecropinXJ, a bioactive antimicrobial peptide from Bombyx mori (Bombycidae, Lepidoptera) in *Escherichia coli*. *Exp. Ther. Med.* **2013**, *5*, 1745–1751. [CrossRef] [PubMed]

37. Akita, M.; Sasaki, S.; Matsuyama, S.-I.; Mizushima, S. SecA interacts with secretory proteins by recognizing the positive charge at the amino terminus of the signal peptide in *Escherichia coli*. *J. Biol. Chem.* **1990**, *265*, 8164–8169. [PubMed]

38. Hartl, F.-U.; Lecker, S.; Schiebel, E.; Hendrick, J.P.; Wickner, W. The binding cascade of SecB to SecA to SecYE mediates preprotein targeting to the *E. coli* plasma membrane. *Cell* **1990**, *63*, 269–279. [CrossRef]

39. Lill, R.; Cunningham, K.; Brundage, L.A.; Ito, K.; Oliver, D.; Wickner, W. SecA protein hydrolyzes ATP and is an essential component of the protein translocation ATPase of *Escherichia coli*. *EMBO J.* **1989**, *8*, 961. [PubMed]

40. Lill, R.; Dowhan, W.; Wickner, W. The ATPase activity of SecA is regulated by acidic phospholipids, SecY, and the leader and mature domains of precursor proteins. *Cell* **1990**, *60*, 271–280. [CrossRef]

41. Collier, D.N. SecB: A molecular chaperone of *Escherichia coli* protein secretion pathway. *Adv. Protein Chem.* **1993**, *44*, 151–193. [PubMed]

42. Duong, F.; Wickner, W. Distinct catalytic roles of the SecYE, SecG and SecDFyajC subunits of preprotein translocase holoenzyme. *EMBO J.* **1997**, *16*, 2756–2768. [CrossRef] [PubMed]

43. Dalbey, R.E.; Wickner, W. Leader peptidase catalyzes the release of exported proteins from the outer surface of the *Escherichia coli* plasma membrane. *J. Biol. Chem.* **1985**, *260*, 15925–15931. [PubMed]

44. Sambrook, J.; Fritsch, E.F.; Maniatis, T. *Molecular Cloning: A Laboratory Manual*; Cold Spring Harbor Laboratory Press: Cold Spring Harbor, NY, USA, 1989.

© 2017 by the authors. Licensee MDPI, Basel, Switzerland. This article is an open access article distributed under the terms and conditions of the Creative Commons Attribution (CC BY) license (http://creativecommons.org/licenses/by/4.0/).

toxins

MDPI

Article

The Influence of Resiniferatoxin (RTX) and Tetrodotoxin (TTX) on the Distribution, Relative Frequency, and Chemical Coding of Noradrenergic and Cholinergic Nerve Fibers Supplying the Porcine Urinary Bladder Wall

Ewa Lepiarczyk [1,*], Agnieszka Bossowska [1], Jerzy Kaleczyc [2], Agnieszka Skowrońska [1], Marta Majewska [1] ⓘ, Michal Majewski [3] and Mariusz Majewski [1] ⓘ

[1] Department of Human Physiology, Faculty of Medical Sciences, University of Warmia and Mazury in Olsztyn, Warszawska 30, 10-082 Olsztyn, Poland; agnboss@uwm.edu.pl (A.B.); agnieszka.skowronska@uwm.edu.pl (A.S.); marta.majewska@uwm.edu.pl (M.M.); mariusz.majewski@uwm.edu.pl (M.M.)

[2] Department of Animal Anatomy, Faculty of Veterinary Medicine, University of Warmia and Mazury in Olsztyn, 10-719 Olsztyn, Poland; jerzy.kaleczyc@uwm.edu.pl

[3] Department of Pharmacology and Toxicology, Faculty of Medical Sciences, University of Warmia and Mazury, Warszawska 30, 10-082 Olsztyn, Poland; michal.majewski@uwm.edu.pl

* Correspondence: ewa.lepiarczyk@uwm.edu.pl; Tel.: +48-895-245-304; Fax: +48-895-245-307

Academic Editor: Steve Peigneur
Received: 29 August 2017; Accepted: 1 October 2017; Published: 3 October 2017

Abstract: The present study investigated the influence of intravesically instilled resiniferatoxin (RTX) or tetrodotoxin (TTX) on the distribution, number, and chemical coding of noradrenergic and cholinergic nerve fibers (NF) supplying the urinary bladder in female pigs. Samples from the bladder wall were processed for double-labelling immunofluorescence with antibodies against cholinergic and noradrenergic markers and some other neurotransmitter substances. Both RTX and TTX caused a significant decrease in the number of cholinergic NF in the urinary bladder wall (in the muscle coat, submucosa, and beneath the urothelium). RTX instillation resulted in a decrease in the number of noradrenergic NF in the submucosa and urothelium, while TTX treatment caused a significant increase in the number of these axons in all the layers. The most remarkable changes in the chemical coding of the NF comprised a distinct decrease in the number of the cholinergic NF immunoreactive to CGRP (calcitonin gene-related peptide), nNOS (neuronal nitric oxide synthase), SOM (somatostatin) or VIP (vasoactive intestinal polypeptide), and an increase in the number of noradrenergic NF immunopositive to GAL (galanin) or nNOS, both after RTX or TTX instillation. The present study is the first to suggest that both RTX and TTX can modify the number of noradrenergic and cholinergic NF supplying the porcine urinary bladder.

Keywords: resiniferatoxin; tetrodotoxin; nerve fibers; urinary bladder; immunohistochemistry; pig

1. Introduction

Overactive bladder (OAB) is a condition characterized by the presence of urinary urgency, typically accompanied by nocturia and frequency, in the absence of urinary tract infection or another evident pathology. This disease is highly prevalent, especially among women, and significantly reduces patient's quality of life [1]. It is believed that OAB symptoms result from sudden, inappropriate contractions of the muscle in the wall of the bladder during the filling phase of the micturition cycle. Since normal and abnormal contractions of the bladder are mediated by stimulation of muscarinic receptors by acetylcholine (ACh) released from parasympathetic/cholinergic neurons supplying the organ (for review see: [2]), antimuscarinic drugs are the treatment of choice in OAB. Clinically, however, their use is often unsatisfactory because of many associating side effects, such as dry mouth, cognitive changes, constipation, urinary retention, or blurred vision, resulting from the widespread blockade of cholinergic activity [3]. Therefore, new therapeutic agents, such as neurotoxins, are being investigated for the therapy of OAB. Currently, the most well-known neurotoxins, which have been successfully applied in urology, are botulinum toxin (BTX) and resiniferatoxin (RTX) (for review see: [4]).

The urine storage process largely depends upon the undisturbed and coordinated control of sympathetic-parasympathetic innervation of the urinary bladder (for review see: [2]). In our previous study [5], we found that BTX strongly influences immunohistochemical characteristics of noradrenergic and cholinergic axons distributed in the porcine urinary bladder wall. The present contribution was aimed at comparing the influence of two other neurotoxins, RTX and tetrodotoxin (TTX), on the distribution, relative frequency, and chemical coding of cholinergic and noradrenergic nerve fibers (NF) supplying this organ.

RTX is a capsaicin analogue produced by spurge Euphorbia resinifera, and it is thousand times stronger than tetrodotoxin. The toxin acts by inhibiting the transient receptor potential of vanilloid type 1 (TRPV1) nonspecific Ca^{2+} channels located mainly on the primary afferent sensory neurons involved in nociceptive signaling (for review see: [6]). Thus, in urology, RTX is predominantly used in patients suffering from detrusor overactivity as it weakens or even blocks an exaggerated C-fiber input-dependent sacral micturition reflex [7]. TTX is a neurotoxin generally found in the liver and ovaries of several marine organisms, for instance pufferfish. The toxin acts by blocking voltage-gated sodium channels in nerve cell membranes and, consequently, similar to RTX, is able to impair nociceptive transmission (for review see: [8,9]). Because of its mechanism of action, TTX has been considered as a potential therapeutic agent in the treatment of certain pains associated with diseases such as leprosy or rheumatoid arthritis (for review see: [9]). Moreover, TTX has been often applied in scientific studies concerning the normal and abnormal functioning of mammalian urinary bladder [10–12].

We decided to perform the present experiment on domestic pigs, as they are considered to be one of the best animal models used in biomedical research because of their anatomical and physiological resemblance to humans in terms of the urinary, cardiovascular, integumentary, and digestive systems [13–16].

2. Results

In the present study, we used two different control groups. Six pigs served as controls for RTX-instilled animals and were treated with intravesical instillation of 5% aqueous solution of ethyl alcohol (60 mL). Another six pigs served as the controls for TTX-instilled animals and they were treated with intravesical instillation of 20 mM citrate buffer (pH 4.9, 60 mL per animal). The above mentioned procedures were performed in the control pigs to ensure that changes in the chemical coding of NF after the toxin treatment (RTX or TTX, respectively) were caused by the toxins themselves, not due to factors associated with the administration processes. Nevertheless, it should be emphasized that no significant differences in the distribution and relative frequency of either vesicular acetylcholine transporter-immunoreactive (VAChT-IR) or dopamine β-hydroxylase-immunoreactive (DβH-IR) NF were found between the two control groups. Therefore, these results will be presented together and described in the next section.

2.1. The Distribution and Relative Frequency of VAChT-Immunoreactive Nerve Fibers

2.1.1. Control Animals

In the control pigs, the muscle layer was densely supplied with VAChT-IR NF. Many cholinergic axons were found around blood vessels. A moderate number of VAChT- IR NF was observed in the submucosa and only a few axons penetrated into the urothelium (Table 1; Figure 1A–C).

Table 1. The distribution and relative frequency of vesicular acetylcholine transporter-immunoreactive (VAChT-IR) nerve fibers supplying the porcine urinary bladder wall. RTX = resiniferatoxin (RTX); TTX = tetrodotoxin; − nerve fibers not found; +/− single fibers; + few fibers; ++ moderate number of fibers; +++ many fibers; ++++ a very dense meshwork of fibers; ↓ a decrease in the nerve fibers density.

Part of the Urinary Bladder Wall	Control Pigs	RTX-Treated Pigs	TTX-Treated Pigs
Muscle layer	++++	++ ↓	++ ↓
Submucosa	++	++	+/− ↓
Urothelium	+	+	− ↓
Blood vessels	+++	+ ↓	+/− ↓

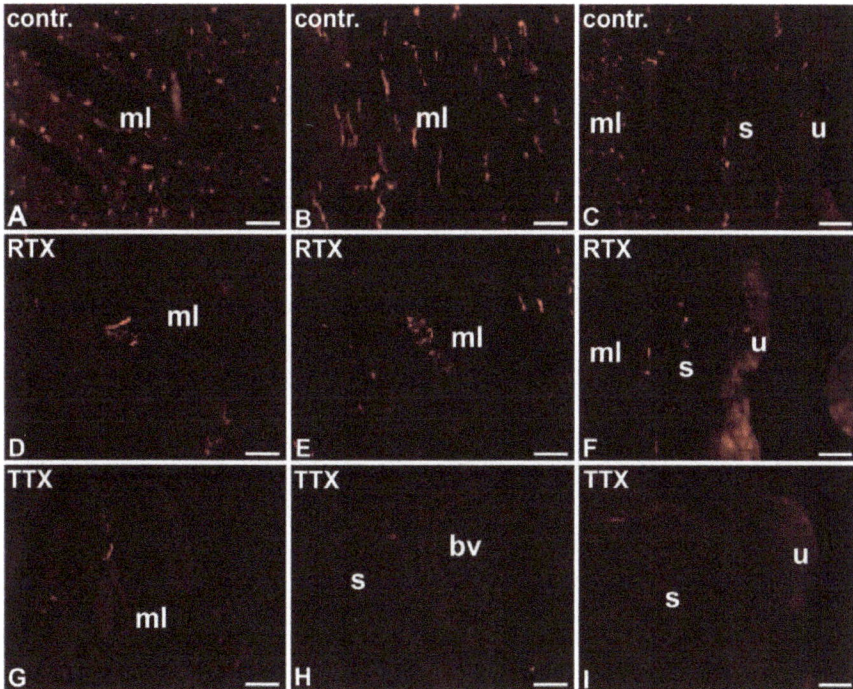

Figure 1. The distribution and relative frequency of VAChT-IR nerve terminals in the control (contr.; **A–C**), RTX-treated (RTX; **D–F**), or TTX-treated (TTX; **G–I**) pigs; muscle layer (mL), blood vessel (bv), submucosa (s), urothelium (u); 20×.

Most of the cholinergic NF supplying the muscle coat were immunopositive to somatostatin (SOM) and many stained for neuropeptide Y (NPY) or neuronal nitric oxide synthase (nNOS). Only single VAChT-IR NF observed in the muscle layer were immunoreactive (IR) to calcitonin gene-related peptide (CGRP) or vasoactive intestinal polypeptide (VIP). The majority of VAChT-IR NF surrounding blood vessels were immunopositive to SOM and single VAChT-IR nerve terminals revealed also immunoreactivity to CGRP, nNOS, NPY, or VIP. VAChT-IR NF observed in the submucosa were predominantly IR to SOM. A moderate number of these nerves expressed immunoreactivity to VIP and single VAChT-IR axons stained for CGRP or nNOS. Few cholinergic nerve terminals penetrating beneath the urothelium were CGRP-, SOM-, or VIP-positive. The cholinergic NF were galanin- (GAL)-, Leu5–enkephalin- (L-ENK)-, pituitary adenylate cyclase-activating polypeptide- (PACAP)- or substance P- (SP)–immunonegative (Table 2; Figure 2A,D,G,J,M).

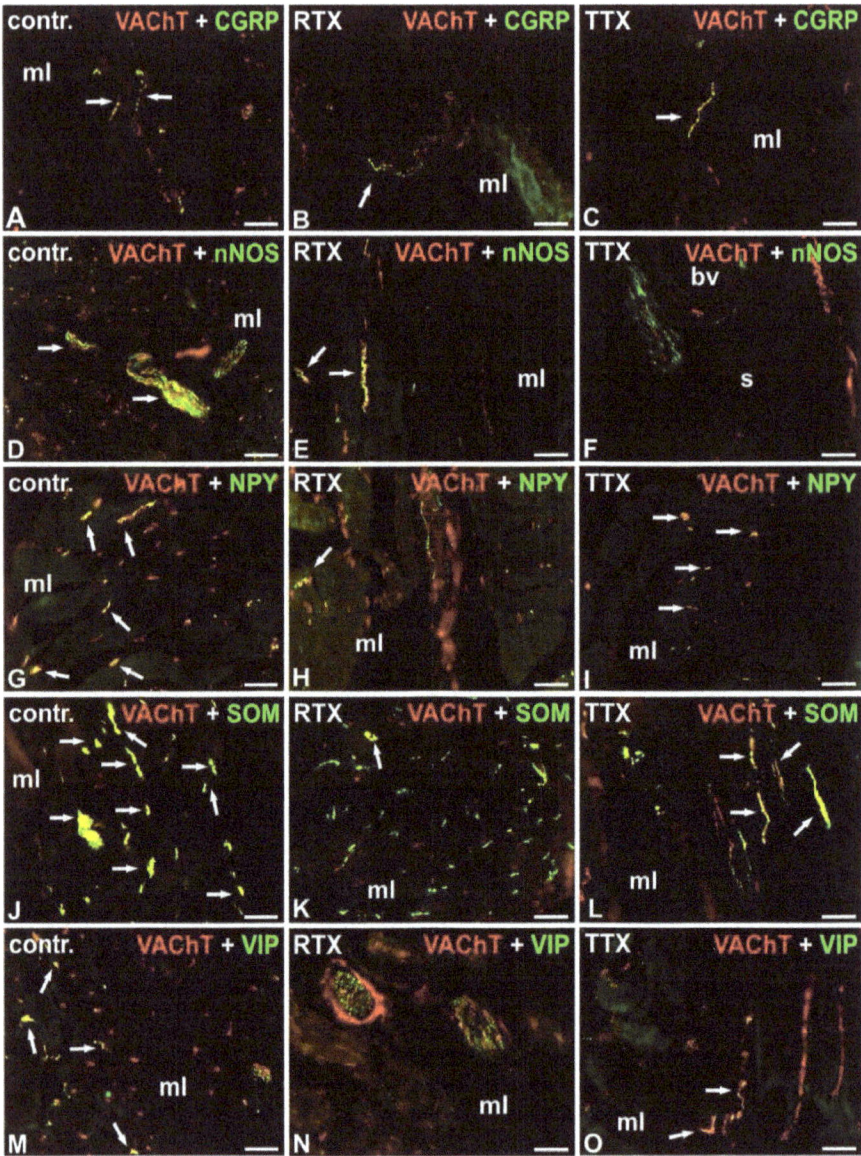

Figure 2. Distribution of VAChT-(red; labelled with CY3) and calcitonin gene-related peptide-(CGRP)- (**A–C**), neuronal nitric oxide synthase- (nNOS)- (**D–F**), neuropeptide Y- (NPY)- (**G–I**), somatostatin- (SOM)- (**J–L**) or vasoactive intestinal polypeptide- (VIP)- (**M–O**) positive (green; labelled with fluorescein isothiocyanate (FITC)) nerve fibers in the urinary bladder wall in the normal (**A,D,G,J,M**), RTX-treated (**B,E,H,K,N**), or TTX-treated (**C,F,I,L,O**) pigs. Red and green channels were digitally superimposed. Double-labelled fibers are yellow to orange, and most of them are indicated with arrows; muscle layer (mL), blood vessel (bv), submucosa (s), urothelium (u); 20×.

Table 2. The degree of colocalization of VAChT with other immunoreactive substances within the nerve fibers supplying the urinary bladder wall. mL—muscle layer; bv—blood vessels; s – submucosa; u – urothelium; − nerve fibers not found; +/− single fibers; + few fibers; ++ moderate number of fibers; +++ many fibers; ++++ a very dense meshwork of fibers. GAL= galanin; L-ENK = Leu5–enkephalin; PACAP = pituitary adenylate cyclase-activating polypeptide; SP = substance P; ↓ a decrease in the nerve fibers density.

Substances	Control Pigs				RTX-Treated Pigs				TTX-Treated Pigs			
	mL	bv	s	u	mL	bv	s	u	mL	bv	s	u
VAChT/CGRP	+	+	+	+	+/−↓	−↓	−↓	−↓	+	−↓	−↓	−↓
VAChT/GAL	-	-	-	-	-	-	-	-	-	-	-	-
VAChT/L-ENK	-	-	-	-	-	-	-	-	-	-	-	-
VAChT/nNOS	+++	+	+	-	+↓	+/−↓	-	-	−↓	−↓	−↓	-
VAChT/NPY	+++	+	-	-	++↓	+/−↓	-	-	+++	+	-	-
VAChT/PACAP	-	-	-	-	-	-	-	-	-	-	-	-
VAChT/SOM	++++	++++	++++	+	++↓	+/−↓	+↓	-	++↓	+↓	++↓	+/−↓
VAChT/SP	-	-	-	-	-	-	-	-	-	-	-	-
VAChT/VIP	+	+	++	+	+/−↓	−↓	−↓	−↓	+	−↓	+↓	−↓

2.1.2. RTX-Treated Pigs

After RTX intravesical treatment, the number of VAChT-IR NF in the smooth muscle layer and around blood vessels was smaller than that observed in the control pigs; a moderate number of these nerve terminals were found in the smooth muscle layer, and only a few VAChT-IR nerve endings were distributed around blood vessels. As in the control group, a moderate number of cholinergic NF were distributed in the submucosa and a few axons penetrated to the urothelium (Table 1; Figure 1D–F).

Double-labeling immunofluorescence revealed that RTX treatment caused a decrease in the number of cholinergic NF immunoreactive to CGRP, nNOS, NPY, SOM, or VIP in the smooth muscle layer and around the blood vessels, as a moderate number of VAChT-IR NF were NPY or SOM-IR, and single cholinergic axons stained also for CGRP, nNOS, or VIP. RTX instillation was followed by a decrease in the number of VAChT-IR axons stained for SOM in the submucosa. Moreover, in contrast to the findings obtained in the control group, no cholinergic NF immunoreactive to CGRP or VIP were found in the submucosa and beneath the urothelium. As in the control group, the cholinergic NF were GAL-, L-ENK-, PACAP-, and SP-immunonegative (Table 2; Figure 2B,E,H,K,N).

2.1.3. TTX-Treated Pigs

After TTX instillation, the number of VAChT-IR NF was smaller in all the layers of the urinary bladder wall than that observed in the control pigs; a moderate number of these axons were found in the smooth muscle layer, and single cholinergic NF were distributed in the submucosa and around blood vessels. No VAChT-IR NF were found beneath the urothelium (Table 1; Figure 1G–I).

Double-labeling immunofluorescence revealed that in the smooth muscle layer, similar to the control group, many VAChT-IR axons were also NPY-IR and few cholinergic axons stained for CGRP or VIP. Again, few cholinergic NF associated with the blood vessels were NPY-IR. However, TTX treatment caused a distinct decrease in the number of cholinergic NF immunoreactive to nNOS or SOM in all the investigated layers of the urinary bladder wall. TTX instillation was also followed by a significant decrease in the number of cholinergic NF surrounding blood vessels and revealing immunoreactivity to CGRP or VIP in the submucosa and beneath the urothelium. As in the control group, the cholinergic NF were GAL-, L-ENK-, PACAP-, and SP-immunonegative (Table 2; Figure 2C,F,I,L,O).

2.2. The Distribution and Relative Frequency of DβH-Immunoreactive Nerve Fibers

2.2.1. Control Animals

In the control pigs, the urinary bladder smooth muscle layer was sparsely supplied with noradrenergic NF, however, many DβH-IR axons surrounded blood vessels. A moderate number of DβH-IR NF was distributed in the submucosa and some of them penetrated beneath the urothelium (Table 3; Figure 3A–C).

Table 3. The distribution and relative frequency of dopamine β-hydroxylase-immunoreactive (DβH-IR) nerve fibers supplying the porcine urinary bladder wall; − nerve fibers not found; +/− single fibers; + few fibers; ++ moderate number of fibers; +++ many fibers; ++++ a very dense meshwork of fibers; ↓ a decrease in the nerve fibers density; ↑ an increase in the nerve fibers density.

Part of the Urinary Bladder Wall	Control Pigs	RTX-Treated Pigs	TTX-Treated Pigs
Muscle layer	+	+	++ ↑
Submucosa	++	+ ↓	+++ ↑
Urothelium	+/−	− ↓	+ ↑
Blood vessels	++++	++++	++++

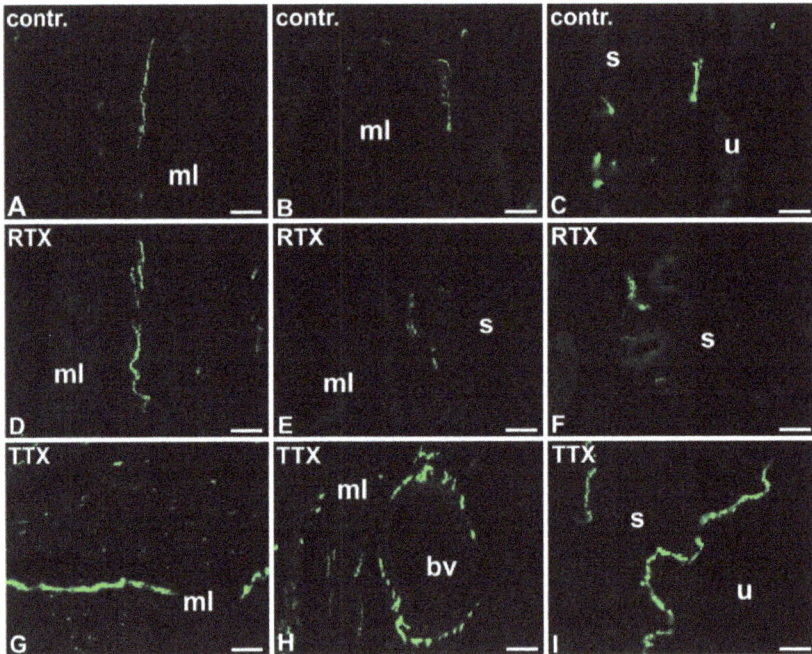

Figure 3. The distribution and relative frequency of DβH-immunoreactive nerve fibers in control (contr.; A–C), RTX-treated (RTX; D–F), or TTX-treated (TTX; G–I) pigs; muscle layer (mL), blood vessel (bv), submucosa (s), urothelium (u); 20×.

Double-labeling investigations revealed that many noradrenergic NF supplying the muscle coat were IR to NPY and a moderate number of DβH-IR axons stained for L-ENK or SOM. Solitary DβH-IR NF were immunopositive to CGRP. Most noradrenergic NF observed around blood vessels were IR to NPY, and single DβH-IR axons exhibited immunoreactivity to L-ENK or SOM. In the submucosa and beneath the urothelium, only a few DβH-IR NF were L-ENK-, NPY-, or SOM-IR. The noradrenergic NF were GAL, nNOS, PACAP, SP, or VIP-immunonegative (Table 4; Figure 4A,D,G,J,M).

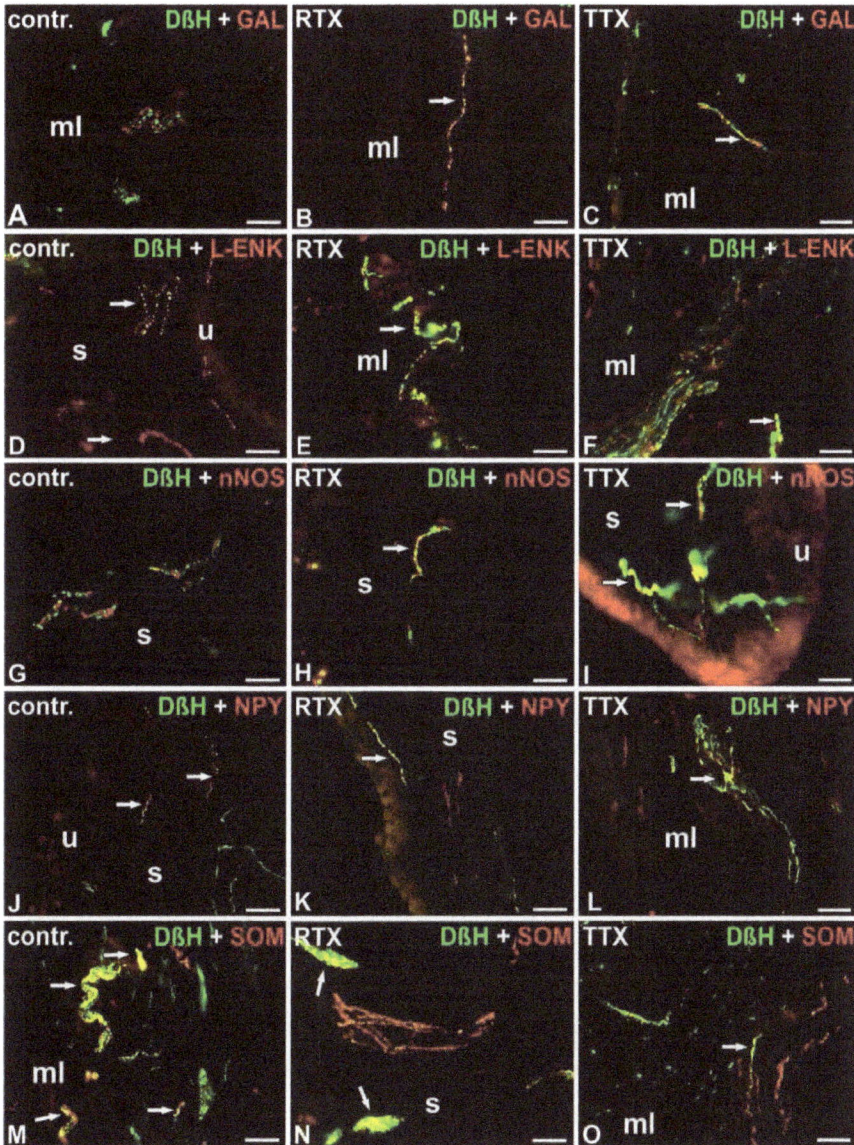

Figure 4. Distribution of DβH-(green; labelled with FITC) and GAL- (**A–C**), L-ENK- (**D–F**), nNOS- (**G–I**), NPY- (**J–L**) or SOM- (**M–O**) positive (red; labelled with CY3) nerve fibers in the urinary bladder wall in the normal (**A,D,G,J,M**), RTX-treated (**B,E,H,K,N**), or TTX-treated (**C,F,I,L,O**) pigs. Red and green channels were digitally superimposed. Double-labelled fibers are yellow to orange and most of them are indicated with arrows; muscle layer (mL), blood vessel (bv), submucosa (s), urothelium (u); 20× (**A–O**); 40× (**I**).

Table 4. The degree of colocalization of DβH with other immunoreactive substances within the nerve fibers supplying the urinary bladder wall. Muscle layer (mL); blood vessels (bv); submucosa (s); urothelium (u); − nerve fibers not found; +/− single fibers; + few fibers; ++ moderate number of fibers; +++ many fibers; ++++ a very dense meshwork of fibers. ↓ a decrease in the nerve fibers density; ↑ an increase in the nerve fibers density.

Substances	Control Pigs				RTX-Treated Pigs				TTX-Treated Pigs			
	mL	bv	s	u	mL	bv	s	u	mL	bv	s	u
DβH/CGRP	+	-	-	-	+	-	-	-	+	-	-	-
DβH/GAL	-	-	-	-	+ ↑	+ ↑	-	-	+ ↑	-	-	-
DβH/L-ENK	++	+/−	+	+	++	+/−	+	+	+++ ↑	++ ↑	+	+
DβH/nNOS	-	-	-	-	-	-	+ ↑	+ ↑	-	-	++ ↑	++ ↑
DβH/NPY	++++	++++	+	+	++ ↓	+ ↓	+/− ↓	− ↓	++ ↓	++ ↓	+/− ↓	− ↓
DβH/PACAP	-	-	-	-	-	-	-	-	-	-	-	-
DβH/SOM	++	+/−	+	+	++	+/−	+	+	+++ ↑	+ ↑	+	+
DβH/SP	-	-	-	-	-	-	-	-	-	-	-	-
DβH/VIP	-	-	-	-	-	-	-	-	-	-	-	-

2.2.2. RTX-Treated Pigs

In the urinary bladder wall of pigs treated with RTX, the number of DβH-positive NF in the smooth muscle layer and around blood vessels was similar to that observed in the control pigs; a small number of DβH-IR axons were found in the smooth muscle layer, while blood vessels were very densely supplied by these NF. However, a smaller number of the NF were distributed in the submucosa as only a few axons were found there, and no axons were observed to penetrate into the urothelium (Table 3; Figure 3D–F).

The chemical profile of the noradrenergic NF after RTX instillation was generally comparable to that observed in the control pigs. However, both in the muscle coat and around the blood vessels the number of noradrenergic NF IR to GAL was slightly higher. Moreover, a lower number of DβH-IR NF containing immunoreactivity to NPY were found in the muscle layer. Unlike in the control animals, some noradrenergic axons supplying the submucosa and urothelium were nNOS-IR. The noradrenergic NF were PACAP-, SP-, and VIP-immunonegative (Table 4; Figure 4B,E,H,K,N).

2.2.3. TTX-Treated Pigs

In the TTX treated pigs, the number of DβH-IR NF was significantly higher in all the layers of the urinary bladder wall than in that of the control animals; a moderate number of DβH-IR NF was distributed in the muscle coat, many of these nerve terminals were observed in the submucosa and few noradrenergic axons penetrated into the urothelium. Similar to the control pigs, blood vessels were densely supplied with DβH-IR NF (Table 3; Figure 3G–I).

Double-labeling investigations revealed some distinct changes in the chemical coding of the noradrenergic NF after RTX treatment. In the muscle layer, in contrast to the findings obtained from the control group, single DβH-IR NF revealed immunoreactivity to GAL. In the muscle layer and around the blood vessels, the number of DβH-IR nerve terminals containing immunoreactivity to L-ENK or SOM was also higher. The number of DβH/NPY-IR axons was definitely lower in all the investigated areas of the urinary bladder wall. Additionally, again, in contrast to the findings obtained from the control group, a moderate number of DβH-IR NF expressed immunoreactivity to nNOS in the submucosa and beneath the urothelium. DβH-positive NF were PACAP-, SP-, and VIP-immunonegative (Table 4; Figure 4C,F,I,L,O).

3. Discussion

The results of the present study clearly indicate, that application of either TTX or RTX is followed by meaningful changes in the distribution, relative frequency, and chemical coding of noradrenergic and cholinergic NF supplying the wall of the porcine urinary bladder.

A thorough discussion regarding the distribution, relative frequency, and chemical coding of noradrenergic and cholinergic NF supplying the wall of the female porcine urinary bladder was already presented in our previous paper [5]. For each of the toxins investigated, a different control group was used. In the earlier study [5], in the control pigs, no medical procedures were applied. In the present experiment, the animals included to the first control group were intravesically instilled with a 5% aqueous solution of ethyl alcohol, while the pigs assigned to the second control group were instilled with a citrate buffer. The aim of the above mentioned procedures was to guarantee that changes in the distribution and chemical coding of NF after the toxin treatment (BTX in the previous study, and RTX or TTX in the present experiment) were caused by the toxins themselves, not due to factors associated with the technique and route of their administration. It needs to be highlighted, that the number, sex, body weight, and age of the animals used as controls in both experiments as well as the surgical and immunohistochemical procedures applied were entirely corresponding. Accordingly, no significant differences in the distribution or chemical coding of NF were observed between the animals in the control groups. Therefore, in the present discussion we are concentrating on the data dealing with the changes caused by either RTX or TTX treatment.

The present findings suggest that both RTX and TTX, like BTX [5], are factors evoking very strong adaptation changes in autonomic neurons supplying the urinary bladder wall. These changes include modifications of the chemical phenotype and/or alterations in the density of NF. This seems to be an interesting finding, especially in the light of the information that RTX as well as TTX exert their primary therapeutic effect by inhibiting the noxious sensory transmission [6,8,9]. Because of their main mechanisms of action, scarce studies on the effect of either RTX or TTX on the innervation of the urinary bladder wall pertain to the influence of these toxins on sensory neurons. However, the present results suggest that the therapeutic effect observed after the toxin treatment can be a result of not only the inhibitory influence on the C-fibers but also involves changes (presumably beneficial) in the distribution and chemical coding of the cholinergic and adrenergic axons. The present data are partially supported by our previous findings, as we already have revealed that intravesically instilled RTX influences immunohistochemical characteristics of sympathetic chain ganglia urinary bladder-projecting neurons [17] and that both RTX and TTX induce plastic changes in caudal mesenteric ganglia urinary bladder-projecting neurons [18,19].

The present study suggests that both investigated toxins distinctly decrease the number of cholinergic nerve terminals in all the layers of the urinary bladder wall. Interestingly, in this respect, they seem to act similarly to BTX [5]. As was mentioned in the introduction section, anticholinergic agents are first-line pharmacotherapy for OAB because of their ability to reduce bladder contractility. Therefore, the anticholinergic effect observed after the application of the investigated toxins may suggest that their mode of action may be based on more than just the ability to block an exaggerated sensory C-fiber input from the urinary bladder wall. This bidirectional action of the investigated toxins should be considered as an advantageous effect as it has been found that antimuscarinic agents which decrease urine volume through C-fibers in the bladder are beneficial for treatment of urinary bladder disorders such as nocturia with nocturnal polyuria [20].

Furthermore, the present study has revealed that both RTX and TTX induce changes in the chemical phenotype of the investigated cholinergic nerve terminals, as the treatment with either of the toxins was followed by a decrease in the number of VAChT-IR NF that were immunopositive to CGRP, nNOS, SOM, or VIP. Additionally, RTX intravesical instillation, in contrast to TTX instillation, resulted in a decrease in the number of VAChT-IR/NPY-IR NF. The reason for the decreased expression of some of the investigated neurotransmitters or their markers in the cholinergic axons after the toxins instillation is not clear, as their functions when co-expressed with ACh are still not fully understood in the urinary tract. However, it may be expected that all these substances somehow influence the contractility of the bladder's smooth musculature.

The toxins investigated in the present study act differently on the distribution and relative frequency of DβH-IR NF. RTX seems to have little impact on the noradrenergic innervation of the urinary bladder wall (only a small decrease in the number of DβH-IR axons was observed in the submucosal layer and beneath the urothelium). TTX, on the other hand, visibly increases the number of DβH-IR NF in the muscle layer, submucosal layer, and beneath the urothelium. In this regard, TTX seems to exert a similar effect to that of BTX, as it has been found that intravesical injections of BTX are also followed by an increase in the number of noradrenergic nerve terminals in the porcine urinary bladder wall [5]. The function of sympathetic innervation of the urinary bladder is opposite to that of parasympathetic innervation. The sympathetic neurotransmitter norepinephrine (NA) inhibits the detrusor muscle by β3-adrenergic receptors and leads to a tonicization of the bladder neck and the smooth-muscular urethra by α-adrenergic receptors, thus ensuring continence (for review see: [21]). For that reason, except of anticholinergic drugs, one of the treatments approved for use in OAB includes the β3-receptor agonists [21]. Therefore, the observed increase in the number of noradrenergic NF after TTX application could be a factor which additionally decreases the spasticity of the overreactive bladder and thus improves the therapy. The present study indicates that both RTX and TTX intravesical instillations induce a decrease in the number of noradrenergic nerve terminals immunopositive to NPY. Additionally, TTX intravesical instillation, in contrast to RTX instillation, was followed by an increase in the number of DβH-IR/L-ENK-IR or DβH-IR/SOM-IR NF. Surprisingly, after both RTX or TTX treatment, some noradrenergic nerve terminals in the urinary bladder wall revealed immunofluorescence to either GAL or nNOS, while such colocalization was not observed in the control animals. Interestingly, both GAL and nitric oxide (NO) seem to exert an inhibitory role in the micturition reflex. It has been found, that galanin delays the onset of micturition through activation of the opioid mechanism [22] while NO relaxes isolated urinary bladder smooth muscle preparations (for review see: [23]).

In conclusion, the present study has revealed the existence of profound differences in the distribution, relative frequency, and chemical coding of cholinergic and noradrenergic NF supplying the wall of urinary bladders in normal female pigs and in female pigs after intravesical RTX or TTX injections. Therefore, it should be assumed that the therapeutic effects of either of this toxins on the mammalian urinary bladder can be partly mediated by the autonomic innervation of this organ.

4. Materials and Methods

4.1. Laboratory Animals

According to the guidelines of the Local Ethics Committee for Animal Experimentation in Olsztyn (affiliated to the National Ethics Commission for Animal Experimentation, Polish Ministry of Science and Higher Education; decision No. 94/2011 from 23 November 2011), the study was carried out on 24 female pigs (8–12 weeks old, 15–20 kg body weight, b.w.) of the Large White Polish race. The animals were kept under standard laboratory conditions. They were fed standard fodder (Grower Plus, Wipasz, Wadąg, Poland) and had free access to water.

Before performing intravesical instillations, all the pigs were pretreated with atropine (Polfa, Warsaw, Poland, 0.04 mg/kg b.w., s.c.) and azaperone (Stresnil, Janssen Pharmaceutica, Belgium, 0.5 mg/kg b.w., i.m.). Thirty minutes later, sodium pentobarbital (Tiopental, 0.5 g per animal) was given intravenously in a slow, fractionated infusion.

The pigs were assigned into four groups. Six animals served as controls for RTX-instilled animals and were treated with an intravesical instillation of a 5% aqueous solution of ethyl alcohol (60 mL). Another six pigs served as controls for TTX-administered animals, and they were treated with an intravesical instillation of a 20 mM citrate buffer pH 4.9 (60 mL per animal). A further group of six pigs was treated with RTX by an intravesical instillation of the toxin (500 nmol per animal in 60 mL of 5% aqueous solution of ethyl alcohol) in order to mimic the route of its administration practiced in humans. The last group of six pigs was treated with an intravesical instillation of TTX (12 µg of TTX dissolved in 60 mL of 20 mM citrate buffer, pH 4.9). In the case of all the intravesical instillations, 10 min after the infusion, the contents of the bladder were evacuated and the catheter was removed.

One week after the administration of the 5% aqueous solution of ethyl alcohol, citrate buffer, RTX, or TTX, all the pigs were deeply anaesthetized with sodium pentobarbital and transcardially perfused with 4% buffered paraformaldehyde (pH 7.4). All the urinary bladders were collected, postfixed in the same fixative (10 min at room temperature), washed several times in 0.1 M phosphate buffer (pH 7.4), and stored in 18% buffered sucrose (pH 7.4) at 4 °C until sectioning.

4.2. Sectioning of the Urinary Bladder Wall Samples and Immunohistochemical Procedure

The tissue samples analyzed in the present experiment were collected (taken) from the trigone region of the urinary bladder wall. Ten-micrometer-thick cryostat sections of the samples were processed for double-labelling immunofluorescence (according to an earlier described method [24]), using antibodies (listed in Table 5) against VAChT (marker of cholinergic fibers), DβH (marker of noradrenergic fibers), CGRP, GAL, L-ENK, nNOS, NPY, PACAP, SOM, SP, or VIP. The application of antisera raised in different species allowed investigation of the coexistence of VAChT or DβH with other substances. Each mixture of primary antibodies applied contained VAChT-antiserum or DβH-antiserum and the antiserum against one of the other biologically active substances mentioned.

Table 5. List of primary antisera and secondary reagents used in the study (CGRP = calcitonin gene-related peptide, DβH = dopamine β-hydroxylase, GAL = galanin, L-ENK = Leu5–enkephalin, nNOS = neuronal nitric oxide synthase, NPY = neuropeptide Y, PACAP = pituitary adenylate cyclase-activating polypeptide, SOM = somatostatin, SP = substance P, VAChT = vesicular acetylcholine transporter, VIP = vasoactive intestinal polypeptide, FITC = fluorescein isothiocyanate.

Antigen	Code	Dilution	Species	Supplier
		Primary antibodies		
CGRP	T-5027	1:400	Guinea pig	Peninsula; San Carlos; CA; USA
	AB5920	1:8000	Rabbit	Millipore; Temecula; CA; USA
DβH	MAB 308	1:300	Mouse	Millipore; Temecula; CA; USA
	D9010-07A.50	1:4000	Rabbit	Biomol; Hamburg; Germany
GAL	T-5036	1:1000	Guinea pig	Peninsula; San Carlos; CA; USA
	AB 5909	1:4000	Rabbit	Millipore; Temecula; CA; USA
L-ENK	4140-0355	1:800	Mouse	Bio-Rad; Kidlington; UK
	AB5024	1:600	Rabbit	Merck; Darmstadt; Germany
nNOS	N2280	1:400	Mouse	Sigma; MSU; USA
	AB 5380	1:17000	Rabbit	Millipore; Temecula; CA; USA
NPY	NA1233	1:8000	Rabbit	Enzo Life Sciences; Farmingdale; NY; USA
	sc-133080	1:100	Mouse	Santa Cruz Biotechnology; TX; USA
PACAP	T-5039	1:300	Guinea pig	Peninsula; San Carlos; CA; USA
	T-4465	1:20000	Rabbit	Peninsula; San Carlos; CA; USA
SOM	11180	1:30	Rabbit	Icn-Cappel; Aurora; OH; USA
	T-1608	1:30	Rat	Peninsula; San Carlos; CA; USA
SP	8450-0505	1:100	Rat	Bio-Rad; Kidlington; UK
VAChT	H-V006	1:6000	Rabbit	Phoenix Pharmaceuticals Inc; Burlingame; CA; USA
VIP	VA 1285	1:6000	Rabbit	Enzo Life Sciences; Farmingdale; NY; USA
	T-5030	1:1000	Guinea pig	Peninsula; San Carlos; CA; USA
		Secondary reagents		
Biotinylated anti-rabbit immunoglobulins	E 0432	1:800	Goat	Dako; Hamburg; Germany
CY3-conjugated streptavidin	711-165-152	1:8000	-	Jackson I.R.; West Grove; PA; USA
FITC-conjugated anti-mouse IgG	715-096-151	1:400	Donkey	Jackson I.R.; West Grove; PA; USA
FITC-conjugated anti-rat IgG	712-095-153	1:400	Donkey	Jackson I.R.; West Grove; PA; USA
FITC-conjugated anti-guinea pig IgG	706-095-148	1:600	Donkey	Jackson I.R.; West Grove; PA; USA

The labeled sections were viewed under an Olympus BX51 microscope equipped with epi-fluorescence and an appropriate filter set for CY3-conjugated streptavidin and fluorescein isothiocyanate (FITC). The images were taken with an Olympus XM10 digital camera (Tokyo, Japan). The microscope was equipped with cellSens Dimension 1.7 Image Processing software (Olympus Soft Imaging Solutions, Münster, Germany). The distribution and relative frequencies of labeled NF were assessed semi-quantitatively [25,26] in 10 sections per one animal (5 fields per section). The evaluation of these structures in the same preparations was performed independently by two investigators (number of the NF immunoreactive to each substance was evaluated subjectively, based on a scale from - (when the NF were not found) to ++++ (a very dense meshwork of fibers).

4.3. Control of Specificity of the Immunohistochemical Procedures

The specificity of the staining reaction was verified by preincubation tests performed on the sections from the urinary bladder wall. Overnight preincubation of 1 mL of the primary antiserum at working dilution with 20 µg/mL of the respective peptide completely eliminated the immunoreaction (Table 6). No detectable fluorescence was exhibited by the specimens after omission and replacement of the respective primary antiserum with the corresponding non-immune serum.

Table 6. List of antigens used in pre-absorption test.

Antigens Used in Pre-Absorption Test		
CGRP	C0292	Sigma; MSU; USA
DβH-blocking peptide	MBS9218238	MyBioSource; CA; USA
GAL	G5773	Sigma; MSU; USA
L-ENK	ab142314	Abcam; UK
nNOS	N3033	Sigma; MSU; USA
NPY	N3266	Sigma; MSU; USA
PACAP	A9808	Sigma; MSU; USA
SOM	S9129	Sigma; MSU; USA
SP	S6883	Sigma; MSU; USA
VAChT	V007	Phoenix Pharmaceuticals Inc; CA; USA
VIP	V6130	Sigma; MSU; USA

Acknowledgments: This study was supported by the Polish Ministry of Science and Higher Education, grant No. N40112931/2865.

Author Contributions: Ewa Lepiarczyk, Agnieszka Bossowska and Mariusz Majewski conceived and designed the experiments; Ewa Lepiarczyk, Agnieszka Bossowska, Agnieszka Skowrońska, Marta Majewska, Michał Majewski and Mariusz Majewski performed surgical procedures and contributed reagents/materials/analysis tools; Ewa Lepiarczyk and Agnieszka Bossowska performed immunohistochemical procedures; Ewa Lepiarczyk analyzed the data; Ewa Lepiarczyk wrote the paper; Jerzy Kaleczyc performed paper revision.

Conflicts of Interest: The authors declare that they have no conflict of interest.

References

1. Haylen, B.T.; de Ridder, D.; Freeman, R.M.; Swift, S.E.; Berghmans, B.; Lee, J.; Monga, A.; Petri, E.; Rizk, D.E.; Sand, P.K.; et al. An International Urogynecological Association (IUGA)/International Continence Society (ICS) joint report on the terminology for female pelvic floor dysfunction. *Neurourol. Urodyn.* **2010**, *29*, 4–20. [CrossRef] [PubMed]

2. De Groat, W.C.; Griffiths, D.; Yoshimura, N. Neural control of the lower urinary tract. *Compr. Physiol.* **2015**, *5*, 327–396. [CrossRef] [PubMed]

3. Fonseca, A.M.; Meinberg, M.F.; Monteiro, M.V.; Roque, M.; Haddad, J.M.; Castro, R.A. The Effectiveness of Anticholinergic Therapy for Overactive Bladders: Systematic Review and Meta-Analysis. *Rev. Bras. Ginecol. Ostet.* **2016**, *38*, 564–575. [CrossRef] [PubMed]

4. Cruz, F.; Dinis, P. Resiniferatoxin and botulinum toxin type A for treatment of lower urinary tract symptoms. *Neurourol. Urodyn.* **2007**, *26*, 920–927. [CrossRef] [PubMed]

5. Lepiarczyk, E.; Bossowska, A.; Kaleczyc, J.; Majewski, M. The influence of botulinum toxin type A (BTX) on the immunohistochemical characteristics of noradrenergic and cholinergic nerve fibers supplying the porcine urinary bladder wall. *Pol. J. Vet. Sci.* **2011**, *14*, 181–189. [CrossRef] [PubMed]
6. Brown, D.C. Resiniferatoxin: The evolution of the "molecular scalpel" for chronic pain relief. *Pharmaceuticals* **2016**, *9*, 47. [CrossRef] [PubMed]
7. Cruz, F.; Guimarães, M.; Silva, C.; Reis, M. Suppression of bladder hyperreflexia by intravesical resiniferatoxin. *Lancet* **1997**, *350*, 640–641. [CrossRef]
8. Lago, J.; Rodríguez, L.P.; Blanco, L.; Vieites, J.M.; Cabado, A.G. Tetrodotoxin, an extremely potent marine neurotoxin: Distribution, toxicity, origin and therapeutical uses. *Mar. Drugs* **2015**, *13*, 6384–6406. [CrossRef] [PubMed]
9. Magarlamov, T.Y.; Melnikova, D.I.; Chernyshev, A.V. Tetrodotoxin-producing bacteria: Detection, distribution and migration of the toxin in aquatic systems. *Toxins* **2017**, *9*, 166. [CrossRef] [PubMed]
10. Gray, S.M.; McGeown, J.G.; McMurray, G.; McCloskey, K.D. Functional innervation of Guinea-pig bladder interstitial cells of cajal subtypes: Neurogenic stimulation evokes in situ calcium transients. *PLoS ONE* **2013**, *8*, e53423. [CrossRef] [PubMed]
11. Ramos-Filho, A.C.; Shah, A.; Augusto, T.M.; Barbosa, G.O.; Leiria, L.O.; de Carvalho, H.F.; Antunes, E.; Grant, A.D. Menthol inhibits detrusor contractility independently of TRPM8 activation. *PLoS ONE* **2014**, *9*, e111616. [CrossRef]
12. Kuga, N.; Tanioka, A.; Hagihara, K.; Kawai, T. Modulation of afferent nerve activity by prostaglandin E2 upon urinary bladder distension in rats. *Exp. Physiol.* **2016**, *101*, 577–587. [CrossRef] [PubMed]
13. Dalmose, A.L.; Hvistendahl, J.J.; Olsen, L.H.; Eskild-Jensen, A.; Djurhuus, J.C.; Swindle, M.M. Surgically induced urologic models in swine. *J. Invest. Surg.* **2000**, *13*, 133–145. [CrossRef] [PubMed]
14. Kuzmuk, K.N.; Schook, L.B. Pigs as a model for biomedical sciences. In *The Genetics of the Pig*, 2nd ed.; Rothschild, M.F., Ruvinsky, A., Eds.; CAB International: Oxford Shire, UK, 2011; pp. 426–444.
15. Swindle, M.M.; Makin, A.; Herron, A.J.; Clubb, F.J., Jr.; Frazier, K.S. Swine as models in biomedical research and toxicology testing. *Vet. Pathol.* **2012**, *49*, 344–356. [CrossRef] [PubMed]
16. Bassols, A.; Costa, C.; Eckersall, P.D.; Osada, J.; Sabrià, J.; Tibau, J. The pig as an animal model for human pathologies: A proteomics perspective. *Proteom. Clin. Appl.* **2014**, *8*, 715–731. [CrossRef] [PubMed]
17. Lepiarczyk, E.; Majewski, M.; Bossowska, A. The influence of intravesical administration of resiniferatoxin (RTX) on the chemical coding of sympathetic chain ganglia (SChG) neurons supplying the porcine urinary bladder. *Histochem. Cell Biol.* **2015**, *144*, 479–489. [CrossRef] [PubMed]
18. Lepiarczyk, E.; Dudek, A.; Kaleczyc, J.; Majewski, M.; Markiewicz, W.; Radziszewski, P.; Bossowska, A. The influence of resiniferatoxin on the chemical coding of caudal mesenteric ganglion neurons supplying the urinary bladder in the pig. *J. Physiol. Pharmacol.* **2016**, *67*, 625–632. [PubMed]
19. Lepiarczyk, E.; Bossowska, A.; Kaleczyc, J.; Majewska, M.; Gonkowski, S.; Majewski, M. The Influence of Tetrodotoxin (TTX) on the Distribution and Chemical Coding of Caudal Mesenteric Ganglion (CaMG) Neurons Supplying the Porcine Urinary Bladder. *Mar. Drugs* **2017**, *15*, 101. [CrossRef] [PubMed]
20. Watanabe, N.; Akino, H.; Kurokawa, T.; Taga, M.; Yokokawa, R.; Tanase, K.; Nagase, K.; Yokoyama, O. Antidiuretic effect of antimuscarinic agents in rat model depends on C-fibre afferent nerves in the bladder. *BJU Int.* **2013**, *112*, 131–136. [CrossRef] [PubMed]
21. Andersson, K.E. On the Site and Mechanism of Action of β3-Adrenoceptor Agonists in the Bladder. *Int. Neurourol. J.* **2017**, *21*, 6–11. [CrossRef] [PubMed]
22. Honda, M.; Yoshimura, N.; Inoue, S.; Hikita, K.; Muraoka, K.; Saito, M.; Chancellor, M.B.; Takenaka, A. Inhibitory role of the spinal galanin system in the control of micturition. *Urology* **2013**, *82*, 1188.e9–1188.e14. [CrossRef] [PubMed]
23. Andersson, K.E.; Persson, K. Nitric oxide synthase and nitric oxide-mediated effects in lower urinary tract smooth muscles. *World J. Urol.* **1994**, *12*, 274–280. [CrossRef] [PubMed]
24. Bossowska, A.; Majewski, M. Botulinum toxin type A-induced changes in the chemical coding of dorsal root ganglion neurons supplying the porcine urinary bladder. *Pol. J. Vet. Sci.* **2012**, *15*, 345–353. [CrossRef] [PubMed]

25. Kaleczyc, J.; Timmermans, J.P.; Majewski, M.; Lakomy, M.; Scheuermann, D.W. Immunohistochemical properties of nerve fibres supplying accessory male genital glands in the pig. A colocalisation study. *Histochem. Cell Biol.* **1999**, *111*, 217–228. [CrossRef] [PubMed]
26. Gonkowski, S.; Kamińska, B.; Landowski, P.; Całka, J. Immunohistochemical distribution of cocaine- and amphetamine-regulated transcript peptide—Like immunoreactive (CART-LI) nerve fibers and various degree of co-localization with other neuronal factors in the circular muscle layer of human descending colon. *Histol. Histopathol.* **2013**, *28*, 851–858. [CrossRef] [PubMed]

© 2017 by the authors. Licensee MDPI, Basel, Switzerland. This article is an open access article distributed under the terms and conditions of the Creative Commons Attribution (CC BY) license (http://creativecommons.org/licenses/by/4.0/).

toxins

MDPI

Article

Antiallodynic Effects of Bee Venom in an Animal Model of Complex Regional Pain Syndrome Type 1 (CRPS-I)

Sung Hyun Lee [1], Jae Min Lee [2], Yun Hong Kim [1], Jung Hyun Choi [2], Seung Hwan Jeon [3], Dong Kyu Kim [2], Hyeon Do Jeong [2], You Jung Lee [2] and Hue Jung Park [2,*]

[1] Department of Anesthesiology and Pain Medicine, Kangbuk Samsung Hospital, Sungkyunkwan University School of Medicine, Seoul 03181, Korea; 4321hoho@naver.com (S.H.L.); yhkim12.kim@samsung.com (Y.H.K.)

[2] Department of Anesthesiology and Pain Medicine, College of Medicine, Seoul St. Mary's Hospital, The Catholic University of Korea, Seoul 06591, Korea; jmlee@catholic.ac.kr (J.M.L.); tzim2000@naver.com (J.H.C.); ramsgate.dk@gmail.com (D.K.K.); paranpisr@naver.com (H.D.J.); dasaki7@gmail.com (Y.J.L.)

[3] Department of Urology, College of Medicine, Seoul St. Mary's Hospital, The Catholic University of Korea, Seoul 06591, Korea; shwan52@naver.com

* Correspondence: huejung@catholic.ac.kr; Tel.: +82-2-2258-2236 (ext. 6157); Fax: +82-2-537-1951

Academic Editor: Steve Peigneur
Received: 25 August 2017; Accepted: 13 September 2017; Published: 15 September 2017

Abstract: Neuropathic pain in a chronic post-ischaemic pain (CPIP) model mimics the symptoms of complex regional pain syndrome type I (CRPS I). The administration of bee venom (BV) has been utilized in Eastern medicine to treat chronic inflammatory diseases accompanying pain. However, the analgesic effect of BV in a CPIP model remains unknown. The application of a tight-fitting O-ring around the left ankle for a period of 3 h generated CPIP in C57/Bl6 male adult mice. BV (1 mg/kg; 1, 2, and 3 times) was administered into the SC layer of the hind paw, and the antiallodynic effects were investigated using the von Frey test and by measuring the expression of neurokinin type 1 (NK-1) receptors in dorsal root ganglia (DRG). The administration of BV dose-dependently reduced the pain withdrawal threshold to mechanical stimuli compared with the pre-administration value and with that of the control group. After the development of the CPIP model, the expression of NK-1 receptors in DRG increased and then decreased following the administration of BV. SC administration of BV results in the attenuation of allodynia in a mouse model of CPIP. The antiallodynic effect was objectively proven through a reduction in the increased expression of NK-1 receptors in DRG.

Keywords: allodynia; bee venom; chronic post-ischaemic pain; complex regional pain syndrome

1. Introduction

Bee venom (BV) has been used in traditional eastern medicine to relieve pain and to treat chronic inflammatory diseases. Various studies have demonstrated the analgesic and anti-inflammatory, as well as anti-cancer, effects of BV. BV contains various peptides, amines, nonpeptide components, and free amino acids, which are presumed to have anti-inflammatory, analgesic, and anti-cancer effects. Recent studies have revealed diverse mechanisms underlying the analgesic and anti-inflammatory effects of BV. The suppression of the expression of inflammation regulatory factors such as cyclooxygenase 2 (COX-2) and phospholipase A2 (PLA2), in addition to the generation of mediators such as tumour necrosis factor-α (TNF-α), interleukin (IL)-1, IL-6, nitric oxide (NO), and reactive oxygen species (ROS), have been reported to be related to the analgesic and anti-arthritic effects of BV [1–3]. Previous studies have demonstrated that BV treatment has analgesic effects in neuropathic pain animal models, with possible mechanisms including the activation of alpha 2-adrenoceptors, the reduction in c-Fos expression in the spinal cord, and the suppression of N-methyl-D-aspartate receptors in the spinal dorsal horn [4–6]. Although diverse effects and mechanisms have been demonstrated, unrevealed mechanisms likely still remain.

Complex regional pain syndrome type I (CRPS I) is one of the most refractory and distressing pain syndromes without a definite nerve injury. Symptoms of CRPS I include sensory changes such as allodynia or hyperalgesia, edema, abnormal vasomotor and sudomotor function, motor dysfunction, and trophic changes. CRPS I occurs following injuries such as sprains, fractures, crush injuries, and minor trauma that are not recognized. The symptoms typically start in the distal part of the affected limb and spread to the unaffected or opposite limb [7,8]. The exact pathophysiology of CRPS has not yet been fully revealed. Various studies have presented several consistent pathophysiological mechanisms that show neurogenic inflammatory responses and central sensitization [8–11]. Several kinds of neurotransmitters, such as substance P (SP), have been implicated in a series of neurogenic inflammatory responses. SP acts through stimulation of neurokinin receptors, especially type 1 (NK-1) receptors. Some studies have shown that SP activation of upregulated NK-1 receptors in the peripheral neuron, dorsal root ganglion, and spinal cord suggests the development of nociceptive and inflammatory changes considered to be an important pathophysiological pathway of CRPS [12–15].

The effect of BV on CRPS I and its mechanism of action have not been studied yet, even though the effects have been demonstrated in other types of pain models. We postulated that BV suppresses the features of CRPS I and conducted behavioural tests in a chronic post-ischaemic pain (CPIP) model produced after a 3 h-ischaemia/reperfusion (I/R) injury in the hind paws of mice induced under general anaesthesia through the application of an O-ring around the mouse's left hind limb just proximal to the ankle joint. Such a chronic post-ischemic pain (CPIP) model had already shown similar features to those described in patients with CRPS-I in previous studies [16,17]. We measured the change in NK-1 receptor expression in dorsal root ganglia (DRG) to verify the antiallodynic effects of BV.

2. Results

2.1. CPIP Mice Exhibited Prominent Mechano-Allodynia

CPIP mice developed mechano-allodynia over a prolonged period in both the ipsilateral and contralateral hind leg, with more prominent effects on the ipsilateral side (Figure 1). Ipsilateral mechano-allodynia was exhibited within 8 h following reperfusion; it peaked at 2 days and was maintained for at least 30 days after reperfusion. Contralateral mechano-allodynia was also present within 8 h following reperfusion; it peaked at 2 days and was maintained for 15 days after reperfusion. Those features were observed on four of six mice tightly-fitted with O-ring mice.

2.2. BV Attenuated Mechanical Allodynia in CPIP Mice

Intrapaw BV injections dose-dependently reduced mechanical allodynia in CPIP mice when compared with that in the control group. In all of the BV-injected groups, the paw withdrawal thresholds (PWTs) were demonstrated to first increase and then decrease. The variance in the PWT among the BV-injected groups was different. Among the three groups injected with BV, injection in triplicate had the greatest effect on the mechanical withdrawal thresholds, indicating that it was the most effective at attenuating allodynia. The effect presented within 30 min after injection and peaked at 1 h in the groups injected with BV two and three times. The effect persisted for different lengths of time in the different BV-injected groups: 90 min for the single injection group, 120 min for the double injection group, and 180 min for the triple injection group (Figure 2). Repeated injections were suggested to amplify the anti-mechano-allodynic effect in CPIP mice.

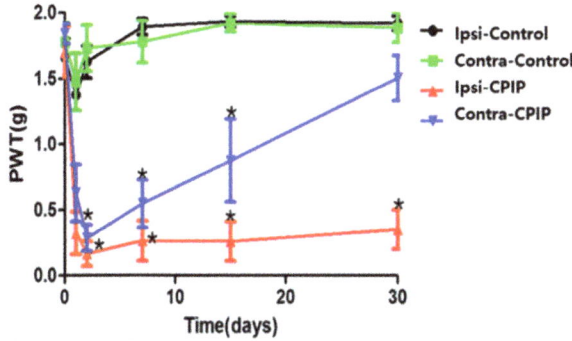

	Ipsi-Vehicle	Contra-Vehicle	Ipsi-CPIP	Contra-CPIP
0	1.66±0.35	1.77±0.31	1.70±0.48	1.84±0.18
1 day	1.37±0.30	1.47±0.50	0.32±0.40	0.63±0.53
3 day	1.63±0.27	2.97±0.43	0.16±0.23	0.29±0.25
7 day	1.90±0.14	1.78±0.39	0.26±0.37	0.55±0.47
15 day	1.93±0.10	1.92±0.16	0.26±0.37	0.88±0.77
30 day	1.92±0.13	1.88±0.28	0.35±0.37	1.50±0.42

Figure 1. Time course of tactile allodynia in the ipsilateral and contralateral hind paw of CPIP and control mice, as shown via von Frey testing. The contralateral withdrawal thresholds of control mice were not meaningfully altered throughout the one month of testing. The withdrawal thresholds of CPIP mice were significantly reduced 30 days after reperfusion ipsilaterally and 15 days contralaterally. Asterisk (∗) indicates $p < 0.05$ at each time point between control and CPIP mice.

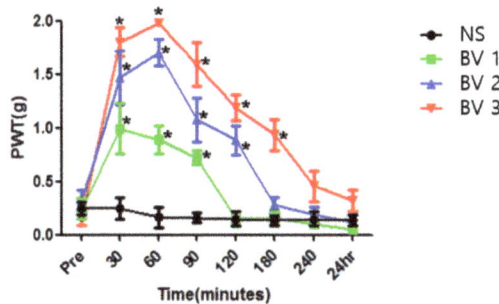

Figure 2. The effect of the administration of bee venom (BV) on the tactile threshold in chronic post-ischaemic pain (CPIP) mice. BV injections dose-dependently reduced mechanical allodynia in CPIP mice when compared with that in the control group. The triple injection group (BV3) showed the most effective attenuation of mechanical allodynia. Asterisk (∗) indicates $p < 0.05$ at each time point compared to that in the saline group.

2.3. BV Attenuated the Increased Expression of NK-1 Receptors in CPIP Mice

The CPIP group showed higher NK-1 receptor expression than the sham group, as mentioned (Figure 3), and as indicated by the higher optical densities measured in the CPIP group ($p = 0.04$).

After the triple injection of BV, 11 days after I/R injury, DRG were harvested and examined for the immunohistochemical expression of NK-1 receptors. In the BV-injected group, the increased expression of NK-1 receptors was significantly reduced, as exhibited by the lower optical densities measured in the BV-injected group than in the CPIP group ($p = 0.013$) (Figure 4). The change in NK-1 receptor expression demonstrated that BV might be effective in CPIP models.

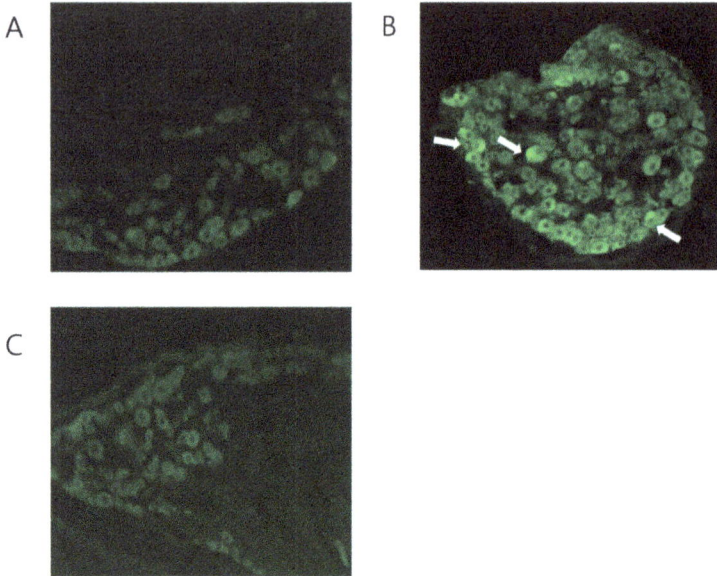

Figure 3. The effect of subcutaneous BV on NK-1 receptor expression in dorsal root ganglia (DRG). Immunostaining for NK-1 receptors in a control mouse. Original magnification: ×200. (**A**); chronic post-ischaemic pain (CPIP) mouse. Original magnification: ×200. (**B**); and BV-injected mouse. Original magnification: ×20. (**C**).

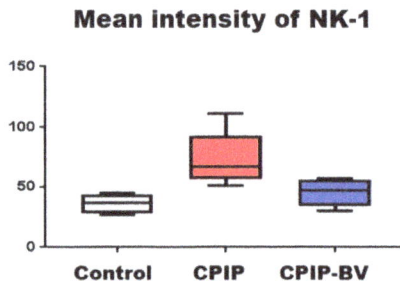

Figure 4. Histogram representing the optical density of NK-1 receptors in DRG from sham ($n = 4$), CPIP ($n = 8$), and BV-treated CPIP mice ($n = 6$). The CPIP group (73.61 ± 20.92 optical density) showed a higher expression of NK-1 receptors than the control group (36.39 ± 8.32 optical density). The lower expression of NK-1 receptors in the BV-treated group (45.57 ± 11.46 optical density) than in the CPIP group demonstrated that BV significantly suppressed the expression of NK-1 receptors (white arrow: immunostaining for the NK-1 receptor).

3. Discussion

Our findings reveal that a novel animal model of complex regional pain syndrome type I (CRPS I), a chronic post-ischaemic pain (CPIP) model, developed mechanical allodynia, which was then attenuated by the administration of bee venom (BV). Histologically, the increased expression of neurokinin type 1 (NK-1) receptors and the decline in NK-1 expression after BV injection in dorsal root ganglia (DRG) validated the effect of BV in CPIP mice.

In previous studies, the effect of BV has been demonstrated on nerve injury models, such as a spinal cord injury model, and neuropathic pain models, such as an oxaliplatin-induced neuropathic pain model and a chronic constrictive injury model [4–6]. However, the analgesic effect of BV on CRPS has not yet been studied. This series of experiments verified the effect of BV on CRPS in a CPIP model. In previous studies, the injection route was usually intraperitoneal or acupoint, which has been employed in traditional medicine. Subcutaneous BV injections, specifically in ischaemia/reperfusion (I/R)-injured paws, were chosen in this study. Intrapaw BV injections attenuated mechanical allodynia in injected paws and decreased NK-1 expression in DRG, suggesting that BV had not only a topical effect but also a systemic and spinal effect.

The specific analgesic mechanisms of BV are unclear, but several mechanisms have been suggested. Activation of spinal α2-adrenoceptors, decreased c-fos expression, and the *N*-methyl-D-aspartate receptor blockade are mechanisms that have been suggested in previous studies [6,18–21]. We found that BV injection significantly reduced NK-1 expression in DRG, potentially suggesting a novel analgesic mechanism of BV, in which suppressed NK-1 expression results in a decrease in Substance P (SP) signalling. Even though all of the CRPS pathophysiological pathways are not understood, neurogenic inflammation has been suggested to cause primary afferent nociceptor sensitization followed by central sensitization. Neurogenic inflammation is mediated by neuropeptides, especially calcitonin gene-related peptide (CGRP) and SP. In the rat fracture/cast model that exhibits the symptoms of CRPS, SP and CGRP expression was increased in the sciatic nerve and serum, and NK-1 receptor expression was upregulated in the skin of the hind paw [14]. Infusion with SP further exaggerated the extravasation responses to an increase in protein leakage in the affected hind-paw skin [12]. Similar to the results observed in the animal models, the infusion of SP through a microdialysis membrane in CPRS volunteers accelerated plasma protein extravasation, an effect that was also present in the contralateral unaffected limb [22]. These findings indicate that the effect of SP is not only regional at the affected lesion but also systemic at the contralateral lesion. Apart from its peripheral actions, SP has distinct effects on the central nervous system. In the rat fracture/cast model of CRPS, the NK-1 receptor signalling in the spinal cord was increased. This upregulation in the spinal cord was sustained through 16 weeks but only lasted 4 weeks in the skin [23]. This study showed a shift in the location of this neuro-inflammatory mediator, leading the CRPS symptoms from the periphery to the central spinal cord. Thus, SP might be an important neuropeptide in CRPS. The findings in the present study suggest that BV injection might be used as a therapeutic treatment for CRPS via the suppression of NK-1 signalling. We could presume that the suppression of NK-1 signalling might occur through the inhibition of nuclear factor-κB (NF-κB) activity. In recent studies, melittin, among a variety of peptides, is an important constituent of the anti-inflammatory, anti-analgesic pathway. Melittin inhibits the DNA binding activity of NF-κB, resulting in a decline in the expression of this inflammation-related gene [1,24]. NF-κB activity is stimulated by many inflammatory stimuli, and activated NF-κB dimers enter into the nucleus, where they bind to DNA binding sites, resulting in the expression of proinflammatory genes. Reduced NF-κB activity induces a decrease in SP production and NK-1 receptor expression [25,26]. BV reduced NK-1 receptor expression and showed an anti-inflammatory or anti-analgesic effect via these pathways.

In a previous study, an injection of a high dose of BV (2.5 mg/kg) into an acupoint induced a motor function deficit at 60 and 120 min [6], as well as skin hypersensitivity; adverse effects such as itching, but not severe effects such as an anaphylactic reaction, have been documented [27,28]. These side effects were not observed in this study.

We conclude that BV given subcutaneously attenuates allodynia in mice models of CPIP without notable adverse effects. The antiallodynic effects were closely associated with a significant decrease in NK-1 receptor expression in DRG. These findings suggest that repetitive BV therapy could be a useful therapeutic modality for the treatment of CRPS. Henceforward, more subjects and clinical studies will be needed to determine the clinical use of BV in CRPS. In addition, the antiallodynic effects of BV in this study were demonstrated during the acute phase of CRPS, 7 days after reperfusion injury. The acute phase of CRPS commonly presents with signs of acute neurogenic inflammation, such as erythema, warmth, oedema, and hyperalgesia. Even if the anti-inflammatory effect of BV might attenuate the symptoms of the acute phase of CRPS, such as neurogenic inflammation, the effect of BV is unlikely to diminish the symptoms of the chronic phase of CRPS. As time passes, the warmth and erythaematous symptoms change to cold and atrophic symptoms. Moreover, signs and symptoms of central sensitization present increasingly in the chronic phase of CRPS. Further studies of a chronic CRPS model will be needed to show the effect of BV in the chronic phase of CRPS. More research on the antiallodynic effect of BV could provide an alternative therapeutic tool to treat neuropathic pain, especially CRPS, for which there is a lack of effective and safe therapeutic regimens [8–10,29].

4. Materials and Methods

4.1. Animals

The study protocol was approved by the Institutional Animal Care and Use Committee (IACUC) of the College of Medicine, Catholic University of Korea. The approval code is 2014-0055 and the date of approval is 5 February 2015. Male adult C57/Bl6 mice (25–30 g) were used in this study and were housed in groups of five, with free access to food and water under a 12:12-h light:dark cycle. All animals were allowed to adapt to their envelopment for 7 days before the experiment.

4.2. CPIP Model

The CPIP model was induced in mice under general anaesthesia with isoflurane by placing a tight-fitting O-ring (O-rings West, Seattle, WA, USA) with a 5/64 inch internal diameter around the left ankle for 3 h, as described by Coderre et al. [17]. The O-rings were removed while mice were still under general anaesthesia, allowing for reperfusion. Mice in the control group were placed under general anaesthesia, but their ankle was loosely rather than tightly surrounded by cutted O-ring.

4.3. Measurement of Tactile Allodynia

The plantar surfaces of the ipsilateral and contralateral hind legs of CPIP and control mice were tested for tactile allodynia 1 day and 30 days after hind leg I/R injury. To determine the threshold of the response, the floors of the cages for the two groups of mice were replaced with mesh floors to easily access the plantar surfaces of their hind legs with a filament. After a 20-min acclimation period, tactile hyperalgesia of the hind leg was assessed using von Frey hairs (Stoelting Co., Wood Dale, IL, USA) ranging from 2.44 to 4.31 (0.03–2.00 g) using the up-down method. The 50% response threshold (grams) was measured based on the response pattern and the value (in log units) of the final von Frey hair [30,31].

4.4. Drug Administration

The effects of BV were evaluated in CPIP mice that exhibited tactile allodynia. BV was delivered 7 days after I/R injury. Before the administration of BV, CPIP mice were acclimated to an observation cage for 20 min, and mechanical allodynia was measured using von Frey hairs. CPIP mice that showed distinct mechanical allodynia were selected. Saline or BV (1 mg/kg, subcutaneous (SC)) was administered into the dorsum of the ipsilateral hind paw that showed an allodynic response in the von Frey test. After injection, mechanical allodynia was assessed through the same process 30, 60, 90, 120, 180, 240 min, and 24 h after BV administration. At the same time on the following day, the same dosage of BV was injected using the same procedure, and mechanical allodynia was assessed again. On the third day, the identical experiment was carried out.

4.5. Assessment of NK-1 Receptor Expression in DRG

Each group of mice administered BV (1, 2, and 3 times) was sacrificed, and DRG were collected 60 min after BV administration, considering the tactile allodynia results. Mice in the control group and CPIP group and mice treated with BV were anaesthetized and transcardially perfused with 50 mL of 4% paraformaldehyde dissolved in 0.01 M phosphate-buffered saline (PBS) with pH 7.2–7.4. The DRG of the mice were then dissected, postfixed, and immersed in a 30% sucrose solution overnight. DRG segments were cut into 10-μm-thick slices on a freezing microtome. The slices were incubated with a rabbit antibody against the NK-1 SP receptor (1:1000; Chemicon, Temecula, CA, USA). After the sections were washed with buffer, they were exposed to the secondary antibody, an anti-rabbit IgG antibody conjugated with Alexa-488 (1:500; Invitrogen, Carlsbad, CA, USA). Digital images were obtained using a Zeiss LSM 510 Meta confocal microscope (Zeiss, Oberkochen, Germany), and the mean intensity was calculated using using Image-Pro Plus v. 6.0 (Media Cybernetics, Inc., Rockville, MD, USA).

4.6. Statistics

The data are presented as the mean ± SEM. Statistical analyses were performed using IBM SPSS Statistics ver. 24. (IBM Co., Armonk, NY, USA). A repeated measures 2-way ANOVA was performed to identify overall differences in the 50% von Frey threshold at each time point under different conditions, followed by Bonferroni post hoc tests. Comparisons between pre-injection and post-injection values were made at each time point using Student's *t*-tests. A two-sided *p* value of less than 0.05 was considered to indicate statistical significance. A Kruskal-Wallis test was used for the comparison of the immunohistochemical expression of NK-1 receptors among the control, CPIP, and treatment groups. The statistical analysis was verified by the Division of Biostatistics, Department of R&D Management, Kangbuk Samsung Hospital, Sungkyunkwan University School of Medicine.

Author Contributions: Hue Jung Park and Jae Min Lee conceived and designed the experiments; Jung Hyun Choi, Sung Hyun Lee, You Jung Lee, and Dong Kyu Kim performed the experiments; Sung Hyun Lee and Seung Hwan Jeon analyzed the data; Hyeon Do Jeong and Yun Hong Kim contributed analysis tools; Sung Hyun Lee and Hue Jung Park wrote the paper.

Conflicts of Interest: The authors declare no conflict of interest.

References

1. Son, D.J.; Lee, J.W.; Lee, Y.H.; Song, H.S.; Lee, C.K.; Hong, J.T. Therapeutic application of anti-arthritis, pain-releasing, and anti-cancer effects of bee venom and its constituent compounds. *Pharmacol. Ther.* **2007**, *115*, 246–270. [CrossRef] [PubMed]
2. Lee, S.H.; Choi, S.M.; Yang, E.J. Bee Venom Acupuncture Augments Anti-Inflammation in the Peripheral Organs of hSOD1G93A Transgenic Mice. *Toxins* **2015**, *7*, 2835–2844. [CrossRef] [PubMed]
3. Lee, J.D.; Kim, S.Y.; Kim, T.W.; Lee, S.H.; Yang, H.I.; Lee, D.I.; Lee, Y.H. Anti-inflammatory effect of bee venom on type II collagen-induced arthritis. *Am. J. Chin. Med.* **2004**, *32*, 361–367. [CrossRef] [PubMed]
4. Li, D.; Lee, Y.; Kim, W.; Lee, K.; Bae, H.; Kim, S.K. Analgesic Effects of Bee Venom Derived Phospholipase A(2) in a Mouse Model of Oxaliplatin-Induced Neuropathic Pain. *Toxins* **2015**, *7*, 2422–2434. [CrossRef] [PubMed]
5. Kang, S.Y.; Roh, D.H.; Yoon, S.Y.; Moon, J.Y.; Kim, H.W.; Lee, H.J.; Beitz, A.J.; Lee, J.H. Repetitive treatment with diluted bee venom reduces neuropathic pain via potentiation of locus coeruleus noradrenergic neuronal activity and modulation of spinal NR1 phosphorylation in rats. *J. Pain* **2012**, *13*, 155–166. [CrossRef] [PubMed]

6. Kang, S.Y.; Roh, D.H.; Park, J.H.; Lee, H.J.; Lee, J.H. Activation of Spinal alpha2-Adrenoceptors Using Diluted Bee Venom Stimulation Reduces Cold Allodynia in Neuropathic Pain Rats. *Evid.-Based Complement. Altern. Med.* **2012**, *2012*, 784713. [CrossRef] [PubMed]
7. Iolascon, G.; de Sire, A.; Moretti, A.; Gimigliano, F. Complex regional pain syndrome (CRPS) type I: Historical perspective and critical issues. *Clin. Cases Miner. Bone Metab.* **2015**, *12* (Suppl. 1), 4–10. [CrossRef] [PubMed]
8. Coderre, T.J.; Bennett, G.J. A hypothesis for the cause of complex regional pain syndrome-type I (reflex sympathetic dystrophy): Pain due to deep-tissue microvascular pathology. *Pain Med.* **2010**, *11*, 1224–1238. [CrossRef] [PubMed]
9. Nahm, F.S.; Park, Z.Y.; Nahm, S.S.; Kim, Y.C.; Lee, P.B. Proteomic identification of altered cerebral proteins in the complex regional pain syndrome animal model. *BioMed Res. Int.* **2014**, *2014*, 498410. [CrossRef] [PubMed]
10. Kortekaas, M.C.; Niehof, S.P.; Stolker, R.J.; Huygen, F.J. Pathophysiological Mechanisms Involved in Vasomotor Disturbances in Complex Regional Pain Syndrome and Implications for Therapy: A Review. *Pain Prac.* **2015**, *16*, 905–914. [CrossRef] [PubMed]
11. Daehyun Jo, R.C.; Alan, R. Light: Glial Mechanisms of Neuropathic Pain and Emerging Interventions. *Korean J. Pain* **2009**, *22*, 1–15.
12. Wei, T.; Li, W.W.; Guo, T.Z.; Zhao, R.; Wang, L.; Clark, D.J.; Oaklander, A.L.; Schmelz, M.; Kingery, W.S. Post-junctional facilitation of Substance P signaling in a tibia fracture rat model of complex regional pain syndrome type I. *Pain* **2009**, *144*, 278–286. [CrossRef] [PubMed]
13. Newby, D.E.; Sciberras, D.G.; Ferro, C.J.; Gertz, B.J.; Sommerville, D.; Majumdar, A.; Lowry, R.C.; Webb, D.J. Substance P-induced vasodilatation is mediated by the neurokinin type 1 receptor but does not contribute to basal vascular tone in man. *Br. J. Clin. Pharmacol.* **1999**, *48*, 336–344. [CrossRef] [PubMed]
14. Guo, T.Z.; Wei, T.; Li, W.W.; Li, X.Q.; Clark, J.D.; Kingery, W.S. Immobilization contributes to exaggerated neuropeptide signaling, inflammatory changes, and nociceptive sensitization after fracture in rats. *J. Pain Off. J. Am. Pain Soc.* **2014**, *15*, 1033–1045. [CrossRef] [PubMed]
15. Marchand, J.E.; Wurm, W.H.; Kato, T.; Kream, R.M. Altered tachykinin expression by dorsal root ganglion neurons in a rat model of neuropathic pain. *Pain* **1994**, *58*, 219–231. [CrossRef]
16. Millecamps, M.; Laferriere, A.; Ragavendran, J.V.; Stone, L.S.; Coderre, T.J. Role of peripheral endothelin receptors in an animal model of complex regional pain syndrome type 1 (CRPS-I). *Pain* **2010**, *151*, 174–183. [CrossRef] [PubMed]
17. Coderre, T.J.; Xanthos, D.N.; Francis, L.; Bennett, G.J. Chronic post-ischemia pain (CPIP): A novel animal model of complex regional pain syndrome-type I (CRPS-I; reflex sympathetic dystrophy) produced by prolonged hindpaw ischemia and reperfusion in the rat. *Pain* **2004**, *112*, 94–105. [CrossRef] [PubMed]
18. Yoon, S.Y.; Yeo, J.H.; Han, S.D.; Bong, D.J.; Oh, B.; Roh, D.H. Diluted bee venom injection reduces ipsilateral mechanical allodynia in oxaliplatin-induced neuropathic mice. *Biol. Pharm. Bull.* **2013**, *36*, 1787–1793. [CrossRef] [PubMed]
19. Koh, W.U.; Choi, S.S.; Lee, J.H.; Lee, S.H.; Lee, S.K.; Lee, Y.K.; Leem, J.G.; Song, J.G.; Shin, J.W. Perineural pretreatment of bee venom attenuated the development of allodynia in the spinal nerve ligation injured neuropathic pain model; an experimental study. *BMC Complement. Altern. Med.* **2014**, 14. [CrossRef] [PubMed]
20. Kwon, Y.B.; Kang, M.S.; Kim, H.W.; Ham, T.W.; Yim, Y.K.; Jeong, S.H.; Park, D.S.; Choi, D.Y.; Han, H.J.; Beitz, A.J.; et al. Antinociceptive effects of bee venom acupuncture (apipuncture) in rodent animal models: A comparative study of acupoint versus non-acupoint stimulation. *Acupunct. Electrother. Res.* **2001**, *26*, 59–68. [CrossRef] [PubMed]
21. Lee, M.J.; Jang, M.; Choi, J.; Lee, G.; Min, H.J.; Chung, W.S.; Kim, J.I.; Jee, Y.; Chae, Y.; Kim, S.H.; et al. Bee Venom Acupuncture Alleviates Experimental Autoimmune Encephalomyelitis by Upregulating Regulatory T Cells and Suppressing Th1 and Th17 Responses. *Mol. Neurobiol.* **2016**, *53*, 1419–1445. [CrossRef] [PubMed]
22. Leis, S.; Weber, M.; Isselmann, A.; Schmelz, M.; Birklein, F. Substance-P-induced protein extravasation is bilaterally increased in complex regional pain syndrome. *Exp. Neurol.* **2003**, *183*, 197–204. [CrossRef]
23. Wei, T.; Guo, T.Z.; Li, W.W.; Kingery, W.S.; Clark, J.D. Acute versus chronic phase mechanisms in a rat model of CRPS. *J. Neuroinflamm.* **2016**, 13. [CrossRef] [PubMed]

24. Darwish, S.F.; El-Bakly, W.M.; Arafa, H.M.; El-Demerdash, E. Targeting TNF-alpha and NF-kappaB activation by bee venom: Role in suppressing adjuvant induced arthritis and methotrexate hepatotoxicity in rats. *PLoS ONE* **2013**, *8*, e79284. [CrossRef] [PubMed]
25. Mashaghi, A.; Marmalidou, A.; Tehrani, M.; Grace, P.M.; Pothoulakis, C.; Dana, R. Neuropeptide substance P and the immune response. *Cell. Mol. Life Sci.* **2016**, *73*, 4249–4264. [CrossRef] [PubMed]
26. Weinstock, J.V.; Blum, A.; Metwali, A.; Elliott, D.; Arsenescu, R. IL-18 and IL-12 signal through the NF-kappa B pathway to induce NK-1R expression on T cells. *J. Immunol.* **2003**, *170*, 5003–5007. [CrossRef] [PubMed]
27. Lee, M.S.; Pittler, M.H.; Shin, B.C.; Kong, J.C.; Ernst, E. Bee venom acupuncture for musculoskeletal pain: A review. *J. Pain* **2008**, *9*, 289–297. [CrossRef] [PubMed]
28. Lim, S.M.; Lee, S.H. Effectiveness of bee venom acupuncture in alleviating post-stroke shoulder pain: A systematic review and meta-analysis. *J. Integr. Med.* **2015**, *13*, 241–247. [CrossRef]
29. Jeon, Y. Cell based therapy for the management of chronic pain. *Korean J. Anesthesiol.* **2011**, *60*, 3–7. [CrossRef] [PubMed]
30. Bonin, R.P.; Bories, C.; De Koninck, Y. A simplified up-down method (SUDO) for measuring mechanical nociception in rodents using von Frey filaments. *Mol. Pain* **2014**, *10*. [CrossRef] [PubMed]
31. Chaplan, S.R.; Bach, F.W.; Pogrel, J.W.; Chung, J.M.; Yaksh, T.L. Quantitative assessment of tactile allodynia in the rat paw. *J. Neurosci. Methods* **1994**, *53*, 55–63. [CrossRef]

© 2017 by the authors. Licensee MDPI, Basel, Switzerland. This article is an open access article distributed under the terms and conditions of the Creative Commons Attribution (CC BY) license (http://creativecommons.org/licenses/by/4.0/).

Article

Cobra Venom Factor and Ketoprofen Abolish the Antitumor Effect of Nerve Growth Factor from Cobra Venom

Alexey V. Osipov [1,†], Tatiana I. Terpinskaya [2,†], Tatiana E. Kuznetsova [2], Elena L. Ryzhkovskaya [2], Vladimir S. Lukashevich [2], Julia A. Rudnichenko [2], Vladimir S. Ulashchyk [2], Vladislav G. Starkov [1] and Yuri N. Utkin [1,*]

[1] Shemyakin-Ovchinnikov Institute of Bioorganic Chemistry, Russian Academy of Sciences, ul. Miklukho-Maklaya 16/10, Moscow 117997, Russia; osipov@mx.ibch.ru (A.V.O.); vladislavstarkov@mail.ru (V.G.S.)

[2] Institute of Physiology, National Academy of Sciences of Belarus, ul. Akademicheskaya, 28, Minsk 220072, Belarus; terpinskayat@mail.ru (T.I.T.); tania_k@mail.ru (T.E.K); ryzhkovskaya@mail.ru (E.L.R.); lukashvs@rambler.ru (V.S.L); link060619@list.ru (J.A.R); ulashchikv@mail.ru (V.S.U.)

* Correspondence: utkin@mx.ibch.ru; Tel.: +7-495-336-6522

† These authors contributed equally to this work.

Academic Editor: Steve Peigneur
Received: 17 August 2017; Accepted: 2 September 2017; Published: 6 September 2017

Abstract: We showed recently that nerve growth factor (NGF) from cobra venom inhibited the growth of Ehrlich ascites carcinoma (EAC) inoculated subcutaneously in mice. Here, we studied the influence of anti-complementary cobra venom factor (CVF) and the non-steroidal anti-inflammatory drug ketoprofen on the antitumor NGF effect, as well as on NGF-induced changes in EAC histological patterns, the activity of lactate and succinate dehydrogenases in tumor cells and the serum level of some cytokines. NGF, CVF and ketoprofen reduced the tumor volume by approximately 72%, 68% and 30%, respectively. The antitumor effect of NGF was accompanied by an increase in the lymphocytic infiltration of the tumor tissue, the level of interleukin 1β and tumor necrosis factor α in the serum, as well as the activity of lactate and succinate dehydrogenases in tumor cells. Simultaneous administration of NGF with either CVF or ketoprofen abolished the antitumor effect and reduced all other effects of NGF, whereas NGF itself significantly decreased the antitumor action of both CVF and ketoprofen. Thus, the antitumor effect of NGF critically depended on the status of the immune system and was abolished by the disturbance of the complement system; the disturbance of the inflammatory response canceled the antitumor effect as well.

Keywords: Ehrlich carcinoma; immune system; nerve growth factor; cobra venom; cobra venom factor; ketoprofen

1. Introduction

Nerve growth factor (NGF) belongs to a protein family of neurotrophins, which are important agents affecting the cell cycle. Usually they are mitogens; mammalian NGF has evident regenerative (primarily neuroprotective) functions. Among prevailing highly toxic ingredients, snake venoms contain non-toxic NGF. NGF from snake venom is a polypeptide possessing homology with more than 80% of identical amino acid residues to the beta-chain of NGF from mammalian saliva [1]; This phenomenon is explained by the origin of the snake venom glands from the salivary glands.

NGF can bind to two classes of receptors on the cell surface: the TrkA receptor, a tyrosine kinase with high specificity and high affinity to NGF, and p75, a low-affinity neurotrophin receptor which can bind to all members of the neurotrophin family with approximately equal affinity. NGF binding to the TrkA receptor triggers a number of metabolic reactions through a cascade of protein kinases. There are some data about the involvement of NGF in carcinogenesis [2], and mammalian NGF may promote or suppress tumor growth, depending on the tumor type. Snake venom NGF exerts some activities characteristic of mammalian NGF; however, the data concerning pro- or anti-oncogenic activity of snake NGF are very limited. We have recently shown that NGF from cobra venom may exert an inhibitory effect on the growth of Ehrlich's adenocarcinoma in vivo [3]. This effect is mediated via a tyrosine kinase cascade, because it can be blocked by the tyrosine kinase inhibitor K252a [4].

Considering the tumor as a disease at the level of the whole organism, we should take into consideration the role of the tumor microenvironment and the overall condition of the homeostasis systems, primarily the immune system. Tumor growth in the body is accompanied by an inflammatory reaction that, on the one hand, is aimed to eliminate tumor cells, and on the other hand, induces a tolerance to the tumor maintenance and growth [5].

One of the systems involved in the development of tumor-associated inflammation is complement. The accumulation of the complement C5 component with the subsequent generation of C5A anaphylatoxin by complement C3/C5 convertase contributes to the immunosuppressive properties of the tumor microenvironment [6]. Cobra venoms contain an activator of alternative pathway of the complement system, the so-called cobra venom factor (CVF), which is similar to the complement C3b component—forming C3/C5 convertase. However, CVF is not regulated by complement components, and activates the system in such a way that depletes it rapidly and completely. We have shown earlier that CVF from the Thailand cobra venom affects subcutaneous Ehrlich carcinoma in mice [7].

The activity of cyclooxygenases, in particular, cyclooxygenase-2 (COX-2), has proved to be essential for the development of chronic inflammation, tumor growth and metastasis during carcinogenesis [8,9]. Use of nonsteroidal anti-inflammatory drugs (NSAIDs) leads to a decrease of tumor size [8,9]. The NSAID ketoprofen has been shown to inhibit tumor cell proliferation in vitro [10] and in vivo [11,12].

Thus, CVF and ketoprofen, through various mechanisms, counteract the development of an inflammatory tumor microenvironment. To check if these drugs act synergistically with NGF, we have studied the influence of CVF and ketoprofen on the antitumor effects of cobra venom NGF in vivo. To see the effects at the tissue level, we have undertaken histological investigations of tumor samples. The concentration of tumor growth factor $\beta 1$ (TGF-$\beta 1$), interleukin 1β (IL-1β) and tumor necrosis factor α (TNF-α) in serum of the experimental animals was also determined.

An anomalous characteristic of energy metabolism in cancer cells is accelerated glycolysis, even in the presence of oxygen, rather than mitochondrial oxidative phosphorylation (the Warburg effect). It is accompanied by an elevated activity of lactate dehydrogenase (LDH, EC 1.1.1.27), which catalyzes the conversion of pyruvate to lactate. However, some sub-populations of cancer cells display high mitochondrial respiration and low glycolysis rates [13]. One of the key enzymes in aerobic oxidation is succinate dehydrogenase (SDH, EC 1.3.99.1). To determine the energy processes ongoing in EAC, the influence of NGF, CVF and ketoprofen on the activity of LDH and SDH have been studied.

Here, we report that simultaneous administration of either CVF or ketoprofen with NGF cancels antitumor effects and reduces other effects of NGF in Ehrlich ascites carcinoma (EAC)-bearing mice.

2. Results

To study effects of NGF, CVF, ketoprofen and combinations thereof, the mice were grafted subcutaneously with EAC, as described earlier [4]. Three experimental inoculated groups were injected intraperitoneally (i.p.) with either NGF, CVF or ketoprofen. Two other experimental inoculated groups were injected with a combination of NGF with either CVF or ketoprofen.

2.1. The Effects on Ehrlich Carcinoma Growth

The effects were analyzed at days eight and twelve after inoculation (Figure 1). NGF inhibited the tumor growth by about 72% (columns two in Figure 1A,B), which was in accordance with our previously published data [4].

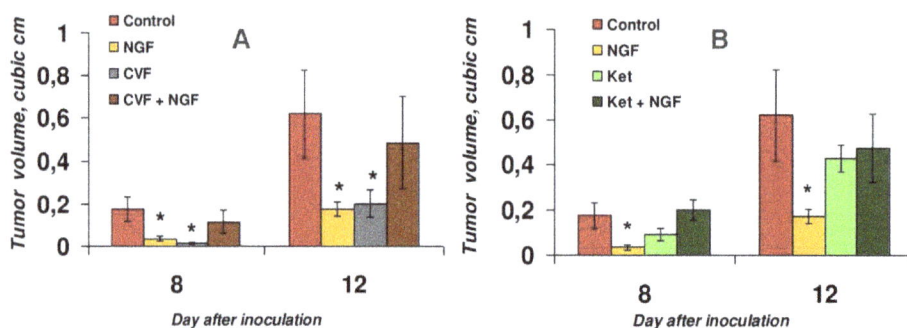

Figure 1. The effects of NGF and its combination with either CVF (**A**) or ketoprofen (Ket) (**B**) on Ehrlich ascites carcinoma (EAC) growth. * $p < 0.05$ compared to control (Mann–Whitney test).

When administered alone, CVF had an in vivo antitumor effect comparable to that of NGF (Figure 1A, column three) which resulted at day 12 in a reduction of mean tumor size by 68% compared to the control. However, combined administration of NGF and CVF critically reduced the antitumor effect of both drugs. Although a decrease in the tumor volume by 22% was observed at day twelve after inoculation (and by 35% at day eight), it was statistically insignificant according to the Mann–Whitney test (Figure 1A, column four). This points to the importance of the normal functioning of the immune system, in particular, the humoral system, for the antitumor effect of NGF. These data correlate well with our preliminary results [7].

Ketoprofen showed a tendency to decrease the tumor size by 48% (statistically insignificant according to the Mann–Whitney test) at day eight (Figure 1B, column three) after inoculation. This tendency was in agreement with the data of da Silveira et al. [12] obtained on glioma-bearing rats, and to the data of Sakayama et al. [11] on pulmonary metastasis of LM8 cells observed in male nude mice. In our work, at day twelve, the average tumor size reduction was only 30%, with a wide variation of this parameter in individual animals. Ketoprofen abolished the antitumor effect of NGF on both day eight and day twelve (Figure 1B, column four). Given that COX-1 and COX-2 are the main targets for ketoprofen [14], our data suggested that, at the level of regulation by cyclooxygenases, the normal manifestation of the inflammatory response is necessary for the antitumor effect of NGF.

2.2. Histological Studies of Tumor Samples

To determine the effects produced by NGF, CVF and ketoprofen, as well as NGF in combination with CVF or ketoprofen, in EAC at the tissue level, histological studies of tumor samples were carried out. Slices of tumor tissues, obtained at day twelve, were stained with hematoxylin–eosin and analyzed by microscopy.

Evident cellular polymorphism was observed in the tumor tissue. Cells with destructive changes, with vacuolization of the nuclei and cytoplasm, and giant multinucleated cells, were found in most of the tumor nodules. In tumor sections from the untreated control series (Figure 2), both minor structural changes and the destruction of a significant number of tumor cells were observed. Immune cell infiltration was absent or was poorly expressed (single scattered lymphocytes and small foci of lymphocytic infiltration).

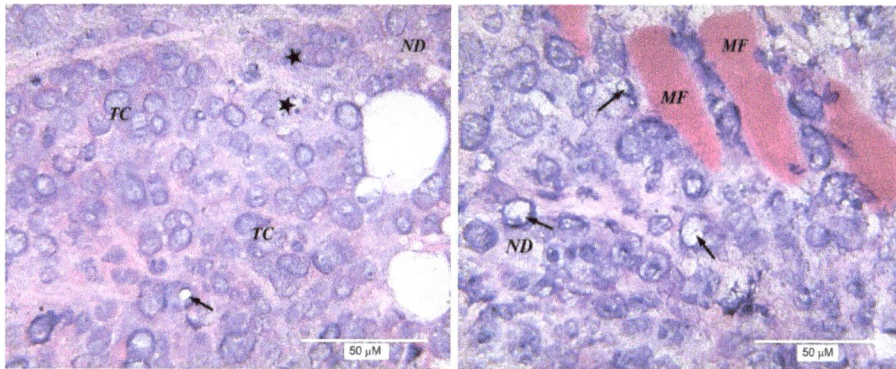

Figure 2. Morphological structure of EAC. For representation of different structures, two panels displaying separate parts of the same tumor are shown. Here, and in Figures 3 and 4, stars indicate lymphocytes; arrows: karyorrhexis or vacuolization of nuclei; MF: muscle fibers (in control: smooth, when exposed to NGF: striated); TC: tumor cells; and ND: necrotic detritus.

When the inoculated mice were treated with NGF, reduction of tumor volume was accompanied by an increase in the number of giant multinucleated cells and areas of unstructured necrotic detritus, as well as by germination of muscle fibers and the presence of large blood vessels in tumor tissue (Figure 3A). The latter indicates that the antitumor effect of NGF cannot be explained by an anti-angiogenic activity. NGF increased the lymphocytic infiltration in the tumor compared to control samples (Figure 3A).

In tumors from mice treated with CVF, the histological analysis revealed no significant differences from the controls. Weak immune cell infiltration was seen (Figure 3B).

The combined administration of NGF and CVF resulted in proliferation of connective tissue. Large blood vessels among nodes of tumor cells (Figure 3C) were typical for the experimental group in which animals received NGF only. Histological analysis of the tumor revealed no significant differences in the pathomorphological features and lymphocytic infiltration from the group that received CVF only, or the control group.

A distinctive feature of the group in which the animals received ketoprofen was a slightly more-pronounced lymphocytic infiltration than in the control—from mild to moderate, with the appearance of medium foci of inflammatory infiltration in the presence of single neutrophils (Figure 4A).

Combined administration of ketoprofen and NGF to inoculated mice resulted in the outgrowth of connective tissue between carcinoma cells. Significant areas of necrotic detritus were seen in the majority of samples (Figure 4B). Similarly to the group treated with NGF only, germination of muscle fibers was detected. Immune cell infiltration was variable, ranging from non-detectable in some tumor slices to weak or moderate in others. In general, lymphocytic infiltration was slightly more pronounced than in the control, but weaker than in the series where mice received ketoprofen or NGF only.

Thus, the increased degree of lymphocytic infiltration was the histological feature of tumors from mice treated with NGF. The application of either CVF or ketoprofen simultaneously with NGF reduced this effect. Moreover, the antitumor activity of CVF was not associated with lymphocytic infiltration, while that of ketoprofen was. Application of NGF alone or in combination with other drugs contributed to the vascularization of tumor nodes, the germination of muscle fibers and the proliferation of connective tissue. In the series where the animals received NGF or NGF with ketoprofen, there was a proliferation of not only smooth muscle, but also striated muscle fibers.

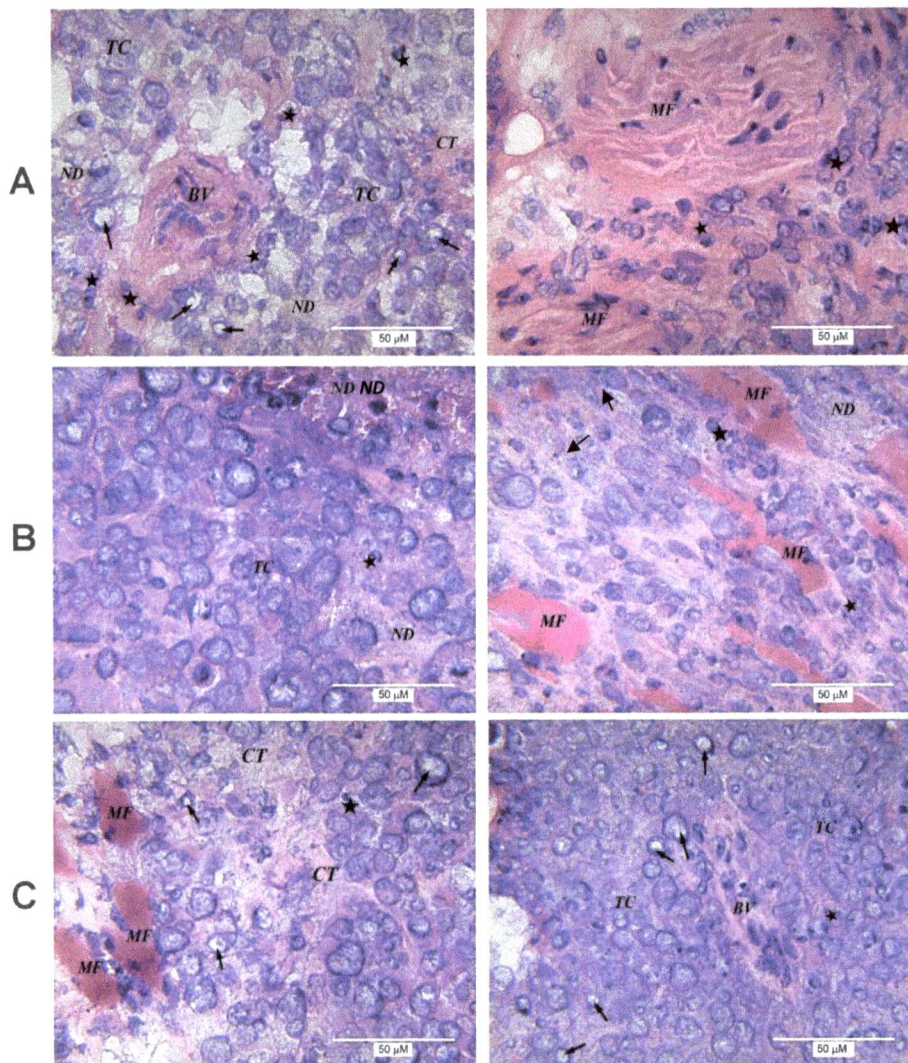

Figure 3. Morphological structure of EAC from mice treated with NGF (**A**) or CVF (**B**) only and with combination of both factors (**C**). CT: connective tissue; BV: blood vessel. Left and right panels show different parts of the same tumor.

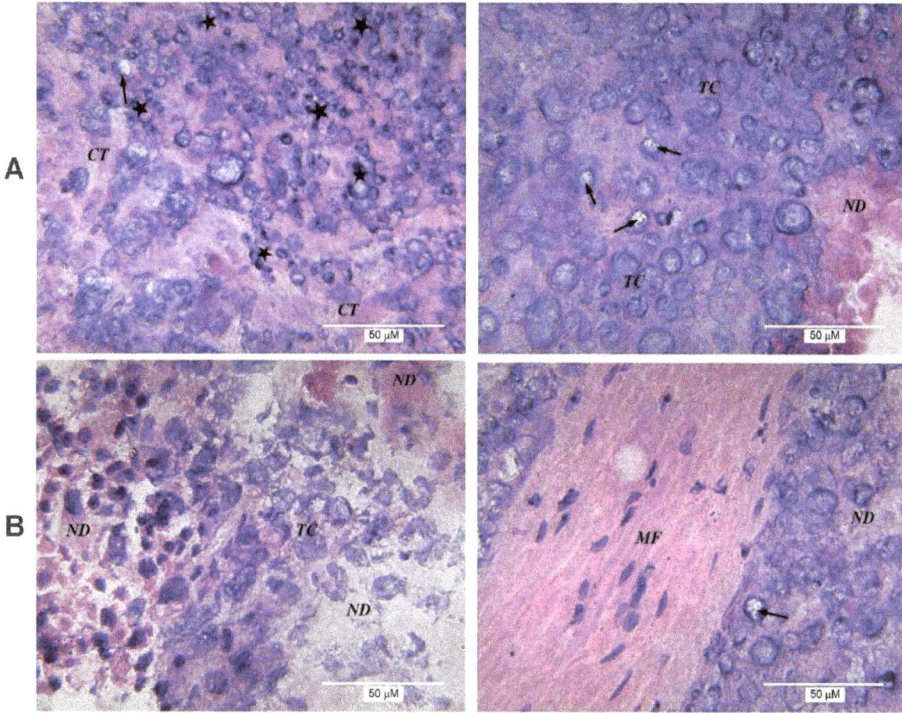

Figure 4. Morphological structure of EAC from mice treated with ketoprofen only (**A**) or with combination of NGF and ketoprofen (**B**). Left and right panels show different parts of the same tumor.

2.3. The Effects on TNF-α, IL-1β and TGF-β1 in Serum

NGF caused a considerable increase in the level of TNF-α and IL-1β in the serum of the tumor-bearing mice. CVF significantly (and ketoprofen insignificantly) reduced the NGF-induced increase in TNF-α, and similar trends were observed for the level of IL-1β (Figure 5A,B). Therefore, it could be assumed that the antitumor effect of NGF was, to a certain degree, related to the increase of IL-1β and THF-α. Mice with EAC had an increased level of serum TGF-β1; CVF or ketoprofen significantly reduced this level (Figure 5C). NGF preserved the increased TGF-β1 level and canceled the decreasing action of ketoprofen, but not that of CVF (Figure 5C).

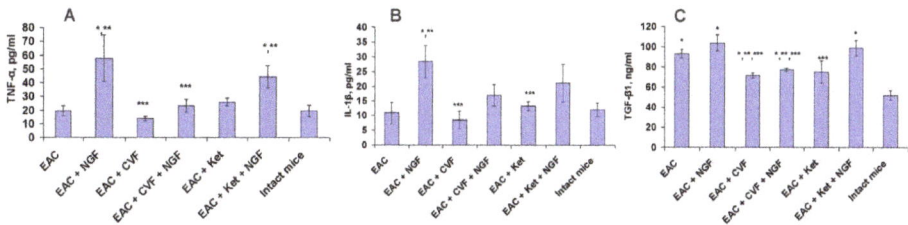

Figure 5. Effect of NGF, CVF or ketoprofen (Ket) on serum levels of TNF-α (**A**), IL-1β (**B**) and TGF-β1 (**C**) in mice bearing EAC. * $p < 0.05$ (compared to intact mice); ** $p < 0.05$ (compared to EAC-bearing mice); *** $p < 0.05$ (compared to EAC-bearing mice treated with NGF).

2.4. Effects on Glycolysis

A characteristic feature of tumor cells is increased glycolysis, which is accompanied by a high activity of LDH; this takes place even under normal oxygenation conditions (aerobic glycolysis). SDH is an enzyme that takes part in the process of aerobic oxidation. Considering these facts, we measured the activity of LDH and SDH as enzymes involved in the carbohydrate and energy metabolism.

Treatment of tumor-bearing mice with NGF led to an increase in the activity of both LDH and SDH (5.9% and 7.4%, respectively, $p < 0.05$) (Figure 6), i.e., both anaerobic and aerobic oxidation were increased under NGF treatment.

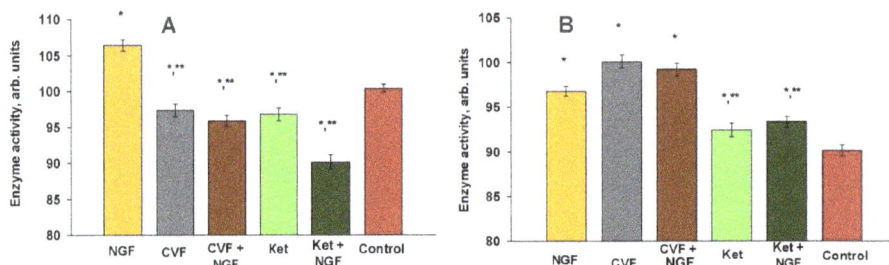

Figure 6. Effect of NGF, CVF and ketoprofen (Ket) on LDH (**A**) and SDH (**B**) activities in EAC (at day 12 after tumor inoculation). * $p < 0.05$ (compared to control), ** $p < 0.05$ (compared to NGF). Control: EAC-bearing mice.

CVF treatment resulted in a decrease of LDH activity ($p < 0.05$) and an increase of SDH activity by 11.1%, $p < 0.05$ (Figure 6). Applied simultaneously, NGF and CVF reduced LDH activity compared to the control (4.5%, $p < 0.05$) and to the group treated with NGF only (9.9%) (column 3 in Figure 6A), and increased SDH activity by 10.1% ($p < 0.05$) compared to the control and by 2.5% ($p < 0.05$) and to the group treated with NGF only (column three in Figure 6B).

Ketoprofen alone reduced LDH activity (3.6%) and increased the activity of SDH (2.4%), shifting the metabolism towards aerobic oxidation (column four in Figure 6A,B). When combined with NGF, ketoprofen neutralized the stimulatory effect of NGF on both SDH and LDH activities (column five versus column one in Figure 6A,B). The reduction in LDH activity was greater by 10.3% than that induced by ketoprofen only (column five versus column four in Figure 6A). The SDH activity after simultaneous application of NGF and ketoprofen was not significantly different from that observed in the presence of ketoprofen only (column five versus column four in Figure 6B).

3. Discussion

Long-term intensive studies have shown that the relationship between immunity and cancer is complex [15]. The immune system can excrete factors promoting survival, growth and invasion of tumor cells. Thus, on the one hand, the immune system can act as an extrinsic tumor suppressor, but on the other hand, it facilitates cancer initiation, promotion and progression [15]. The complement system is generally recognized as a protective mechanism against the formation of tumors, but recent studies also indicated a pro-tumorigenic potential of the complement system in certain cancers and under certain conditions [16]. Nevertheless, treatment of tumor-bearing mice with CVF results in a significant growth retardation of B16 melanoma tumors [17] and EAC [7]. It is also known that different NSAIDs, including inhibitors of COX-1 and COX-2 such as ketoprofen, can restrain the development of tumors [8,9]. We have found that ketoprofen suppresses the growth of EAC as well (Figure 1). Earlier, we showed that NGF from cobra venom exerts the same effect on the subcutaneous form of EAC [4]. Based on these data, we decided to check if CVF and ketoprofen would exhibit a synergistic effect with NGF. However, the results obtained showed that neither CVF nor ketoprofen enhanced the antitumor effect of NGF (Figure 1). Instead, both compounds abolished the effect of NGF on EAC (Figure 1). This data indicates that the normal function of the immune system is a prerequisite for the antitumor effect of NGF.

3.1. The Inflammatory Infiltration and EAC Growth

Inflammatory reactions play a key role at different stages of tumor development [18–20]. At present, most data underscore the benefits of inflammatory tissue infiltration for tumor progression. One of the potential mechanisms is that chronic inflammation can generate an immunosuppressive microenvironment that benefits tumor formation and progression [21]. However, the balance between antitumor and tumor-promoting immunity can be shifted either to protect against the neoplasia development, or to support tumor growth.

In our experiments, the immune cell infiltrate is mainly composed of lymphocytes. In clinical studies, the increase in lymphocytic infiltration in many (although not all) cases is considered a favorable prognostic sign [22,23]. Our data show that retardation of the EAC growth under NGF treatment is accompanied with an increase in the lymphocyte infiltration (Figure 3). CVF or ketoprofen abolish the NGF effect on EAC, and reduce NGF-induced local lymphocytic infiltration (Figures 3 and 4). Based on these data, one may suggest that the antitumor effect of NGF is mediated by immune cells. At the same time, NGF promotes the sprouting of blood vessels, i.e., it does not display anti-angiogenic properties when exerting its antitumor effect. The promotion of breast cancer angiogenesis by NGF was observed earlier [24]. Such a property is generally considered pro-oncogenic, and in this case might facilitate access of immune cells to the tumor.

In our work, CVF alone has practically no effect on the immune cell infiltration (Figure 3). This finding is in some contradiction with the data of [17], wherein a significant slowdown of the B16 melanoma growth under CVF treatment was accompanied by a significant increase in the number of tumor-infiltrating immune cells. This might be explained by the different schemes of CVF administration used by [17] and in our experiments.

Our data show that lymphocyte infiltration plays some role in tumor development. This issue deserves further study, and may provide new information on the mechanisms of the antitumor effects of NGF, CVF and NSAIDs.

3.2. The Levels of TNF-α, IL-1β and TGF-β1 in the Serum of Tumor-Bearing Mice

An increase in the levels of proinflammatory cytokines IL-1β and TNF-α indicates that the antitumor effect of NGF is realized under conditions of increased inflammation. Ketoprofen, and especially CVF, attenuate or cancel this effect (Figure 5). A decrease in the level of IL-1β and TNF-α by CVF points to the latter's pronounced anti-inflammatory effect. In turn, the antitumor activity of CVF and ketoprofen, but not NGF, is possibly associated with a decrease in the level of serum TGF-β1, which directs the cells of the tumor microenvironment to the pro-tumor phenotype, and negatively regulates the cytotoxic function of immune cells [25,26].

The antitumor effect of NSAIDs is associated with inhibition of cyclooxygenase activity and a decrease in the synthesis of PGE2, which has a tumor-promoting effect [9,27], and is also associated with a series of cyclooxygenase-independent events [28].

It may be possible that suppression of inflammation prevents the antitumor effect of NGF. Conversely, inflammation induced by NGF may cancel the antitumor effect of CVF.

3.3. Glycolysis and EAC Growth

Oxidative stress is considered one of the inducers and markers of carcinogenesis. It contributes to the shift of energy metabolism to enhanced aerobic glycolysis, which is associated with malignant cell proliferation [29], as well as to the enhancement of pro-carcinogenic properties of the tumor cell microenvironment [30]. On the other hand, there is evidence for the "reverse Warburg effect", consisting of a metabolic interaction between glycolytic stroma cells and cancer cells with enhanced oxidative metabolism. Thus, aerobic glycolysis can be characteristic of stromal cells surrounding the tumor, while the tumor cells convert lactate, produced by the stromal cells, into pyruvate, and use the latter in oxidative phosphorylation [13].

To study the influence of our compounds on energy metabolism in EAC, we determined the activity of LDH and SDH in the tumor cells (Figure 6). Our data show that the retardation of EAC growth by NGF is accompanied by an increase in both anaerobic and aerobic oxidation in EAC cells. Ketoprofen and CVF influenced NGF-stimulated aerobic oxidation to different extents; ketoprofen significantly reduced the NGF-induced stimulation of aerobic oxidation, while CVF did not. Suppression of the NGF antitumor effect by both CVF and ketoprofen coincided with the suppression by these drugs of the NGF-induced enhancement of glycolysis represented by LDH activity.

Thus, the inhibition of tumor growth after the NGF treatment was accompanied by increased activity of both LDH and SDH. CVF and ketoprofen had different effects on energy metabolism in EAC. Ketoprofen, but not CVF, reduced the increase of aerobic oxidation induced by NGF. However, both significantly reduced the stimulating effect of NGF on anaerobic oxidation (glycolysis).

It should be noted that the activation and cytotoxic response of lymphocytes are associated with an increase in the intensity of glycolysis [31]. One of the possible mechanisms for the antitumor effect of NGF may represent its action on energy metabolism in immune cells. CVF and ketoprofen might also act on energy metabolism in immune cells, however, in a way opposite to that of NGF. Direct experimental confirmation of this suggestion is outside of the scope of this work, and further studies should be performed to address this matter.

3.4. The Possible Role of TrkA Receptors

It is known that NGF participates in some inflammatory processes [32]. Thus, various myeloid cells are capable of expressing the NGF receptor TrkA and responding to NGF. For example, about 20% of fresh mouse natural killer (NK) cells express the TrkA receptor, and this fraction reaches 100% when NK cells are activated by interleukin 2. NGF does not influence the expression of surface molecules important for NK cell activation or inhibition. In contrast to TrkA, the other NGF receptor, p75, is not expressed by NK cells (neither resting nor activated) [33]. NGF is involved in eosinophil or B cell survival; its effect may be completely abolished in the presence of K252a, the TrkA receptor antagonist [34,35]. NGF can also act as a chemotactic factor for lymphocytes, considering that it mediates attraction of monocytes—other leukocytes from the agranulocyte group—without modifying the production of proinflammatory cytokines [36].

We have earlier shown that K252a can block the NGF inhibitory effect on the growth of EAC [4]. In the present paper, we demonstrate that treatment of mice by NGF leads to increased lymphocyte infiltration of the tumor. Considering these facts, one could suggest that the antitumor effect of NGF is mediated through the TrkA receptor, activation of which on lymphocytes might be more important than on the tumor cells.

4. Conclusions

The treatment of tumor-bearing mice with NGF resulted in a reduction of tumor volume by 72%, increased lymphocytic infiltration of the tumor tissue, elevated levels of serum IL-1β and TNF-α and an increase in the activities of both SDH and LDH in EAC cells. Under the same conditions, CVF inhibited the tumor growth by 68%; however, the degree of inflammatory infiltration did not differ significantly from controls. A decrease in the TGF-β1 level in the serum of CVF-treated mice, and a tendency to decrease levels of IL-1β and TNF-α, were observed. With ketoprofen administration, the lymphocytic infiltration was slightly more pronounced; however, the inhibition of tumor growth was less apparent than that induced by CVF. CVF and ketoprofen increased the activity of SDH and decreased LDH activity in tumor cells.

Unlike NGF administration only, the application of NGF together with either ketoprofen or CVF resulted in abrogation of its inhibitory effect on EAC. CVF and ketoprofen decreased lymphocytic infiltration and levels of IL-1β and TNF-α, stimulated by NGF. They also prevented the NGF-induced increase in LDH activity.

Thus, the data obtained in this work suggest that the antitumor effect of NGF in vivo depends critically on the normal status of the immune system. The substances that disturb immunity neutralize the inhibitory effect of NGF on the development of EAC in mice. The NGF antitumor mechanism may cause an increase of lymphocytic infiltration in the tumor, a rise in the levels of IL-1β and TNF-α in the serum of tumor-bearing mice, and an increase in aerobic glycolysis.

5. Materials and Methods

5.1. Materials

NGF and CVF were isolated from *Naja kaouthia* cobra venom as described in [37,38], respectively. The purity of samples was more than 98%, as confirmed by analytical reversed-phase HPLC and MALDI mass spectrometry. Ketoprofen in the form of 4% solution for injection was obtained from Pharmaceuticals Lek (Ljubljana, Slovenia).

5.2. Mice

Female Af/WySnMv mice were inbreeded at the Institute of Physiology, National Academy of Sciences of Belarus (Minsk, Belarus). EAC was purchased from Blokhin Russian Cancer Research Center, Russian Academy of Medical Sciences (Moscow, Russia). Tumor cells were obtained from the EAC maintained by intraperitoneal passages. All the appropriate actions were taken to minimize discomfort to mice. The World Health Organization's International Guiding Principles for Biomedical Research Involving Animals were followed during experiments on animals.

5.3. Carcinoma Growth and Application of Drugs

Viable EAC cells (20×10^6) were inoculated into the left flank of the mice.

The mice with EAC were divided into groups of 5–6 animals. NGF at a dose of 104 µg/kg, CVF at a dose of 250 µg/kg and ketoprofen at a dose of 40 mg/kg were administered intraperitoneally each 3rd–4th day over 12 days; two groups of animals received NGF and ketoprofen or NGF and CVF at doses indicated above. Drugs were dissolved in saline. The control group received saline. Each animal received three treatments. The volume of developed subcutaneous tumor was assessed as in [3]. The tumor tissue was taken from animals at day 12 after inoculation; cryostat sections for histochemical studies were prepared, stained by hematoxylin–eosin and analyzed as in [39], and the blood samples were collected for preparation of serum. Sera of intact mice were also obtained.

5.4. Immunoenzyme Analysis (ELISA)

Immunoenzyme analysis (ELISA) was performed with DuoSet ELISA Mouse IL-1β/IL-1F2, DuoSet ELISA TGF-β1 and DuoSet ELISA TNF-α assay kits (R & D Systems, Minneapolis, MN, USA). Serum levels of IL-1β, TGF-β1 and TNF-α were determined according to the manufacturer's protocol. Optical density values at 450 nm were measured using a BioTek ELx808 microplate reader (BioTek, Winooski, VT, USA). The concentration of samples was calculated according to the corresponding OD value and the concentration of the standard substance.

5.5. Enzymatic Activity Determination

The activities of LDH and SDH were determined by the tetrazolium method according to [40], using the computer data-processing program Image J (National Institutes of Health, Bethesda, MD, USA). Each measurement was performed on more than 300 individual tumor cells obtained from three to five animals from each group.

5.6. Statistical Analysis

Statistical analysis of the tumor volume, cytokine serum levels and enzymatic activity was performed with the Mann–Whitney test. The differences were considered significant for *p* values <0.05. All results are presented as the mean ± SEM (standard error of the mean).

All studies involving animals were approved by the Commission on Bioethics of the Institute of Physiology of the National Academy of Sciences of Belarus, established in accordance with Order No. 44 of 07.06.2013, consisting of: Chairman—Head of the Laboratory A. Yu. Molchanova; Deputy Chairman—Deputy Director for Scientific and Innovation S. V. Mankovskaya; Members—head of the laboratory of physiotherapy and balneology, E.I. Kalinovskaya; chief scientific worker, L.I. Archakova; research worker, A.E. Pyzh; senior research worker, V.S. Lukashevich; research worker, S.B. Kohan; scientific secretary of the Institute of Physiology—N.F. Pavlova; secretary—T.E. Kuznetsova. The ethical approval code is No. 2. The date of approval is 7 August, 2014.

Acknowledgments: This study was supported by the Russian Foundation for Basic Research (project No. 16-54-00199) and by the Belarusian Republican Foundation for Fundamental Research (projects No. M14P-101 and M16P-171).

Author Contributions: T.I.T. and V.S.U. conceived and designed the experiments; T.I.T., T.E.K., E.L.R., V.S.L. and J.A.R. performed the experiments; T.I.T., A.V.O., T.E.K., Y.N.U. and V.S.U. analyzed the data; A.V.O. and V.G.S. contributed reagents/materials/analysis tools; A.V.O., T.I.T. and Y.N.U. wrote the paper.

Conflicts of Interest: The authors declare no conflict of interest. The funding sponsors had no role in the design of the study; in the collection, analyses, or interpretation of data; in the writing of the manuscript, and in the decision to publish the results.

References

1. Koh, D.C.; Armugam, A.; Jeyaseelan, K. Sputa nerve growth factor forms a preferable substitute to mouse 7S-beta nerve growth factor. *BioChem. J.* **2004**, *383*, 149–158. [CrossRef] [PubMed]
2. Wang, W.; Chen, J.; Guo, X. The role of nerve growth factor and its receptors in tumorigenesis and cancer pain. *Biosci. Trends* **2014**, *8*, 68–74. [CrossRef] [PubMed]
3. Osipov, A.V.; Terpinskaya, T.I.; Ulaschik, V.S.; Tsetlin, V.I.; Utkin, Y.N. Nerve growth factor suppresses Ehrlich carcinoma growth. *Dokl. BioChem. Biophys.* **2013**, *451*, 207–208. [CrossRef] [PubMed]
4. Osipov, A.V.; Terpinskaya, T.I.; Kryukova, E.V.; Ulaschik, V.S.; Paulovets, L.V.; Petrova, E.A.; Blagun, E.V.; Starkov, V.G.; Utkin, Y.N. Nerve growth factor from cobra venom inhibits the growth of Ehrlich tumor in mice. *Toxins* **2014**, *6*, 784–795. [CrossRef] [PubMed]
5. Crusz, S.M.; Balkwil, L.F.R. Inflammation and cancer: Advances and new agents. *Nat. Rev. Clin. Oncol.* **2015**, *12*, 584–596. [CrossRef] [PubMed]
6. Piao, C.; Cai, L.; Qiu, S.; Jia, L.; Song, W.; Du, J. Complement 5a enhances hepatic metastases of colon cancer via monocyte chemoattractant protein-1-mediated inflammatory cell infiltration. *J. Biol. Chem.* **2015**, *290*, 10667–10676. [CrossRef] [PubMed]
7. Terpinskaya, T.I.; Ulashchik, V.S.; Osipov, A.V.; Tsetlin, V.I.; Utkin, Y.N. Suppression of Ehrlich carcinoma growth by cobra venom factor. *Dokl. Biol. Sci.* **2016**, *470*, 240–243. [CrossRef] [PubMed]
8. Guan, X. Cancer metastases: Challenges and opportunities. *Acta Pharm. Sin. B* **2015**, *5*, 402–418. [CrossRef] [PubMed]
9. Liu, B.; Qu, L.; Yan, S. Cyclooxygenase-2 promotes tumor growth and suppresses tumor immunity. *Cancer Cell Int.* **2015**, *15*, 106. [CrossRef] [PubMed]
10. Damnjanovic, I.; Najman, S.; Stojanovic, S.; Stojanovic, D.; Veljkovic, A.; Kocic, H.; Langerholc, T.; Damnjanovic, Z.; Pesic, S. Crosstalk between possible cytostatic and antiinflammatory potential of ketoprofen in the treatment of culture of colon and cervix cancer cell lines. *Bratisl. Lek. Listy* **2015**, *116*, 227–232. [CrossRef] [PubMed]
11. Sakayama, K.; Kidani, T.; Miyazaki, T.; Shirakata, H.; Kimura, Y.; Kamogawa, J.; Masuno, H.; Yamamoto, H. Effect of ketoprofen in topical formulation on vascular endothelial growth factor expression and tumor growth in nude mice with osteosarcoma. *J. Orthop. Res.* **2004**, *22*, 1168–11674. [CrossRef] [PubMed]

12. Da Silveira, E.F.; Chassot, J.M.; Teixeira, F.C.; Azambuja, J.H.; Debom, G.; Beira, F.T.; Del Pino, F.A.; Lourenço, A.; Horn, A.P.; Cruz, L.; et al. Ketoprofen-loaded polymeric nanocapsules selectively inhibit cancer cell growth in vitro and in preclinical model of glioblastoma multiforme. *Invest. New Drugs* **2013**, *31*, 1424–1435. [CrossRef] [PubMed]

13. Lee, M.; Yoon, J.H. Metabolic interplay between glycolysis and mitochondrial oxidation: The reverse Warburg effect and its therapeutic implication. *World J. Biol. Chem.* **2015**, *6*, 148–161. [CrossRef] [PubMed]

14. Brune, K.; Patrignani, P. New insights into the use of currently available non-steroidal anti-inflammatory drugs. *J. Pain Res.* **2015**, *8*, 105–118. [CrossRef] [PubMed]

15. Vesely, M.D.; Kershaw, M.H.; Schreiber, R.D.; Smyth, M.J. Natural innate and adaptive immunity to cancer. *Annu. Rev. Immunol.* **2011**, *29*, 235–271. [CrossRef] [PubMed]

16. Mamidi, S.; Höne, S.; Kirschfink, M. The complement system in cancer: Ambivalence between tumour destruction and promotion. *Immunobiology* **2017**, *222*, 45–54. [CrossRef] [PubMed]

17. Janelle, V.; Langlois, M.P.; Tarrab, E.; Lapierre, P.; Poliquin, L.; Lamarre, A. Transient complement inhibition promotes a tumor-specific immune response through the implication of natural killer cells. *Cancer Immunol. Res.* **2014**, *2*, 200–206. [CrossRef] [PubMed]

18. Finn, O.J. Cancer immunology. *N. Engl. J. Med.* **2008**, *358*, 2704–2715. [CrossRef] [PubMed]

19. Grivennikov, S.I.; Greten, F.R.; Karin, M. Immunity, inflammation, and cancer. *Cell* **2010**, *140*, 883–899. [CrossRef] [PubMed]

20. Morrison, W.B. Inflammation and cancer: A comparative view. *J. Vet. Intern. Med.* **2012**, *26*, 18–31. [CrossRef] [PubMed]

21. Wang, D.; DuBois, R.N. Immunosuppression associated with chronic inflammation in the tumor microenvironment. *Carcinogenesis* **2015**, *36*, 1085–1093. [CrossRef] [PubMed]

22. Simonson, W.T.N.; Allison, K.H. Tumour-infiltrating lymphocytes in cancer: Implications for the diagnostic pathologist. *Diagn. Histopathol.* **2011**, *17*, 80–90. [CrossRef]

23. Lee, N.; Zakka, L.R.; Mihm, M.C., Jr.; Schatton, T. Tumour-infiltrating lymphocytes in melanoma prognosis and cancer immunotherapy. *Pathology* **2016**, *48*, 177–187. [CrossRef] [PubMed]

24. Romon, R.; Adriaenssens, E.; Lagadec, C.; Germain, E.; Hondermarck, H.; Le Bourhis, X. Nerve growth factor promotes breast cancer angiogenesis by activating multiple pathways. *Mol. Cancer* **2010**, *9*, 157. [CrossRef] [PubMed]

25. Gigante, M.; Gesualdo, L.; Ranieri, E. TGF-beta: A master switch in tumor immunity. *Curr. Pharm. Des.* **2012**, *18*, 4126–4134. [CrossRef] [PubMed]

26. Caja, F.; Vannucci, L. TGFβ: A player on multiple fronts in the tumor microenvironment. *J. Immunotoxicol.* **2015**, *12*, 300–307. [CrossRef] [PubMed]

27. Nakanishi, M.; Rosenberg, D.W. Multifaceted roles of PGE2 in inflammation and cancer. *Semin. Immunopathol.* **2013**, *35*, 123–137. [CrossRef] [PubMed]

28. Liggett, J.L.; Zhang, X.; Eling, T.E.; Baek, S.J. Anti-tumor activity of non-steroidal anti-inflammatory drugs: Cyclooxygenase-independent targets. *Cancer Lett.* **2014**, *346*, 217–224. [CrossRef] [PubMed]

29. Shi, D.; Xie, F.; Zhai, C.; Stern, J.S.; Liu, Y.; Liu, S. The role of cellular oxidative stress in regulating glycolysis energy metabolism in hepatoma cells. *Mol. Cancer* **2009**, *8*, 32. [CrossRef] [PubMed]

30. Lisanti, M.P.; Martinez_Outschoorn, U.E.; Chiavarina, B.; Pavlides, S.; Whitaker-Menezes, D.; Tsirigos, A.; Witkiewicz, A.; Lin, Z.; Balliet, R.; Howell, A.; et al. Understanding the "lethal" drivers of tumor-stroma co-evolution: Emerging role(s) for hypoxia, oxidative stress and autophagy/mitophagy in the tumor micro-environment. *Cancer Biol. Therap.* **2010**, *10*, 537–542. [CrossRef] [PubMed]

31. Park, B.V.; Pan, F. Metabolic regulation of T cell differentiation and function. *Mol. Immunol.* **2015**, *68*, 497–506. [CrossRef] [PubMed]

32. Aloe, L. Nerve growth factor and neuroimmune responses: Basic and clinical observations. *Arch. Physiol. Biochem.* **2001**, *109*, 354–356. [CrossRef] [PubMed]

33. Ralainirina, N.; Brons, N.H.; Ammerlaan, W.; Hoffmann, C.; Hentges, F.; Zimmer, J. Mouse natural killer (NK) cells express the nerve growth factor receptor TrkA, which is dynamically regulated. *PLoS ONE* **2010**, *5*, e15053. [CrossRef] [PubMed]

34. Kronfeld, I.; Kazimirsky, G.; Gelfand, E.W.; Brodie, C. NGF rescues human B lymphocytes from anti-IgM induced apoptosis by activation of PKCzeta. *Eur. J. Immunol.* **2002**, *32*, 136–143. [CrossRef]

35. Hahn, C.; Islamian, A.P.; Renz, H.; Nockher, W.A. Airway epithelial cells produce neurotrophins and promote the survival of eosinophils during allergic airway inflammation. *J. Allergy Clin. Immunol.* **2006**, *117*, 787–794. [CrossRef] [PubMed]
36. Samah, B.; Porcheray, F.; Gras, G. Neurotrophins modulate monocyte chemotaxis without affecting macrophage function. *Clin. Exp. Immunol.* **2008**, *151*, 476–486. [CrossRef] [PubMed]
37. Kukhtina, V.V.; Tsetlin, V.I.; Utkin, Y.N.; Inozemtseva, L.S.; Grivennikov, I.A. Two forms of nerve growth factor from cobra venom prevent the death of PC12 cells in serum-free medium. *J. Nat. Toxins* **2001**, *10*, 9–16. [PubMed]
38. Osipov, A.V.; Mordvintsev, D.Y.; Starkov, V.G.; Galebskaya, L.V.; Ryumina, E.V.; Bel'tyukov, P.P.; Kozlov, L.V.; Romanov, S.V.; Doljansky, Y.; Tsetlin, V.I.; et al. Naja melanoleuca cobra venom contains two forms of complement-depleting factor (CVF). *Toxicon* **2005**, *46*, 394–403. [CrossRef] [PubMed]
39. Terpinskaya, T.I.; Osipov, A.V.; Kuznetsova, T.E.; Ryzhkovskaya, E.L.; Ulaschik, V.S.; Ivanov, I.A.; Tsetlin, V.I.; Utkin, Y.N. α-conotoxins revealed different roles of nicotinic cholinergic receptor subtypes in oncogenesis of Ehrlich tumor and in the associated inflammation. *Dokl. BioChem. Biophys.* **2015**, *463*, 216–219. [CrossRef] [PubMed]
40. Lojda, Z.; Gossrau, R.; Schibler, T.H. *Enzyme Histochemistry: A Laboratory Manual*; Springer-Verlag: Berlin, Germany, 1979; p. 278.

© 2017 by the authors. Licensee MDPI, Basel, Switzerland. This article is an open access article distributed under the terms and conditions of the Creative Commons Attribution (CC BY) license (http://creativecommons.org/licenses/by/4.0/).

Article

High Throughput Identification of Antimicrobial Peptides from Fish Gastrointestinal Microbiota

Bo Dong [1,†], Yunhai Yi [1,2,†] (ORCID), Lifeng Liang [1] and Qiong Shi [1,2,3,*]

1   BGI Education Center, University of Chinese Academy of Sciences, Shenzhen 518083, China; dongbo@genomics.cn (B.D.); yiyunhai@genomics.cn (Y.Y.); lianglifeng@genomics.cn (L.L.)
2   Shenzhen Key Lab of Marine Genomics, Guangdong Provincial Key Lab of Molecular Breeding in Marine Economic Animals, BGI Academy of Marine Sciences, BGI Marine, BGI, Shenzhen 518083, China
3   Laboratory of Aquatic Genomics, College of Life Sciences and Oceanography, Shenzhen University, Shenzhen 518060, China
*   Correspondence: shiqiong@genomics.cn; Tel.: +86-185-6627-9826
†   These authors contributed equally to this work.

Academic Editor: Steve Peigneur
Received: 10 August 2017; Accepted: 28 August 2017; Published: 30 August 2017

Abstract: Antimicrobial peptides (AMPs) are a group of small peptides, which are secreted by almost all creatures in nature. They have been explored in therapeutic and agricultural aspects as they are toxic to many bacteria. A considerable amount of work has been conducted in analyzing 16S and metagenomics of the gastrointestinal (GI) microbiome of grass carp (*Ctenopharyngodon idellus*). However, these datasets are still untapped resources. In this present study, a homologous search was performed to predict AMPs from our newly generated metagenome of grass carp. We identified five AMPs with high similarities to previously reported bacterial toxins, such as lantibiotic and class II bacteriocins. In addition, we observed that the top abundant genus in the GI microbiota of the grass carp was generally consistent with the putative AMP-producing strains, which are mainly from *Lactobacillales*. Furthermore, we constructed the phylogenetic relationship of these putative AMP-producing bacteria existing in the GI of grass carp and some popular commercial probiotics (commonly used for microecologics), demonstrating that they are closely related. Thus, these strains have the potential to be developed into novel microecologics. In a word, we provide a high-throughput way to discover AMPs from fish GI microbiota, which can be developed as alternative pathogen antagonists (toxins) for microecologics or probiotic supplements.

Keywords: antimicrobial peptide (AMP); fish gastrointestinal microbiota; high throughput identification; AMP-producing bacteria

---

1. Introduction

Antimicrobial peptides (AMPs) are a group of small peptides, which are secreted by almost all creatures in nature. They are being explored with regards to their potential therapeutic and agricultural uses as they are toxic to many bacteria. These AMPs synthesized in the ribosomes of bacteria are also called bacteriocins, which normally present antibacterial activity towards closely-related strains, although it has been reported that a broad range of antimicrobial activity occurs in some bacteriocins [1]. Those bacteriocins of lactic acid bacteria (LAB) have raised a considerable amount of attention nowadays [2]. Nisin, the first bacteriocin applied to therapeutics without any side effects, can effectively inhibit the growth of Gram-positive (G+) bacteria, including many members of *Lactobacillus*, *Pediococcus*, *Leuconostoc*, *Micrococcus*, *Staphylococcus*, *Listeria*, and *Clostridium* [3].

It is well known that the majority of G+ bacteria are beneficial to intestinal health. *Lactobacillus* is the largest genus of the LAB group (with over 50 species in total), contributing important metabolic reactions to the production of cheese, yogurt, and other dairy products [4]. They are commonly found in the oral, vaginal, and intestinal regions of many animals. These organisms have also been shown to stimulate the immune system and have antibacterial activity against intestinal pathogens, which indicates that they may be useful as probiotics. *Pediococcus acidilactici* is usually used to regulate gastrointestinal (GI) floras as they balance microecology by acid production [5]. *Bacillus subtilis* is a commonly applied microecologic in agriculture feed additives and food products [6]. *Streptococcus* contains many commensal floras located at the epidermis, respiratory tract, and gut, such as beneficial *S. thermophilus* and pathogenic *S. pyogenes* [7].These probiotics are active against microorganisms by antagonism in vivo, particularly against those closely-related competing bacteria. Some of them have a broader inhibitory spectrum in certain conditions, while some of them have a rather narrow spectrum. Additionally, their metabolites improve the activity of acidic proteases and prevent the production of harmful molecules.

The grass carp, *Ctenopharyngodon idellus*, is an herbivorous freshwater fish of the Cyprinidae family and a commercial species widely cultivated in China. Many previous works have focused on employing the 16S rRNA method in analyzing the GI microbiota of grass carp [8]. However, in order to obtain a more accurate phylogenetic composition, the whole metagenomic data should be put to use. Meanwhile, since antibiotic abuse in aquaculture has caused many problems, such as pathogen resistance and food contamination, we are trying to screen new bacterial species with a high efficacy of AMP production as potential probiotics to combat aquatic pathogens.

In this research, we sequenced the whole metagenome of the GI microbiota in grass carp and a high-throughput homologous search was performed to predict AMPs. In addition, the phylogenetic relationship of some genera in grass carp GI microbiota, putative AMP-producing strains and commonly used probiotics for microecologics were described. Interestingly, these newly identified AMP-producing strains have the potential to be developed into novel microecologics or probiotic supplements.

2. Results

2.1. Summary of Achieved Metagenome Datasets

In total, 11 gigabase (Gb) of raw data were generated from a pool of sequencing data of eight GI samples (see more details in Section 4.1). After filtering the host contamination, 218 megabase (Mb) of metagenome data were assembled. Finally, from these acquired data, a total of 4966 species, 1453 genera, 378 families, 178 orders, 76 classes, and 54 phyla were annotated (Table S1). The GI microbiome of grass carp are dominated by members of phyla Proteobacteria (36.12%), Firmicutes (7.14%), Bacteroidetes (5.16%), Fusobacteria (3.82%), and Actinobacteria (1.31%; Figure 1). The well represented genera are *Aeromonas* (19.02%), *Shewanella* (3.79%), *Cetobacterium* (2.96%), *Bacteroides* (1.60%), and *Clostridium* (0.81%), respectively (Figure 1, Table S1).

2.2. AMPs Identified in the Metagenome

Based on sequence similarity, we identified five AMPs by homologous alignment (Figure 2). However, we could not assert the exact strains for producing them, although four of them seem to be incomplete. Meanwhile, we employed BLASTP to search against public databases in NCBI and found that they are novel bacteriocins except for lantibiotic (Table 1). These putative AMP-producing bacteria, summarized in Table 1, will be useful for further screening and validation of potential probiotic supplements.

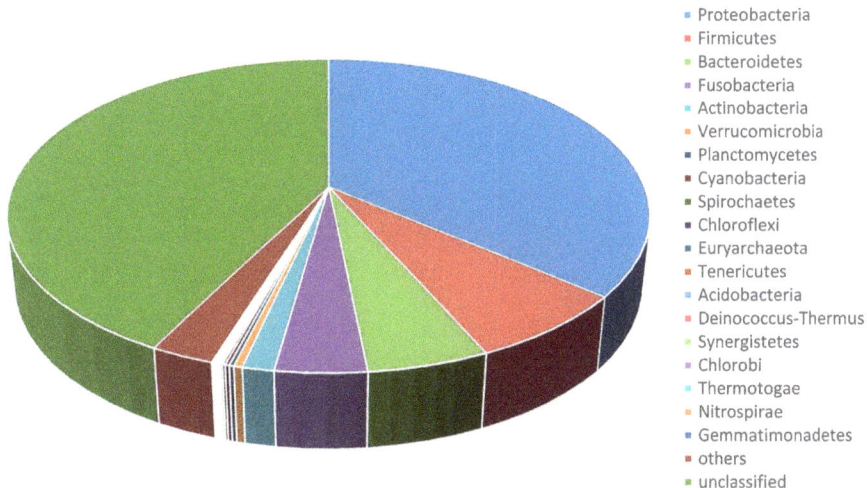

Figure 1. Relative abundance of phyla annotated in the metagenome of grass carp.

A 39 amino acid (aa) fragments identified from filtered reads of current metagenome are exactly the same as the lantibiotics of *Streptococcus* spp., including *S. macedonicus*, *S. pyogenes*, and *S. agalactiae*. This post-translationally modified peptide with characteristic thioether amino acids represents a large group of promising candidates for future applications as it can be a good template for improving lanthipeptide analogs or introducing thioether cross-links into reported therapeutics [9]. The mature peptide GKNGVFKTISHECHLNTWAFLATCCS belongs to lacticin 481-like peptides of class I of lantibiotics, which are LanM modified globular peptides with a classification into class I type B lantibiotic [10]. However, their structures require further confirmation.

The predicted complete sequence of a pediocin-like bacteriocin belongs to class IIa as it has a highly conserved consensus sequence $YYGNGX_2CX_4CXVX_n$ [11] within its N-terminal domain. This 49-aa mature peptide seems to be produced primarily by strains of genus *Streptococcus* [12]. Lactococcin 972 (lcn972) was first isolated from *Lactococcus lactis* subsp. *lactis* IPLA 972 [13]. Our predicted lcn972-like peptide also has a high similarity to lcn972 and thus, we speculate that it may come from strains of the genus *Lactococcus*. Furthermore, members of this family tend to be associated with transmembrane putative immunity proteins related to cellular processes, toxin production, and resistance. It has been proven that lcn972 was the first non-lantibiotic bacteriocin that specifically interacts with the cell wall precursor lipid II [14], which was categorized into subclass IIc [11].

The predicted subtilosin A-like bacteriocin is another class IIc member identified by us. Subtilosin A is a group of antilisterial bacteriocins previously reported only produced by *Bacillus subtilis*, before being found in many other species [15]. Aureocin-like type II bacteriocin is a small family usually encoded on a plasmid. Characteristically, the members are small, cationic, rich in Lys and Trp in addition of bringing about a generalized membrane permeabilization leading to leakage of ions. The family includes aureocin A, lacticins Q and Z, BhtB, as well as an archaeal member [16,17]. However, our predicted BhtB-like seems to be very special compared with others.

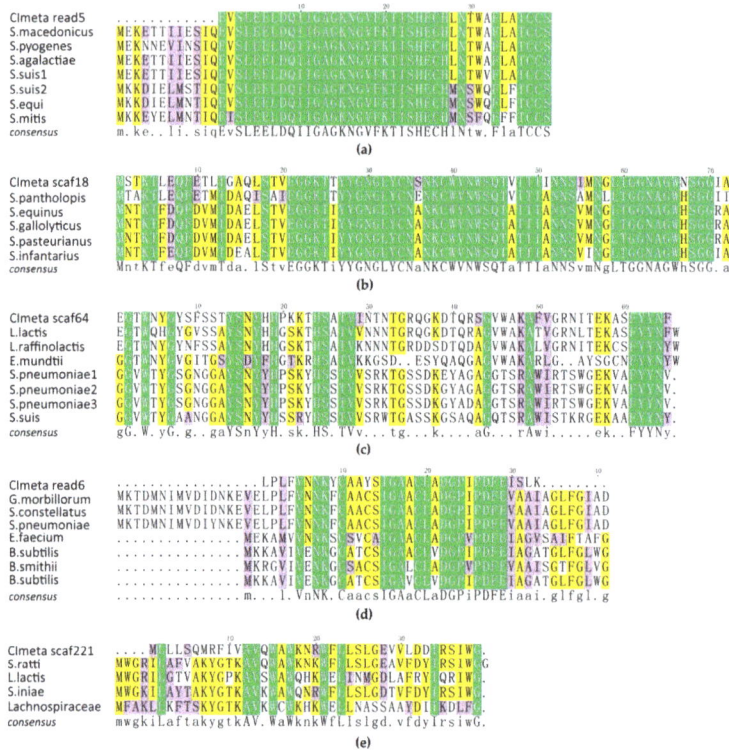

Figure 2. Multiple sequence alignment of identified AMPs with public bacteriocins in the NCBI. (**a**) lantibiotic; (**b**) Pediocin-like (Class IIa); (**c**) Lactococcin 972 (Class IIc); (**d**) SubtilosinA (Class IIc); (**e**) Aureocin-like (Class IId). Please note that with the exception of (**b**), the sequences seem to be incomplete, which may be due to missing reads in the assembly procedure. Similar amino acids of at least 50% were shaded in purple, 80% were shaded in yellow, and those with same characteristics were shaded in green.

Table 1. Summary of known bacteriocins in the NCBI with the highest ranked hits.

Category	Identified AMPs [1]	NCBI AMPs	GenBank Accession No.	NCBI AMP-Producing Bacterial Species
Class I (Lantibiotic)	a	McdA	ABI30227.1	*Streptococcus macedonicus*
		Streptococcin A-FF22	P36501.1	*Streptococcus pyogenes*
		Nukacin	CZT39525.1	*Streptococcus agalactiae*
		Macedocin	CZA89639.1	*Streptococcus suis*
		type-A lantibiotic	EEF65507.1	*Streptococcus suis 89/1591*
		type A2 lantipeptide	WP_041790396.1	*Streptococcus equi*
		type A2 lantipeptide	WP_033685037.1	*Streptococcus mitis*
		lantibiotic nukacin	KEO43205.1	*Streptococcus salivarius*
Class IIa (Pediocin-like)	b	Hypothetical	AND78905.1	*Streptococcus pantholopis*
		piscicolin-126	EFM26697.1	*Streptococcus equinus*
		Bacteriocin	KUE92317.1	*Streptococcus gallolyticus*
		putative piscicolin-126	KXI11412.1	*Streptococcus pasteurianus*
		infantaricin E	AHW46171.1	*Streptococcus infantarius*
		MundKS	ACI25616.1	*Enterococcus mundtii*
		leucocin C	AEY55410.1	*Leuconostoc carnosum*
		SakX	AAP44569.1	*Lactobacillus sakei*

Table 1. *Cont.*

Category	Identified AMPs [1]	NCBI AMPs	GenBank Accession No.	NCBI AMP-Producing Bacterial Species
Class IIc	c	lactococcin 972	CAA05247.1	*Lactococcus lactis*
		lactococcin 972 family	WP_061775386.1	*Lactococcus raffinolactis*
		lactococcin 972 family	WP_065096983.1	*Enterococcus mundtii*
		lactococcin 972	CTL98394.1	*Streptococcus pneumoniae*
		lactococcin 972 family	SNP59245.1	*Streptococcus pneumoniae*
		lactococcin 972 family	CWJ26067.1	*Streptococcus pneumoniae*
		lactococcin 972 family	CYW17154.1	*Streptococcus suis*
	d	sboA protein	EFV34710.1	*Gemella morbillorum* M424
		subtilosin A	EGV07582.1	*Streptococcus constellatus*
		putative subtilosin A	CVX48913.1	*Streptococcus pneumoniae*
		Hypothetical	ELB10075.1	*Enterococcus faecium*
		subtilosin A	CAD23198.1	*Bacillus subtilis*
		subtilosin A	AKP46487.1	*Bacillus smithii*
		subtilosin A	WP_087992738.1	*Bacillus subtilis*
Class IId	e	Mutacin BhtB	AAZ76605.1	*Streptococcus ratti*
		lactolisterin BU	SDR48784.1	*Lactococcus lactis*
		Hypothetical	WP_081348647.1	*Streptococcus iniae*
		Aureocin-like	SFG15527.1	*Lachnospiraceae* C7

[1] See more details in Figure 2.

2.3. Phylogenetic Analysis

We could not determine the unique strains for these predicted AMP sequences. However, in order to evaluate the utility for potential probiotics in grass carp according to close relations between known probiotics and these putative AMP-producing bacteria, we constructed a phylogenetic tree (Figure 3) using microbiota from annotated top abundant species in class *Bacilli*, top bacterial hits in the NCBI by searching from our predicted AMP sequences, and those commonly-used commercial probiotics in microecologics.

Three *Lactococcus* spp. exist in the GI of grass carp, with both *L. raffinolactis* and *L. lactis* being putative AMP-producing probiotics. However, several *Lactococcus* strains are reported as pathogens. Thus, only some strains have the potential to become probiotics in aquaculture as agonists of *S. aureus*, while *L. garvieae* SYP-B-301 has been employed as a feed additive [18–21]. In *Streptococcus* spp., *S. iniae* I1 was selected as predominant LAB in the intestines of young common carp, which is consistent with what we observed. Streptococcaceae was the most plentiful family under the order *Lactobacillales*, followed by Lactobacillaceae and Enterococcaceae families. *Streptococcus* was the 7th most abundant genus in the identified 1453 genera, while *Bacillus* was the 11th, *Lactobacillus* was the 42nd, and *Enterococcus* was the 46th. It seems that a large amount of *Streptococcus* spp. in the GI are predicted to be AMP-producing species (Figure 3) and *S. agalactiae* with a closest relationship to *S. iniae* (a previously applied probiotics) can be a probiotic candidate.

Meanwhile, *Gemella morbillorum* M424, a subtilosin A-producing strain closely related to some probiotics, may be beneficial to the intestine. In addition, while probiotics *Bacillus coagulans*, *B. firmus*, *B. licheniformis*, and *B. subtilis*, were found at relatively low levels, the predominant species *B. cellulosilyticus*, *B. clausii*, and *B. mannanilyticus* (Table 1) may play an important alternative role in the GI. A similar situation occurred for *Lactobacillus* spp. as *L. sakei*, *L. fructivorans*, and *L. ceti* may be helpful. *Enterococcus* spp. are another superior resource, with their functions waiting to be explored.

In summary, based on our phylogenetic tree (Figure 3), we speculate that these bacteria closely related to some known probiotics are promising candidates for novel AMPs or applicable probiotics in the grass carp, such as *L. garvieae*, *S. agalactiae*, *G. morbillorum*, and *L. sakei*. Furthermore, these predominant species of *Lactobacillus* and *Bacillus* are also valuable genetic resources for further investigation.

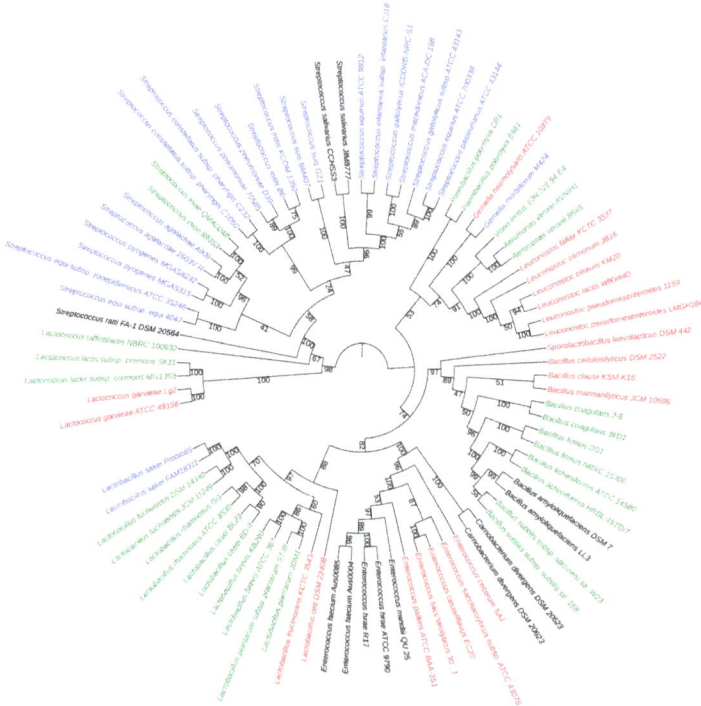

Figure 3. Phylogenetic tree of AMP-producing microbiota. Commercial probiotics are green, top abundant bacteria of class *Bacilli* in grass carp GI are red, and the putative strains producing our predicted AMPs are blue.

3. Discussion

In commercial aquaculture, the stressful conditions fish are exposed to usually result in a decrease in production and increase the risk of infectious diseases. However, resistance from antibiotic abuse has encouraged the scientific community to search for alternatives to antibiotics. Our present strategy to identify the potential AMPs in fish GI microbiota is an alternative way to overcome the deficiency of antibiotics.

In this study, we identified five AMPs (pediocin, lactococcin 972, subtilosin A, and aureocin-like bacteriocins) in the GI of grass carp. However, it is still necessary to confirm the exact strains producing these AMPs. Our analysis suggested that these bacteriocins were encoded by rare members of the GI microbiota or those that have not been previously identified as important bacteria. Furthermore, we predicted the microbiota that may produce our five identified AMPs from the GI metagenomics of *C. idellus* and constructed an interesting phylogenetic tree for them. It revealed that many commensal organisms in grass carp, including three species of *Lactococcus* (*L. raffinolactis*, *L. lactis*, and *L. garvieae*) as well as many members of *Streptococcus*, *Bacillus*, *Lactobacillus*, and *Enterococcus*, are potential probiotics in grass carp, as they are closely related to the known strains used in microecologics and consistent with other characterized probiotics [22,23].

Interestingly, all the predicted bacterial species belong to the class *Bacilli*. In fact, antimicrobial compounds produced by *Bacilli* and isolated from GI of Japanese costal fish [24] and Indian major carp [25] have been suggested as bio-control agents, while antimicrobial peptides or bacteriocins have received attention as an alternative tool to control colonization of pathogenic bacteria in fish intestine [26]. It can be inferred that these *Bacilli* are dominant strains in many animals and other metabolites certainly have irreplaceable functions in GI. Actually, apart from G+ bacteria derived bacteriocins, antimicrobial metabolites from G− bacteria have a rather high concentration in antimicrobial providers to protect their host. In *Pseudoalteromonas*, bacteriocins are almost large proteins (>100 kDa) and their strains are of great ecological significance as part of the resident microbiota or antimicrobial metabolite producers [27]. That may explain why class γ-Proteobacteria is the largest group in the GI of grass carp (Table S1).

GI bacteria can be variably changed by components of food or cultural media, interaction with enterocytes, water temperatures, or seasons [28–30]. A healthy gut is always kept in a dynamic equilibrium, which requires the coordination of many factors [29]. It seems that each bacterial species has its own natural enemies in GI and it restricts other competitors in a reasonable condition by secreting toxins such as AMPs. Their abundance in GI flora is surely in accord with their functions in this complex bacteria community. Once their balance is disrupted, diseases will occur in the host. Antibiotics are commonly abused in animal agriculture, thus causing the emergence of antibiotic-resistant bacteria. An alternative way for applying bacteriocins is to deliver some AMP-producing microbiota directly through prophylactic administration of probiotics, as long as they can stably survive in the intestine and maintain physiological homeostasis.

The rapid advancement of next-generation sequencing (NGS) has accelerated our pace of identifying novel probiotics in fish GI. Previously reported bacteriocins are usually purified by traditional isolation [31]. There are more and more approaches being developed to dig out more bacteriocins in nature, including Bacteriocin Operon and gene block Associator (BOA; [32]), Hidden Markov Model (HMM) [33], and BAGEL2 [34]. These genome mining bioinformatic tools facilitate our work in discovering more useful functional genes by fully utilizing the big data in the post-genomics era [35]. They have a great potential to look into the indicators of many diseases or the prediction of novel AMPs in metagenomes of animals. Thus, this improves the digestive conditions of fish and feed efficiency. This also provides opportunities to better determine the specific types of bacteriocins and bacterial species can be found in different environmental niches through the investigation of metagenomic data. Exploiting antimicrobial-producing resources to shape the host-associated microbiota along with high-throughput sequencing may elucidate the roles of different strains for host defense.

This is the first big-data-based study looking for AMPs and AMP-producing bacteria in the GI metagenome of grass carp. There is a great temptation to connect related microbiota that may produce bacteriocins and abundant bacteria groups with commonly acknowledged probiotics. Future studies should focus on AMPs in fish GI and bacterial strains producing them, thus offering efficient probiotic supplements in agriculture and feed industry. Their basic characteristics are worthy of further investigation and their physiological roles in interacting microbiota remain to be elucidated. Finally, in vivo experiments are required to determine the ecological significance of the association between predicted potential probiotics and their host macro-organisms. The major issue is to determine whether AMP-producing microbiota provide a real benefit to their host, especially in the context of pathogenic events. In such a case, they may stand out as the next generation of probiotics of economic importance in aquaculture.

4. Materials and Methods

4.1. Sample Collection

Pond-cultured grass carps were fed twice a day with a commercial feed (crude protein ≥ 30.0%; crude fiber ≥ 12.0%; crude ash ≤ 15.0%; calcium 0.4–2.5%; phosphate ≥ 0.7%; salt 0.3–1.2%; moisture ≤ 12.5%; and lysine ≥ 1.2%) from Shenzhen Alphafeed Co. Ltd., Shenzhen, China. They were originally collected from a local hatchery in Shenzhen, Guangdong, China by trawl netting. These cultured fish, with an approximate size of 1–2 kg, were selected. The body surface of these fish was rinsed with sterile distilled water and subsequently, 70% ethanol was used to reduce contamination.

The GI was dissected aseptically from their abdominal cavity, while the GI content and the epithelial GI mucosa were squeezed out for a separate harvest. The GI contents from eight individuals were obtained individually and stored at $-80\,^\circ$C before use.

All experiments were performed in accordance with the guidelines of the Animal Ethics Committee and were approved by the Institutional Review Board on Bioethics and Biosafety of BGI (No. FT15103) on 6 July 2016.

4.2. Assembly and Annotation of the Metagenome

Eight samples were separately sequenced by an Illumina Hiseq PE150 platform at BGI, Shenzhen, China. The pool of metagenomic data were assembled by MegaHIT software [36]. Analysis was implemented by Guangdong Magigene Biotechnology Co., Ltd, Guangzhou, China. Annotation of deduced amino acid sequences was performed through BLASTP against the public Nr database with $E \leq 1 \times 10^{-3}$ by KAIJU [37]. To determine the accurate phylogenetic composition of GI microbiota, we assigned all the metagenomic reads to prokaryotic reference genomes that have been submitted to the genome database of NCBI using BLASTN with default parameters. The aligned reads with sequence similarity $\geq$75% were filtered by BLAST against the metagenome of grass carp.

4.3. Prediction and Identification of AMPs

We applied a homology search with the reported AMPs from the newly established *C. idellus* metagenome datasets. These previously validated AMPs were retrieved from online Antimicrobial Peptides Database (APD3) [38] and were used as queries. Assembled metagenome and raw reads of grass carp GI microbiota were used as a local index database. Subsequently, we applied TBLASTN with the threshold of E-value $\leq 10^{-5}$ to run the queries against the database. Following this, the hits of nucleotides were translated into peptide sequences and submitted to NCBI using BLASTP tool for further verification. However, those AMPs only exist in eukaryotes were removed.

4.4. Alignment and Homology of AMPs

The obtained peptide sequences and downloaded proteins from NCBI (Table 1) were aligned by ClustalW of BioEdit [39]. Amino acid residues with conservation among sequences were shaded by TexShade [40].

4.5. Construction of the Phylogenetic Tree

We downloaded the metagenomes of 84 strains and applied their 16S rRNA sequences to construct a phylogenetic tree. First of all, MUSCLE v3.8.31 [41] was employed for multiple alignment, in which all positions containing gaps and missing data were eliminated. Subsequently, an UPGMA tree was constructed using MEGA7 [42]. Bootstrapping was carried out using 100 replicates and values are indicated on the nodes of the phylogeny. The rate variation among sites was modeled with a gamma distribution (shape parameter = 1). Finally, the phylogenetic tree was shown in iTOL [43].

Supplementary Materials: The following is available online at www.mdpi.com/2072-6651/9/9/266/s1. Table S1: Bacterial components in the gastrointestinal samples of grass carp.

Acknowledgments: This work was supported by Shenzhen Special Program for Development of Emerging Strategic Industries (No. JSGG20170412153411369), and Zhenjiang Leading Talent Program for Innovation and Entrepreneurship.

Author Contributions: Q.S. conceived and designed the project; B.D. and Y.Y. analyzed the data; B.D., Y.Y. and Q.S. wrote the paper; L.L. participated in the data analysis and figure preparation.

Conflicts of Interest: The authors declare no conflict of interest.

References

1. Rodney, H.P.; Takeshi, Z.; Kenji, S. Novel bacteriocins from lactic acid bacteria (LAB): Various structures and applications. *Microb. Cell Fact.* **2014**, *13* (Suppl. 1), S3.
2. Zacharof, M.P.; Lovitt, R.W. Bacteriocins produced by lactic acid bacteria a review article. *APCBEE Procedia* **2012**, *2*, 50–56. [CrossRef]
3. Field, D.; Cotter, P.D.; Ross, R.P.; Hill, C. Bioengineering of the model lantibiotic nisin. *Bioengineered* **2015**, *6*, 187–192. [CrossRef] [PubMed]
4. Sun, Z.; Harris, H.M.; McCann, A.; Guo, C.; Argimon, S.; Zhang, W.; Yang, X.; Jeffery, I.B.; Cooney, J.C.; Kagawa, T.F.; et al. Expanding the biotechnology potential of lactobacilli through comparative genomics of 213 strains and associated genera. *Nat. Commun.* **2015**, *6*, 8322. [CrossRef] [PubMed]
5. Papagianni, M.; Anastasiadou, S. Pediocins: The bacteriocins of Pediococci. Sources, production, properties and applications. *Microb. Cell Fact.* **2009**, *8*, 3. [CrossRef] [PubMed]
6. Qi, Z.; Zhang, X.H.; Boon, N.; Bossier, P. Probiotics in aquaculture of China—Current state, problems and prospect. *Aquaculture* **2009**, *290*, 15–21. [CrossRef]
7. Bassis, C.M.; Erb-Downward, J.R.; Dickson, R.P.; Freeman, C.M.; Schmidt, T.M.; Young, V.B.; Beck, J.M.; Curtis, J.L.; Huffnagle, G.B. Analysis of the upper respiratory tract microbiotas as the source of the lung and gastric microbiotas in healthy individuals. *mBio* **2015**, *6*, e00037-15. [CrossRef] [PubMed]
8. Wu, S.; Wang, G.; Angert, E.R.; Wang, W.; Li, W.; Zou, H. Composition, diversity, and origin of the bacterial community in grass carp intestine. *PLoS ONE* **2012**, *7*, e30440. [CrossRef] [PubMed]
9. Dischinger, J.; Basi, C.S.; Bierbaum, G. Lantibiotics: Promising candidates for future applications in health care. *Int. J. Med. Microbiol.* **2014**, *304*, 51–62. [CrossRef] [PubMed]
10. Vaillancourt, K.; LeBel, G.; Frenette, M.; Fittipaldi, N.; Gottschalk, M.; Grenier, D. Purification and characterization of Suicin 65, a novel class I type B lantibiotic produced by *Streptococcus suis*. *PLoS ONE* **2015**, *10*, e0145854. [CrossRef] [PubMed]
11. Drider, D.; Fimland, G.; Hechard, Y.; McMullen, L.M.; Prevost, H. The continuing story of class IIa bacteriocins. *Microbiol. Mol. Biol. Rev.* **2006**, *70*, 564–582. [CrossRef] [PubMed]
12. Cui, Y.; Zhang, C.; Wang, Y.; Shi, J.; Zhang, L.; Ding, Z.; Qu, X.; Cui, H. Class IIa bacteriocins: Diversity and new developments. *Int. J. Mol. Sci.* **2012**, *13*, 16668–16707. [CrossRef] [PubMed]
13. Martinez, B.; Suárez, J.E.; Rodríguez, A. Lactococcin 972: A hornodimeric lactococcal bacteriocin whose primary target is not the plasma membrane. *Microbiology* **1996**, *142*, 2393–2398. [CrossRef] [PubMed]
14. Martinez, B.; Bottiger, T.; Schneider, T.; Rodriguez, A.; Sahl, H.G.; Wiedemann, I. Specific interaction of the unmodified bacteriocin Lactococcin 972 with the cell wall precursor lipid II. *Appl. Environ. Microbiol.* **2008**, *74*, 4666–4670. [CrossRef] [PubMed]
15. Parveen, R.R.; Anandharaj, M.; Hema, S.; Deepika, R.; David, R.A. Purification of antilisterial peptide (Subtilosin A) from novel *Bacillus tequilensis* FR9 and demonstrate their pathogen invasion protection ability using human carcinoma cell line. *Front. Microbiol.* **2016**, *7*, 1910. [CrossRef] [PubMed]
16. Netz, D.A.; Pohl, R.; Beck-Sickinger, A.G.; Selmer, T.; Pierik, A.J.; Bastos, M.F.; Sahl, H.G. Biochemical characterisation and genetic analysis of aureocin A53, a new, a typical bacteriocin from *Staphylococcus aureus*. *J. Mol. Biol.* **2002**, *319*, 745–756. [CrossRef]
17. Netz, D.A.; Bastos, M.C.; Sahl, H.G. Mode of action of the antimicrobial peptide aureocin A53 from *Staphylococcus aureus*. *Appl. Environ. Microbiol.* **2002**, *68*, 5274–5280. [CrossRef] [PubMed]
18. Vendrell, D.; Balcázar, J.L.; Ruiz, Z.I.; De, B.I.; Gironés, O.; Múzquiz, J.L. *Lactococcus garvieae* in fish: A review. *Comp. Immunol. Microbiol. Infect. Dis.* **2006**, *29*, 177–198. [CrossRef] [PubMed]
19. Delpech, P.; Rifa, E.; Ball, G.; Nidelet, S.; Dubois, E.; Gagne, G.; Montel, M.C.; Delbes, C.; Bornes, S. New insights into the anti-pathogenic potential of *Lactococcus garvieae* against *Staphylococcus aureus* based on RNA sequencing profiling. *Front. Microbiol.* **2017**, *8*, 359. [CrossRef] [PubMed]
20. Sugimura, Y.; Hagi, T.; Hoshino, T. Correlation between in vitro mucus adhesion and the in vivo colonization ability of lactic acid bacteria: Screening of new candidate carp probiotics. *Biosci. Biotechnol. Biochem.* **2011**, *75*, 511–515. [CrossRef] [PubMed]
21. Hagi, T.; Hoshino, T. Screening and characterization of potential probiotic lactic acid bacteria from cultured common carp intestine. *Biosci. Biotechnol. Biochem.* **2009**, *73*, 1479–1483. [CrossRef] [PubMed]

22. Pandiyan, P.; Balaraman, D.; Thirunavukkarasu, R.; George, E.J.; Subaramaniyan, K.; Manikkam, S.; Sadayappan, B. Probiotics in aquaculture. *Drug Invent. Today* **2013**, *5*, 55–59. [CrossRef]

23. Merrifield, D.L.; Dimitroglou, A.; Foey, A.; Davies, S.J.; Baker, R.M.; Bøgwald, J.; Castex, M.; Ringø, E. The current status and future focus of probiotic and prebiotic applications for salmonids. *Aquaculture* **2010**, *302*, 1–18. [CrossRef]

24. Sugita, H.; Hirose, Y.; Matsuo, N.; Deguchi, Y. Production of the antibacterial substance by *Bacillus* sp. strain NM 12, an intestinal bacterium of Japanese coastal fish. *Aquaculture* **1998**, *165*, 269–280. [CrossRef]

25. Giri, S.S.; Sukumaran, V.; Sen, S.S.; Vinumonia, J.; Banu, B.N.; Jena, P.K. Antagonistic activity of cellular components of potential probiotic bacteria, isolated from the gut of *Labeo rohita*, against *Aeromonas hydrophila*. *Probiotics Antimicrob. Proteins* **2011**, *3*, 214–222. [CrossRef] [PubMed]

26. Ghanbari, M.; Jami, M.; Domig, K.J.; Kneifel, W. Seafood biopreservation by lactic acid bacteria—A review. *LWT Food Sci. Technol.* **2013**, *54*, 315–324. [CrossRef]

27. Offret, C.; Desriac, F.; Le, C.P.; Mounier, J.; Jegou, C.; Fleury, Y. Spotlight on antimicrobial metabolites from the marine bacteria *Pseudoalteromonas*: Chemodiversity and ecological significance. *Mar. Drugs* **2016**, *14*, 129. [CrossRef] [PubMed]

28. Harikrishnan, R.; Balasundaram, C.; Heo, M.S. Potential use of probiotic- and triherbal extract-enriched diets to control *Aeromonas hydrophila* infection in carp. *Dis. Aquat. Org.* **2010**, *92*, 41–49. [CrossRef] [PubMed]

29. Ni, J.; Yan, Q.; Yu, Y.; Zhang, T. Factors influencing the grass carp gut microbiome and its effect on metabolism. *FEMS Microbiol. Ecol.* **2014**, *87*, 704–714. [CrossRef] [PubMed]

30. Hooper, L.V.; Littman, D.R.; Macpherson, A.J. Interactions between the microbiota and the immune system. *Science* **2012**, *336*, 1268–1273. [CrossRef] [PubMed]

31. Zendo, T.; Nakayama, J.; Fujita, K.; Sonomoto, K. Bacteriocin detection by liquid chromatography/mass spectrometry for rapid identification. *J. Appl. Environ. Microbiol.* **2008**, *104*, 499–507. [CrossRef] [PubMed]

32. Morton, J.T.; Freed, S.D.; Lee, S.W.; Friedberg, I. A large scale prediction of bacteriocin gene blocks suggests a wide functional spectrum for bacteriocins. *BMC Bioinform.* **2015**, *16*, 381. [CrossRef] [PubMed]

33. Walsh, C.J.; Guinane, C.M.; O'Toole, P.W.; Cotter, P.D. A Profile Hidden Markov Model to investigate the distribution and frequency of LanB-encoding lantibiotic modification genes in the human oral and gut microbiome. *PeerJ* **2017**, *5*, e3254. [CrossRef] [PubMed]

34. De Jong, A.; van Heel, A.J.; Kok, J.; Kuipers, O.P. BAGEL2: Mining for bacteriocins in genomic data. *Nucleic Acids Res.* **2010**, *38*, W647–W651. [CrossRef] [PubMed]

35. Scheffler, R.J.; Colmer, S.; Tynan, H.; Demain, A.L.; Gullo, V.P. Antimicrobials, drug discovery, and genome mining. *Appl. Microbiol. Biotechnol.* **2013**, *97*, 969–978. [CrossRef] [PubMed]

36. Li, D.; Luo, R.; Liu, C.M.; Leung, C.M.; Ting, H.F.; Sadakane, K.; Yamashita, H.; Lam, T.W. MEGAHIT v1.0: A fast and scalable metagenome assembler driven by advanced methodologies and community practices. *Methods* **2016**, *102*, 3–11. [CrossRef] [PubMed]

37. Menzel, P.; Ng, K.L.; Krogh, A. Fast and sensitive taxonomic classification for metagenomics with Kaiju. *Nat. Commun.* **2016**, *7*, 11257. [CrossRef] [PubMed]

38. Wang, G.; Li, X.; Wang, Z. APD3: The antimicrobial peptide database as a tool for research and education. *Nucleic Acids Res.* **2016**, *44*, D1087–D1093. [CrossRef] [PubMed]

39. Hall, T.A. BioEdit: A user-friendly biological sequence alignment editor and analysis program for Windows 95/98/NT. *Nucleic Acids Symp. Ser.* **1999**, *41*, 95–98.

40. Beitz, E. TeXshade: Shading and labeling of multiple sequence alignments using LaTeX2e. *Bioinformatics* **2000**, *16*, 135–139. [CrossRef] [PubMed]

41. Edgar, R.C. MUSCLE: Multiple sequence alignment with high accuracy and high throughput. *Nucleic Acids Res.* **2004**, *32*, 1792–1797. [CrossRef] [PubMed]

42. Kumar, S.; Stecher, G.; Tamura, K. MEGA7: Molecular Evolutionary Genetics Analysis version 7.0 for bigger datasets. *Mol. Biol. Evol.* **2016**, *33*, 1870–1874. [CrossRef] [PubMed]

43. Letunic, I.; Bork, P. Interactive tree of life (iTOL) v3: An online tool for the display and annotation of phylogenetic and other trees. *Nucleic Acids Res.* **2016**, *44*, W242–W245. [CrossRef] [PubMed]

© 2017 by the authors. Licensee MDPI, Basel, Switzerland. This article is an open access article distributed under the terms and conditions of the Creative Commons Attribution (CC BY) license (http://creativecommons.org/licenses/by/4.0/).

toxins

MDPI

Article

Lengths of the C-Terminus and Interconnecting Loops Impact Stability of Spider-Derived Gating Modifier Toxins

Akello J. Agwa, Yen-Hua Huang, David J. Craik, Sónia T. Henriques ⬤ and Christina I. Schroeder *

Institute for Molecular Bioscience, the University of Queensland, Brisbane, Queensland 4072, Australia; joanna.agwa@uqconnect.edu.au (A.J.A.); y.huang@imb.uq.edu.au (Y.-H.H.); d.craik@imb.uq.edu.au (D.J.C.); s.henriques@imb.uq.edu.au (S.T.H.)
* Correspondence: c.schroeder@imb.uq.edu.au; Tel.: +61-7-334-62021

Academic Editor: Steve Peigneur
Received: 17 July 2017; Accepted: 8 August 2017; Published: 12 August 2017

Abstract: Spider gating modifier toxins (GMTs) are potent modulators of voltage-gated ion channels and have thus attracted attention as drug leads for several pathophysiological conditions. GMTs contain three disulfide bonds organized in an inhibitory cystine knot, which putatively confers them with high stability; however, thus far, there has not been a focused study to establish the stability of GMTs in physiological conditions. We examined the resistance of five GMTs including GpTx-1, HnTx-IV, HwTx-IV, PaurTx-3 and SgTx-1, to pH, thermal and proteolytic degradation. The peptides were stable under physiological conditions, except SgTx-1, which was susceptible to proteolysis, probably due to a longer C-terminus compared to the other peptides. In non-physiological conditions, the five peptides withstood chaotropic degradation, and all but SgTx-1 remained intact after prolonged exposure to high temperature; however, the peptides were degraded in strongly alkaline solutions. GpTx-1 and PaurTx-3 were more resistant to basic hydrolysis than HnTx-IV, HwTx-IV and SgTx-1, probably because a shorter interconnecting loop 3 on GpTx-1 and PaurTx-3 may stabilize interactions between the C-terminus and the hydrophobic patch. Here, we establish that most GMTs are exceptionally stable, and propose that, in the design of GMT-based therapeutics, stability can be enhanced by optimizing the C-terminus in terms of length, and increased interactions with the hydrophobic patch.

Keywords: $Na_V1.7$; nuclear magnetic resonance; pain; rational drug design; serum stability; spider venom

---

1. Introduction

Gating modifier toxins (GMTs), a class of disulfide-rich peptides expressed in the venom of spiders, modulate the gating mechanism (opening and closing of an ion conduction pore), of voltage-gated ion channels [1–3]. Human voltage-gated ion channels are involved in several pathophysiological conditions, including chronic pain, epilepsy, and cardiovascular conditions, and, accordingly, GMTs have potential as drug leads [4–11].

Spider GMTs are classified within the inhibitory cystine knot (ICK) family of peptides because of the presence of a conserved disulfide bridge connectivity consisting of Cys I–Cys IV, Cys II–Cys V and Cys III–Cys VI [12–16]. Furthermore, some ICK GMTs including GpTx-1, HnTx-IV, PaurTx-3 and SgTx-1 contain two to three antiparallel β-sheets as is expected for ICK peptides (Figure 1A) [13–19]. However, this is not a conserved structural feature, as some GMTs, including HwTx-IV, do not display β-sheets, exemplifying the diversity in structures of peptides containing the ICK motif

(Figure 1A) [12,20,21]. In addition to the ICK motif, a second conserved feature of the structures of GMTs is a hydrophobic patch surrounded by a charged ring of amino acids (Figure 1B) [17,19,21–24]. This amphipathic characteristic, similar to that of membrane active antimicrobial peptides [25], is thought to facilitate GMT interactions with both the voltage-gated ion channels and the lipid membrane in which the channels are embedded [12,20,26–31].

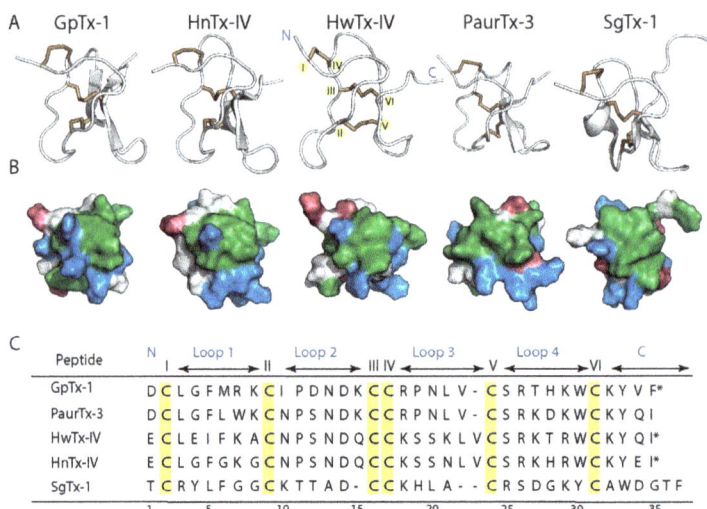

Figure 1. Conserved structural features of spider-derived GMTs. (**A**) ribbon representations of GpTx-1 [17], HnTx-IV (PDB ID: 1NIY) [19], HwTx-IV (PDB ID: 2M4X) [21], PaurTx-3 (PDB ID: 5WE3, this study) and SgTx-1 (PDB ID: 1LA4) [18]. The backbones of the GMTs are shown in white and the disulfide bridges forming the inhibitory cystine knot are shown in brown. GpTx-1, HnTx-IV and PaurTx-3 each have two anti-parallel β-sheets, SgTx-1 has three, and HwTx-IV has no anti-parallel β-sheets. Locations of N- and C-termini and of Cys I–VI (highlighted in yellow) are identified on HwTx-IV for clarity; (**B**) surface representations of the GMTs showing the conserved hydrophobic patch and charged ring of spider toxins, where hydrophobic residues are green, positively charged residues are blue and negatively charged residues are red. The disulfide bridges are buried within the hydrophobic patch; (**C**) sequences of the peptides are shown (aligned to cysteine residues of HwTx-IV and HnTx-IV) with cysteines highlighted in yellow, residues making up interconnecting loops are identified with arrows, N- and C-termini are shown and * denotes amidated C-terminal.

It is generally assumed that GMTs have high stability; however, the ability of these ICK peptides to maintain their structural integrity in physiologically relevant conditions has not been studied in a systematic manner, though a recent study looked at the stability of Hv1a in the context of the development of insecticides [32]. The current study was designed to determine whether the ICK motif engenders GMTs with stability against thermal degradation, pH dependent hydrolysis, proteolysis and chemical degradation using GpTx-1, HnTx-IV, HwTx-IV, PaurTx-3 and SgTx-1 as model GMTs (see Figure 1C for sequences). GpTx-1, HnTx-IV and HwTx-IV are potent inhibitors of the voltage-gated sodium type 1.7 channel (Na$_V$1.7), a channel that has been implicated in chronic pain and PaurTx-3 and SgTx-1 are modulators of Na$_V$1.2 [33,34], a sodium channel associated with epilepsy and ataxia [1].

Here we show that GpTx-1, HnTx-IV, HwTx-IV, and PaurTx-3 are indeed stable in physiologically relevant conditions; however, SgTx-1 is susceptible to proteolysis. Our results show that, although a majority of GMTs provide excellent scaffolds for drug development, the length of the C-terminus may affect stability and needs to be considered if GMTs are to be used as templates in drug design.

2. Results

2.1. Solution NMR Structure of PaurTx-3

Three-dimensional structures of the peptides were used to evaluate the relationship between peptide structure and stability. The structures of HnTx-IV (PDB ID: 1NIY) [19], HwTx-IV (PDB ID: 2M4X) [21] and SgTx-1 (PDB ID: 1LA4) [18] were available from the protein data bank (PDB), and PDB coordinates for GpTx-1 were made available to us as a gift [17]. The three-dimensional solution structure of PaurTx3 was not available from the PDB website and therefore calculated using nuclear magnetic resonance (NMR) spectroscopy (Figure 2) including 350 distance restraints, comprising 134 intraresidue ($i - j = 0$), 120 sequential ($i - j = 1$), 42 medium range ($i - j < 5$), 40 long range ($i - j > 5$) and 14 hydrogen bond restraints and 41 dihedral restraints including 18 ϕ, 18 ψ and 5 $\chi 1$ angle restraints. Root Mean Square Deviation (RMSD) values for the 20 lowest energy structures superimposed where global backbone = 0.88 ± 0.23 Å; global heavy = 1.67 ± 0.20 Å (residues 1–34) and global backbone = 0.71 ± 0.21 Å; global heavy = 1.51 ± 0.21 Å (residues 2–31, excluding flexible termini) (Table 1).

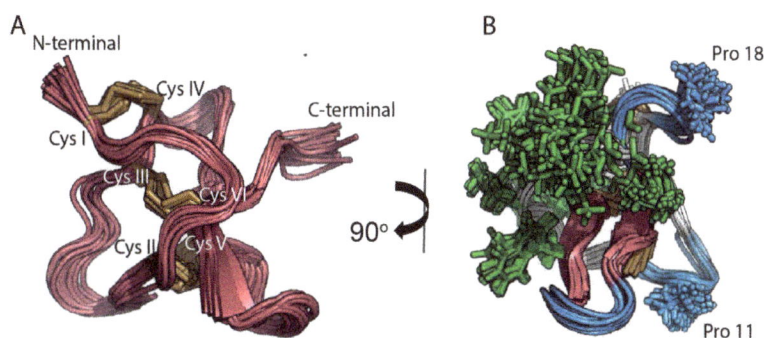

Figure 2. Solution NMR structure of PaurTx-3 (PDB ID: 5WE3, this study). The 20 best conformers selected from lowest energy and best MolProbity scores are shown [35]. (**A**) Backbone of the structures is red and disulfide bridges are yellow with N- and C-termini and cysteines I-VI labeled; (**B**) two antiparallel β-sheets are shown in red, the residues forming the hydrophobic patch are shown in green and the turns on the structures are shown in blue with proline residues in position 11 and 18.

Table 1. Energies and structural statistics [1] for the final 20 [2] structures of PaurTx-3.

Energies (kcal/mol)	
Overall	−1232.31 ± 50.68
Bonds	23.83 ± 2.03
Angles	64.83 ± 6.52
Improper	19.89 ± 2.47
Dihedral	164.27 ± 1.66
Van der Waals	−131.51 ± 7.15
Electrostatic	−1374.68 ± 52.43
NOE	0.38 ± 0.04
Constrained dihedral (cDih)	0.66 ± 0.40
MolProbity Statistics	
Clash score (>0.4 Å/1000 atoms)	11.10 ± 3.94
Poor rotamers (%)	2.88 ± 2.30
Ramachandran outliers (%)	0.47 ± 1.14
Ramachandran favoured (%)	87.50 ± 3.36

Table 1. Cont.

Energies (kcal/mol)	
MolProbity score	2.45 ± 0.26
MolProbity percentile [3]	50.40 ± 14.91
Atomic RMSD (Å)	
Mean global backbone (2–31) [4]	0.71 ± 0.21
Mean global heavy (2–31)	1.51 ± 0.21
Mean global backbone (1–34)	0.88 ± 0.23
Mean global heavy (1–34)	1.67 ± 0.20
Distance Restraints	
Intraresidue ($i − j = 0$)	134
Sequential ($|i − j| = 1$)	120
Medium range ($|i − j| < 5$)	42
Long range ($|i − j| > 5$)	40
Hydrogen bonds [5]	14
Total	350
Dihedral Angle Restraints	
ϕ	18
ψ	18
$\chi 1$	5
Total	41
Violations from Experimental Restraints	
Total NOE violations exceeding 0.2 Å	1
Total dihedral violations exceeding 2.0°	2

[1] ±St Dev. [2] Based on structures with highest overall MolProbity score [35]. [3] 100th percentile is the best among structures of comparable resolution; 0th percentile is the worst. [4] RMSD calculated in MOLMOL (Version 2k.2, Institute of Molecular Biology and Biophysics, ETH, Zürich, Switzerland) [36]. [5] Two restraints were used per hydrogen bond.

The structure comprises the expected ICK disulfide bond connectivity of Cys I–Cys IV, CysII–Cys V and Cys III–Cys VI (Figure 2A) and two antiparallel β-sheets between Val 22–Ser 24 and Lys 28–Lys 31 (Figure 2B). The structure also includes a high density of solvent exposed hydrophobic residues flanked by a series of loops and turns (Figure 2B). Two of the turns are stabilized by proline residues, whereas the third loop lacks a proline residue, probably facilitating flexibility for interactions with target voltage-gated ion channels (Figure 2B).

2.2. Thermal Stability

Effects of varying temperature on peptide structure were examined using one-dimensional (1D) NMR spectra at temperatures ranging from 20 to 80 °C. The spectra (Figure 3) show peaks with narrow line widths and minimal loss in signal intensity up to 80 °C, at which temperature HnTx-IV and SgTx-1 showed peak broadening and loss of intensity. GpTx-1, HwTx-IV and PaurTx-3 maintained narrow peaks but also showed some loss in intensity at 80 °C. Changes in the chemical shifts for all the peptides were reversible, as cooling back to 20 °C resulted in spectra identical to the initial ones, at 20 °C, with narrow peaks and no loss of signal (Figure 3). Notably, for PaurTx-3, the $\varepsilon 1$ proton shifts on the two imidazole side chains of Trp 7 and Trp 29 begin to split with increasing temperature (Figure 3 inset). Therefore, nuclear Overhauser effect (NOE) correlation peaks of PaurTx-3 were compared at 20 °C and 50 °C to confirm that the appearance of a second peak was due to a separation of Trp 7 and Trp 29 and not a result of cis-trans proline isomerization of the peptide (Figure 3 inset). In summary, the reversibility of the chemical shift changes and the well-dispersed peaks in the amide regions of the spectra of all five GMTs at each temperature level indicate that the peptides remain structured and have good thermal stability.

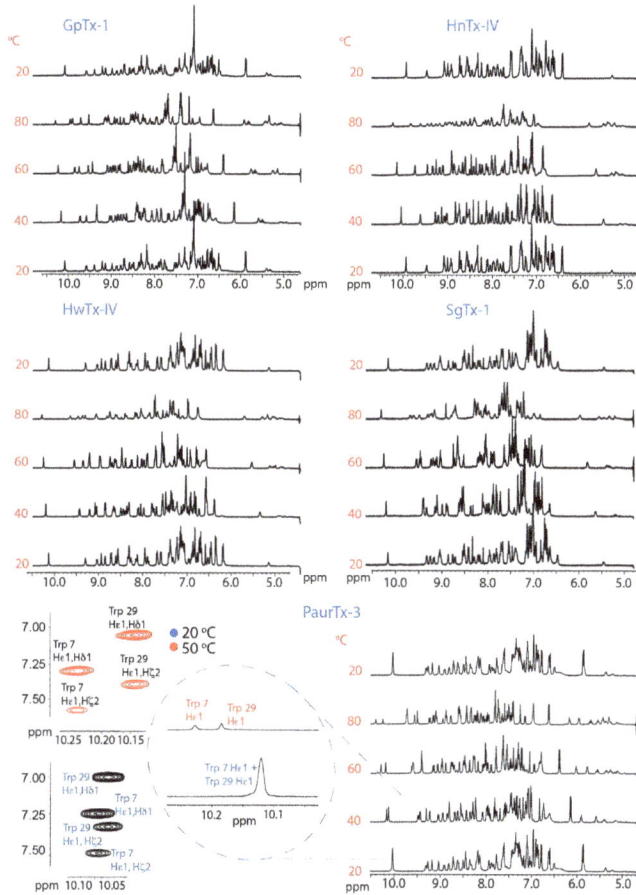

Figure 3. Reversible thermal denaturation of GMTs. 1D ^{1}H solution NMR was used to monitor the structural changes to the five peptides at temperatures ranging from 20 to 80 °C. GpTx-1, HwTx-IV and PaurTx-3 maintained narrow peaks up to 80 °C, although there was some loss in peak intensity in the amide region of the spectra. SgTx-1 and HnTx-IV showed both a loss in intensity and broadening of peaks at 80 °C. NOE correlations confirm that the appearance of two peaks in the 10 ppm region of the 1D ^{1}H spectra of PaurTx-3 are a result of the separation of the ε1 proton on the imidazole rings of Trp 29 and Trp 7.

The peptides were also subjected to more extreme temperature by boiling (100 °C) for 30 min and their stability was examined using analytical reverse-phase high performance liquid chromatography (RP-HPLC). GpTx-1, HnTx-IV, HwTx-IV and PaurTx-3 could withstand the high temperatures, as was demonstrated by identical analytical traces with and without heat treatment (Figure 4), whereas analytical traces of SgTx-1 showed a minor degradation product (Figure 5A). The degradation product from SgTx-1 had an 18 Da loss from the molecular weight of native SgTx-1 (3776.7 Da) as determined by matrix-assisted laser desorption/ionization mass spectrometry (MALDI-MS) (Figure 5A). We suspected that the longer C-terminal chain (six amino acids from the last cysteine in the sequence) of SgTx-1 made this GMT susceptible to degradation; therefore, ProTx-1, a GMT that is one amino acid residue longer than SgTx-1 was similarly subjected to 100 °C for 30 min resulting in two prominent degradation product masses, one of which was equivalent to the loss of 390.2 Da from the molecular weight of native ProTx-1 (3987.6 Da) (Figure 5C). This most likely involved the sequential hydrolytic cleavage of three ProTx-1 C-terminal residues, Thr 33, Phe 34 and Ser 35 (combined Mw = 389.4 Da) (see Figure 5C for sequence).

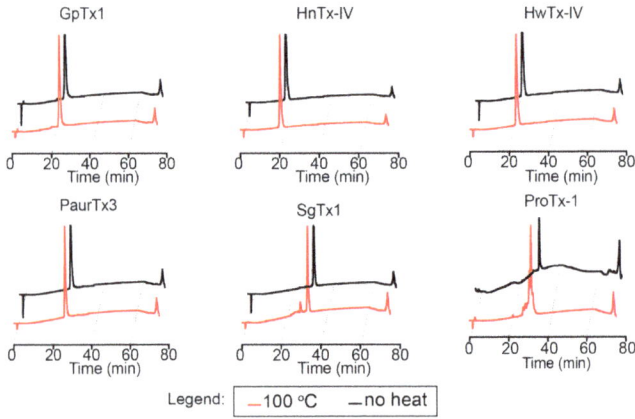

Figure 4. Analytical RP-HPLC traces of the GMTs following heating to 100 °C. All of the peptides except SgTx-1 and ProTx-1 showed remarkable stability after the thermal assault.

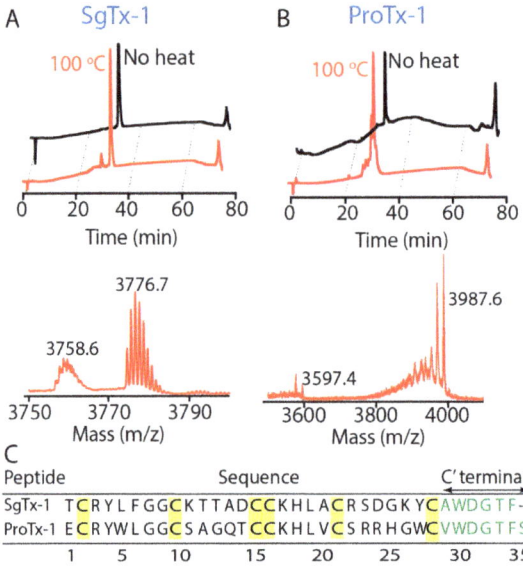

Figure 5. Irreversible thermal degradation of SgTx-1 and ProTx-1. Analytical RP-HPLC was used to examine the peptides after heating to 100 °C and subsequent cooling to room temperature. Degradation products were further characterized using MALDI-MS (**A**) SgTx-1 lost 18 Da from the parent peptide; (**B**) ProTx-1 underwent more extensive degradation, showing a 389.4 Da loss; (**C**) sequences of SgTx-1 and ProTx-1 are shown with cysteines highlighted in yellow and C-terminal residues shown in green.

2.3. pH Dependent Hydrolysis

All five peptides in the current study were stable when incubated at 37 °C in phosphate buffer adjusted to pH 2, 4, 7.4 and 9, with more than 75% peptide remaining after the 24 h incubation (reduced PaurTx-3 was used as a control) (Figure 6A). At pH 12, all of the peptides were partially or fully degraded, as shown in the analytical RP-HPLC traces of the peptides at pH 12 compared to pH 4 (Figure 6B). GpTx-1 and PaurTx-3 had a larger proportion of the folded peptide remaining in strong alkaline conditions compared to the HnTx-IV, HwTx-IV and SgTx-1 (Figure 6B). Analysis of the degradation products of SgTx-1, HwTx-IV and PaurTx-3 are shown as examples (Figure 6C), and reveal that the masses of the native for SgTx-1 (3776.7 Da) and HwTx-IV (4106.6 Da) were absent, whereas the mass of parent PaurTx-3 (4059.5 Da) was still observable. Analysis of the degradation products for PaurTx-3 shows a mass loss of 112.9 Da corresponding to the loss of Ile 34, and HwTx-IV shows a loss of 461 Da, within a range corresponding to the sequential hydrolytic cleavage of Ile 35, Gln 34 and Tyr 33 (458.5 Da).

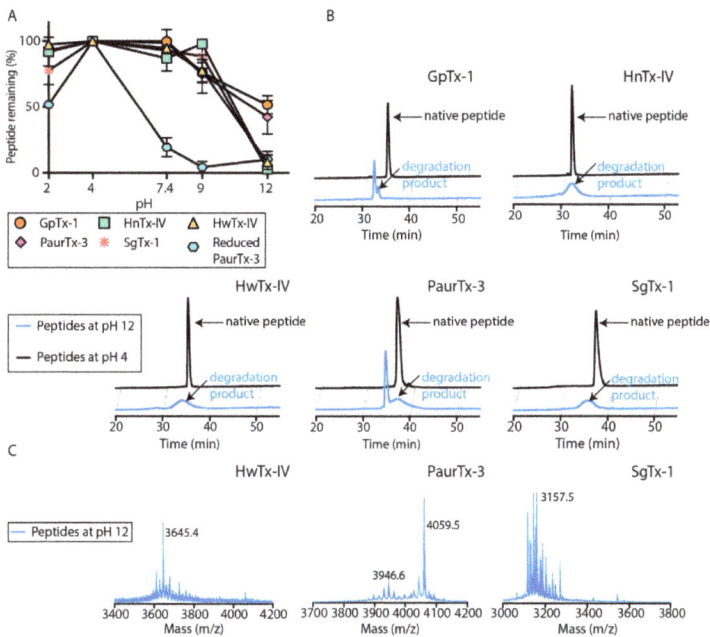

Figure 6. (**A**) pH dependent hydrolysis of the peptides was monitored at pH 2, 4, 7.4, 9 and 12 using analytical RP-HPLC following a 24 h incubation at 37 °C. Reduced PaurTx-3 was used as a control. Data points are relative to amount of peptide at pH 4 and error bars are ± SE for n = 3; (**B**) comparisons of the analytical RP-HPLC traces of the peptides at pH 4 and pH 12 are shown and (**C**) MALDI-MS spectra for HwTx-IV, PaurTx-3 and SgTx-1 at pH 12 are also shown.

2.4. Proteolytic Degradation

Stability of the peptides against proteolytic degradation in human serum was examined at 0 h, 1 h, 8 h and 24 h upon incubation at 37 °C. R-BP100 [37], a linear peptide rich in positively charged residues that does not contain disulfide bonds and is therefore likely to be susceptible to proteolyic degradation, underwent rapid proteolytic degradation to less than 20% peptide remaining after 1 h, and was completely degraded in 8 h (Figure 7). In contrast, HwTx-IV, HnTx-IV, GpTx-1 and PaurTx-3 were resistant to proteolysis with more than 90% peptide remaining after a 24 h incubation in human serum (Figure 7). SgTx-1 was degraded to approximately 70% peptide remaining in 8 h, and to approximately 50% peptide remaining in 24 h.

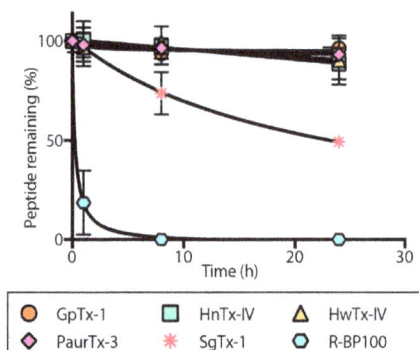

Figure 7. Proteolytic degradation of GMTs. The five GMTs were incubated in human serum and degradation was monitored using RP-HPLC at 0 h, 1 h, 8 h and 24 h. R-BP100 was used as a control. Data points are relative to amount of peptide at 0 h and error bars are ± SE for $n = 3$.

2.5. Chaotropic Degradation

Resistance of the peptides to chemical degradation was examined by incubating the GMTs in 6 M guanidine hydrochloride (GdHCl) at 25 °C for 16 h. Each peptide displayed an identical analytical trace in the presence or absence of 6 M GdHCl upon monitoring using RP-HPLC (Figure 8).

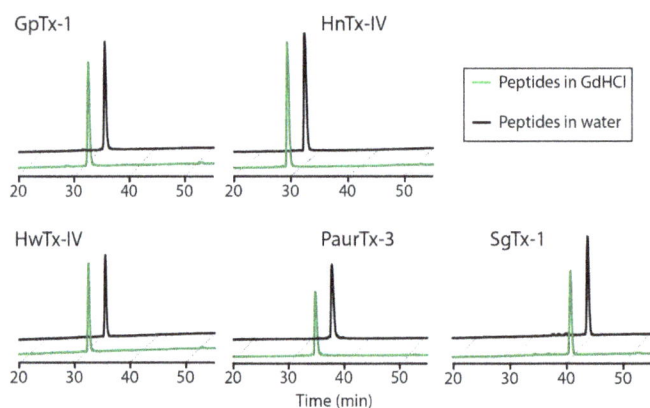

Figure 8. GMT stability in 6 M GdHCl following 16 h incubation at 25 °C, as monitored using RP-HPLC.

2.6. Root-Mean-Square Deviation at the C-Termini GMTs

Global RMSDs of the backbones of the amino acids forming the C-terminus (comprising amino acids from the last cysteine to the C-terminal residue) of the GMTs were calculated using MOLMOL [36]. GpTx-1 and PaurTx-3 had relatively less disordered C-terminal backbones than HwTx-IV, HnTx-IV and SgTx-1 (Table 2). This low disorder is believed to reflect decreased flexibility in this region of GpTx-1 and PaurTx-3, since other factors that can contribute to disorder, such as paucity of NOEs, do not apply for these two peptides.

Table 2. GMT C-terminal RMSD [1] values.

Peptide	Backbone RMSD (Å) (Pre C-Term) [2]	Backbone RMSD (Å) (C-Term) [3]
GpTx-1	0.18 ± 0.09 (Asp 1–Trp 29)	0.17 ± 0.09 (Lys 31–Phe 34)
HnTx-IV	0.60 ± 0.13 (Glu 1–Trp 30)	0.51 ± 0.22 (Lys 32–Ile 35)
HwTx-IV	0.34 ± 0.09 (Glu 1–Trp 30)	0.39 ± 0.17 (Lys 32–Ile 35)
PaurTx-3	0.79 ± 0.22 (Asp 1–Trp 29)	0.30 ± 0.14 (Lys 31–Ile 34)
SgTx-1	0.38 ± 0.11 (Thr 1–Tyr 27)	1.58 ± 0.43 (Asp 29–Phe 34)

[1] Global RMSD calculated using MOLMOL [36]. [2] Residues prior to the final cysteine in the sequence ± standard deviation (SD) from the mean. [3] Residues after the final cysteine in the sequence ± SD from the mean.

3. Discussion

This study set out to examine the stability of spider-derived GMTs under thermal, pH-dependent, proteolytic and chemical conditions. We focused on GpTx-1, HnTx-IV, HwTx-IV, PaurTx-3 and SgTx-1, which like other peptides in their class, contain the conserved ICK motif that stabilizes a solvent exposed hydrophobic patch surrounded by a charged ring of amino acid residues [15–17,19,21–24,38]. We were primarily interested in examining whether GMTs withstand physiologically relevant conditions; however, we also subjected the peptides to extreme thermal, pH and chemical assault to gauge the extent of the overall stability of spider-derived GMTs.

The five peptides in the current study display high thermal stability, as is illustrated by the reversible changes to their 1D ^{1}H NMR chemical shifts after heating to 80 °C in the instrument and subsequent cooling back to 20 °C. Only SgTx-1 was degraded when the peptides were exposed to prolonged, extreme heating and on comparing the structure of this GMT to the remaining four peptides, we hypothesized that the longer, more flexible C-terminus of SgTx-1 (six amino acids for SgTx-1 compared to four amino acids for the remaining GMTs) may have an impact on the lower stability of the peptide. To further examine this hypothesis, ProTx-1, a GMT with one additional amino acid residue at the C-terminus compared to SgTx-1, was exposed to the same prolonged heat treatment, and we observed an even greater extent of degradation for ProTx-1 compared to SgTx-1. Furthermore, C-terminal NOEs for SgTx-1 and ProTx-1 are broad and less intense, compared to the other peptides included in this study, suggesting a more disordered C-terminal. The C-terminal residues in SgTx-1 and ProTx-1 also and have fewer long-distance interactions to hydrophobic residues in other loops compared to GpTx-1, PaurTx-3, HnTx-IV and HwTx-IV (as observed by the lack of long-range NOEs), which may explain their susceptibility to thermal degradation. Similarly, recent work on Hm3a and PcTx1, two spider ICK peptides with high sequence homology, found that Hm3a, which contains four C-terminal amino acid residues, has higher thermal stability compared to PcTx1, which has seven C-terminal amino acid residues [39]. OAIP, another spider ICK peptide containing three C-terminal amino acid residues, previously showed thermal stability [40]; however, PVIIA, an ICK toxin extracted from cone snail venom containing only one C-terminal residue, was irreversibly denatured at 56 °C [38]; therefore, it appears that a C-terminus of 3–4 amino acids may be optimal for the stability of ICK peptides against thermal assault.

We examined the stability of the peptides at pH values representing various sites in the body including pH 2 (stomach), pH 4 (jejunum), pH 7.4 (plasma) and pH 9 (approximate ileum pH). pH 12 was used to represent the peptides in an extreme (non-physiological) environment. The results suggest that the peptides can withstand physiological pH ranges, with SgTx-1 perhaps being the least favorable because of the small loss of peptide at pH 2 (Figure 6A). The peptides began to show degradation at pH 9, but the most substantial degradation occurred at pH 12. Previous studies on spider-derived ICK peptides have reported that disulfide bond shuffling occurs in alkaline conditions [32,40]. However, the degradation products of the GMTs in the present study contained mass losses most probably from the C-terminus of the peptides (Figure 6B,C), and disulfide bond shuffling was not observed. Assuming that SgTx-1 was susceptible to the alkaline hydrolysis because of the length of the C-terminal on this GMT, we were curious to find out why the four remaining peptides each containing four C-terminal amino acids showed different behavior from each other when incubated at pH 12. There are three key structural features on the peptides that could withstand the alkaline assault within GpTx-1 and PaurTx-3, which are absent on HwTx-IV and HnTx-IV (which were completely degraded). First, a shorter loop 3, second, the presence of Pro 18 in the turn between loop 2 and loop 3 and, third, less disorder across the C-terminal residues (Figure 9, Table 2). It is possible that this combination of structural features facilitates a stronger attraction of the hydrophobic C-terminal residues of PaurTx-3 and GpTx-1 to the hydrophobic patch of the peptides, providing some protection from alkaline hydrolysis (Figure 9), whereas peptides like HwTx-IV and HnTx-IV, lacking this additional structural rigidity and subsequent hydrophobic attraction, become more susceptible to alkaline hydrolysis.

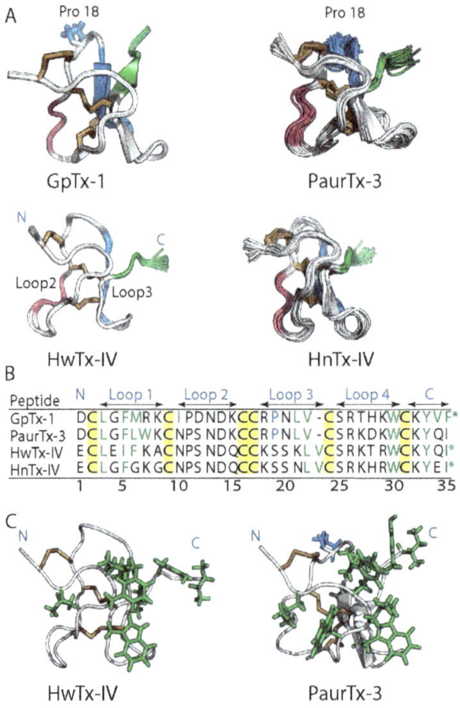

Figure 9. Structural features of the loops, turns and C-termini of GMTs. (**A**) ribbon representations of 20 structures of PaurTx-3 (PDB ID: 5W3E, this study), HwTx-IV (PDB ID: 2M4X) [21], and HnTx-IV (PDB ID: 1NIY) [19], and 10 structures of GpTx-1 [17], are shown where loop 2 is highlighted in red, loop 3 is in blue and the C-terminal is in green (labels on HwTx-IV for clarity). The side chain of Pro 18 is shown for GpTx-1 and PaurTx-3 in blue; (**B**) sequences of the peptides are also shown and the residues forming each loop and the C-terminal are identified by arrows. Sequences of hydrophobic amino acid residues and residues forming the hydrophobic patch are colored green and * denotes amidated C-terminal; (**C**) green sticks show the side chains of hydrophobic residues on HwTx-IV and PaurTx-3.

To consider GMTs as viable drug leads, stability of the peptides in human serum is essential. GpTx-1, PaurTx-3, HwTx-IV and HnTx-IV show exceptional serum stability (Figure 8); however, SgTx-1 is susceptible to proteolytic degradation. This degradation of SgTx-1 is most likely due to the relatively longer C-terminal, making the peptide more susceptible to proteolysis in comparison to the other GMTs in the current study.

As a final analysis of the overall stability of disulfide-rich GMTs, we examined the ability of the peptides to withstand the chaotropic effects of GdHCl. GdHCl has the potential to interfere with the integrity of peptide structures via interactions either with the backbone of the peptides or through π-cation interactions with the solvent exposed aromatic side chains of the solvent exposed hydrophobic amino acid residues [41,42]. The five GMTs in the current study were unaffected by the chemical assault, most likely because of the presence of the ICK motif. A previous study on AS-48, a globular cyclic peptide also containing a hydrophobic surface patch, but lacking disulfide bridges, showed that the peptide was degraded by 6.3 M GdHCl at 25 °C [43]. Cyclic peptides containing the ICK motif were subsequently found to be stable when subjected to GdHCl [38]. Therefore, we conclude that disulfide bridges confer chemical stability to spider-derived ICK GMTs.

4. Conclusions

In conclusion, GpTx-1, HwTx-IV, HnTx-IV and PaurTx-3 are stable in physiologically relevant environments. SgTx-1 underwent partial degradation when subjected to proteolytic, acidic (pH 2) and extreme temperatures, probably because of the length of the C-terminal of this particular peptide. To avoid similar degradation in the use of GMTs as templates for drug design, a three to four residue C-terminal appears to be preferable for stability. Hydrophobic interactions between the C-terminal chain and the hydrophobic patch on the surface of the peptide may provide additional stability to GMTs against hydrolysis in strongly alkaline solutions, and these hydrophobic interactions can be augmented by designing GMT analogues with a shorter loop 3 stabilized by a proline on the turn between loop 2 and 3.

Spider-derived GMTs are undeniably among nature's more interesting pharmacological probes in the study of voltage-gated ion channels [6,12,34]. The current work has provided additional insight into the potential of these ICK peptides as templates for drugs designed to target ailments linked to the voltage-gated ion channels.

5. Materials and Methods

5.1. Peptide Synthesis

Peptides were synthetically assembled using automated solid phase peptide synthesis on a Symphony peptide synthesizer (Protein Technologies Inc., Tucson, AZ, USA), as previously described [26]. Side chain protecting groups were removed and the reduced peptides were released from the resin using 96% (*v/v*) trifluoroacetic acid (TFA), 2% (*v/v*) water and 2% (*v/v*) triisopropylsaline (TIPS) as before [26].

Oxidative folding of PaurTx-3 and GpTx-1 was achieved in 16 h at room temperature in 7.5% (*v/v*) acetonitrile (ACN), 0.1 M Tris(hydroxymethyl)aminomethane, 0.81 mM reduced glutathione (GSH) and 0.81 mM oxidized glutathione (GSSG) at pH 7.7 [17]. ProTx-1 and SgTx-1 were oxidized over 72 h at 4 °C in a buffer containing 0.1 M ammonium acetate, 2 M urea, 2.5 mM GSH and 0.25 mM GSSG at pH 7.8 [18]. HnTx-IV and HwTx-IV were oxidized at room temperature for 16 h in a buffer containing 0.1 M Tris, 5 mM GSH, 0.5 mM GSSG ph 8, oxidation of HnTx-IV additionally required 0.1 M NaCl in the buffer solution [26,44,45]. The peptides were purified using RP-HPLC on a C18 column using preparatory (8 mL/min) and semi-preparatory (3 mL/min) flow rates on gradients obtained using solvent A (0.05% TFA in water) and solvent B (0.05% TFA in 90% ACN) as previously described [26].

5.2. NMR Structure Calculation for PaurTx-3

Data used for PaurTx-3 structure calculations was acquired on a Bruker Avance 600 MHz NMR spectrometer (Bruker, Billerica, MA, USA) equipped with a cryoprobe. The solution NMR structure for PaurTx-3 was calculated using previously described protocols [26], on a Bruker Avance 600 MHz NMR spectrometer equipped with a cryoprobe. CCPNMR Analysis 2.4.1 (CCPN, University of Cambridge, Cambridge, UK) was used for amino acid assignment [46,47]. The solution NMR structure for PaurTx-3 was calculated as previously described [26] using the AUTO and ANNEAL functions in CYANA 3.97 (Güntert Group, Goethe-Univerity Frankfurt, Frankfurt, Germany) [48] to refine peak assignments. Dihedral angle restraints were generated using TALOS-N (Bax Group, NIH, Pike Bethseda, MD, USA) [49]. After initial structure determination on CYANA, protocols on the RECOORD database [50], were used to generate 50 structures which were then refined in a water shell [51], and a final set of 20 structures was chosen based on the lowest energy, best MolProbity scores [35], and fewest distance and dihedral angle violations (Table 1). PaurTx-3 structure and restraints have been submitted to the Protein data bank (PDB ID: 5WE3) and the Biomagnetic Resonance Data bank (BMRB: 30317).

5.3. Peptide Quantification

Stock concentrations of peptides were quantified using Nanodrop at 280 nm using extinction coefficient (ε_{280}) values as follows: GpTx-1 ε_{280} = 7365 M^{-1} cm^{-1}; HnTx-IV ε_{280} = 7365 M^{-1} cm^{-1}; HwTx-IV ε_{280} = 7365 M^{-1} cm^{-1}; PaurTx-3 ε_{280} = 12,865 M^{-1} cm^{-1}; SgTx-1 ε_{280} = 8855 M^{-1} cm^{-1}. Unless otherwise stated, stock concentrations of the peptides were 300 μM in water.

5.4. Analytical RP-HPLC

For experiments monitored using RP-HPLC, solvent A (0.05% TFA in water) and solvent B (0.05% TFA in 90% ACN) were used on a C18 phenomenex column at a flow rate of 0.3 mL/min using a 1% gradient of 0–50% solvent B and monitored at 215 nm.

5.5. Thermal Stability

A Bruker 500 MHz Avance nuclear magnetic resonance (NMR) spectrometer was used to heat the peptides (1 mg/mL in 9:1 v/v H_2O/D_2O) and to monitor structural changes at temperatures ranging from 20 to 80 °C. One-dimensional (1D) ^1H NMR spectra and two-dimensional (2D) nuclear Overhauser effect spectroscopy (NOESY) (200 ms mixing time) spectra were acquired and processed using TopSpin 3.5 (Bruker, Billerica, MA, USA) and CCPNMR Analysis 2.4.1 was used in the assignment of the NOE spectra [46,47].

Peptides were also subjected to boiling (100 °C) in water for 30 min using a heating block, cooled to 25 °C and analyzed using analytical RP-HPLC. The mass of the degradation products from SgTx-1 and ProTx-1 were determined using matrix-assisted laser desorption/ionization mass spectrometry (MALDI-MS).

5.6. pH Dependent Hydrolysis

The ability of the peptides to withstand varying pH conditions was monitored by incubating the peptides (30 μM) at 37 °C in phosphate buffer adjusted to pH 2, 4, 7.4, 9 and 12 using phosphoric acid and/or sodium hydroxide. After 24 h, each sample was adjusted to pH 2 using phosphoric acid and analytical RP-HPLC was used to quantify the percentage of peptide remaining by examining the height of the peaks at each pH relative to pH 4 (the pH where most samples showed highest stability). Reduced PaurTx-3 was used as a control.

5.7. Proteolytic Degradation

Human serum isolated from male AB plasma was used to examine the stability of the peptides to proteolysis. Peptides (30 μM) were incubated with human serum or phosphate buffered saline (PBS) at 37 °C and reactions were stopped at 0, 1, 8 or 24 h by placing the samples on ice, 6 M urea was used to denature the proteases and 20% trichloroacetic acid was used to precipitate the proteases. Peptides were separated from the proteases by centrifugation at 17,000 g for 10 min. Effects of proteolysis were examined using RP-HPLC, whereby retention times of the peptides were determined using the PBS controls at 0 h and percentage of peptide remaining in human serum was obtained from the heights of the peaks at each time point relative to 0 h. RBP-100 was the control.

5.8. Chaotropic Stability

Chemical stability of the peptides (12 µM) was studied by incubating them in 6 M GdHCl at 25 °C for 16 h. Analytical RP-HPLC was used to compare the peptides in GdHCl to peptides in water. RBP-100 was used as a control.

5.9. RMSD Calculation

MOLMOL (Version 2k.2, Institute of Molecular Biology and Biophysics, ETH, Zürich, Switzerland) [36] was used to calculate the global atomic RMSDs of the backbone of the GMTs including the amino acid residue following the last cysteine to the C-terminal of each sequence. RMSD values of the amino acids prior to the final cysteine were also calculated.

Acknowledgments: The authors would like to thank Olivier Cheneval and Thomas Dash, Institute for Molecular Bioscience, University of Queensland, Australia, for the assembly of the peptides used in this study and for assistance with MALDI-MS, respectively, and Les Miranda, Amgen, Thousand Oaks, CA, USA, for providing PDB coordinates for GpTx-1. This work was funded by an Australian National Health and Medical Research Council (NHMRC) grant awarded to C.I.S. and S.T.H. (APP1080405). C.I.S. (FT160100055) and S.T.H. (FT150100398) are Australian Research Council (ARC) Future Fellows, D.J.C. is an ARC Australian Laureate Fellow (FL150100146) and A.J.A. holds a University of Queensland International postgraduate student scholarship. No additional funds were received for covering the costs to publish in open access.

Author Contributions: A.J.A. and C.I.S. conceived and designed the experiments; A.J.A. performed the experiments and analyzed the data and Y.-H.H. provided expertise on the proteolysis experiments. A.J.A. wrote the paper and C.I.S., S.T.H., Y.-H.H. and D.J.C. provided critical reviews of the paper.

Conflicts of Interest: The authors declare no conflicts of interest.

References

1. Smith, J.J.; Lau, C.H.Y.; Herzig, V.; Ikonomopoulou, M.P.; Rash, L.D.; King, G.F. Therapeutic applications of spider-venom peptides. In *Venoms to Drugs: Venom as a Source for the Development of Human Therapeutics*; King, G.F., Ed.; Royal Society of Chemistry: London, UK, 2015; pp. 221–244.
2. Escoubas, P.; Bosmans, F. Spider peptide toxins as leads for drug development. *Expert Opin. Drug Discov.* **2007**, *2*, 823–835. [CrossRef] [PubMed]
3. Catterall, W.A.; Cestele, S.; Yarov-Yarovoy, V.; Yu, F.H.; Konoki, K.; Scheuer, T. Voltage-gated ion channels and gating modifier toxins. *Toxicon* **2007**, *49*, 124–141. [CrossRef] [PubMed]
4. Vetter, I.; Deuis, J.R.; Mueller, A.; Israel, M.R.; Starobova, H.; Zhang, A.; Rash, L.D.; Mobli, M. Na$_V$1.7 as a pain target—From gene to pharmacology. *Pharmacol. Ther.* **2017**, *172*, 73–100. [CrossRef] [PubMed]
5. Osteen, J.D.; Herzig, V.; Gilchrist, J.; Emrick, J.J.; Zhang, C.; Wang, X.; Castro, J.; Garcia-Caraballo, S.; Grundy, L.; Rychkov, G.Y.; et al. Selective spider toxins reveal a role for the Na$_V$1.1 channel in mechanical pain. *Nature* **2016**, *534*, 494–499. [CrossRef] [PubMed]
6. Lewis, R.J.; Vetter, I.; Cardoso, F.C.; Inserra, M.; King, G. Does nature do ion channel drug discovery better than us? In *Ion Channel Drug Discovery*; Cox, B., Gosling, M., Eds.; Royal Society of Chemistry: London, UK, 2015; pp. 297–319.
7. Catterall, W.A. Voltage-gated sodium channels at 60: Structure, function and pathophysiology. *J. Physiol.* **2012**, *590*, 2577–2589. [CrossRef] [PubMed]
8. Dib-Hajj, S.D.; Cummins, T.R.; Black, J.A.; Waxman, S.G. Sodium channels in normal and pathological pain. *Annu. Rev. Neurosci.* **2010**, *33*, 325–347. [CrossRef] [PubMed]
9. Flinspach, M.; Xu, Q.; Piekarz, A.D.; Fellows, R.; Hagan, R.; Gibbs, A.; Liu, Y.; Neff, R.A.; Freedman, J.; Eckert, W.A.; et al. Insensitivity to pain induced by a potent selective closed-state Na$_V$1.7 inhibitor. *Sci. Rep.* **2017**, *7*, 39662. [CrossRef] [PubMed]
10. Netirojjanakul, C.; Miranda, L.P. Progress and challenges in the optimization of toxin peptides for development as pain therapeutics. *Curr. Opin. Chem. Biol.* **2017**, *38*, 70–79. [CrossRef] [PubMed]
11. Osteen, J.D.; Sampson, K.; Iyer, V.; Julius, D.; Bosmans, F. Pharmacology of the Nav1.1 domain IV voltage sensor reveals coupling between inactivation gating processes. *Proc. Natl. Acad. Sci. USA* **2017**, *114*, 6836–6841. [PubMed]

12. Agwa, A.J.; Henriques, S.T.; Schroeder, C.I. Gating modifier toxin interactions with ion channels and lipid bilayers: is the trimolecular complex real? *Neuropharmacology* **2017**, in press. [CrossRef] [PubMed]
13. King, G.F.; Tedford, H.W.; Maggio, F. Structure and function of insecticidal neurotoxins from Australian funnel-web spiders. *J. Toxicol. Toxin Rev.* **2002**, *21*, 361–389. [CrossRef]
14. Craik, D.J.; Daly, N.L.; Waine, C. The cystine knot motif in toxins and implications for drug design. *Toxicon* **2001**, *39*, 43–60. [CrossRef]
15. Norton, R.S.; Pallaghy, P.K. The cystine knot structure of ion channel toxins and related polypeptides. *Toxicon* **1998**, *36*, 1573–1583. [CrossRef]
16. Pallaghy, P.K.; Nielsen, K.J.; Craik, D.J.; Norton, R.S. A common structural motif incorporating a cystine knot and a triple-stranded beta-sheet in toxic and inhibitory polypeptides. *Protein Sci.* **1994**, *3*, 1833–1839. [CrossRef] [PubMed]
17. Murray, J.K.; Ligutti, J.; Liu, D.; Zou, A.; Poppe, L.; Li, H.; Andrews, K.L.; Moyer, B.D.; McDonough, S.I.; Favreau, P.; et al. Engineering potent and selective analogues of GpTx-1, a tarantula venom peptide antagonist of the Na$_V$1.7 sodium channel. *J. Med. Chem.* **2015**, *58*, 2299–2314. [CrossRef] [PubMed]
18. Lee, C.W.; Kim, S.; Roh, S.H.; Endoh, H.; Kodera, Y.; Maeda, T.; Kohno, T.; Wang, J.M.; Swartz, K.J.; Kim, J.I. Solution structure and functional characterization of SGTx1, a modifier of K$_V$2.1 channel gating. *Biochemistry* **2004**, *43*, 890–897. [CrossRef] [PubMed]
19. Li, D.; Xiao, Y.; Xu, X.; Xiong, X.; Lu, S.; Liu, Z.; Zhu, Q.; Wang, M.; Gu, X.; Liang, S. Structure-activity relationships of hainantoxin-IV and structure determination of active and inactive sodium channel blockers. *J. Biol. Chem.* **2004**, *279*, 37734–37740. [CrossRef] [PubMed]
20. Henriques, S.T.; Deplazes, E.; Lawrence, N.; Cheneval, O.; Chaousis, S.; Inserra, M.; Thongyoo, P.; King, G.F.; Mark, A.E.; Vetter, I.; et al. Interaction of tarantula venom peptide ProTx-II with lipid membranes is a prerequisite for its inhibition of human voltage-gated sodium channel Na$_V$1.7. *J. Biol. Chem.* **2016**, *29*, 17049–17065. [CrossRef] [PubMed]
21. Minassian, N.A.; Gibbs, A.; Shih, A.Y.; Liu, Y.; Neff, R.A.; Sutton, S.W.; Mirzadegan, T.; Connor, J.; Fellows, R.; Husovsky, M.; et al. Analysis of the structural and molecular basis of voltage-sensitive sodium channel inhibition by the spider toxin huwentoxin-IV (mu-TRTX-Hh2a). *J. Biol. Chem.* **2013**, *288*, 22707–22720. [CrossRef] [PubMed]
22. Deuis, J.R.; Dekan, Z.; Wingerd, J.S.; Smith, J.J.; Munasinghe, N.R.; Bhola, R.F.; Imlach, W.L.; Herzig, V.; Armstrong, D.A.; Rosengren, K.J.; et al. Pharmacological characterisation of the highly Na$_V$1.7 selective spider venom peptide Pn3a. *Sci. Rep.* **2017**, *7*, 40883. [CrossRef] [PubMed]
23. Klint, J.K.; Smith, J.J.; Vetter, I.; Rupasinghe, D.B.; Er, S.Y.; Senff, S.; Herzig, V.; Mobli, M.; Lewis, R.J.; Bosmans, F.; et al. Seven novel modulators of the analgesic target Na$_V$1.7 uncovered using a high-throughput venom-based discovery approach. *Br. J. Pharmacol.* **2015**, *172*, 2445–2458. [CrossRef] [PubMed]
24. Bosmans, F.; Rash, L.; Zhu, S.; Diochot, S.; Lazdunski, M.; Escoubas, P.; Tytgat, J. Four novel tarantula toxins as selective modulators of voltage-gated sodium channel subtypes. *Mol. Pharmacol.* **2006**, *69*, 419–429. [CrossRef] [PubMed]
25. Mandard, N.; Bulet, P.; Caille, A.; Daffre, S.; Vovelle, F. The solution structure of gomesin, an antimicrobial cysteine-rich peptide from the spider. *Eur. J. Biochem.* **2002**, *269*, 1190–1198. [CrossRef] [PubMed]
26. Agwa, A.J.; Lawrence, N.; Deplazes, E.; Cheneval, O.; Chen, R.; Craik, D.J.; Schroeder, C.I.; Henriques, S.T. Spider peptide toxin HwTx-IV engineered to bind to lipid membranes has an increased inhibitory potency at human voltage-gated sodium channel hNa$_V$1.7. *Biochim. Biophys. Acta* **2017**, *1859*, 835–844. [CrossRef] [PubMed]
27. Deplazes, E.; Henriques, S.T.; Smith, J.J.; King, G.F.; Craik, D.J.; Mark, A.E.; Schroeder, C.I. Membrane-binding properties of gating modifier and pore-blocking toxins: Membrane interaction is not a prerequisite for modification of channel gating. *Biochim. Biophys. Acta* **2016**, *1858*, 872–882. [CrossRef] [PubMed]
28. Lau, C.H.Y.; King, G.F.; Mobli, M. Molecular basis of the interaction between gating modifier spider toxins and the voltage sensor of voltage-gated ion channels. *Sci. Rep.* **2016**, *6*, 34333. [CrossRef] [PubMed]
29. Revell Phillips, L.; Milescu, M.; Li-Smerin, Y.; Mindell, J.A.; Kim, J.I.; Swartz, K.J. Voltage-sensor activation with a tarantula toxin as cargo. *Nature* **2005**, *436*, 857–860. [CrossRef] [PubMed]
30. Lee, S.-Y.; MacKinnon, R. A membrane-access mechanism of ion channel inhibition by voltage sensor toxins from spider venom. *Nature* **2004**, *430*, 232–235. [CrossRef] [PubMed]

31. Salari, A.; Vega, B.S.; Milescu, L.S.; Milescu, M. Molecular interactions between tarantula toxins and low-voltage-activated calcium channels. *Sci. Rep.* **2016**, *6*, 23894. [CrossRef] [PubMed]
32. Herzig, V.; King, G.F. The cystine knot is responsible for the exceptional stability of the insecticidal spider toxin omega-hexatoxin-Hv1a. *Toxins* **2015**, *7*, 4366–4380. [CrossRef] [PubMed]
33. Bosmans, F.; Martin-Eauclaire, M.-F.; Swartz, K.J. Deconstructing voltage sensor function and pharmacology in sodium channels. *Nature* **2008**, *456*, 202–208. [CrossRef] [PubMed]
34. Bosmans, F.; Swartz, K.J. Targeting sodium channel voltage sensors with spider toxins. *Trends Pharmacol. Sci.* **2010**, *31*, 175–182. [CrossRef] [PubMed]
35. Davis, I.W.; Leaver-Fay, A.; Chen, V.B.; Block, J.N.; Kapral, G.J.; Wang, X.; Murray, L.W.; Arendall, W.B., 3rd; Snoeyink, J.; Richardson, J.S.; et al. MolProbity: all-atom contacts and structure validation for proteins and nucleic acids. *Nucleic Acids Res.* **2007**, *35*, W375–383. [CrossRef] [PubMed]
36. Koradi, R.; Billeter, M.; Wüthrich, K. MOLMOL: A program for display and analysis of macromolecular structures. *J. Mol. Graph.* **1996**, *14*, 51–55, 29–32. [CrossRef]
37. Torcato, I.M.; Huang, Y.H.; Franquelim, H.G.; Gaspar, D.; Craik, D.J.; Castanho, M.A.; Troeira Henriques, S. Design and characterization of novel antimicrobial peptides, R-BP100 and RW-BP100, with activity against Gram-negative and Gram-positive bacteria. *Biochim. Biophys. Acta* **2013**, *1828*, 944–955. [CrossRef] [PubMed]
38. Colgrave, M.L.; Craik, D.J. Thermal, chemical, and enzymatic stability of the cyclotide kalata B1: The importance of the cyclic cystine knot. *Biochemistry* **2004**, *43*, 5965–5975. [CrossRef] [PubMed]
39. Er, S.Y.; Cristofori-Armstrong, B.; Escoubas, P.; Rash, L.D. Discovery and molecular interaction studies of a highly stable, tarantula peptide modulator of acid-sensing ion channel 1. *Neuropharmacology* **2017**. [CrossRef] [PubMed]
40. Hardy, M.C.; Daly, N.L.; Mobli, M.; Morales, R.A.V.; King, G.F. Isolation of an orally active insecticidal toxin from the venom of an australian tarantula. *PLoS ONE* **2013**, *8*, e73136. [CrossRef] [PubMed]
41. Smith, J.S.; Scholtz, J.M. Guanidine hydrochloride unfolding of peptide helices: separation of denaturant and salt effects. *Biochemistry* **1996**, *35*, 7292–7297. [CrossRef] [PubMed]
42. Mason, P.E.; Dempsey, C.E.; Neilson, G.W.; Kline, S.R.; Brady, J.W. Preferential interactions of guanidinium ions with aromatic groups over aliphatic groups. *J. Am. Chem. Soc.* **2009**, *131*, 16689–16696. [CrossRef] [PubMed]
43. Cobos, E.S.; Filimonov, V.V.; Gálvez, A.; Valdivia, E.; Maqueda, M.; Martinez, J.C.; Mateo, P.L. The denaturation of circular enterocin AS-48 by urea and guanidinium hydrochloride. *Biochim. Biophys. Acta* **2002**, *1598*, 98–107. [CrossRef]
44. Liu, Y.; Li, D.; Wu, Z.; Li, J.; Nie, D.; Xiang, Y.; Liu, Z. A positively charged surface patch is important for hainantoxin-IV binding to voltage-gated sodium channels. *J. Pept. Sci.* **2012**, *18*, 643–649. [CrossRef] [PubMed]
45. Xiao, Y.; Bingham, J.P.; Zhu, W.; Moczydlowski, E.; Liang, S.; Cummins, T.R. Tarantula huwentoxin-IV inhibits neuronal sodium channels by binding to receptor site 4 and trapping the domain II voltage sensor in the closed configuration. *J. Biol. Chem.* **2008**, *283*, 27300–27313. [CrossRef] [PubMed]
46. Vranken, W.F.; Boucher, W.; Stevens, T.J.; Fogh, R.H.; Pajon, A.; Llinas, M.; Ulrich, E.L.; Markley, J.L.; Ionides, J.; Laue, E.D. The CCPN data model for NMR spectroscopy: development of a software pipeline. *Proteins* **2005**, *59*, 687–696. [CrossRef] [PubMed]
47. Wüthrich, K. *NMR of Proteins and Nucleic Acids*; Wiley Interscience: New York, NY, USA, 1986.
48. Güntert, P. Automated NMR structure calculation with CYANA. *Methods Mol. Biol.* **2004**, *278*, 353–378. [PubMed]
49. Shen, Y.; Bax, A. Protein backbone and sidechain torsion angles predicted from NMR chemical shifts using artificial neural networks. *J. Biomol. NMR* **2013**, *56*, 227–241. [CrossRef] [PubMed]
50. Nederveen, A.J.; Doreleijers, J.F.; Vranken, W.; Miller, Z.; Spronk, C.A.; Nabuurs, S.B.; Güntert, P.; Livny, M.; Markley, J.L.; Nilges, M.; et al. RECOORD: A recalculated coordinate database of 500+ proteins from the PDB using restraints from the BioMagResBank. *Proteins* **2005**, *59*, 662–672. [CrossRef] [PubMed]
51. Brünger, A.T.; Adams, P.D.; Clore, G.M.; DeLano, W.L.; Gros, P.; Grosse-Kunstleve, R.W.; Jiang, J.S.; Kuszewski, J.; Nilges, M.; Pannu, N.S.; et al. Crystallography & NMR system: A new software suite for macromolecular structure determination. *Acta Crystall. D Biol. Crystall.* **1998**, *54*, 905–921.

© 2017 by the authors. Licensee MDPI, Basel, Switzerland. This article is an open access article distributed under the terms and conditions of the Creative Commons Attribution (CC BY) license (http://creativecommons.org/licenses/by/4.0/).

Review

Cone Snails: A Big Store of Conotoxins for Novel Drug Discovery

Bingmiao Gao [1,†], Chao Peng [2,†], Jiaan Yang [3], Yunhai Yi [2,4] ⓘ, Junqing Zhang [1,*] and Qiong Shi [2,4,*]

1   Hainan Provincial Key Laboratory of Research and Development of Tropical Medicinal Plants,
     Hainan Medical University, Haikou 571199, China; gaobingmiao1982@163.com or hy0207083@hainmc.edu.cn
2   Shenzhen Key Lab of Marine Genomics, Guangdong Provincial Key Lab of Molecular Breeding in Marine
     Economic Animals, BGI Academy of Marine Sciences, BGI Marine, BGI, Shenzhen 518083, China;
     pengchao@genomics.cn (C.P.); yiyunhai@genomics.cn (Y.Y.)
3   Micro Pharmtech, Ltd., Wuhan 430075, China; jyang@micropht.com
4   BGI Education Center, University of Chinese Academy of Sciences, Shenzhen 518083, China
*   Correspondence: jqzhang2011@163.com or hy0207002@hainmc.edu.cn (J.Z.); shiqiong@genomics.cn (Q.S.);
     Tel.: +86-898-6689-5337 (J.Z.); +86-755-3630-7807 (Q.S.)
†   These authors contributed equally to this work.

Academic Editor: Steve Peigneur
Received: 26 October 2017; Accepted: 4 December 2017; Published: 7 December 2017

Abstract: Marine drugs have developed rapidly in recent decades. Cone snails, a group of more than 700 species, have always been one of the focuses for new drug discovery. These venomous snails capture prey using a diverse array of unique bioactive neurotoxins, usually named as conotoxins or conopeptides. These conotoxins have proven to be valuable pharmacological probes and potential drugs due to their high specificity and affinity to ion channels, receptors, and transporters in the nervous systems of target prey and humans. Several research groups, including ours, have examined the venom gland of cone snails using a combination of transcriptomic and proteomic sequencing, and revealed the existence of hundreds of conotoxin transcripts and thousands of conopeptides in each *Conus* species. Over 2000 nucleotide and 8000 peptide sequences of conotoxins have been published, and the number is still increasing quickly. However, more than 98% of these sequences still lack 3D structural and functional information. With the rapid development of genomics and bioinformatics in recent years, functional predictions and investigations on conotoxins are making great progress in promoting the discovery of novel drugs. For example, ω-MVIIA was approved by the U.S. Food and Drug Administration in 2004 to treat chronic pain, and nine more conotoxins are at various stages of preclinical or clinical evaluation. In short, the genus *Conus*, the big family of cone snails, has become an important genetic resource for conotoxin identification and drug development.

Keywords: conotoxin; cone snail; transcriptome; proteome; drug development

---

1. Introduction

Forming the biggest single genera of living marine invertebrates [1], cone snails are composed of various carnivorous predators. They are usually classified into three groups depending on their feeding habits: worm hunters (vermivorous), mollusk hunters (molluscivorous), and fish hunters (piscivorous) [1–3]. There are now around 700 *Conus* species, with the majority distributed throughout tropical and subtropical waters, such as the South China Sea, Australia, and the Pacific Ocean [4].

The venom gland of cone snails can secrete large amounts of unique neurotoxic peptides, commonly referred to as conopeptides or conotoxins, and most conotoxins are rich in disulfide bridges with many pharmacological activities [5,6]. Each *Conus* species typically possesses an average of 100–200 conotoxins as potential pharmacological targets [7]. More than 80,000 natural conotoxins have been estimated to exist in various cone snails around the world [7–9]. Therefore, the cone snails construct the largest library of natural drug candidates for the development of marine drugs.

Usually, conotoxins are categorized into many different families based on the types of their molecular targets and corresponding pharmacological activities [9,10]. Their structures and functions are highly diverse and mainly target membrane proteins, particularly ion channels, membrane receptors, and transporters. Some toxins that have specific targets and short sequences and are easy to synthesize have been developed as drug leads and effective research agents for distinguishing subtypes of molecular targets [10,11]. The most well-known commercial conotoxin is ω-MVIIA (ziconotide), which has been derived from the venom of a fish-hunting *C. magus* species, and approved by the U.S. Food and Drug Administration (FDA) to treat chronic pain in serious cancer and AIDS patients [12,13]. Conotoxins are now increasingly undergoing development for the treatment of multiple diseases including pain, Alzheimer's disease, Parkinson's disease, cardiac infarction, hypertension, and various neurological diseases [14–16]. Therefore, conotoxins, being important potential therapeutic targets due to the physiological roles that they play, are a practical prospect with wide implications in the neuroscience research field.

2. Diversity of Cone Snails

2.1. Various Phenotypes

Cone snails have been of interest as collector's items for a long time due to their beautiful patterned shells (Figure 1), and the identification of *Conus* species is mainly based on their morphology and color of shell. The genus *Conus* is a member of the most diverse and taxonomically complex superfamily, Conoidea. According to recent estimations, it includes around 700 species, and most of them still remain undescribed [17]. However, species determination of live cone snails using shell characteristics poses difficulties due to regional and intra-specific variations [18].

C. coronatus	*C. striatus*	*C. quercinus*	*C. emaciatus*	*C. lividus*	*C. betulinus*	*C. capitaneus*	*C. imperialis*	*C. marmoreus*	*C.generalis*
Gmelin	Linnaeus	Lightfoot	Reeve	Hwass	Linnaeus	Linnaeus	Linnaeus	Linnaeus	Linnaeus

C. virgo	*C. chaldaeus*	*C.litteratus*	*C. vexillum*	*C.caracteris-*	*C. pulicar-*	*C. leopardus*	*C. miles*	*C. consors*	*C. textile*
Linnaeus	Roding	Linnaeus	Gmelin	*ticus* Fischer	*ius* Hwass	Roding	Linnaeus	Sowerby	Linnaeus

Figure 1. Twenty most abundant *Conus* species in the South China Sea.

The rapid development of molecular biology, the application of the mitochondrial genome, and the partial sequences of COI (cytochrome c oxidase subunit I), 16S rRNA, 12S rRNA, and calmodulin genes have all enabled a greater understanding of the diversity of gastropods at various levels, in terms of population, varieties, species, and so on [19–25]. Until recently, efforts on molecular taxonomy have been focused on the higher taxonomic categories above the species level. A phylogenetic tree of 72–138 *Conus* species with known diets was obtained on the basis of mitochondrial 16S rRNA and nuclear calmodulin gene sequences, and a distinctive clustering of species with similar diets was observed [26,27]. More details on a phylogenetic basis were established [28] to evaluate morphological criteria and characterize the genetic discontinuity so that *Conus* members can be identified based on gene sequence data from 16S rRNA, COI, and a four-loop conotoxin gene. Monophyly of the Conoidea, characterized by a venom apparatus, has not been questioned; however, subdivisions within the Conoidea and the relationships among them are controversial, mostly because of the uncertainty around the extensive morphological and anatomical variations [29,30]. In summary, a molecular perspective can aid in the phylogenetic classification of Conoidea. Hence, phylogenetic analyses are still essential in determining patterns of speciation and divergence.

2.2. Diverse Conotoxins and Targets

Predatory cone snails have long been of interest because of their highly evolved hunting strategies that employ conotoxins to paralyze prey [2]. Cone snails move slowly in an environment of fast-moving prey, which presents a major survival challenge to these predators. However, they have overcome this problem by developing a highly sophisticated venomous apparatus, which is responsible for the synthesis, storage, and delivery of a huge diversity of conotoxins [31].

About 1800 mature conotoxin sequences are available to date, and this number is increasing rapidly as the costs of transcriptome and proteome sequencing continue to reduce [7,8,32]. These diverse conotoxins were originally organized into various superfamilies with the help of two sequence elements, namely the conserved signal sequence and the characteristic cysteine framework. Currently, conotoxins can be classified into 26 gene superfamilies (A, B1, B2, B3, C, D, E, F, G, H, I1, I2, I3, J, K, L, M, N, O1, O2, O3, P, S, T, V, and Y) [33]. Each superfamily can be further divided into several families according to the array of cysteine frameworks. For example, the A-superfamily conotoxins include four cysteine frameworks (I, II, IV, and XIV) and are categorized into α, αA, and κA families; the M-superfamily includes five cysteine frameworks (II, XIV, III, VI, and VII), and is separated into μ and ψ families; the O-superfamily is composed of four cysteine frameworks (XII, XV, VI, and VII), and is classified into δ, μ, O, ω, κ, and γ families [11,14]. A schematic illustration is drawn in Figure 2 to summarize the updated 19 major gene superfamilies, frameworks, families, and ion channel-target networks identified to date [34,35].

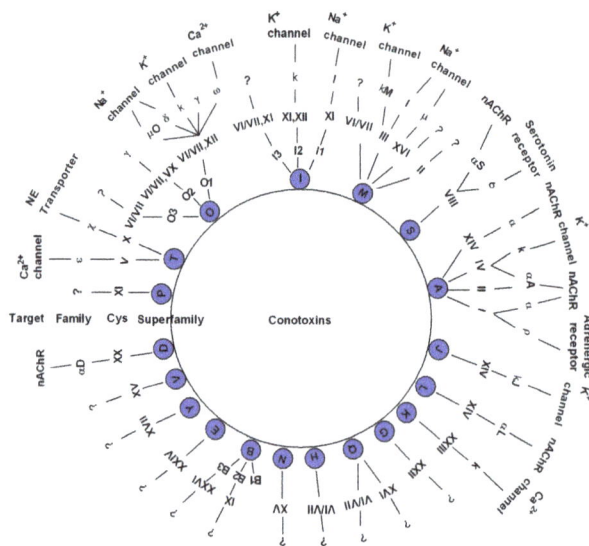

Figure 2. Classification of conotoxins (modified from [34,35]). On the basis of their conserved signal sequence homology, framework, and target receptor, conotoxins are classified into various superfamilies and families. NE: norepinephrine; nAChR: nicotinic acetylcholine receptor.

The molecular variety of conotoxins mirrors the diversity of their molecular targets [11,14–16]. Due to their high specificity and affinity to ion channels, various conotoxins can also be categorized into nicotinic acetylcholine receptor conotoxins (nAChR-conotoxins), sodium channel-targeted conotoxins (Na$^+$-conotoxins), potassium channel-targeted conotoxins (K$^+$-conotoxins), and calcium channel-targeted conotoxins (Ca^{2+}-conotoxins) [11]. Among them, α-, μ-, and ω-conotoxins are the most characterized families so far. Not only is the spectrum of the molecular targets expanding, but the diversity of different sites in a given molecular target also continues to surprise researchers.

2.3. Different Distribution and Ecology

Cone snails are the most diverse genus of marine invertebrates and contribute substantially to the great biodiversity in the tropical Indo-Pacific reef environments [36]. Most cone snails are widely distributed throughout all tropical oceans comprising a quarter of the earth's ocean area, yet more than 60% of their habitation occurs in the Indo-Pacific region (Figure 3). A few species have adapted to cooler temperate ocean environments, such as *C. californicus*, which is found on the North American Pacific coast [37].

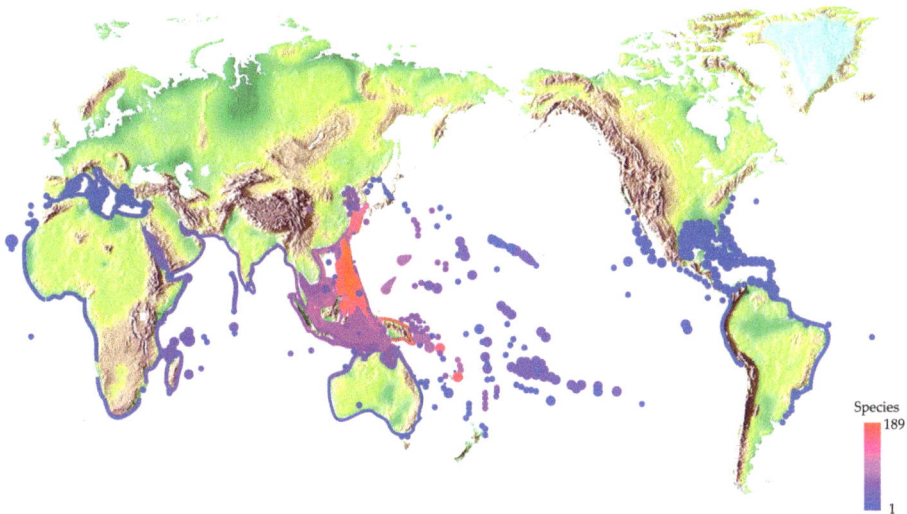

Figure 3. Worldwide distribution of cone snails. Spot colors stand for various species number.

Over 20 species have been observed to co-occur on certain reef platforms, with a maximum of 27 species in Indonesia [38,39]. In more recent papers, 36 *Conus* species were reported on the reef platform fringing Laing Island and 32 species on the four small reefs near Madang of the Northeast Papua Guinea [27]. *C. anemone* and *C. victoriae* are the dominant species in intertidal habitats along the inner region of the Dampier Archipelago, and both reside predominantly under rocks, on sand or limestone substrates [40]. Inter- and sub-tidal regions of the Indian coasts contain nearly 100 *Conus* species, but 16 of the reported species are still currently placed on the list of unverified species due to a lack of sufficient information [41]. However, some considered as unverified species, such as *C. generalis* and *C. litoglyphus*, have been confirmed as a species native to Indian Coastal waters [27,40]. China's coastal waters have more than 60 species, mainly distributed in the Xisha Islands, Hainan Island, Taiwan Island, and other tropical areas, and the vermivorous *C. betulinus* is the dominant *Conus* species inhabiting the South China Sea [7].

3. Multi-Omics Sequencing for High-Throughput Identification of New Conotoxins

Cone snails aroused the interest of some biochemists in the mid-20th century, because of the numerous cases of human injury; several fatalities were recorded due to stings inflicted by these snails [42,43]. After subsequent investigations, a correlation was established between the toxicity of their venoms to vertebrates and their prey type [44]. The first conotoxin was isolated from the venom of the piscivorous *C. geographus* in 1978 [45]. To date, more than 100 natural conotoxin peptides have been purified from the crude venom of cone snails via multi-step chromatography [14].

Cone snails are precious biological resources for marine medicine acquisition. However, researchers face incompatibility problems in the collection and execution of living individuals of the endangered *Conus* species. Moreover, there are many obvious disadvantages to extracting and purifying conotoxins, such as its time-consuming, laborious, high-cost, and low-yield nature, and it could be a substantial waste of bioresources and even cause serious ecological damage.

PCR technology was invented in 1983 [46], with the first conotoxin gene obtained in 1992 [47]; subsequently, PCR has become an important clue for screening novel conotoxin genes. With primes designed from the conservative sequence of each superfamily, PCR amplification has been employed to deal with genomic DNAs, cDNAs, or cDNA libraries in subsequent decades [48–50].

In recent years, transcriptomics has developed rapidly with the application of next-generation sequencing technology (Figure 4A) in a cost-effective manner on account of its high throughput sequencing and massive bioinformation analysis capacity [51]. A large number of new conotoxin genes from different species were obtained quickly and efficiently when this technology was used in studies of transcriptomes of the *Conus* venom duct [7]. Transcriptome analysis can provide a comprehensive understanding of mRNA information, including almost all types of the mRNA and their transcription quantity data, from tissue(s) or cell(s) at specific developmental stages or functional status, and can provide a comprehensive reflection of gene transcription within a dynamic scope.

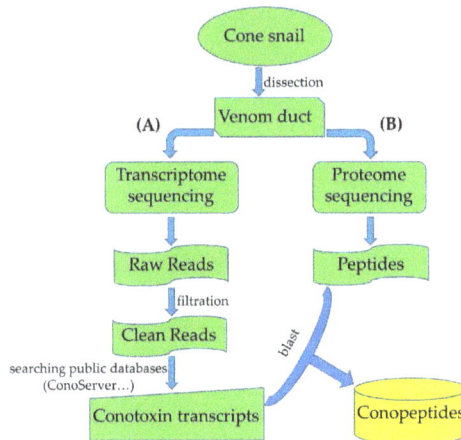

Figure 4. High-throughput identification of new conotoxin transcripts (**A**) and conopeptides (**B**) by transcriptome and proteome sequencing, respectively. More details about sequencing and data analysis can be found in several recent papers [7,9].

The first report of a *Conus* transcriptome in 2011 was achieved by a team at the University of Utah, which was led by Dr. Olivera BM, a pioneer in the study of conotoxins [52]. The study was the first to show that conotoxins are highly expressed within the venom duct of *Conus* species (*C. bullatus*), and described the first bioinformatics pipeline for high-throughput discovery and characterization of conotoxins. The study also identified 30 putative conotoxin sequences.

From 2013 to 2016, Professor Alewood PF and his team at the University of Queensland completed multi-omics studies of six different *Conus* species, which have improved our understanding of the diversity of conotoxins [1,2,9,32,53–55]. Their most representative research in 2013 employed an integrated approach, combining the next-generation transcriptome sequencing with high sensitivity proteomics (Figure 4), to investigate how *Conus* can generate impressive diversity of conotoxins, and a total of 105 conopeptide precursor sequences from 13 gene superfamilies (including five novel superfamilies) were identified from the venom duct of *C. marmoreus* [9]. Interestingly, for the first time, they observed that an average of 20 conopeptides were generated from each conotoxin precursor through a procedure of variable peptide processing. Hence, they estimated that over 2000 conotoxins could be generated in the venom by a single *C. marmoreus* specimen, given that 105 conotoxin precursors were identified from the transcriptome sequencing.

Soon afterwards, Alewood's team performed an analysis of venom duct transcriptome of *C. marmoreus* with the application of a new algorithm, ConoSorter, and they identified 158 novel conotoxin transcripts and another 13 novel conotoxin gene superfamilies. However, only 106 of these 158 transcripts were confirmed by peptide mass spectrometry, indicating that the effectiveness of ConoSorter is still necessary to be proved [53].

Over the past three years, the algorithm ConoSorter has been improved continuously and has been applied to identify conotoxin precursors from the transcriptome data of five different *Conus* species. From the transcriptomic data of *C. miles*, ConoSorter retrieved 662 putative conotoxin encoded sequences, comprising 48 conotoxin sequences validated at both transcript and peptide levels [32]. In 2015, the team presented a study of transcriptomes and proteomes of the radular sac, salivary gland, and venom duct of a single *C. episcopatus* specimen and discovered 3305 novel toxin sequences from a single *Conus* specimen—the highest number of conotoxins ever found in one specimen [54]. In another work, by the same team, on the *C. catus* transcriptome, 557 putative conotoxin sequences were identified using the 454 sequencing via programs ConoSorter, SignalP, and ConoServer, but only 104 precursors were ultimately recovered because the majority were rare isoforms and excluded from further analysis [2]. Similar sequencing and analysis strategies were applied in subsequent studies of venom duct transcriptomes of *C. planorbis* [55] and *C. vexillum* [1], and a final list of 182 and 220 transcripts, respectively, were obtained.

To explore as many novel conotoxins as possible, we employed an integrated approach to combine next-generation transcriptome sequencing with traditional Sanger sequencing and established a set of efficient methods for the high-throughput identification and validation of novel conotoxins from cone snails [7]. Based on the transcriptomes of the venom duct and venom bulb of vermivorous *C. betulinus*, a dominant *Conus* species inhabiting the South China Sea, we identified a total of 215 conotoxin transcripts within 38 superfamilies or groups, of which nine superfamilies were reported for the first time. We also performed a transcriptomic survey of ion-channel-based conotoxins [56] for the development of conotoxins as potential drugs to treat ion-channel related human diseases. Interestingly, more than 20 conotoxins with potential insecticidal activity were screened out [57] from our transcriptome-based dataset [7] by a homologous search with a reported positive control (IMI from *C. imperialis*) as the query. Two of them were further validated as presenting high insecticidal activity [57], which supports their further study in field investigations as a potential insecticide.

Studies of conotoxins using multi-omics methods have been developing rapidly in recent years. Seventeen articles about venom gland transcriptome research of cone snails have been published to date [1,2,7–10,32,52–55,58–63]. The related data of 67 transcriptomes, involving a total of 30 *Conus* species, are available from the NCBI. Massive new genes and their encoded toxin peptides can be discovered from this treasure house of marine drugs.

4. Structural Prediction with Protein Structure Fingerprinting for Novel Conformations

Over the last few decades, conotoxins have been an important subject of pharmacological interest. As we know, the biological activities of conotoxins are affected by their sequences as well as their structural conformations. A knowledge of conformation is critical to exploiting drug development for conotoxins. As conotoxin peptides usually consist of 10–30 amino acid residues, the conformations are mainly determined by nuclear magnetic resonance (NMR) spectroscopy, X-ray crystallography, or computational prediction approaches. However, except for the more complex procedures, these approaches offer limited conformations under certain circumstances. To date, a large amount of conotoxin data has been collected and compiled in the ConoServer database [64,65]. About 160 conotoxins with determined 3D structures are now available in the Protein Data Bank (PDB); however, for over 8000 conotoxins, only sequence information has been recorded, and 3D conformations have not yet been obtained. Therefore, the challenge is determining how this information can be collected.

One innovative approach—protein folding shape codes (PFSCs) [66], which we established—is able to provide comprehensive conformations for conotoxin peptides. A set of 27 PFSCs completely covers the folding shapes of five successive amino acids, simply represented by the 26 alphabet letters and the "\$" symbol as a digitized expression. Consequently, any conformation can be expressed by a string in alphabetical letters comprising a protein structure fingerprint, which can be aligned to directly display the similarity or dissimilarity of conotoxins. For example, the letter A represents a typical alpha-helical fold, and the letters H, D, V, L, Y, and P are for folds with partial alpha-helical similarity. The letter B represents a typical beta-strand fold, and E, G, V, J, M, and S are for folds with partial beta-strand similarity. C, F, L O, \$, N, Q, R, I, T, K, X, U, Z, and W mostly relate to irregular folds of the elements of the tertiary structure fragment. The beauty of this approach is that a set of 27 PFSCs covers the complete folding space for the five amino acid residues. Meanwhile, each PFSC vector can be transformed from each other by a skeleton relationship according to the partial sharing of folding similarity.

Here, examples of the conformations of three different conotoxins are described using protein structure fingerprinting with the 27 PFSCs. These examples demonstrate how the conformations are presented for a given 3D structure, and how the predicted conformations are obtained from the sequences. The first structure is conotoxin pl14a with 25 amino acid residues (left column in Table 1). Isolated from vermivorous cone snails, it has potent activity in both nAChR and a voltage-gated potassium channel subtypes [67]. Its structure with 20 conformations is available in PDB (ID: 2FQC), which was determined by NMR spectroscopy. The second structure is alpha-Conotoxin EI with 25 amino acid residues (middle column), originally purified from the venom of *C. ermineus*. Alpha-Conotoxin EI targets neuronal nicotinic acetylcholine receptors but antagonizes neuromuscular receptors [68]. Its structure with 13 conformations is available in PDB (ID: 1K64), which was determined by NMR spectroscopy. The third structure is Omega-conotoxin MVIIA with 26 amino acid residues (right column). It has a range of selectivity for different subtypes of the voltage-sensitive calcium channel [69]. Its structure with 17 conformations is available in PDB (ID: 1K64), which was also determined by NMR spectroscopy. The detailed conformation images, the conformation descriptions for given 3D structures, as well as the conformation predictions from the sequences for these three structures are summarized in Table 1.

The conformations with a given structure can be well described. Section A in Table 1 displays the protein structure fingerprints of 20 conformations of conotoxin pl14a, 13 conformations of Alpha-conotoxin EI, and 17 conformations of Omega-conotoxins MVIIA. Usually, the folding changes are difficult to observe in related structural images; however, the fine differentiations of folding shapes between multiple isomers are easily revealed using a protein structure fingerprint consisting of the 27 PFSCs.

Table 1. Conformation descriptions of three conotoxin examples.

PDB ID	2FQC	1K64	1CNN
Images			
Info	Conotoxin pl14a; Inhibitor of Kv1.6 Channel and nAChR	Alpha-conotoxin EI; torpedo nicotinic acetylcholine receptor	Omega-conotoxins MVIIA; N and P/Q-type calcium channels
Tenth	0000000001111111111222222	00000000011111111112222222	00000000011111111112222222
Digital	1234567890123456789012345	12345678901234567890123456	12345678901234567890123456
Sequence	FPRPRICNLACRAGIGHKYPFCHCR	CKGKGAKCSRLMYDCCTGSCRSGKC	CKGKGAPCRKTMYDCCSGSCGRRGKC

Table 1. *Cont.*

PDB ID	2FQC	1K64	1CNN
1	. . C A A A A A A A R A A A A J D B .	. . C A H . W C C J J R A J J Q C C W \$ Z . .	. . C J A A N . C . V J C J J . W E Q . .
2	. . W S . B B J D J Q J A J Y N . .	. . A Z \$ W B F J . B Y A B Y J F E R Q . .	. . A Z A . W B J A F P A J Y Z C . C Z A . .
3	. . C J . I S . L \$ J C D B . .	. . W C J A . B R A W B P A B C C A . .	. . W C . W . C . F Q . J Q C Q B R A J .
4	. . L . . J \$ L . .	. . W Z B . L W . Y J Y . C J J R J Z . .	. . W P . . Q . W C J W B . A .
5	. . A C W . .	. . V J C . J F . J C . A J J W . D . .	. . A P Y . B . J . D R A . J U .
6	. . P b . .	. . A A P . . R J J . Z J R C . L . .	. . Y A . R . J . Y Y P B . .
7	. . L . .	. . Z J Q . J F . R . .	. . Z Y . C . Y L J C .
B 8	. . C . .	. . D S . . D . .	. . D . . \$ C J . .
9	. . B . .	. . J F V . L . .	. . J F . . . J . .
10	. . J . .	. . P J D . J . .	. . J P J . Z . .
11	. . D . .	. . Q W . Z . .	. . Q . .
12	. .	. . J D . A . .	. . J D . .
13	. .	. . L \$. U . .	. . L \$ . .
14	. .	. . U L . X . .	. . U L . .

Notes: The first row is the structure entity (PDB ID). The following rows, respectively, are conformation images presented by a solid ribbon format, conotoxin information, rules of position, and amino acid sequence. Section A lists the conformation descriptions of the structure fingerprint by PFSCs according to the given 3D structures. Section B displays the predicted folding variations according to the conotoxin sequences, which are ensembles of folding shapes for five successive amino acids. The PFSC folding shapes are marked by different colors: red is for a typical helix fold; blue is for a typical beta fold; pink and light blue are for folds with a partial helix or beta; black is for irregular folds.

The NMR spectroscopy approach, indeed, measures the nature of conotoxin conformations in fluctuation. Furthermore, with protein structure fingerprinting, the conformation alignment of the isomers for each structure reveals the locations of folding fluctuations or stability along sequences. In the 2FQC structure, it is apparent that in 20 isomers most of the local folds are alpha helices alike, and more folding changes happen in the N-terminus while stable conformations appear in the C-terminus. In the 1K64 structure, a fragment (10–13) of the 13 isomers has a stable conformation, but other parts showed more fluctuation in conformations. In the 1CNN structure, a fragment (7, 9, 13–19) of the 17 isomers has a stable conformation, but other parts showed more changeable conformations. Together, 2FQC has relatively longer alpha helices in conformation; 1CNN has a stable fragment with alike alpha helices; 1K64 has many folding variations with shorter fragments.

The comprehensive folding variations in each conformation can be predicted directly according to the sequence. Section B at the bottom of Table 1 displays the complete variations of local folding shapes for three structures, which are actually the ensembles of folding shapes for five successive amino acids along the conotoxin sequences. In fact, the complete variations provide rich information to cover all possible changes in local folding shapes. It is noted that the possible types and numbers of folding shapes are altered for each of the five successive amino acids along the conotoxin sequence. For example, the predicted folding variations for conotoxin pl14a indicate stability in the C-terminus. However, the fragment with the sequence "RAGIG" (12–16) has the highest number of folding shape variations, indicating the location with the most flexibility of folding changes in the conotoxin pl14a. The predicted folding variations in Section B of Table 1 are significant, suggesting which locations have a more flexible selection in the folds and less selection for local folds. Meanwhile, all folding changes in the isomers for given 3D structures are encompassed by the complete folding variations found from prediction. Although each structure in Section A of Table 1 has different conformations with altered folds, all these folds are totally covered by the folding variations in Section B and are marked in yellow. In other words, any conformation of isomer for each conotoxin structure in Section A of Table 1 can be well predicted using the comprehensive folding variations in Section B.

The binding sites for conotoxins in any protein receptor can be described by its protein structure fingerprint with the 27 PFSCs. For example, alpha-conotoxin is a peptide antagonist of nAChRs that has been used as a pharmacological probe and investigated as a drug lead for nAChR-related disorders [70]. The co-crystal structure (PDB ID: 5T90) is an alpha-conotoxin binding to human $\alpha 3 \beta 4$ nAChR. A detailed image of the structure and its binding sites using protein structure fingerprinting is presented in Table 2. In this structure, the alpha-conotoxin is defined as Chain F (yellow color), which is surrounded by Chain A and Chain C of nAChR (images on the left side of Table 2). The binding sites were determined by an 8 Å distance of interaction of all atoms from alpha-conotoxin. Hence, it was revealed that the binding site is formed by five fragments from Chains A and C. It is hard to study the conformation of binding sites with an image or computational molecule modeling approach. However, the topological space of binding sites can be explicitly described using PFSCs. The five fragments with sequences and folding descriptions are displayed in the bottom section of Table 2, which may better illustrate how alpha-conotoxin peptides act as antagonist leads for nAChR-related disorders. Furthermore, to query the similar fingerprint of the known binding sites with other proteins, multiple protein targets may be discovered with high-throughput screening of protein databases. This approach may be better for understanding various interactions between conotoxins and protein receptors, such as nAChR in nerves and muscles [71], voltage-dependent sodium channels [72], potassium channels [73], sodium channels in muscles [74], and N-type voltage-dependent calcium channels [75].

In summary, the protein structure fingerprint with 27 PFSC vectors is able to provide rich information for studying conotoxin conformations. First, based on a given 3D structure, the protein structure fingerprint is able to provide a complete description of conotoxin conformations. Second, based on sequences, the protein structure fingerprint can provide comprehensive folding variations to predict the conotoxin conformations of unknown 3D structures. Third, the protein structure fingerprint may provide another means of exploring the mechanisms of interactions between various conotoxins and related protein targets.

Table 2. Binding site description using a protein structure fingerprint for an alpha-conotoxin (LsIA) binding with human $\alpha3\beta4$ nicotinic acetylcholine receptor.

Alpha-Conotoxin in Structure (PDB ID: 5T90 for LsIA)				
Chain	Start	End	Sequence	PFSC
F	1	17	SGCCSNPACRVNNPNIC	..AAAJVAAAAAJVA..

Image of LsIA (yellow; PDB ID: 5T90) binding with human $\alpha3\beta4$ nicotinic acetylcholine receptor		
	Chain A (red); Chain C (green); Conotoxin as Chain F (yellow)	Conotoxin binding site with five fragments

Binding Fragments of nAChR for Alpha-Conotoxin (PDB ID = 5T90)				
Chain	Start	End	Sequence	PFSC
A	53	57	WQQTT	RBBEE
A	102	116	LARVVSDGEVLYMPS	BLREWYQSBBEBEWR
A	155	165	TTENSDDSEYF	CYJVAJVAAAP
C	141	149	GSWTHHSRE	WSVAJWZAD
C	183	193	VTYSCCPEAYE	BEWYAJVPREE

Notes: The images are presented in a solid ribbon format. The image on the left is colored to distinguish each chain; the image on the right presents the binding site formed by five fragments. The binding fragments were determined by the 8 Å distance of interaction of all atoms from the alpha-conotoxin. The PFSC folding shapes are marked as various colors: red is for a typical helix fold; blue is for a typical beta fold; pink and light blue are for folds with a partial helix or beta; black is for irregular folds.

5. Recent Advances in Conotoxins for Drug Development

The therapeutic potential of conotoxins is ascribed to their special ion-channel targets in nervous systems [76]. Thus, conotoxins have potentially wide applications in the fields of neuroscience research and development. Many conotoxins are proving to be valuable as research tools, drug leads, and drugs [14,15,77]. We performed a detailed survey of US patent literature covering conotoxins to determine their potential therapeutic applications.

The State Intellectual Property Office of China (SIPO; [78]) was used as the primary search, while the United States Patent and Trademark Office (USPTO; [79]) Patent Office for Europe (EPO; [80]) and World Intellectual Property Organization (WIPO; [81]) assisted the search. All these databases were searched with the following keywords: conotoxin, conopeptide, conantokin, contryphan, and contulakin. The search period was set from 1998 to 2017. From this search, 811 patents, of which 243 were authorized, were obtained. These patents were classified into different families based on continuity relationships. Among these patent families, the majority (451) refer to conotoxin compositions of matter and the remaining 360 primarily cover processes and methods for their applications [82]. We observed that the number of patent applications has fluctuated in the past 20 years, and the maximum number occurred in 2014 (Figure 5). Patent applicants are mainly distributed in the United States, China, and Australia (Figure 6).

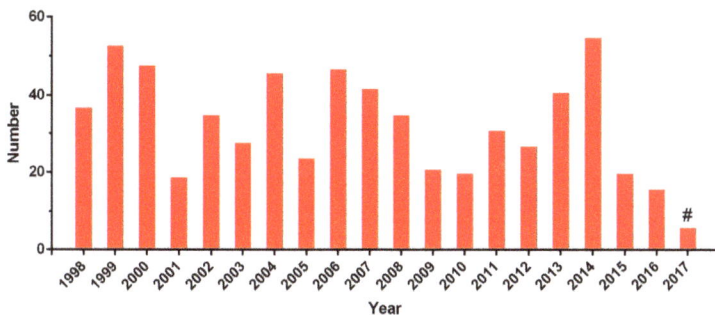

Figure 5. The number of patents for conotoxins per year. Note that # indicates the incomplete number counted to 2017.

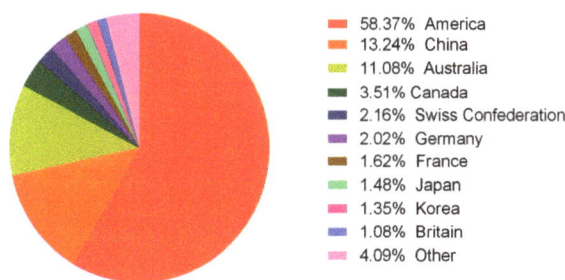

- 58.37% America
- 13.24% China
- 11.08% Australia
- 3.51% Canada
- 2.16% Swiss Confederation
- 2.02% Germany
- 1.62% France
- 1.48% Japan
- 1.35% Korea
- 1.08% Britain
- 4.09% Other

Figure 6. The national distribution of patents for conotoxins.

To date, several conotoxins have already demonstrated potential therapeutic effects in preclinical or clinical trials (Table 3). The most well-known commercial conotoxin is ω-MVIIA (ziconotide), which has been approved by the America FDA to treat intractable chronic pain in cancer and AIDS patients [12,13,83]. Its introduction into the market not only demonstrated the therapeutic potential of conotoxins but also stimulated more interest from biotechnology companies to support conotoxin research. Those conotoxins currently in clinical trials include an analog of the conotoxin χ-MrIA, which noncompetitively inhibits noradrena-line transporter and is undergoing phase II clinical trials as a treatment for neuropathic pain [84]. Other ω-conotoxins in the clinical trial pipeline include ω-CVID, which successfully completed preclinical studies, but high cytotoxic effects were observed during phase IIa trials [85]. In addition, contulakin-G and conantokin G are, respectively, the specific antagonist against the Neurotensin receptor and NR2B subunit of the NMDA receptor, and are currently in human clinical trials for pain and intractable epilepsy [86,87]. Therefore, an increasing number of conopeptides are undergoing development for the treatment of pathologies including pain, cancer, cardiac infarction, hypertension, Parkinson's disease, Alzheimer's disease, epilepsy, and various neurological diseases [3,6,11,14,15]. Several conopeptides (Table 3) reaching human clinical trials have already established the *Conus* pharmacopoeia as a rich source of therapeutics for neurological disorders [14].

Table 3. Therapeutic applications of conotoxins [14,15,77].

Clinical Application	Conopeptide	Molecular Target	Clinical Status	Reference
Pain	ω-MVIIA (Ziconitide)	Ca^{2+} channel (CaV2.2) N-type calcium channels/blocker	FDA-approved	[83]
Pain	χ-MrIA (Xen2174)	Norepinephrine transporter	Phase IIa *	[84]
Pain	ω-CVID (AM336)	Ca^{2+} channel (CaV2.2) N-type calcium channels/blocker	Phase IIa *	[85]
Pain	Contulakin-G (CGX-1160)	Neurotensin receptor	Phase Ib *	[86]
Pain/Neuro protection	Conantokin-G (CGX-1007)	NMDA receptor (NR2B)	Preclinical *	[87]
Pain	α-Vc1.1 (ACV1)	nAChR (α9α10)	Phase II *	[88]
Myocardial infarction	κ-PVIIA (CGX-1051)	K^+ channel (K_V1)	Preclinical	[89]
Neuropathic pain	μO-MrVIB (CGX-1002)	Sodium channels/subtype selective blocker	Preclinical *	[90]

Note: * indicates that development of these conotoxins has been terminated.

6. Conclusive Remarks

The diversity of cone snails offers a promising prospect for drug discovery, with the rapid development of genomic/proteomic data and bioinformatics methods. However, despite the great advances of conotoxins in drug development, the incapability of these conotoxins to cross the blood–brain barrier results in their dependence on intrathecal administration. This remains a major challenge in the therapeutic application of conotoxins. Elevation of in vivo stability and efficient absorption, as well as 3D structural modifications as described above, will also greatly boost their clinical success.

Acknowledgments: This work was supported by Natural Science Foundation of Hainan Province (No. 317170), International Cooperation Project of Shenzhen Science and Technology (No. GJHZ20160229173052805), National Natural Science Foundation of China (No. 81560611), Shenzhen Dapeng Special Program for Industrial Development (No. KY20150207), and State High-Tech Research and Development Project (863) of the Ministry of Science and Technology of China (No. 2012AA021706).

Author Contributions: Bingmiao Gao, Qiong Shi, and Junqing Zhang designed the review; Bingmiao Gao, Chao Peng, and Jiaan Yang wrote the manuscript; Yunhai Yi participated in figure preparation; Bingmiao Gao, Chao Peng, and Qiong Shi revised the manuscript.

Conflicts of Interest: The authors declare no conflict of interest.

References

1. Prashanth, J.R.; Dutertre, S.; Jin, A.H.; Lavergne, V.; Hamilton, B.; Cardoso, F.C.; Griffin, J.; Venter, D.J.; Alewood, P.F.; Lewis, R.J. The role of defensive ecological interactions in the evolution of conotoxins. *Mol. Ecol.* **2016**, *25*, 598–615. [CrossRef] [PubMed]
2. Himaya, S.W.; Jin, A.H.; Dutertre, S.; Giacomotto, J.; Mohialdeen, H.; Vetter, I.; Alewood, P.F.; Lewis, R.J. Comparative Venomics Reveals the Complex Prey Capture Strategy of the Piscivorous Cone Snail *Conus catus*. *J. Proteom. Res.* **2015**, *14*, 4372–4381. [CrossRef] [PubMed]
3. Lewis, R.J.; Dutertre, S.; Vetter, I.; Christie, M.J. *Conus* venom peptide pharmacology. *Pharmacol. Rev.* **2012**, *64*, 259–298. [CrossRef] [PubMed]
4. Kumar, P.S.; Kumar, D.S.; Umamaheswari, S. A perspective on toxicology of *Conus* venom peptides. *Asian Pac. J. Trop. Med.* **2015**, *8*, 337–351. [CrossRef]
5. Prashanth, J.R.; Brust, A.; Jin, A.H.; Alewood, P.F.; Dutertre, S.; Lewis, R.J. Cone snail venomics: From novel biology to novel therapeutics. *Future Med. Chem.* **2014**, *6*, 1659–1675. [CrossRef] [PubMed]
6. Vetter, I.; Lewis, R.J. Therapeutic potential of cone snail venom peptides (conopeptides). *Curr. Top. Med. Chem.* **2012**, *12*, 1546–1552. [CrossRef] [PubMed]

7.   Peng, C.; Yao, G.; Gao, B.M.; Fan, C.X.; Bian, C.; Wang, J.; Cao, Y.; Wen, B.; Zhu, Y.; Ruan, Z.; et al. High-throughput identification of novel conotoxins from the Chinese tubular cone snail (*Conus betulinus*) by multi-transcriptome sequencing. *Gigascience* **2016**, *5*, 17. [CrossRef] [PubMed]

8.   Barghi, N.; Concepcion, G.P.; Olivera, B.M.; Lluisma, A.O. High conopeptide diversity in *Conus tribblei* revealed through analysis of venom duct transcriptome using two high-throughput sequencing platforms. *Mar. Biotechnol.* **2015**, *17*, 81–98. [CrossRef] [PubMed]

9.   Dutertre, S.; Jin, A.H.; Kaas, Q.; Jones, A.; Alewood, P.F.; Lewis, R.J. Deep venomics reveals the mechanism for expanded peptide diversity in cone snail venom. *Mol. Cell. Proteom.* **2013**, *12*, 312–329. [CrossRef] [PubMed]

10. Robinson, S.D.; Safavi-Hemami, H.; McIntosh, L.D.; Purcell, A.W.; Norton, R.S.; Papenfuss, A.T. Diversity of conotoxin gene superfamilies in the venomous snail, *Conus victoriae*. *PLoS ONE* **2014**, *9*, e87648. [CrossRef] [PubMed]

11. Lewis, R.J.; Garcia, M.L. Therapeutic potential of venom peptides. *Nat. Rev. Drug Discov.* **2003**, *2*, 790–802. [CrossRef] [PubMed]

12. Rigo, F.K.; Dalmolin, G.D.; Trevisan, G.; Tonello, R.; Silva, M.A.; Rossato, M.F.; Klafke, J.Z.; Cordeiro Mdo, N.; Castro Junior, C.J.; Montijo, D.; et al. Effect of omega-conotoxin MVIIA and Phalpha1beta on paclitaxel-induced acute and chronic pain. *Pharmacol. Biochem. Behav.* **2013**, *114*, 16–22. [CrossRef] [PubMed]

13. Eisapoor, S.S.; Jamili, S.; Shahbazzadeh, D.; Ghavam Mostafavi, P.; Pooshang Bagheri, K. A New, High Yield, Rapid, and Cost-Effective Protocol to Deprotection of Cysteine-Rich Conopeptide, Omega-Conotoxin MVIIA. *Chem. Biol. Drug Des.* **2016**, *87*, 687–693. [CrossRef] [PubMed]

14. Han, T.S.; Teichert, R.W.; Olivera, B.M.; Bulaj, G. *Conus* venoms-a rich source of peptide-based therapeutics. *Curr. Pharm. Des.* **2008**, *14*, 2462–2479. [CrossRef] [PubMed]

15. Olivera, B.M.; Teichert, R.W. Diversity of the neurotoxic *Conus* peptides: A model for concerted pharmacological discovery. *Mol. Interv.* **2007**, *7*, 251–260. [CrossRef] [PubMed]

16. Fedosov, A.E.; Moshkovskii, S.A.; Kuznetsova, K.G.; Olivera, B.M. Conotoxins: From the biodiversity of gastropods to new drugs. *Biomed. Khim.* **2013**, *59*, 267–294. [CrossRef] [PubMed]

17. Fallon, P.J. Taxonomic review of tropical western Atlantic shallow water Drilliidae (Mollusca: Gastropoda: Conoidea) including descriptions of 100 new species. *Zootaxa* **2016**, *4090*, 1–363. [CrossRef] [PubMed]

18. Marshall, J.; Kelley, W.P.; Rubakhin, S.S.; Bingham, J.P.; Sweedler, J.V.; Gilly, W.F. Anatomical correlates of venom production in *Conus californicus*. *Biol. Bull.* **2002**, *203*, 27–41. [CrossRef] [PubMed]

19. Chen, P.W.; Hsiao, S.T.; Huang, C.W.; Chen, K.S.; Tseng, C.T.; Wu, W.L.; Hwang, D.F. The complete mitochondrial genome of *Conus tulipa* (Neogastropoda: Conidae). *Mitochondrial DNA Part A* **2016**, *27*, 2738–2739.

20. Bandyopadhyay, P.K.; Stevenson, B.J.; Ownby, J.P.; Cady, M.T.; Watkins, M.; Olivera, B.M. The mitochondrial genome of *Conus textile*, coxI-coxII intergenic sequences and Conoidean evolution. *Mol. Phylogenet. Evol.* **2008**, *46*, 215–223. [CrossRef] [PubMed]

21. Barghi, N.; Concepcion, G.P.; Olivera, B.M.; Lluisma, A.O. Characterization of the complete mitochondrial genome of *Conus tribblei* Walls, 1977. *Mitochondrial DNA Part A* **2016**, *27*, 4451–4452. [CrossRef] [PubMed]

22. Cunha, R.L.; Grande, C.; Zardoya, R. Neogastropod phylogenetic relationships based on entire mitochondrial genomes. *BMC Evol. Biol.* **2009**, *9*, 210. [CrossRef] [PubMed]

23. Chen, P.W.; Hsiao, S.T.; Chen, K.S.; Tseng, C.T.; Wu, W.L.; Hwang, D.F. The complete mitochondrial genome of *Conus capitaneus* (Neogastropoda: Conidae). *Mitochondrial DNA Part B* **2016**, *1*, 520–521. [CrossRef]

24. Boore, J.L. Animal mitochondrial genomes. *Nucleic Acids Res.* **1999**, *27*, 1767–1780. [CrossRef] [PubMed]

25. Pastukh, V.M.; Gorodnya, O.M.; Gillespie, M.N.; Ruchko, M.V. Regulation of mitochondrial genome replication by hypoxia: The role of DNA oxidation in D-loop region. *Free Radic. Biol. Med.* **2016**, *96*, 78–88. [CrossRef] [PubMed]

26. Remigio, E.A.; Duda, T.F., Jr. Evolution of ecological specialization and venom of a predatory marine gastropod. *Mol. Ecol.* **2008**, *17*, 1156–1162. [CrossRef] [PubMed]

27. Kohn, A.J. Maximal species richness in *Conus*: Diversity, diet and habitat on reefs of northeast Papua New Guinea. *Coral Reefs* **2001**, *20*, 25–38.

28. Duda, T. Origins of diverse feeding ecologies within *Conus*, a genus of venomous marine gastropods. *Biol. J. Linn. Soc.* **2001**, *73*, 391–409. [CrossRef]

29. Puillandre, N.; Bouchet, P.; Duda, T.F., Jr.; Kauferstein, S.; Kohn, A.J.; Olivera, B.M.; Watkins, M.; Meyer, C. Molecular phylogeny and evolution of the cone snails (Gastropoda, Conoidea). *Mol. Phylogenet. Evol.* **2014**, *78*, 290–303. [CrossRef] [PubMed]

30. Guindon, S.; Gascuel, O. A simple, fast, and accurate algorithm to estimate large phylogenies by maximum likelihood. *Syst. Boil.* **2003**, *52*, 696–704. [CrossRef]

31. Olivera, B.M.; Seger, J.; Horvath, M.P.; Fedosov, A.E. Prey-Capture Strategies of Fish-Hunting Cone Snails: Behavior, Neurobiology and Evolution. *Brain Behav. Evol.* **2015**, *86*, 58–74. [CrossRef] [PubMed]

32. Jin, A.H.; Dutertre, S.; Kaas, Q.; Lavergne, V.; Kubala, P.; Lewis, R.J.; Alewood, P.F. Transcriptomic messiness in the venom duct of *Conus miles* contributes to conotoxin diversity. *Mol. Cell. Proteom.* **2013**, *12*, 3824–3833. [CrossRef] [PubMed]

33. Robinson, S.D.; Norton, R.S. Conotoxin gene superfamilies. *Mar. Drugs* **2014**, *12*, 6058–6101. [CrossRef] [PubMed]

34. Halai, R.; Craik, D.J. Conotoxins: Natural product drug leads. *Nat. Prod. Rep.* **2009**, *26*, 526–536. [CrossRef] [PubMed]

35. Akondi, K.B.; Muttenthaler, M.; Dutertre, S.; Kaas, Q.; Craik, D.J.; Lewis, R.J.; Alewood, P.F. Discovery, synthesis, and structure-activity relationships of conotoxins. *Chem. Rev.* **2014**, *114*, 5815–5847. [CrossRef] [PubMed]

36. Kohn, A.J. Superfamily conoidea. In *Mollusca: The Southern Synthesis. Fauna of Australia*; Beesley, P.L., Ross, G.J.B., Wells, A., Eds.; CSIRO Publishing: Melbourne, Australia, 1998; Volume 5, pp. 846–854.

37. Briggs, J.C. *Marine Zoogeography*; McGraw-Hill: New York, NY, USA, 1974.

38. Kohn, A.J.; Nybakken, J.W. Ecology of *Conus* on eastern Indian Ocean fringing reefs: Diversity of species and resource utilization. *Mar. Biol.* **1975**, *29*, 211–234. [CrossRef]

39. Kohn, A.J. Tempo and mode of evolution in Conidae. *Malacologia* **1990**, *32*, 55–67.

40. Kohn, A.J. The feeding process in *Conus victoriae*. In *The Marine Flora and Fauna of Dampier, Western Australia*; Wells, F.E., Walker, D.I., Jones, D.S., Eds.; Western Australian Museum: Perth, Australia, 2003.

41. Kohn, A.J. The conidae (Mollusca: Gastropoda) of India. *J. Nat. Hist.* **1978**, *12*, 295–335. [CrossRef]

42. Clench, W.J.; Kondo, Y. The poison cone shell. *Am. J. Trop. Med. Hyg.* **1943**, *23*, 105–121. [CrossRef]

43. Kohn, A.J. Cone shell stings; recent cases of human injury due to venomous marine snails of the genus *Conus*. *Hawaii Med. J.* **1958**, *17*, 528–532. [PubMed]

44. Endean, R.; Rudkin, C. Studies of the venoms of some Conidae. *Toxicon* **1963**, *1*, 49–64. [CrossRef]

45. Cruz, L.J.; Gray, W.R.; Olivera, B.M. Purification and properties of a myotoxin from *Conus geographus* venom. *Arch. Biochem. Biophys.* **1978**, *190*, 539–548. [CrossRef]

46. Mullis, K.B.; Faloona, F. Specific synthesis of DNA in vitro via a polymerase catalyzed chain reaction. *Meth. Enzymol.* **1987**, *155*, 335–350. [PubMed]

47. Hillyard, D.R.; Monje, V.D.; Mintz, I.M.; Bean, B.P.; Nadasdi, L.; Ramachandran, J.; Miljanich, G.; Azimi-Zoonooz, A.; McIntosh, J.M.; Cruz, L.J.; et al. A new *Conus* peptide ligand for mammalian presynaptic Ca2+ channels. *Neuron* **1992**, *9*, 69–77. [CrossRef]

48. Duda, T.F., Jr.; Chang, D.; Lewis, B.D.; Lee, T. Geographic variation in venom allelic composition and diets of the widespread predatory marine gastropod *Conus ebraeus*. *PLoS ONE* **2009**, *4*, e6245. [CrossRef] [PubMed]

49. Liu, Z.; Xu, N.; Hu, J.; Zhao, C.; Yu, Z.; Dai, Q. Identification of novel I-superfamily conopeptides from several clades of *Conus* species found in the South China Sea. *Peptides* **2009**, *30*, 1782–1787. [CrossRef] [PubMed]

50. Conticello, S.G.; Gilad, Y.; Avidan, N.; Ben-Asher, E.; Levy, Z.; Fainzilber, M. Mechanisms for evolving hypervariability: The case of conopeptides. *Mol. Biol. Evol.* **2001**, *18*, 120–131. [CrossRef] [PubMed]

51. Ansorge, W.J. Next-generation DNA sequencing techniques. *Nat. Biotechnol.* **2009**, *25*, 195–203. [CrossRef] [PubMed]

52. Hu, H.; Bandyopadhyay, P.K.; Olivera, B.M.; Yandell, M. Characterization of the *Conus bullatus* genome and its venom-duct transcriptome. *BMC Genom.* **2011**, *12*, 60. [CrossRef] [PubMed]

53. Lavergne, V.; Dutertre, S.; Jin, A.H.; Lewis, R.J.; Taft, R.J.; Alewood, P.F. Systematic interrogation of the *Conus marmoreus* venom duct transcriptome with ConoSorter reveals 158 novel conotoxins and 13 new gene superfamilies. *BMC Genom.* **2013**, *14*, 708. [CrossRef] [PubMed]

54. Lavergne, V.; Harliwong, I.; Jones, A.; Miller, D.; Taft, R.J.; Alewood, P.F. Optimized deep-targeted proteotranscriptomic profiling reveals unexplored *Conus* toxin diversity and novel cysteine frameworks. *Proc. Natl. Acad. Sci. USA* **2015**, *112*, E3782–E3791. [CrossRef] [PubMed]

55. Jin, A.H.; Vetter, I.; Himaya, S.W.; Alewood, P.F.; Lewis, R.J.; Dutertre, S. Transcriptome and proteome of *Conus planorbis* identify the nicotinic receptors as primary target for the defensive venom. *Proteomics* **2015**, *15*, 4030–4040. [CrossRef] [PubMed]

56. Huang, Y.; Peng, C.; Yi, Y.; Gao, B.; Shi, Q. A Transcriptomic Survey of Ion Channel-Based Conotoxins in the Chinese Tubular Cone Snail (*Conus betulinus*). *Mar. Drugs* **2017**, *15*, 228. [CrossRef] [PubMed]

57. Gao, B.; Peng, C.; Lin, B.; Chen, Q.; Zhang, J.; Shi, Q. Screening and Validation of Highly-Efficient Insecticidal Conotoxins from a Transcriptome-Based Dataset of Chinese Tubular Cone Snail. *Toxins* **2017**, *9*, 214. [CrossRef] [PubMed]

58. Terrat, Y.; Biass, D.; Dutertre, S.; Favreau, P.; Remm, M.; Stöcklin, R.; Piquemal, D.; Ducancel, F. High-resolution picture of a venom gland transcriptome: Case study with the marine snail *Conus consors*. *Toxicon* **2012**, *59*, 34–46. [CrossRef] [PubMed]

59. Lluisma, A.O.; Milash, B.A.; Moore, B.; Olivera, B.M.; Bandyopadhyay, P.K. Novel venom peptides from the cone snail *Conus pulicarius* discovered through next-generation sequencing of its venom duct transcriptome. *Mar. Genom.* **2012**, *5*, 43–51. [CrossRef] [PubMed]

60. Hu, H.; Bandyopadhyay, P.K.; Olivera, B.M.; Yandell, M. Elucidation of the molecular envenomation strategy of the cone snail *Conus geographus* through transcriptome sequencing of its venom duct. *BMC Genom.* **2012**, *13*, 284. [CrossRef] [PubMed]

61. Barghi, N.; Concepcion, G.P.; Olivera, B.M.; Lluisma, A.O. Comparison of the Venom Peptides and Their Expression in Closely Related *Conus* Species: Insights into Adaptive Post-Speciation Evolution of *Conus* Exogenomes. *Genome Biol. Evol.* **2015**, *7*, 1797–1814. [CrossRef] [PubMed]

62. Phuong, M.A.; Mahardika, G.N.; Alfaro, M.E. Dietary breadth is positively correlated with venom complexity in cone snails. *BMC Genom.* **2016**, *17*, 401. [CrossRef] [PubMed]

63. Robinson, S.D.; Li, Q.; Lu, A.; Bandyopadhyay, P.K.; Yandell, M.; Olivera, B.M.; Safavi-Hemami, H. The Venom Repertoire of *Conus gloriamaris* (Chemnitz, 1777), the Glory of the Sea. *Mar. Drugs* **2017**, *15*, 145. [CrossRef] [PubMed]

64. Kaas, Q.; Yu, R.; Jin, A.H.; Dutertre, S.; Craik, D.J. ConoServer: Updated content, knowledge, and discovery tools in the conopeptide database. *Nucleic Acids Res.* **2012**, *40*, D325–D330. [CrossRef] [PubMed]

65. Kaas, Q.; Westermann, J.C.; Halai, R.; Wang, C.K.; Craik, D.J. ConoServer, a database for conopeptide sequences and structures. *Bioinformatics* **2008**, *24*, 445–446. [CrossRef] [PubMed]

66. Yang, J. Comprehensive description of protein structures using protein folding shape code. *Proteins* **2008**, *71*, 1497–1518. [CrossRef] [PubMed]

67. Imperial, J.S.; Bansal, P.S.; Alewood, P.F.; Daly, N.L.; Craik, D.J.; Sporning, A.; Terlau, H.; López-Vera, E.; Bandyopadhyay, P.K.; Olivera, B.M. A novel conotoxin inhibitor of Kv1.6 channel and nAChR subtypes defines a new superfamily of conotoxins. *Biochemistry* **2006**, *45*, 8331–8340. [CrossRef] [PubMed]

68. Park, K.H.; Suk, J.E.; Jacobsen, R.; Gray, W.R.; McIntosh, J.M.; Han, K.H. Solution conformation of alpha-conotoxin EI, a neuromuscular toxin specific for the alpha 1/delta subunit interface of torpedo nicotinic acetylcholine receptor. *J. Biol. Chem.* **2001**, *276*, 49028–49033. [CrossRef] [PubMed]

69. Nielsen, K.J.; Adams, D.; Thomas, L.; Bond, T.; Alewood, P.F.; Craik, D.J.; Lewis, R.J. Structure-activity relationships of omega-conotoxins MVIIA, MVIIC and 14 loop splice hybrids at N and P/Q-type calcium channels. *J. Mol. Biol.* **1999**, *289*, 1405–1421. [CrossRef] [PubMed]

70. Abraham, N.; Healy, M.; Ragnarsson, L.; Brust, A.; Alewood, P.F.; Lewis, R.J. Structural mechanisms for alpha-conotoxin activity at the human alpha 3 beta 4 nicotinic acetylcholine receptor. *Sci. Rep.* **2017**, *7*, 45466. [CrossRef] [PubMed]

71. Nicke, A.; Wonnacott, S.; Lewis, R.J. Alpha-conotoxins as tools for the elucidation of structure and function of neuronal nicotinic acetylcholine receptor subtypes. *Eur. J. Biochem.* **2004**, *271*, 2305–2319. [CrossRef] [PubMed]

72. Leipold, E.; Hansel, A.; Olivera, B.M.; Terlau, H.; Heinemann, S.H. Molecular interaction of delta-conotoxins with voltage-gated sodium channels. *FEBS Lett.* **2005**, *579*, 3881–3884. [CrossRef] [PubMed]

73. Shon, K.J.; Stocker, M.; Terlau, H.; Stühmer, W.; Jacobsen, R.; Walker, C.; Grilley, M.; Watkins, M.; Hillyard, D.R.; Gray, W.R.; et al. kappa-Conotoxin PVIIA is a peptide inhibiting the shaker K+ channel. *J. Biol. Chem.* **1998**, *273*, 33–38. [CrossRef] [PubMed]

74. Li, R.A.; Tomaselli, G.F. Using the deadly mu-conotoxins as probes of voltage-gated sodium channels. *Toxicon* **2004**, *44*, 117–122. [CrossRef] [PubMed]

75. Nielsen, K.J.; Schroeder, T.; Lewis, R. Structure-activity relationships of omega-conotoxins at N-type voltage-sensitive calcium channels (abstract). *J. Mol. Recognit.* **2000**, *13*, 55–70. [CrossRef]

76. Mir, R.; Karim, S.; Kamal, M.A.; Wilson, C.M.; Mirza, Z. Conotoxins: Structure, Therapeutic Potential and Pharmacological Applications. *Curr. Pharm. Des.* **2016**, *22*, 582–589. [CrossRef] [PubMed]

77. Tosti, E.; Boni, R.; Gallo, A. μ-Conotoxins Modulating Sodium Currents in Pain Perception and Transmission: A Therapeutic Potential. *Mar. Drugs* **2017**, *15*, 295. [CrossRef] [PubMed]

78. SIPO. Available online: http://www.sipo.gov.cn/ (accessed on 15 November 2017).

79. USPTO. Available online: https://www.uspto.gov/patent (accessed on 15 November 2017).

80. EPO. Available online: http://www.epo.org/ (accessed on 15 November 2017).

81. WIPO. Available online: http://www.wipo.int/ (accessed on 15 November 2017).

82. Durek, T.; Craik, D.J. Therapeutic conotoxins: A US patent literature survey. *Expert Opin. Ther. Pat.* **2015**, *25*, 1159–1173. [CrossRef] [PubMed]

83. Miljanich, G.P. Ziconotide: Neuronal calcium channel blocker for treating severe chronic pain. *Curr. Med. Chem.* **2004**, *11*, 3029–3040. [CrossRef] [PubMed]

84. Nielsen, C.K.; Lewis, R.J.; Alewood, D.; Drinkwater, R.; Palant, E.; Patterson, M.; Yaksh, T.L.; McCumber, D.; Smith, M.T. Anti-allodynic efficacy of the chi-conopeptide, Xen2174, in rats with neuropathic pain. *Pain* **2005**, *118*, 112–124. [CrossRef] [PubMed]

85. Adams, D.J.; Smith, A.B.; Schroeder, C.I.; Yasuda, T.; Lewis, R.J. Omega-conotoxin CVID inhibits a pharmacologically distinct voltage-sensitive calcium channel associated with transmitter release from preganglionic nerve terminals. *J. Biol. Chem.* **2003**, *278*, 4057–4062. [CrossRef] [PubMed]

86. Craig, A.G.; Norberg, T.; Griffin, D.; Hoeger, C.; Akhtar, M.; Schmidt, K.; Low, W.; Dykert, J.; Richelson, E.; Navarro, V.; et al. Contulakin-G, an O-glycosylated invertebrate neurotensin. *J. Biol. Chem.* **1999**, *274*, 13752–13759. [CrossRef] [PubMed]

87. Malmberg, A.B.; Gilbert, H.; McCabe, R.T.; Basbaum, A.I. Powerful antinociceptive effects of the cone snail venom-derived subtype-selective NMDA receptor antagonists conantokins G and T. *Pain* **2003**, *101*, 109–116. [CrossRef]

88. Satkunanathan, N.; Livett, B.; Gayler, K.; Sandall, D.; Down, J.; Khalil, Z. Alpha-conotoxin Vc1.1 alleviates neuropathic pain and accelerates functional recovery of injured neurones. *Brain Res.* **2005**, *1059*, 149–158. [CrossRef] [PubMed]

89. Lubbers, N.L.; Campbell, T.J.; Polakowski, J.S.; Bulaj, G.; Layer, R.T.; Moore, J.; Gross, G.J.; Cox, B.F. Postischemic administration of CGX-1051, a peptide from cone snail venom, reduces infarct size in both rat and dog models of myocardial ischemia and reperfusion. *J. Cardiovasc. Pharmacol.* **2005**, *46*, 141–146. [CrossRef] [PubMed]

90. Ekberg, J.; Jayamanne, A.; Vaughan, C.W.; Aslan, S.; Thomas, L.; Mould, J.; Drinkwater, R.; Baker, M.D.; Abrahamsen, B.; Wood, J.N.; et al. muO-conotoxin MrVIB selectively blocks Nav1.8 sensory neuron specific sodium channels and chronic pain behavior without motor deficits. *Proc. Natl. Acad. Sci. USA* **2006**, *103*, 17030–17035. [CrossRef] [PubMed]

© 2017 by the authors. Licensee MDPI, Basel, Switzerland. This article is an open access article distributed under the terms and conditions of the Creative Commons Attribution (CC BY) license (http://creativecommons.org/licenses/by/4.0/).

![toxins logo] *toxins*

MDPI

Review

Direct Fibrinolytic Snake Venom Metalloproteinases Affecting Hemostasis: Structural, Biochemical Features and Therapeutic Potential

Eladio F. Sanchez [1],*, Renzo J. Flores-Ortiz [2], Valeria G. Alvarenga [1] and Johannes A. Eble [3]

[1] Research and Development Center, Ezequiel Dias Foundation, Belo Horizonte 30510-010, MG, Brazil; eladio.flores@funed.mg.gov.br
[2] Graduate Program in Nursing, Federal University of Minas Gerais, Belo Horizonte 30130-100, MG, Brazil; renzojfo@gmail.com
[3] Institute for Physiological Chemistry and Pathobiochemistry, University of Münster, 15, 48149 Muenster, Germany; johannes.eble@uni-muenster.de
* Correspondence: eladio.flores@funed.mg.gov.br; Tel.: +55-31-3314-4784

Academic Editor: Steve Peigneur
Received: 25 October 2017; Accepted: 27 November 2017; Published: 5 December 2017

Abstract: Snake venom metalloproteinases (SVMPs) are predominant in viperid venoms, which provoke hemorrhage and affect hemostasis and thrombosis. P-I class enzymes consist only of a single metalloproteinase domain. Despite sharing high sequence homology, only some of them induce hemorrhage. They have direct fibrin(ogen)olytic activity. Their main biological substrate is fibrin(ogen), whose Aα-chain is degraded rapidly and independently of activation of plasminogen. It is important to understand their biochemical and physiological mechanisms, as well as their applications, to study the etiology of some human diseases and to identify sites of potential intervention. As compared to all current antiplatelet therapies to treat cardiovascular events, the SVMPs have outstanding biochemical attributes: (a) they are insensitive to plasma serine proteinase inhibitors; (b) they have the potential to avoid bleeding risk; (c) mechanistically, they are inactivated/cleared by α2-macroglobulin that limits their range of action in circulation; and (d) few of them also impair platelet aggregation that represent an important target for therapeutic intervention. This review will briefly highlight the structure–function relationships of these few direct-acting fibrinolytic agents, including, barnettlysin-I, isolated from *Bothrops barnetti* venom, that could be considered as potential agent to treat major thrombotic disorders. Some of their pharmacological advantages are compared with plasmin.

Keywords: metalloproteinases; animal toxins; thrombolysis; antithrombotics

1. Introduction

Among the venomous animals, snakes are the best-studied creatures throughout human history; this is partially due to the bad reputation associated with snakes, as many people have experienced that these small and often fragile-looking animals are harmful to man, and can inflict devastating damage in envenomed victims [1]. Indeed, snake venoms, especially those of the Viperidae (pit vipers and true vipers) family, contain extremely complex mixtures of pharmacologically active proteins/peptides that disrupt normal physiological or biochemical processes in line with their function to immobilize, to kill, and to digest their prey, as well as to defend themselves from predators [2,3]. They belong to a few structural classes of major protein families, including proteins with and without enzymatic activity, such as metalloproteinases (SVMPs), serine proteinases (SVSPs), phospholipases A_2 (PLA$_2$s), L-amino acid oxidases (L-AAOs), hyaluronidases, and non-enzymatic proteins: disintegrins, C-type lectin-like proteins/snaclecs, bradykinin-potentiating peptides (BPPs), nerve and vascular endothelial grow factors (VEGF), and Kunitz-type proteinase inhibitors [2–5]. Acting synergistically, various venom proteins/toxins are able to cause severe and detrimental effects on the hemostatic system and lead to cardiovascular shock [5–7]. Moreover, several of these compounds offer interesting and often unique insights into several biological systems [7–9]. In the context of drug discovery, the isolation and characterization of active venom proteins/toxins is carried out for two main purposes: (1) to identify and determine the compound(s) responsible for a specific activity observed in a bioassay; or (2) to survey the complete structural diversity of a venom to discover new sequences with novel structural scaffolds and pharmacological properties. Therefore, after a snake venom constituent has been properly purified, after its molecular structure has been resolved and its specific pharmacological effects have been revealed, the resulting pharmaceutical lead structure and the known molecular mechanism are beneficial to mankind, in contrast to the envenomation with the crude venom [3,7–10]. Underpinning research in the biomedicine field, such as looking for new thrombolytic and/or antithrombotic agents, becomes of increasing medical importance. It is noteworthy that around half of the drugs which are currently in therapeutic use have originated from natural products [10,11].

Snake venom metalloproteinases (SVMPs) are the crucial endopeptidases associated with the pathologies of snake envenoming. Especially coagulopathies commonly associated with viperid (*Serpentes viperidae*) are caused by enzymatic and non-enzymatic proteins in these venoms, and usually lead to incoagulability of the blood [12–14]. Proteomic analysis of snake venoms showed that in some venoms they are the most abundant >50% proteins of the proteome, e.g., see reference [15], and most lethal protein in viper and pit viper venoms [13–15], but are less significant in the venoms of Elapidae, Atractaspididae and Colubridae. In addition, differential proteomic compositions among different families of snakes have been reported recently [16]. SVMPs represent a group of multigene protein families that encode different multidomain protein molecules capable of producing a diverse array of activities, including hemorrhage, pro-coagulant, anticoagulant, fibrinolysis, apoptosis, and antiplatelet effects [1,15,16]. They were inferred to be derived through recruitment, duplication, and neofunctionalization of an ancestral gene, which must have been very similar to the recent genes encoding ADAM7, 28, and ADAMDEC-1 [1]. Consequently, SVMPs/reprolysins are also referred to as snake ADAMs (a disintegrin and metalloprotease). Actually, the large P-III class metalloproteases have a modular structure that is homologous to the ectodomain of membrane-anchored ADAMs [17,18]. Furthermore, SVMPs, ADAMs and ADAMTSs, the latter of which contain an additional thrombospondin (TS) motifs, share a topological similarity to matrix metalloproteinases of tissues (MMPs) within the structural organization of their catalytic center [19]. However, their non-catalytic ancillary domains are distinct from those of MMPs and other metalloproteinases [20]. On the other hand, SVMPs selectively hydrolyze a small number of proteins involved in key reactions in the coagulation cascade, fibrinolysis, and in platelet function. Such mechanism of action leads to either activation or inactivation of the protein that participate in these processes, thereby severely interfering with blood coagulation and platelet function. In the following sections, we provide an overview of several direct-acting fibrinolytic P-I metalloproteinases that affect hemostasis and impair platelet function. Their potential application in therapy of major arterial occlusive disorders is surveyed.

2. Structure and Classification of Snake Venom Metalloproteinases

SVMPs are zinc-dependent endopeptidases also known as adamalysins/reprolysins, based on the name of its structural prototype adamalysin II, isolated from eastern diamondback rattlesnake (*Crotalus adamanteus*) venom, and on the mammalian reproductive tract proteins involved in cellular adhesion [21–23]. These enzymes are also termed as ADAMs (a disintegrin and metallropteinase), MDC (metalloproteinase-like, disintegrin-like and cysteine-rich proteins), and are grouped into three major classes, P-I to P-III, according to their general structural organization, and are subdivided into several subgroups (Figure 1) [19,24,25]. They were initially characterized by their ability or inability to induce hemorrhage in experimental in vivo models [26,27]. Hemorrhage is defined as the escape of blood from the vascular system. This leaking is caused by damage of the vessel wall, which consists of the endothelial cell layer and the subjacent extracellular matrix, such as basement membranes and interstitial stroma. Proteolytic cleavage of extracellular matrix proteins, of blood clotting factors, and of cell adhesion receptors on platelets and endothelial cells by SVMPs are the main reason for venom-induced hemorrhages.

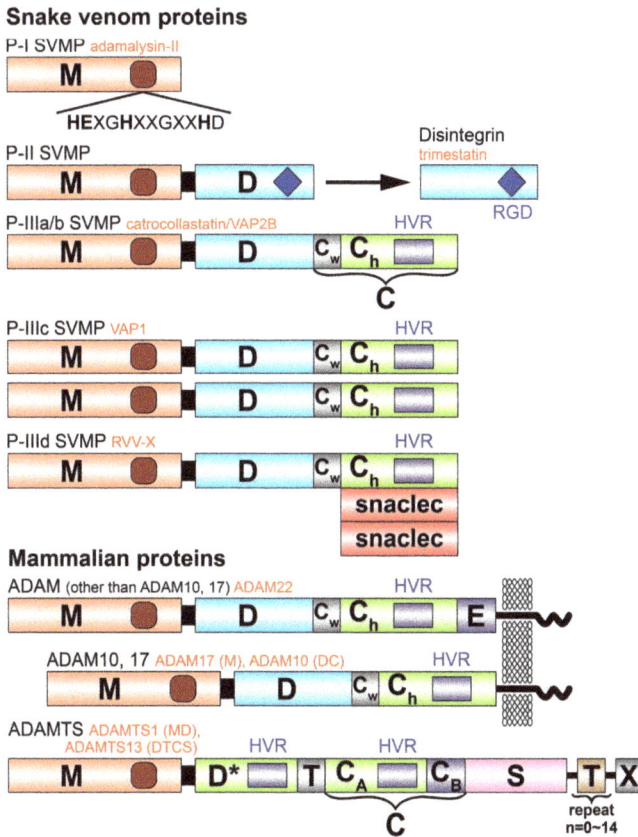

Figure 1. Protein domain structure of snake venom metalloproteinases (SVMPs) and related molecules. Each domain or subdomain is represented by a different color. M, metalloproteinase; D, disintegrin (or disintegrin-like) domain; C, cysteine-rich domain; C_W, cysteine-rich "wrist" subdomain; C_h, the cysteine-rich "hand" subdomain; snaclec, snake venom C-type lectin-like domain; E, epidermal growth factor (EGF)-like domain; T, thrombospondin type-1 (TSP) motif; S, spacer domain; X, domain variable among ADAMTSs. Representatives of each class of SVMPs and ADAM/ADAMTSs, whose crystal structure have been determined, are indicated in red letters. The P-III classes SVMPs are divided into subclasses (IIIa–IIId) based on their distinct post-translation modifications. Recently, it was found that the D domain of ADAMTS family proteinases does not have a disintegrin-like structure but adopt the C_h subdomain fold, and thus, is represented as D*. The previously cysteine-rich domain of ADAMTSs is structurally subdivided into the N-terminal G_h subdomain-fold domain (C_A) and the C-terminal domain (C_B). The ADAMTS family commonly possesses the N-terminal M, D, T, C, S domains whereas the C-terminal is variable among ADAMTSs e.g., ADAMTS13 possess six repeats of TSP and two CUB (complement, uEGF, and bone morphogenesis) domains that follow the S domain. Reproduced from [14], copyright 2012, Elsevier.

Class I (P-I) SVMPs, have a single catalytic metalloproteinase (MP) domain in their mature form [23,28–30]. All SVMPs exhibit an extended zinc-binding consensus sequence HEXXHXXGXXH/D, which comprises three zinc-coordinating histidine side chains, and generally, a glutamate residue. Moreover, these proteins also possess a strictly conserved methionine containing 1,4-β-turn, termed Met-turn, bordering the substrate-binding site, which is a typical feature of the metzincin clan of metalloproteinases [19,21,31]. In general, there are two structural forms of the proteinase domain: a two-disulfide-containing structure e.g., in adamalysin II [19,21] and a three-disulfide-stabilized structure e.g., in mutalysin-II (mut-II) [30,32] and in leucurolysin-a (leuc-a) [29]. Sequence alignment of the P-I enzymes indicate that they possess high sequence homologies (Figure 2).

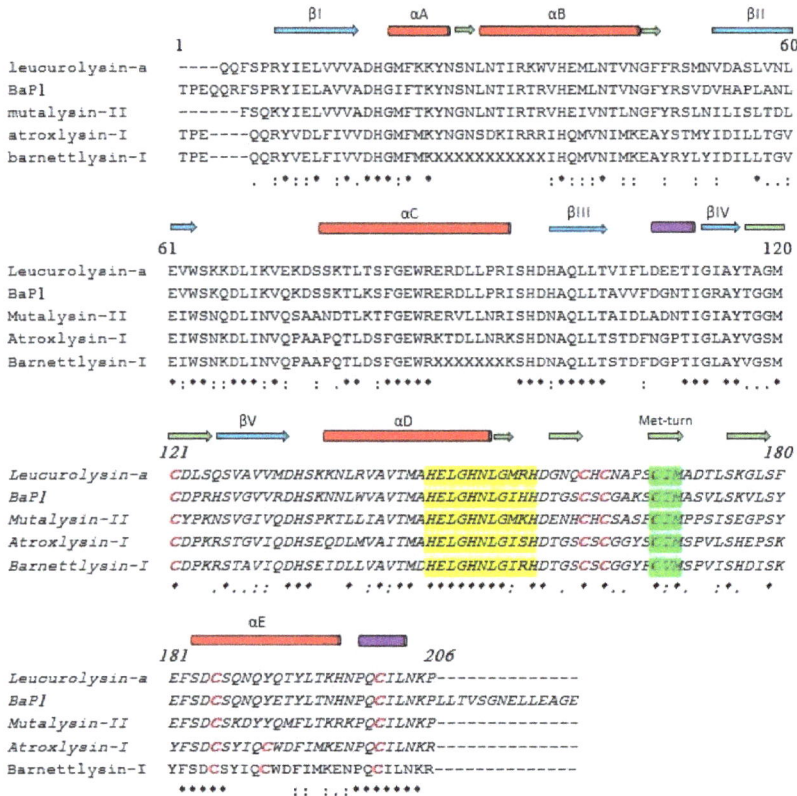

```
                         βI          αA          αB                    βII
                  1                                                           60
leucurolysin-a    ----QQFSPRYIELVVVADHGMFKKYNSNLNTIRKWVHEMLNTVNGFFRSMNVDASLVNL
BaP1              TPEQQRFSPRYIELAVVADHGIFTKYNSNLNTIRTRVHEMLNTVNGFYRSVDVHAPLANL
mutalysin-II      ------FSQKYIELVVVADHGMFTKYNGNLNTIRTRVHEIVNTLNGFYRSLNILISLTDL
atroxlysin-I      TPE----QQRYVDLFIVVDHGMFMKYNGNSDKIRRRIHQMVNIMKEAYSTMYIDILLTGV
barnettlysin-I    TPE----QQRYVELFIVVDHGMFMKXXXXXXXXXXXXXIHQMVNIMKEAYRYLYIDILLTGV
                    . :*::* :*.***:* *                  :*:::* ::   :  : :   *..:

                                    αC            βIII        βIV
                  61                                                          120
Leucurolysin-a    EVWSKKDLIKVEKDSSKTLTSFGEWRERDLLPRISHDHAQLLTVIFLDEETIGIAYTAGM
BaP1              EVWSKQDLIKVQKDSSKTLKSFGEWRERDLLPRISHDNAQLLTAVVFDGNTIGRAYTGGM
Mutalysin-II      EIWSNQDLINVQSAANDTLKTFGEWRERVLLNRISHDNAQLLTAIDLADNTIGIAYTGGM
Atroxlysin-I      EIWSNKDLINVQPAAPQTLDSFGEWRKTDLLNRKSHDNAQLLTSTDFNGPTIGLAYVGSM
Barnettlysin-I    EIWSNKDLINVQPAAPQTLDSFGEWRXXXXXXXXKSHDNAQLLTSTDFDGPTIGLAYVGSM
                  *:**::***:*:   :  .** :*****       ***:***** :     *** **...*

                        βV          αD                       Met-turn
                  121                                                         180
Leucurolysin-a    CDLSQSVAVVMDHSKKNLRVAVTMAHELGHNLGMRHDGNQCHCNAPSC:MADTLSKGLSF
BaP1              CDPRHSVGVVRDHSKNNLWVAVTMAHELGHNLGIHHDTGSCSCGAKSC:MASVLSKVLSY
Mutalysin-II      CYPKNSVGIVQDHSPKTLLIAVTMAHELGHNLGMKHDENHCHCSASFC:MPPSISEGPSY
Atroxlysin-I      CDPKRSTGVIQDHSEQDLMVAITMAHELGHNLGISHDTGSCSCGGYSC:MSPVLSHEPSK
Barnettlysin-I    CDPKRSTAVIQDHSEIDLLVAVTMDHELGHNLGIRHDTGSCSCGGYFC:MSPVSHDISK
                  *   .*..::  ***  * :*;** *********: **  . * *..  *:*.  :*.  *

                      αE
                  181              206
Leucurolysin-a    EFSDCSQNQYQTYLTKHNPQCILNKP--------------
BaP1              EFSDCSQNQYETYLTNHNPQCILNKPLLTVSGNELLEAGE
Mutalysin-II      EFSDCSKDYYQMFLTKRRKPQCILNKP--------------
Atroxlysin-I      YFSDCSYIQCWDFIMKENPQCILNKR--------------
Barnettlysin-I    YFSDCSYIQCWDFIMKENPQCILNKR--------------
                  *****   :: :.;********
```

Figure 2. Sequence comparisons of four P-I class SVMPs. UniProt accession numbers sequences were assigned by using the program ClustalW. Non-hemorrhagic: leuc-a (P84907), mut-II (P22796), bar-I (P86976), and hemorrhagic: atr-I (P85420) and BaP1 (P83512). The sequences of these proteins were determined by the Edman degradation method and the sequences of leuc-a and BaP1 were confirmed by crystallography. Secondary-structure elements were defined by MAFFT V7 (multiple alignment) and PSIPRED V3.3 (predict secondary structure). The blue and dark green arrows indicate the locations of β-strands and turns, respectively, in the crystal structure of leuc-a. The red and purple cylinders represent α-helices and 3₁₀ helices, respectively. Cys residues are highlighted in red; (*) identical residues; (:) strongly similar residues; (.) weakly similar residues. The conserved zinc biding motif and the met-turn are highlighted in yellow and bright green, respectively. (-) indicate gaps.

Based on the functional ability to induce hemorrhage, the P-I SVMPs are further divided into two subgroups: P-IA which induce hemorrhage [28,33], and P-IB with weak (or no) hemorrhagic effect [29,32,34]. SVMPs play important roles in the overall pathophysiology of viperid envenoming by inducing local and systemic hemorrhage, which was primarily attributed to their potential to degrade basement membrane (BM) components surrounding capillaries, like type IV collagen, laminin (LM), nidogen, and fibronectin (FN), as well as to induce other tissue damaging and hemostatic alterations [8,22,25,35–37].

In addition to the MP domain, class II (P-II) SVMPs possess a C-terminal disintegrin domain (Dis). A group of disintegrins released from precursor P-II MPs have an RGD motif, which mediates the interaction with integrins, thereby offering many potentials for pharmacological applications [7,38,39]. Other active tripeptide sequences such as KGD, MDV, MLD, VGD, ECD, MDG, and KTS have been reported [40]. The RGD and KGD tripeptide sequences are the primary recognition sites for the integrin αIIbβ3 receptor. The binding of disintegrin to integrin αIIbβ3, blocks the binding of fibrinogen to the receptor, and hence, platelet aggregation [7,26,39,40]. Two FDA-approved drugs, Eptifibatide (Integrilin®, Millennium Pharmaceuticals, Shering-Plough, Cambridge, MA, USA), and Tirofiban (Aggrastat®, Merck, Darmstadt, Hesse, Germany), antagonists of the platelet receptor glycoprotein αIIbβ3 of human platelets, inhibit platelet aggregation, and are the first rationally designed antiplatelet agents [9,41,42].

Class III (P-III) SVMPs contain the MP, disintegrin-like (containing a disulfide-linked XXCD, mostly SECD, in place of RGD) and cysteine rich (Cys) domains, and are the most mysterious enzymes in terms of complexity and function. Their structure is homologous to a group of membrane bound ADAM, which act in cell–cell and cell–matrix adhesion and signaling [14,24]. P-III SVMPs are further grouped into subclasses based on their different post-translational modifications, such as homo-dimerization (P-IIIc), proteolysis between the MP and Dis domain (P-IIIb), or complexation (P-IIId) with additional snake C-type lectin-like proteins (snaclecs) [43]. All SVMPs have a signal sequence (pre-form) and a zymogenic sequence (pro-form) N-terminal to the MP domain in their gene structures. The signal sequence is cleaved co-translationally in the endoplasmic reticulum, whereas cleavage of the zymogenic sequence occurs extracellularly, is regulated, and activates the proteolytic enzyme. Evolution of viperid SVMPs is characterized by domain loss along the evolutionary timeline, thus, the loss of the Cys domain had preceded the development of the P-II class, which in turn preceded the formation of the P-I SVMPs [17,44].

3. Three-Dimensional Structures of P-I Class SVMPs

Table 1, summarizes the three-dimensional structures currently available for eleven P-I class SVMPs, as well as their main proteolytic activity related to hemorrhage. The molecular structure of P-I SVMP adamalysin II from the eastern diamondback rattlesnake (*Crotalus adamanteus*) venom was the first one of the M12B proteinase to be solved by X-ray crystallography in 1993 [21]. The first P-III SVMP, vascular-apoptosis inducing protein-1 (VAP-1) was reported by 2006 [18]. The 3D structures of a number of P-I SVMPs soon followed, and to date, the structures of eleven P-I proteinases are available in the Protein Data Bank (PDB) [20]. By the 1990s, the Sanchez lab (Biochemistry of Proteins from Animal Venoms) at FUNED, Brazil, started to investigate/identify fibrinolytic activity of the small SVMPs from bushmaster snake (*Lachesis muta muta*). Later, this research was extended to other South American *Bothrops* snakes. Thus, other P-I class enzymes, including leucurolysin-a (leuc-a) from the venom of the Brazilian snake *Bothrops leucurus* (white-tailed jararaca), were discovered and described [29]. The mature leuc-a is composed of 202 amino acid residues, and was crystallized using the hanging-drop vapor-diffusion technique at 1.8 Å. The crystal structure of leuc-a (PDB code 4Q1L) complexed with an endogenous tripeptide (QSW) was solved by molecular replacement technique using the proteinase BaP1 (*B. asper*) structure (PDB code INDI) as template Ferreira et al., unpublished [45]. The crystal structure analysis reveals that leuc-a is an ellipsoidal molecule with a relatively flat active-site cleft that separates two subdomains similar to the two jaws of the oral cavity (Figure 3). The upper jaw is formed by the N-terminal subdomain of the molecule (residues 1-152) and characterized by a β-strand with four parallel and one antiparallel β-strand (strands I, II, III, IV, and V), which is flanked by a long and short surface located helix on its convex side, and by two long helices, one of which represents the central active site helix, on its concave side. The lower jaw, comprising the 50 C-terminal residues, is folded in a more irregular fold, which is organized in multiple turns, with the chain ending in a long C-terminal helix and an extended segment that is linked to the upper subdomain by a disulfide bond. The catalytic zinc ion is located at the active site cleft between the two subdomains (jaws). It is tetrahedrally coordinated by His^{142}, and His^{146} of the upper subdomain, by His^{152} of the lower subdomain, and a water molecule, which is polarized by Glu^{143}, and therefore attacks the scissile peptide bond in a nucleophilic manner. These three His residues and the nearby Glu play a critical role in both the structure and activity of P-I proteinases, and explains their occurrence in the $H^{142}EXXHXXGXXH^{152}D$ consensus sequence. In addition, Asp^{153} is strictly conserved in the SVMPs that establish a hydrogen bond with an invariant serine (Ser^{179}), located in the first turn of αC cleft, and the sequence $C^{164}I^{165}M^{166}$ associated with the characteristic "Met-turn". These structural features are typical of the metzincin superfamily of metalloproteinases [14,18–21,23].

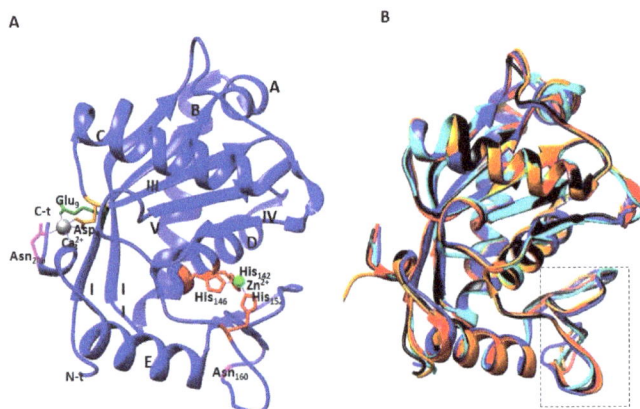

Figure 3. Ribbon plot of the overall structure of metalloproteinase leucurolysin-a (**A**) and its superposition with the structures of other P-I SVMPs (**B**), depicted in standard orientation. (**A**) The molecular structure of leucurolysin-a, a non-hemorrhagic P-I SVMP. The catalytic zinc atom is highlighted in green, together with the three histidines (red), a glutamic acid, and the water molecule forming the active site environment. The localization of α-helices (A–E), β-strands (I–V) and the methionine-turn as well as the N- and C-terminal residues are also indicated; (**B**) Superposition of structures by UCSF CHIMERA system of non-hemorrhagic P-I SVMPs: leuc-a (blue, PDB 4Q1L; mut-II, red, P22796, predicted by I-Tasser program C-score 1.58; bar-I, cyano, P86976, predicted by I-Tasser program C-score 1.49), and hemorrhagic P-I SVMPs: BaP1 (orange, PBD 2W12), and atr-I (black, P85420, predicted by I-Tasser program C-score 1.49). The superposed structures within the Met-turn are highlighted in the insert figure. However, flexibility of this variable motif does not provide relevant details responsible for hemorrhagic activity.

As depicted in Figure 2, primary structures of P-I SVMPs identified in our laboratory: leuc-a [29,45], mut-II [30], atroxlysin-I (atr-I, [28]), and barnettlysin-I (bar-I, [34]), align with other homologous P-I SVMPs, and show their high similarity. Nevertheless, despite their high sequence homology, some SVMPs induce hemorrhage, while others are (almost) inactive, and fail to cause any bleeding. This functional difference is probably related to the structural determinants in the MP domain [19–21,23,28–31]. In connection with this, Wallnoefer and collegues [46] have investigated the protein–protein interfaces of four P-I SVMPs, including: hemorrhagic (BaP1 and acutolysin A) and non-hemorrhagic (leuc-a and H2-proteinase ones).

The P-I SVMPs hydrolyze basement membrane (BM) proteins in good correlation with their ability to bind them and to induce profuse bleeding in vivo. The authors applied computer simulations to obtain information about the backbone flexibility in certain surface regions/loops of these enzymes to carry out their damaging function. The findings indicated that the sequences of these four MPs mainly differ in the loop following the highly conserved active site, which surrounds the so-called Met-turn. For instance, the active hemorrhagic MPs (BaP1 and acutolysin A) both present a GSCSCGA/GKS (residues 154–163) before the Met-turn, whereas the inactive (leuc-a and H2-proteinase) do not show any identical residues in this section, besides the two conserved Cys residues. This added further evidence to the hypothesis that flexibility might play a role in distinguishing between active and inactive enzymes. Thus, a certain combination of flexibility (residues 156–165) and rigidity of the neighboring loop C-terminal of the Met-turn (residues 167–176) provides an appropriate association domain for individual target protein [46,47]. However, despite intense investigation on this topic, detailed structural determinants of hemorrhagic activity have remained unclear, and no experimental data have been provided yet.

Table 1. Three dimensional structures of P-I class SVMPs deposited in the PDB and their main biological activities.

SVMP	Source	Activities	PDB ID	Year	Reference
Adamalysin II	C. adamanteus	non-hemorrhagic	1LAG	1993	[21]
Atrolysin C	C. atrox	hemorrhagic	1ATL, 1HTD	1994	[48]
H2 proteinase	T. Flavoviridis	non-hemorrhagic	1WNI	1996	[49]
Acutolysin A	A. Acutus	hemorrhagic	1BSW,1BUD	1998	[50]
Acutolysin C	A. Acutus	hemorrhagic	1QUA	1999	[51]
TM-3	T. Mucrosquamatus	fibrinogenolytic	1KUF, 1KUI	2002	[52]
BaP1	B. asper	hemorrhagic	1ND1	2003	[53]
FII	A. acutus	non-hemorrhagic	1YP1	2005	[54]
BmooMPα-I	B. moogeni	non-hemorrhagic	3GBO	2010	[55]
TM-1	T. mucrosquamatus	fibrinogenolytic	4J4M	2013	[56]
Leuc-a	B. leucurus	non-hemorrhagic	4Q1L	2015. unpublished	

4. Action on Some Plasma and ECM Protein Substrates

Most of the relevant proteolytic enzymes that act on fibrin (Fb) and fibrinogen (Fbg) belong to one of two families: the metalloproteinases, and the serine proteinases. These proteinases can lead to defibrinogenation of blood, lysis of fibrin clots, and a consequent decrease in blood viscosity. Therefore, they can be regarded as true anticoagulants. The majority of fibrin(ogen)olytic enzymes are metalloproteinases which selectively cleave the α-chains of fibrin(ogen) to a ~44 kDa fragment and thereby are termed as α-fibrinogenases [4,28,29,57]. However, generalizations about chains specificity are not always applicable, since the other chains of fibrinogen can be substantially degraded over time. These P-I SVMPs are direct-acting fibrinolytics, as they are not reliant on components in the blood for activity. Like plasmin, the prototype of direct-acting fibrinolytic enzyme, a number of P-I SVMPs may represent an attractive therapeutic option in thrombolysis to allow reperfusion of ischemic tissue [58–63]. Among other P-I SVMPs described in the literature, alfimeprase, the recombinant form of fibrolase that first was isolated from the venom of the southern copperhead snake (Agkistrodon contortrix contortrix) [4,57,63], made the best progress, and has been investigated as a potential and safe thrombolytic agent, using in vitro, and also, many animal thrombosis models [57,63–66]. Alfimeprase proved to have clinical potential for drug development as a direct thrombolytic compound, however, the enzyme failed to successfully complete phase 3 clinical trials. Thus, Nuvelo (San Carlos, CA, USA) has discontinued further clinical development (for details, see [57]). A possible reason for suboptimal performance of alfimeprase in clinical use is its inability to bind to fibrin, thereby failing to reach a critical concentration of proteolytic activity locally at the thrombus [57]. Based on previous studies with fibrolase [67,68], Prof. Markland and his group at University of Southern California (Los Angeles, CA, USA), have not given up hope for alfimeprase and have constructed a chimeric compound possessing both thrombolytic and antiplatelet properties, and now the potential of this enzyme with bifunctional activity could be investigated in animal models of arterial thrombosis [57].

In studies of the vascular system, blood coagulation, fibrinolysis, and platelet function, snake venom proteins have been crucial in elucidating the complex physiological mechanisms which rule the vascular system, the coagulation cascade, and platelet functions. Moreover, they have been instrumental in elucidating the structure–function relationships of human clotting factors and platelet glycoproteins, because of their potency, selectivity, and high biological efficacy [65–70]. Notably, in the past fifteen years, the field has advanced due to the continued development of new or alternative agents that provide greater hemostatic safety and thrombolytic efficacy, as well as the identification of risk factors for arterial and venous thrombosis [71–73]. In this context, all the thrombolytic agents in current therapeutic use for deep vein thrombosis (DVT) e.g., the variants of tissue type plasminogen activator (tPA) are plasminogen activators (PAs). They efficiently dissolve thrombi, but adversely carry the unavoidable risk of bleeding complications [74,75].

To elucidate the molecular mode of action of four P-I SVMPs, identified in our laboratory: bar-I [34], leuc-a [29,45], mut-II [30], and atr-I [28], we have characterized the peptide bond specificity (Table A1). This is essential to understand the active site preference of the proteinases in correlation to their proteolytic and hemorrhagic activity. Moreover, this offers the opportunity to design peptide substrates and proteinase inhibitors. In addition to oxidized insulin B chain, which was used as the standard to compare peptide bond specificity among SVMPs, we also employed the designed peptides [406]RREYHTEKLVSKGD[420] and [693]GHARLVHVEEPH[704] as model substrates [34]. These sequences mimic the Aα-chain of Fbg and the "bait" region of α2-macroglobulin (α2-M), respectively. The results suggest that the SVMP-mediated cleavage is directed to an X–Leu bond, where X is a small residue at the P1 position, and a bulky hydrophobic residue at P1′, with a clear preference for leucine residue. An interesting study was carried out analyzing a plasma based, proteome-derived peptide library as substrate with mass spectrometry, to investigate the peptide bond specificity of three P-I SVMPs: atrolysin C (*C. atrox*, [75]), BaP1 [53], leuc-a [29], and a P-III bothropasin (*B. jararaca*, [76]). This study revealed the consensus sequence, ETAL–LLLD, that was similar to the other P-I enzymes, except of the acidic aspartate residue at the P4′ position for leuc-a [77,78]. These interesting differences in the peptide bond specificities at the other P and P′ sites, may imply functional differences between these proteases. For instance, the P-I enzymes showed preferences across the full P4 to P4′ range, whereas the P-III bothropasin exhibited narrow preferences across the sites, in accordance with earlier studies related with P-III SVMPs [78]. Furthermore, in the case of the non-hemorrhagic leuc-a, the preference for the acidic residue (Asp) in the P4′ site may have had a negative effect in inducing hemorrhage by this proteinase [77]. This finding merits further investigation for understanding the mechanism by which SVMPs induce hemorrhage. Moreover, the manner by which these enzymes act on various plasma and ECM proteins, including Fbg, FN, LM, fibrin, and collagen I and IV, were also performed. It is known that disruption of capillaries is the result of proteolytic degradation of key BM and ECM components, allowing for the escape of blood components into the stroma, and thus, producing local hemorrhage [22,28,36,37]. A comparative study of two P-I SVMPs: BaP1 (hemorrhagic) and leuc-a (non-hemorrhagic), provided insights into the putative mechanism of bleeding produced by SVMPs [36]. Both enzymes showed differences to degrade BM and associated ECM protein substrates, in vivo, mainly type IV collagen that is degraded by BaP1. To support these findings, further in vivo studies indicated that hydrolysis of type IV collagen by SVMPs, mainly P-II and P-III classes, is crucial in destabilizing microvessel structures and causing hemorrhage [79].

5. Antiplatelet Properties of P-I SVMPs

Blood platelets play a crucial role in hemostasis, and in the development of arterial thrombosis and of cardiovascular diseases. In response to vascular injury they rapidly adhere to exposed subendothelial matrix proteins, mainly von Willebrand factor (vWF) and collagen. As a result, adherent platelets are activated, spread, and release the content of storage vesicles [80,81]. More importantly, the main targets for antithrombotic drugs development are platelets and coagulation proteins [82,83]. According to current knowledge, the pathophysiology of arterial thrombosis differs from that of venous thrombosis as a consequence of higher shear forces in the arterial branch of the circulation, which require especially vWF and its shear-force-dependent conformational change for platelet adhesion [84]. Therefore, arterial thrombosis is treated with drugs that target platelets, while venous thrombosis is treated with drugs that target compounds of the coagulation cascade [71–73]. On the other hand, it has become clear that platelet function can be inhibited to reduce thrombotic tendencies by blocking either surface receptors, key cytoplasmic enzymes, e.g., cyclooxygenase or signaling proteins, including kinases or phosphatases [85,86]. As shown in Figure 4, there are a number of SVMPs that affect platelet aggregation. Notably, the glycoproteins (GPs), GPIb-IX-V and GPVI, receptors for vWF and collagen, respectively, bind to their respective ligands in different ways [85,86]. While at low physiological shear conditions, GPVI binds collagen exposed within the damaged blood vessel walls, and GPIb-IX-V binds vWF which has undergone conformational changes after exposure to high shear rates in arterioles and stenotic arteries [82,86,87]. Notably, only a small number of P-I SVMPs with direct-acting fibrinolytic activity inhibit platelet aggregation. Therefore, it is arguable whether fibrinogen degradation products generated by α-fibrinogenases play a role in inhibiting platelet aggregation [88].

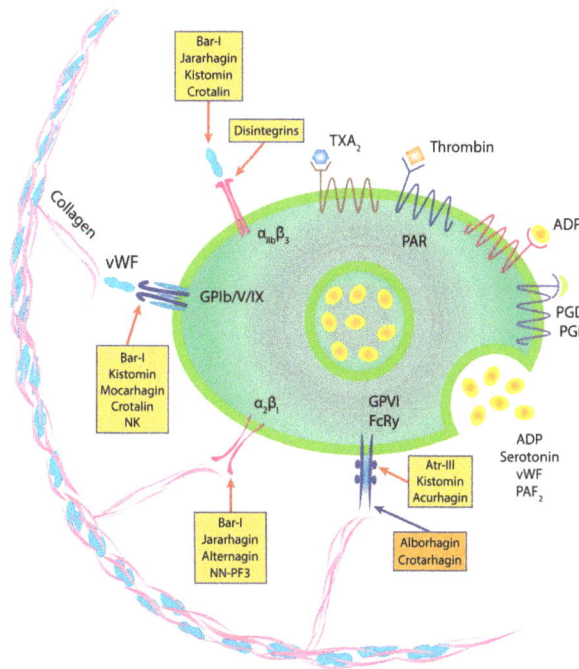

Figure 4. Schematic representation of some SVMPs and their effects on platelet function. Induction or inhibition of platelet aggregation by these proteinases are indicated in blue or red arrows, respectively. The inhibitory activity of disintegrins, the antagonists of αIIbβ3 integrin are indicated in red arrow. The ligands for various receptors are shown. PAR, protease activated receptor; TXA2, thromboxane A$_2$; PGD, prostaglandin D; PGI, prostaglandin I. Modified from [88], copyright 2016 MDPI.

An α-fibrinogenase, termed kistomin was indentified in the venom of *Calloselasma rhodostoma* (formerly *Agkistrodon rhodostoma*) during the 1990s. The enzyme inhibited platelet aggregation induced by low concentrations of thrombin ($\geq$0.2 U/mL). Moreover, it attenuated cytosolic calcium rise and blocked thromboxane B2 formation in platelets stimulated by thrombin (0.1 U/mL). Importantly, the enzyme inhibited ristocetin-induced platelet aggregation in the presence of vWF. Thus, kistomin was the first P-I SVMP that exhibited anti-thrombotic effects [89]. Further investigation demonstrated that kistomin specifically inhibited vWF-induced platelet aggregation through binding and cleavage of platelet GPIbα and vWF [90]. Incubation of human platelets with kistomin resulted in a selective cleavage of platelet membrane glycoprotein GPIbα, with the release of ~45 and 130 kDa soluble fragments [90]. In addition, tail-bleeding time was prolonged in mice which had been injected with kistomin intravenously. The platelet receptor GPIb-IX-V complex plays a dominant role in the first steps of platelet adhesion under high shear stress conditions and arterial thrombus formation. Since GPIb–vWF interaction is very significant for hemostasis/thrombosis, the modulation of GPIbα–vWF interactions during thrombotic events could be beneficial [91,92]. Another P-I metalloproteinase, named crotalin, was purified from *Crotalus atrox* venom by the group of Prof. Huang [93]. Crotalin showed potent antithrombotic effect in vivo by cleaving vWF and GPIb, as shown by Western blotting and flow cytometry. Moreover, the author stated that crotalin, due to its multiple actions, may be a useful tool for investigating the interactions among vWF, ECM proteins, and GPIb-IX-V in static and flow conditions [93].

Based on the preliminary observations that kistomin and crotalin inhibited platelet aggregation induced by ristocetin, which promote vWF binding to GPIb, and taking into consideration that platelet dysfunction is responsible for increased patient morbidity and mortality, platelets represent the major target for therapeutic intervention [82,83,94]. Moreover, as thrombus formation is primarily stabilized by platelets and fibrin [73], we investigated a recently purified P-I metalloproteinase bar-I [34], and focused not only on the interaction of GPIb-IX-V complex with its major ligand, vWF, but also on other platelet surface ligands, such as fibrin. Its antithrombotic effect by targeting this receptor and potential platelet ligands, as well as its potential benefits, are worth discussing. Bar-I (23 kDa) was characterized as a direct-acting fibrinolytic enzyme which does not require the conversion of the zymogen plasminogen to the active form plasmin. Its amino acid sequence shows high sequence similarity with homologous P-I SVMPs (Figure 2). Furthermore, bar-I hydrolyzed several plasma and extracellular matrix (ECM) proteins in vitro [34,94]. Although SVMPs have similar proteolytic activity toward several substrates in vitro, bar-I is devoid of hemorrhagic activity in the mice skin model [34]. More importantly, the enzyme dose-dependently inhibited collagen- and plasma vWF-induced platelet aggregation. This effect was inhibited by treatment bar-I with EDTA, suggesting that the effect of bar-I on platelet activation is due to its enzymatic activity. Furthermore, vWF-induced platelet activation was more efficiently inhibited than collagen-induced platelet activation by this enzyme with (IC_{50}) values of 1.4 and 3.2 μM, respectively. Interesting, studies in which platelets were pretreated with bar-I revealed that the vWF-receptor, GPIb-IX-V complex, is more susceptible to bar-I cleavage than the collagen binding receptor [34]. In addition to the cleavage of the multimeric adhesive protein vWF and its receptor GPIb, bar-I also cleaves the collagen receptor α2β1 integrin, albeit at a slower rate [34]. The essential interaction of GPIb-IX-V complex with vWF for normal hemostasis is well documented by the severe bleeding disorders as a consequence of the lack of either the receptor (Bernard–Soulier syndrome) or the ligand (von Willebrand disease). Thus, interactions of platelets via their receptors GPIb-IX-V and α2β1 integrin with vWF and collagen, respectively, indicates early events in platelet activation, especially under high shear rates of the arterial flow [73,83,95]. Furthermore, we have tested bar-I's antithrombotic effect in vivo, with a tail bleeding assay using mice CF2 strain (18–20 g). When bar-I at doses of 2.5, 5 and 7.5 μg were injected intravenously into mice, the tail bleeding time was not altered (83 ± 4 s, $n = 5$) in comparison with saline (negative control, 78 ± 5 s, $n = 5$), (Sanchez et al., unpublished results). Similar data have been reported for batroxase, a P-I SVMP from *B. atrox* [96]. Moreover, bar-I is an analog of mut-II that disrupts formed thrombi through the hydrolysis of fibrin, rather than by plasminogen activation. We are now much interested in examining the hypothesis that vascular recanalization in the absence of plasmin generation results in improved thrombolysis without inducing bleeding side effects.

6. Biochemical Advantages of P-I SVMPs in Comparison to Plasminogen Activators (PAs)

One approach to treat thrombosis is to infuse thrombolytic agents to dissolve the blood clots and to restore tissue perfusion and oxygenation. This is ultimately accomplished by the serine proteinase plasmin, which is derived from its zymogen plasminogen in a reaction catalyzed by plasminogen activators (PAs), e.g., tissue type plasminogen activator (tPA), urokinase type-PA (u-PA) and staphylokinase. PAs effectively dissolve relatively small clots which occur in the coronary artery of patients with acute myocardial infarction [59,74,97,98]. Importantly, long retracted clots which are found in peripheral arterial occlusion (PAO) and deep-vein thrombosis (DVT), often lack plasminogen. Therefore, plasmin should be more efficacious to treat this kind of thrombus successfully [97–101].

Hemorrhage is a common complication of all the PAs, and is observed with any of the multiple agents and therapeutic indications [101]. An important conceptual framework of fibrinolytic activation and inhibition have been reported, and provided a foundation for understanding the mechanism of action of PAs and plasmin in a thrombus milieu [100–102]. A notable issue of direct-acting thrombolytics, of which plasmin is the prototype, has been well documented in in vitro and in vivo models, such as the use of catheter-delivered plasmin for treatment of DVT and PAO. As reported, plasmin should induce safe and efficacious thrombolysis at the thrombus site, whereas circulating enzyme is rapidly neutralized by α2-antiplasmin and by α2-macroglobulin (α2-M). This regime avoids bleeding complications, for details see [73,101,102]. Direct thrombolytic agents under investigation can be grouped into two categories: (a) plasmin and its derivatives: mini-plasmin, micro-plasmin, and delta-plasmin [74]; and (b) fibrinolytic SVMPs [63,99,100]. Figure 5 shows a schematic representation of the plasminogen/fibrinolytic system, including a number of snake venom proteins. Although plasmin is involved in other normal and pathological conditions, such as cell migration, inflammation and tissue remodeling, the most important function of plasmin is intravascular thrombolysis [58,59,98].

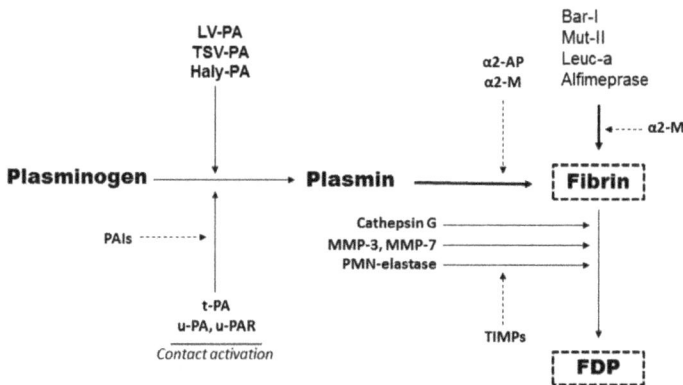

Figure 5. Schematic representation of the plasminogen/fibrinolytic system. Local effects of plasminogen activators (PAs) (physiological and snake venom PAs) and direct-acting fibrinolytic agents. PAs convert plasminogen to active enzyme plasmin, which degrades fibrin. In parallel using fibrin as substrate, direct-acting P-I SVMPs or endogenous plasmin proteolytically degrade fibrin and dissolve the fibrin clot. Inhibition (indicated by dotted lines) occurs either (i) at the level of PAs by plasminogen activator inhibitor (mainly by PAI-1 and PAI-2) or (ii) by plasmin inhibitors (by α2-antiplasmin and α2-macroglobulin). Matrix metalloproteinases (MMPs) degrade fibrin into smaller fragments, termed fibrin degradation products (FDP), and are inhibited by tissue inhibitor of MMPs (TIMPs). Snake venom serine proteinases, PAs (TSV-PA, LV-PA and Haly-PA [62]) and the direct-acting fibrinolytic metalloproteinases (bar-I, leuc-a, alfimeprase and mut-II), are shown at their site of inhibition.

Fibrinolytic activity in viperid snake venoms was described for the first time in 1956 [60], when several pit viper venoms of *Agkistrodon, Bothrops,* and *Crotalus* genus were examined. Furthermore, the fibrinolytic activity in the venom of *A. contortrix contortrix* was reported to act directly on fibrin and independently of activation of the endogenous fibrinolytic system [60,61]. These findings supported a significant clinical potential of the direct thrombolytic agents to degrade fibrin without requiring an intermediate step of plasminogen activation. Therefore, alfimeprase, the recombinant fibrolase from *A. c. contortrix* venom, was used in clinical trials. In contrast to plasmin, this P-I SVMP does not bind directly to fibrin. In addition, the systemic effects of venom fibrinolytic enzymes are inhibited by α2-M, which is the last line of defense against exogenous proteolytic enzymes. As direct-acting thrombolytic P-I SVMPs are not inhibited by the normal blood serine proteinase inhibitors (serpins), they may serve as templates for the development of alternative thrombolytic compounds, and have received special attention due to their possible therapeutic role for dissolution of blood clots [62–70]. These enzymes act on fibrin and Fbg, leading to defibrinogenation of blood, lysis of fibrin clots, and a consequent decrease in blood viscosity [69]. They may be classified as being either α or β chain fibrin(ogen)ases. Due to their broad spectrum of proteolytic activity leading to fibrin(ogen) digestion, they can be regarded as true anticoagulants, and are metalloproteinases or serine proteinases [29,34,69]. Also, it was observed that the level of fibrinolytic effect varies widely between the P-I SVMPs of different species within a particular genus [60–63,65]. Therefore, it was suggested to use a properly purified enzyme(s) from snake venoms as fibrin clot-lysing agent for clinical applications. In the early 1990s, we reported the isolation and complete amino acid sequence of a P-I class metalloproteinase, termed lachesis hemorrhagic factor II (LHF-II), from bushmaster (*Lachesis muta muta*) snake venom [30]. Additionally, we had elucidated that traces of hemorrhagic effect of LHF-II were due to the presence of a minor contaminant by the P-III metalloproteinase named mutalysin-I (mut-I). Therefore, the proteinase was renamed to mutalysin-II (mut-II). Furthermore, its pharmacological properties have been reevaluated, and these reports had indicated that mut-II does not elicit any hemorrhagic response in mice or rabbit [32]. More importantly, we have evaluated by intravital microscopy, the effects of mut-II on the recanalization of microvessels after thrombus induction in the ear of hairless mice. At the doses used (50 μg, 2.0 mg/kg, iv), thrombolytic efficacy was achieved in all animals (*n* = 5) after approximately 12 min, followed by recanalization. A control group (*n* = 5) that received u-PA (250 U/mouse, iv), showed blood flow restoration within the same interval, 12 min. In addition, under the experimental conditions, mut-II does not alter hemostasis or cause bleeding events, as confirmed by histopathology [32]. Based on these data, we have also initiated in vivo studies to assess the thrombolytic potential of a recently isolated bar-I by using intravital microscopy in comparison with recombinant t-PA. This direct-acting fibrinolytic enzyme dissolves fibrin clots in vitro, and also inhibits collagen- and plasma vWF induced platelet aggregation by cleaving not only the vWF and its receptor GPIb, but also the collagen receptor α2β1 integrin. Although the current thrombolytic agents have proved to be of clinical benefit, the failure to rapidly restore reperfusion in some patients, and the continuous risk of bleeding of all PAs, are still setbacks which have to be improved in order to introduce them in the routine of clinical therapy [72,82,83,100]. Therefore, continued development of safer and more efficient thrombolytic agents, in combination with more effective antiplatelet approaches, are the future goals in this research field.

7. Conclusions

Fibrin clot-based vascular occlusion, a life-threatening disorder, has to be treated immediately by dissolving the fibrin clot in the vessel which impairs the blood flow. Treatment with recombinant plasmin or with plasminogen activator is usually the choice of means in the hospital. As an alternative, recombinantly produced P-I SVMPs are investigated. Their extremely high fibrinolytic activity gives them an advantage over current fibrin clot-dissolving agents. However, substrate specificity of such P-I SVMP should be mainly limited to fibrin. Any cleavage or degradation of the blood vessel wall, especially of the basement membrane subjacent to the endothelial cells, must be absent to avoid blood leakage and hemorrhages. Whereas several P-I SVMPs cleave vessel wall components, some non-hemorrhagic P-I SVMPs have been identified. The structural comparison between both groups may reveal characteristics for non-hemorrhagic P-I SVMPs to accelerate the search for such fibrinolytic, non-hemorrhagic P-I SVMPs in the biodiversity of snake venoms. Thereby, lead-structures can be obtained for the design of novel fibrinolytic, non-hemorrhagic proteinases. Some of the recent non-hemorrhagic P-I SVMPs also cleave adhesion receptors on platelets, such as vWF-receptor. Cleavage of those "off-targets" would be tolerable, if not even beneficial, as such a proteinase would have, in addition to their fibrinolytic activity, antithrombotic functions by preventing platelets from adhesion and thrombus formation. Moreover, platelets support leukocyte extravasation. Although the molecular mechanism is not fully understood, neutrophils seem to interact with platelets and use their adhesive potential, likely via their adhesion receptors, GPIb and αIIbβ3, to attach to the vessel wall, especially under higher shear rates and at atheroslerotic lesions [103]. Similarly, blood-borne tumor cells during hematogenic dissemination also interact with platelets and subvert their adhesive potential to the vessel wall, likely also mediated via the vWF-receptor and different integrins [104]. Hence, it is worthwhile to think whether non-hemorrhagic snake venom proteinases which cleave the adhesion receptors on platelets might be useful in reducing platelet-supported extravasation of leukocyte, or disseminating blood-borne tumor cells. This could be a strategy to reduce formation of atherosclerotic plaques or metastasis. Another criterion for the use of fibrinolytic, non-hemorrhagic P-I SVMPs is the restriction of the fibrinolytic activity to the thrombus site and to prevent potential adverse systemic effects. A long way ahead, but the goal of utilizing fibrinolytic, non-hemorrhagic P-I SVMPs in clearing thrombotic occlusions or inhibiting platelet-assisted cell extravasation is promising.

Acknowledgments: This work was supported by the Brazilian agencies Fundação de Amparo a Pesquisa do Estado de Minas Gerais (FAPEMIG, grants number: CBB-AUC-00022-16, APQ 01858-15) and CNPq to E.F.S. J.A.E. is financially supported by the Deutsche Forschungsgemeinschaft through a joint German-Brazilian cooperation project (grant: DFG:EB177/13-1). We apologize to the authors whose work was not cited.

Author Contributions: E.F.S. and J.A.E. wrote the review. R.J.F.-O. and V.G.A. contributed to prepare the figures and revision the manuscript.

Conflicts of Interest: The authors declare no conflict of interest.

Appendix A

Table A1. Cleavage sites of three synthetic substrates by some P-I SVMPs.

Proteinase	Bonds Cleaved	Reference
	Oxidized Insulin B chain	
leuc-a	Ala_{14}-Leu_{15}, Tyr_{16}-Leu_{17}	[29]
atr-I	Ala_{14}-Leu_{15}, Tyr_{16}-Leu_{17}	[28]
BaP1	Ala_{14}-Leu_{15}, Tyr_{16}-Leu_{17}	[36]
mut-II	His_5-Leu_6, His_{11}-Leu_{11}, Ala_{14}-Leu_{15}, Phe_{24}-Phe_{25}	[78]
	Human α2-M (bait region)	
leuc-a	Arg_{696}-Leu_{697}	[105]
atr-I	Arg_{696}-Leu_{697}	[unpublished]
mut-II	Arg_{696}-Leu_{697}	[unpublished]
bar-I	Arg_{696}-Leu_{697}	[34]
	Human fibrinogen Aα-chain	
leuc-a	Lys_{413}-Leu_{414}	[105]
atr-I	Lys_{413}-Leu_{414}	[unpublished]
mut-II	Lys_{413}-Leu_{414}	[unpublished]
bar-I	Lys_{413}-Leu_{414}	[34]

References

1. Casewell, N.R.; Wüster, W.; Vonk, F.J.; Harrison, R.A.; Fry, B.G. Complex cocktails: The evolutionary novelty of venoms. *Trends Ecol. Evol.* **2013**, *28*, 219–229. [CrossRef] [PubMed]
2. Fry, B.G. The structural and functional diversification of the Toxicofera reptile venom system. *Toxicon* **2012**, *60*, 434–448. [CrossRef] [PubMed]
3. King, G.F. Venoms as platform for human drug: Translating toxins into therapeutics. *Expert Opin. Biol. Ther.* **2011**, *11*, 1469–1484. [CrossRef] [PubMed]
4. Markland, F.S. Snake venoms and the hemostatic system. *Toxicon* **1998**, *36*, 1749–1800. [CrossRef]
5. Kang, T.S.; Georgieva, D.; Genov, N.; Murakami, M.T.; Sinha, M.; Kumar, R.P.; Kaur, P.; Kurmar, W.S.; Dey, S.; Sharma, S.; et al. Enzymatic toxins from snake venoms: Structural characterization and mechanism of catalysis. *FEBS J.* **2011**, *278*, 4544–4576. [CrossRef] [PubMed]
6. Sajevic, T.; Leonardi, A.; Krizaj, I. Haemostatically active proteins in snake venoms. *Toxicon* **2011**, *57*, 627–645. [CrossRef] [PubMed]
7. Lu, Q.; Clemetson, J.M.; Clemetson, K.J. Snake venoms and hemostasis. *J. Thromb. Haemost.* **2005**, *3*, 1791–1799. [CrossRef] [PubMed]
8. Yamazaki, Y.; Morita, T. Snake venom components affecting blood coagulation and the vascular system: Structural similarities and marked diversity. *Curr. Pharm. Des.* **2007**, *13*, 2927–2934. [CrossRef]
9. Fox, J.W.; Serrano, S.M.T. Approaching the golden age of natural product pharmaceuticals from venom libraries: An overview of toxins and toxin-derivatives currently involved in therapeutic or diagnostic applications. *Curr. Pharm. Des.* **2007**, *13*, 2927–2934. [CrossRef] [PubMed]
10. Butler, M.S. Natural products to drugs: Natural product derived compounds in clinical trials. *Nat. Prod. Rep.* **2005**, *22*, 162–195. [CrossRef] [PubMed]
11. Paterson, I.; Anderson, E.A. The renaissance of natural products as drug candidates. *Science* **2005**, *310*, 451–453. [CrossRef] [PubMed]
12. White, J. Snake venom and coagulopathy. *Toxicon* **2005**, *45*, 951–967. [CrossRef] [PubMed]
13. Gutierrez, J.M.; Rucavado, A.; Escalante, T.; Dias, C. Hemorrhage induced by snake venom metalloproteinases: Biochemical and biophysical mechanisms involved in microvessel damage. *Toxicon* **2005**, *45*, 997–1011. [CrossRef] [PubMed]
14. Takeda, S.; Takeya, H.; Iwanaga, S. Snake venom metalloproteinases: Structure, function and relevance to the mammalian ADAM/ADAMTS family proteins. *Biochim. Biophys. Acta* **2012**, *1824*, 164–176. [CrossRef] [PubMed]
15. Kohlhoff, M.; Borges, M.H.; Yarleque, A.; Cabezas, C.; Richardson, M.; Sanchez, E.F. Exploring the proteomes of the venoms of the Peruvian pit vipers. *J. Proteom.* **2012**, *75*, 2181–2195. [CrossRef] [PubMed]
16. Calvete, J.J. Venomics: Integrative venom proteomics and beyond. *Biochem. J.* **2017**, *474*, 611–634. [CrossRef] [PubMed]
17. Casewell, N.R.; Wagstaff, S.C.; Harrison, R.A.; Renjifo, C.; Wüster, W. Domain loss facilitates accelerated evolution and neofunctionalization of duplicate snake venom metalloproteinase toxin genes. *Mol. Biol. Evol.* **2011**, *28*, 2637–2649. [CrossRef] [PubMed]
18. Takeda, S.; Igarashi, T.; Mori, H.; Araki, S. Crystal structures of VAP1 reveal ADAMs' MDC domain architecture and its unique C-shaped scaffold. *EMBO J.* **2006**, *25*, 2388–2396. [CrossRef] [PubMed]
19. Gomis-Ruth, F.X. Structural aspects of the metzinzin clan of metalloendopeptidases. *Mol. Biotechnol.* **2003**, *24*, 157–202. [CrossRef]
20. Takeda, S. ADAM and ADAMTS family proteins and snake venom metalloproteinases: A structural overview. *Toxins* **2016**, *8*, 155. [CrossRef] [PubMed]
21. Gomis-Rüth, F.X.; Kress, L.F.; Bode, W. First structure of a snake venom metalloproteinase: A prototype for matrix metalloproteinases/collagenases. *EMBO J.* **1993**, *12*, 4151–4157. [PubMed]
22. Bjarnason, J.B.; Fox, J.B. Hemorrhagic metalloproteinases from snake venoms. *Pharmacol. Ther.* **1994**, *62*, 325–372. [CrossRef]
23. Fox, J.W.; Serrano, S.M. Structural considerations of snake venom metalloproteinases, key members of the M12 reprolysin family of metalloproteinases. *Toxicon* **2005**, *45*, 969–985. [CrossRef] [PubMed]

24. Fox, J.W.; Serrano, S.M. Insights into and speculations about snake venom metalloproteinases (SVMP) synthesis, folding and disulfide bond formation and their contribution to venom complexity. *FEBS J.* **2008**, *275*, 3016–3030. [CrossRef] [PubMed]

25. Sanchez, E.F.; Gabriel, L.M.; Gontijo, S.; Gremski, L.H.; Veiga, S.S.; Evangelista, K.S.; Eble, J.A.; Richardson, M. Structural and functional characterization of a P-III metalloproteinase, leucurolysin-B from *Bothrops leucurus* venom. *Arch. Biochem. Biophys.* **2007**, *468*, 193–204. [CrossRef] [PubMed]

26. Huang, T.F.; Chang, G.H.; Ouyang, C.H. Characterization of hemorrhagic principles from *Trimeresurus gramineous* snake venom. *Toxicon* **1984**, *22*, 45–52. [CrossRef]

27. Sanchez, E.F.; Magalhaes, A.; Diniz, C.R. Purification of a hemorrhagic factor (LHF-I) from the venom of the bushmaster snake (*Lachesis muta muta*). *Toxicon* **1987**, *25*, 611–619. [CrossRef]

28. Sanchez, E.F.; Schneider, F.S.; Yarleque, A.; Borges, M.H.; Richardson, M.; Figueiredo, S.G.; Eble, J.A. A novel metalloproteinase atroxlysin-I from the Peruvian *Bothrops atrox* (jergon) snake venom acts both on ECM and platelets. *Arch. Biochim. Biophys.* **2010**, *496*, 9–20. [CrossRef] [PubMed]

29. Bello, C.; Hermogenes, A.L.; Magalhaes, A.; Veiga, S.S.; Gremski, L.H.; Richardson, M.; Sanchez, E.F. Isolation and characterization of a fibrinolytic proteinase from *Bothrops leucurus* (White-tailed jararaca) snake venom. *Biochimie* **2006**, *88*, 189–200. [CrossRef] [PubMed]

30. Sanchez, E.F.; Diniz, C.R.; Richardson, M. The complete amino acid sequence of the haemorrhagic factor LHFII, a metalloproteinase isolated from the venom of the bushmaster snake (*Lachesis muta muta*). *FEBS Lett.* **1991**, *282*, 178–182. [CrossRef]

31. Bode, W.; Gomis-Rüth, F.X.; Stocker, W. Astacins, serralysins, snake venom and matrix metalloproteinases exhibit identical zinc-binding environment (HEXXHXXGXXH) and Met-turn) and topologies and should be grouped into a common family, the "metzincins". *FEBS Lett.* **1993**, *331*, 134–140. [CrossRef]

32. Aguero, U.; Arantes, R.M.S.; Lazerda-Queiroz, N.; Mesquita, O.N.; Magalhaes, A.; Sanchez, E.F.; Carvalho-Tavares, J. Effect of mutalysin-II on vascular recanalization after thrombosis induction in the ear of hairless mice model. *Toxicon* **2007**, *50*, 698–707. [CrossRef] [PubMed]

33. Rucavado, A.; Nuñez, J.; Gutierrez, J.M. Blister formation and skin damage induced by BaP1, a hemorrhagic metalloproteinase from the venom of the snake *Bothrops asper*. *Int. J. Exp. Pathol.* **1998**, *79*, 245–254. [PubMed]

34. Sanchez, E.F.; Richardson, M.; Gremski, L.H.; Veiga, S.S.; Yarleque, A.; Stephan, N.; Martins-Lima, A.; Estevao-Costa, M.I.; Eble, J.A. A novel fibrinolytic metalloproteinase, barnettlysin-I from *Bothrops barnetti* (barnett's pitviper) snake venom with anti-platelet properties. *Biochim. Biophys. Acta* **2016**, *1860*, 542–556. [CrossRef] [PubMed]

35. Tsai, L.H.; Wang, Y.M.; Chiang, T.Y.; Chen, Y.L.; Huang, R.J. Purification, cloning and sequence analyses for pro-metalloprotease-disintegrin variants from *Deinagkistrodon acutus* venom and subclassification of the small venom metalloproteinases. *Eur. J. Biochem.* **2000**, *267*, 1359–1367. [CrossRef] [PubMed]

36. Escalante, T.; Ortiz, N.; Rucavado, A.; Sanchez, E.F.; Richardson, M.; Fox, J.W.; Gutierrez, J.M. Role of collagens and perlecan in microvascular stability: Exploring the mechanism of capillary vessel damage by snake venom metalloproteinases. *PLoS ONE* **2011**, *6*, e28017. [CrossRef] [PubMed]

37. Gutierrez, J.M.; Escalante, T.; Rucavado, A.; Herrera, C. Hemorrhage caused by snake venom metalloproteinases: A journey of discovering and understanding. *Toxins* **2016**, *8*, 93. [CrossRef] [PubMed]

38. Huang, T.F.; Holt, J.C.; Lukasiewicz, H.; Nieviaroswski, S. Trigramin, a low molecular weight peptide inhibiting fibrinogen interaction with platelet receptors expressed on glycoprotein IIb/IIIa. *J. Biol. Chem.* **1987**, *262*, 16157–16163. [PubMed]

39. Marcinkiewicz, C.; Weinreb, P.H.; Calvete, J.J.; Kisiel, D.G.; Mousa, S.A.; Tuszynski, G.P.; Lobb, R.R. Obtustatin: A potent selective inhibitor of alfa1beta1 integrin in vitro and angiogenesis in vivo. *Cancer Res.* **2003**, *63*, 2020–2023. [PubMed]

40. Calvete, J.J.; Marcinkiewicz, C.; Monleon, D.; Esteve, V.; Celda, B.; Juarez, P.; Sanz, L. Snake venom disintegrins: Evolution of structure and function. *Toxicon* **2005**, *45*, 1063–1074. [CrossRef] [PubMed]

41. Coller, B.S. Platelet GPIIbIIIa antagonists, the first anti-integrin receptor therapeutic. *J. Clin. Investig.* **1997**, *99*, 1467–1471. [CrossRef] [PubMed]

42. Michelson, A.D. Antiplatelet therapies for the treatment of cardiovascular disease. *Nat. Rev. Drug Discov.* **2010**, *9*, 154–169. [CrossRef] [PubMed]

43. Clemetson, K.J. Snaclecs (snake C-type lectins) that inhibit or activate platelets by binding to receptors. *Toxicon* **2010**, *56*, 1236–1246. [CrossRef] [PubMed]

44. Moura-da-Silva, A.M.; Theakston, R.D.G.; Crampton, J.M. Evolution of disintegrin cysteine-rich and mammalian matrix-degrading metalloproteinases: Gene duplication and divergence of a common ancestor rather than convergent evolution. *J. Mol. Evol.* **1996**, *43*, 263–269. [CrossRef] [PubMed]
45. Ferreira, R.N.; Rates, B.; Richardson, M.; Guimaraes, B.G.; Sanchez, E.F.; Pimenta, A.M.; Nagem, R.A.P. Complete amino acid sequence, crystallization and preliminary X-ray diffraction studies of leucurolysin-a, a nonhemorrhagic metalloproteinase from *Bothrops leucurus* snake venom. *Acta Cryst.* **2009**, *F65*, 798–801.
46. Wallnoefer, H.G.; Lingott, T.; Gutierrez, J.M.; Merfort, I.; Liedi, K.R. Backbone flexibility controls the activity and specificity of a protein-protein interface: Specificity in snake venoms metalloproteases. *J. Am. Chem. Soc.* **2010**, *132*, 10330–10337. [CrossRef] [PubMed]
47. Lingott, T.; Schleberger, C.; Gutierrez, J.M.; Merfort, I. High-resolution crystal structure of the snake venom metalloproteinase BaP1 complexed with a peptidomimetic: Insight into inhibitor binding. *Biochemistry* **2009**, *48*, 6166–6174. [CrossRef] [PubMed]
48. Zhang, D.; Botos, I.; Gomis-Rüth, F.X.; Doll, R.; Blood, C.; Njoroge, F.G.; Fox, J.W.; Bode, W.; Meyer, E.F. Structural interaction of natural and synthetic inhibitors with the venom metalloproteinase, atrolysin C (form d). *Proc. Natl. Acad. Sci. USA* **1994**, *91*, 8447–8451. [CrossRef] [PubMed]
49. Kumasaka, T.; Yamamoto, M.; Moriyama, H.; Tanaka, N.; Sato, M.; Katsube, Y.; Yamakawa, Y.; Omori-Satoh, T.; Iwanaga, S.; Ueki, T. Crystal structure of H2-proteinase from the venom of *Trimeresurus flavoviridis*. *J. Biochem.* **1996**, *119*, 49–57. [CrossRef] [PubMed]
50. Gong, W.; Zhu, X.; Liu, S.; Teng, M.; Niu, L. Crystal structures of acutolysin A, a three-disulfide hemorrhagic zinc metalloproteinase from the snake venom of *Agkistrodon acutus*. *J. Mol. Biol.* **1998**, *283*, 657–668. [CrossRef] [PubMed]
51. Zhu, X.Y.; Teng, M.K.; Niu, L.W. Structure of acutolysin-C, a hemorrhagic toxin from the venom of *Agkistrodon acutus*, providing further evidence for the mechanism of the pH-dependent proteolytic reaction of zinc metalloproteinases. *Acta Crystallogr. D Biol. Crystallogr.* **1999**, *55*, 1834–1841. [CrossRef] [PubMed]
52. Huang, K.F.; Chiou, S.H.; Ko, T.P.; Yuann, J.M.; Wang, A.H. The 1.35 A structure of cadmium-substituted TM-3, a snake venom metalloproteinase from Taiwan habu: Elucidation of a TNFalfa-converting enzyme-like active-site structure with a distorted octahedral geometry of cadmium. *Acta Crystallogr. D.* **2002**, *58*, 1118–1128. [CrossRef] [PubMed]
53. Watanabe, L.; Shannon, J.D.; Valente, R.H.; Rucavado, A.; Alape-Giron, A.; Theakston, R.D.; Fox, J.W.; Gutierrez, J.M.; Arni, R.K. Amino acid sequence and crystal structure of BaP1, a metalloproteinase from *Bothrops asper* snake venom that exerts multiple tissue-damaging activities. *Protein Sci.* **2003**, *12*, 2273–2281. [CrossRef] [PubMed]
54. Lou, Z.; Hou, J.; Liang, X.; Chen, J.; Qiu, P.; Liu, Y.; Li, M.; Rao, Z.; Yan, G. Crystal structure of a non-hemorrhagic fibrin(ogen)olytic metalloproteinase complexed with a novel natural tri-peptide inhibitor from venom of *Agkistrodon acutus*. *J. Struct. Biol.* **2005**, *152*, 195–203. [CrossRef] [PubMed]
55. Akao, P.K.; Tonoli, C.C.C.; Navarro, M.S.; Cintra, A.C.O.; Neto, J.R.; Arni, R.K.; Murakami, M.T. Structural studies BmooMPalpha-I, a non-hemorrhagic metalloproteinase from *Bothrops moojeni* venom. *Toxicon* **2010**, *55*, 361–368. [CrossRef] [PubMed]
56. Chou, T.L.; Wu, C.H.; Huang, K.F.; Wang, A.H.J. Crystal structure of a *Trimeresurus mucrosquamatus* venom metalloproteinase providing new insights into the inhibition by endogenous tripeptide inhibitors. *Toxicon* **2013**, *71*, 140–146. [CrossRef] [PubMed]
57. Markland, F.S.; Swenson, S. Snake venom metalloproteinases. *Toxicon* **2013**, *62*, 3–18. [CrossRef] [PubMed]
58. Carmeliet, P.; Collen, D. Gene targeting and gene transfer studies of the biological role of the plasminogen /plasmin system. *Thromb. Haemost.* **1995**, *74*, 429–436. [PubMed]
59. Novokhatny, V. Structure and activity of plasmin and other direct thrombolytic agents. *Thromb. Res.* **2008**, *122*, S3–S8. [CrossRef] [PubMed]
60. Didisheim, P.; Lewis, J.H. Fibrinolytic and coagulant activities of certain snake venoms and proteases. *Proc. Soc. Exp. Biol. Med.* **1956**, *93*, 10–13. [CrossRef] [PubMed]
61. Kornalik, F.; Styblova, Z. Fibrinolytic proteases in snake venoms. *Experientia* **1967**, *23*, 999–1000. [CrossRef] [PubMed]
62. Sanchez, E.F.; Swenson, S. Proteases from South American snake venoms affecting fibrinolysis. *Curr. Pharm. Anal.* **2007**, *3*, 147–157. [CrossRef]

63. Swenson, S.; Toombs, C.F.; Pena, L.; Johanson, J.; Markland, F.S. α-fibrinogenases. *Curr. Drug Targets Cardiovasc. Haematol. Disord.* **2004**, *4*, 417–434. [CrossRef] [PubMed]
64. Deitcher, S.R.; Toombs, C.F. Non-clinical and clinical characterization of a novel acting thrombolytic: Alfimeprase. *Pathophysiol. Haemost. Thromb.* **2005**, *34*, 215–220. [CrossRef] [PubMed]
65. Hong, T.-T.; Huang, J.; Lucchesi, B.R. Effect of thrombolysis on myocardial injury: Recombinant tissue type plasminogen activator vs. alfimeprase. *Am. J. Physiol. Heart Circ. Physiol.* **2006**, *290*, 959–967. [CrossRef] [PubMed]
66. Weaver, H.S.M.; Comerota, A.J.; Perler, B.A.; Joing, M. Efficacy and safety of alfimeprase in patients with acute peripheral arterial occlusion (PAO). *J. Vasc. Surg.* **2010**, *51*, 600–609.
67. Sanchez, E.F.; Bush, L.R.; Swenson, S.; Markland, F.S. Chimeric fibrolase: Covalent attachment of an RGD-like peptide to create a potentially more effective thrombolytic agent. *Thromb. Res.* **1997**, *87*, 289–302. [CrossRef]
68. Swenson, S.; Bush, L.R.; Markland, F.S. Chimeric derivative of fibrolase, a fibrinolytic enzyme from southern copperhead venom, possesses inhibitory activity on platelet aggregation. *Arch. Biochem. Biophys.* **2000**, *384*, 227–237. [CrossRef] [PubMed]
69. Koh, C.Y.; Kini, R.M. From snake venom toxins to therapeutics-Cardiovascular examples. *Toxicon* **2012**, *59*, 497–506. [CrossRef] [PubMed]
70. Shah, A.R.; Scher, L. Drug evaluation: Alfimeprase, a plasminogen-independent thrombolytic. *Drugs* **2007**, *10*, 329–335.
71. Mackman, N. Triggers, targets and treatments for thrombosis. *Nature* **2008**, *451*, 914–918. [CrossRef] [PubMed]
72. Jackson, S.P. Arterial thrombosis-insidious, unpredictable and deadly. *Nat. Med.* **2011**, *17*, 1423–1436. [CrossRef] [PubMed]
73. Marder, V.J. Thrombolytic therapy for deep vein thrombosis: Potential application of plasmin. *Thromb. Res.* **2009**, *123*, 556–561. [CrossRef]
74. Marder, V.J.; Novokhatny, V. Direct fibrinolytic agents: Biochemical attributes, preclinical foundation and clinical potential. *J. Thromb. Haemost.* **2010**, *8*, 433–444. [CrossRef] [PubMed]
75. Bjarnason, J.B.; Tu, A.T. Hemorrhagic toxins from western diamondback rattlesnake (*Crotalus. atrox*) venom; isolation and characterization of five toxins and the role of zinc in hemorrhagic toxin e. *Biochemistry* **1978**, *17*, 395–404. [CrossRef]
76. Oliveira, A.K.; Paes Leme, A.F.; Assakura, M.T.; Menezes, M.C.; Zelanis, A.; Tashima, A.K.; Lopes-Ferreira, M.; Lima, C.; Camargo, A.C.; Fox, J.W.; et al. Simplified procedure for the isolation of HF3, bothropasin, disintegrin-like/cysteine-rich protein and a novel P-I metalloproteinase from *Bothrops jararaca* venom. *Toxicon* **2009**, *53*, 797–801. [CrossRef] [PubMed]
77. Paes Leme, A.F.; Escalante, T.; Pereira, J.G.C.; Oliveira, A.K.; Sanchez, E.F.; Gutierrez, J.M.; Serrano, S.M.T.; Fox, J.W. High resolution analysis of snake venom metalloproteinase (SVMP) peptide bond cleavage specificity using proteome based peptide library and mass spectrometry. *J. Proteom.* **2011**, *74*, 401–410. [CrossRef] [PubMed]
78. Sanchez, E.F.; Cordeiro, M.; DE Oliveira, E.B.; Juliano, L.; Prado, E.S.; Diniz, C.R. Proteolytic specificity of two hemorrhagic factors, LHF-I and LHF-II from the venom of the bushmaster snake (*Lachesis muta muta*). *Toxicon* **1995**, *33*, 1061–1069. [CrossRef]
79. Herrera, C.; Escalante, T.; Voisin, M.-B.; Rucavado, A.; Morazan, D.; Macedo, J.K.A.; Calvete, J.J.; Sanz, L.; Nourshargh, S.; Gutierrez, J.M.; et al. Tissue localization and extracellular matrix degradation by PI, PII and PIII snake venom metalloproteinases: Clues on the mechanisms of venom-induced hemorrhage. *PLoS Negl. Trop. Dis.* **2015**, *9*, e0003731. [CrossRef] [PubMed]
80. Jackson, S.P.; Schoenwaelder, S.M. Antiplatelet therapy: In search of the magic bullet. *Nat. Rev. Drug Discov.* **2003**, *2*, 775–789. [CrossRef] [PubMed]
81. Clemetson, K.J.; Clemetson, J.M. Platelet GPIb complex as a target for anti-thrombotic drug development. *Thromb. Haemost.* **2008**, *99*, 473–479. [CrossRef] [PubMed]
82. Gardiner, E.E.; Andrews, R.K. Platelet receptor expression and shedding: Glycoprotein Ib-IX-V and glycoprotein VI. *Transfus. Med. Rev.* **2014**, *28*, 56–60. [CrossRef] [PubMed]
83. Schneider, S.W.; Nuschele, S.; Wixforth, A.; Gorzelanny, C.; Alexander-Katz, A.; Netz, R.R.; Schneider, M.F. Shear-induced unfolding triggers adhesion of von Willebrand factor fibers. *Proc. Natl. Acad. Sci. USA* **2007**, *104*, 7899–7903. [CrossRef] [PubMed]

84. Choi, S.; Vilaire, G.; Marcinkiewicz, C. Small molecule inhibitors of integrin α2β1. *J. Med. Chem.* **2007**, *50*, 5457–5462. [CrossRef] [PubMed]
85. Clemetson, K.J.; Clemetson, J.M. Collagen receptors as potential targets for novel anti-platelet targets. *Curr. Pharm. Des.* **2007**, *13*, 2673–2683. [CrossRef] [PubMed]
86. Andrews, R.K.; Arthur, J.F.; Gardiner, E.E. Targeting GPVI as a novel antithrombotic strategy. *J. Blood Med.* **2014**, *5*, 59–68. [PubMed]
87. Kroll, M.H.; Hellums, J.D.; McIntire, L.V.; Schafer, A.I.; Moake, J.L. Platelets and shear stress. *Blood* **1996**, *88*, 1525–1541. [PubMed]
88. Kini, M.R.; Koh, C.Y. Metalloproteinases affecting blood coagulation, fibrinolysis and platelet aggregation from snake venoms: Definition and nomenclature of interaction sites. *Toxins* **2016**, *8*, 284. [CrossRef] [PubMed]
89. Huang, T.-F.; Chang, M.-C.; Teng, C.-M. Antiplatelet protease, kistomin, selectively cleaves human platelet glycoprotein Ib. *Biochim. Biopys. Acta* **1993**, *1158*, 293–299. [CrossRef]
90. Hsu, C.-C.; Wu, W.-B.; Chang, Y.-H.; Kuo, H.-L.; Huang, T.-F. Antithrombotic effect of a protein-type I class snake venom metalloproteinase, kistomin, is mediated by affecting glycoprotein Ib-von Willebrand factor interaction. *Mol. Pharmacol.* **2007**, *72*, 984–992. [CrossRef] [PubMed]
91. Vanhoorebeke, K.; Ulrichts, H.; Van de Walle, G.; Fontayne, A.; Deckmyn, H. Inhibition of glycoprotein Ib and its antithrombotic potential. *Curr. Pharm. Des.* **2007**, *13*, 2684–2697. [CrossRef]
92. Bonnefoy, A.; Vermylen, J.; Hoylaerts, M.F. Inhibition of von Willebrand factor-GPIb/IX/V interactions as a strategy to prevent arterial thrombosis. *Expert Rev. Cardiovasc. Ther.* **2003**, *1*, 257–269. [CrossRef] [PubMed]
93. Wu, W.-B.; Peng, H.-C.; Huang, T.-F. Crotalin, a vWF and GPIb cleaving metalloproteinase from venom of *Crotalus. atrox. Thromb. Haemost.* **2001**, *86*, 1501–1511. [PubMed]
94. Sanchez, E.F.; Richardson, M.; Gremski, L.H.; Veiga, S.S.; Yarleque, A.; Niland, S.; Lima, A.M.; Estevao-Costa, M.I.; Eble, J.A. Data for a direct fibrinolytic metalloproteinase, barnettlysin-I from *Bothrops barnetti* (barnett's pit viper) snake venom with anti-thrombotic effect. *Data Brief* **2016**, *7*, 1609–1613. [CrossRef] [PubMed]
95. Collen, D.; Lijnen, H.R. Recent developments in thrombolytic therapy. *Fibrinolysis Proteolysis* **2000**, *14*, 66–72. [CrossRef]
96. Jacob-Ferreira, A.L.; Menaldo, D.L.; Bernardes, C.P.; Sartim, M.A.; de Angelis, C.D.; Tanus-Santos, J.E.; Sampaio, S.V. Evaluation of the in vivo thrombolytic activity of a metalloprotease from *Botrops atrox* venom using a model of venous thrombosis. *Toxicon* **2016**, *109*, 18–25. [CrossRef] [PubMed]
97. Tooms, C.F. New directions in thrombolytic therapy. *Curr. Opin. Pharmacol.* **2001**, *1*, 164–168. [CrossRef]
98. Marder, V.J. Pre-clinical studies of plasmin: Superior benefit-to-risk ratio of plasmin compared to tissue plasminogen activator. *Thromb. Res.* **2008**, *122*, S9–S15. [CrossRef] [PubMed]
99. Swenson, S.; Markland, F.S. Snake venom fibrin(ogen)olytic enzymes. *Toxicon* **2005**, *45*, 1021–1039. [CrossRef] [PubMed]
100. Robbie, L.A.; Bennett, B.; Croll, A.M.; Brown, P.A.; Booth, N.A. Proteins of the fibrinolytic system in human thrombi. *Thromb. Haemost.* **1996**, *75*, 127–133. [PubMed]
101. Marder, V.J.; Landskroner, K.; Novokhatny, V.; Zimmerman, T.P.; Kong, M.; Kanouse, J.J.; Jesmok, G. Plasmin induces local thrombolysis without causing hemorrhage: A comparison with tissue type plasminogen activator in the rabbit. *J. Thromb. Haemost.* **2001**, *86*, 739–746.
102. Marder, V.J.; Jahan, R.; Gruber, T.; Goyal, A.; Arora, V. Thrombolysis with plasmin. *Stroke* **2010**, *41*, S41–S49. [CrossRef] [PubMed]
103. Kral, J.B.; Schrottmaier, W.C.; Salzmann, M.; Assinger, A. Platelet Interaction with Innate Immune Cells. *Transfus. Med. Hemother.* **2016**, *43*, 78–88. [CrossRef] [PubMed]
104. Lavergne, M.; Janus-Bell, E.; Schaff, M.; Gachet, C.; Mangin, P.H. Platelet Integrins in Tumor Metastasis: Do They Represent a Therapeutic Target? *Cancers* **2017**, *9*, 133. [CrossRef] [PubMed]
105. Gremski, L.H.; Veiga, S.S.; Sanchez, E.F. Leucuraolysin-A. In *Handbook of Proteolytic Enzymes*, 3nd ed.; Rawlings, N.D., Salvesen, G.S., Eds.; Publisher: London, UK, 2013; Volume 1, pp. 1013–1016, ISBN 978-0-12-382219-2.

© 2017 by the authors. Licensee MDPI, Basel, Switzerland. This article is an open access article distributed under the terms and conditions of the Creative Commons Attribution (CC BY) license (http://creativecommons.org/licenses/by/4.0/).

toxins

MDPI

Review

Targeting Metastasis with Snake Toxins: Molecular Mechanisms

Félix A. Urra [1,2,*] and Ramiro Araya-Maturana [3,*] ![ORCID]

[1]   Anatomy and Developmental Biology Program, Institute of Biomedical Sciences, University of Chile, Independencia 1027, Casilla 7, Santiago 7800003, Chile
[2]   Geroscience Center for Brain Health and Metabolism, Independencia 1027, Casilla 7, Santiago 7800003, Chile
[3]   Instituto de Química de Recursos Naturales, Universidad de Talca, Casilla 747, Talca 3460000, Chile
*   Correspondence: felix.urra@qf.uchile.cl (F.A.U.); raraya@utalca.cl (R.A.-M.); Tel.: +56-22-978-9580 (F.A.U.); +56-71-220-0285 (R.A.-M.)

Academic Editor: Steve Peigneur
Received: 2 November 2017; Accepted: 28 November 2017; Published: 30 November 2017

Abstract: Metastasis involves the migration of cancer cells from a primary tumor to invade and establish secondary tumors in distant organs, and it is the main cause for cancer-related deaths. Currently, the conventional cytostatic drugs target the proliferation of malignant cells, being ineffective in metastatic disease. This highlights the need to find new anti-metastatic drugs. Toxins isolated from snake venoms are a natural source of potentially useful molecular scaffolds to obtain agents with anti-migratory and anti-invasive effects in cancer cells. While there is greater evidence concerning the mechanisms of cell death induction of several snake toxin classes on cancer cells; only a reduced number of toxin classes have been reported (i.e., disintegrins/disintegrin-like proteins, C-type lectin-like proteins, C-type lectins, serinproteases, cardiotoxins, snake venom cystatins) as inhibitors of adhesion, migration, and invasion of cancer cells. Here, we discuss the anti-metastatic mechanisms of snake toxins, distinguishing three targets, which involve (1) inhibition of extracellular matrix components-dependent adhesion and migration, (2) inhibition of epithelial-mesenchymal transition, and (3) inhibition of migration by alterations in the actin/cytoskeleton network.

Keywords: anti-cancer agents; cancer cells; invasion; migrastatic drugs; snake venom

---

1. Introduction

Currently, anticancer therapies target the uncontrolled clonal proliferation of cancer cells with cytostatic drugs, which are an effective therapeutic strategy for certain cancer types such as hematological malignancies. However, in solid cancers, the proliferation is accompanied by the ability to invade and execute metastasis, involving different molecular mechanisms that are not inhibited or affected by conventional anti-cancer drugs. Therefore, to search for and design specific drugs to inhibit invasion and metastasis for treatment of solid cancers is a highly relevant issue [1].

The composition of solid tumors is heterogeneous, having several cancer cell subpopulations with different tumorigenic properties [2]. In a tumor, cancer cells acquire mutations that confer them with different proliferative capacities and survival advantages. A subpopulation, named metastasis-initiating cells (MICs), exhibits high plasticity to adapt their metabolic and proliferative requirements, ability to enter and exit dormancy state, and resistance to apoptosis and immune evasion, which is responsible for metastatic growth [3]. For example, during the initial steps of tumor growth of cancer cells confined to epithelium, certain colonies of malignant cells can form a carcinoma in situ separated from the stroma. In some cells, mutations provide the ability to establish a physical relationship with stroma and changes in extracellular signals from the microenvironment, triggering the secretion of soluble factors by stromal and hematopoietic cells [4,5] and inducing phenotypic changes in cancer cells known as epithelial–mesenchymal transition (EMT). This process recapitulates properties displayed by tissues during the embrionary development [6] that facilitate the dissociation of cancer cell from the tumor bulk and dissemination to distant organs, being considered a prerequisite for invasion and metastasis [2,7].

Metastasis is a complex process in which cancer cells disseminate from a primary tumor to invade a distant organ, this ability characterizes the tumor malignancy [6]. It has been described that about 90% of cancer-related deaths are caused by a metastatic disease [8]. It is clear that dissemination to specific organs depends upon blood flow patterns and of the relationship of the migrating cells with distant organ microenvironments, the stromal cell content, vascular architecture, presence of growth factors, metabolic substrates, and signaling molecules. These characteristics can be permissive or antagonistic to metastatic colonization, determining whether these cells grow to form secondary tumors [9].

The detailed mechanistic insight of the metastatic process contrast with the minimal progress in the identification of effective therapeutic targets and in the design of new anti-metastatic drugs [1]. Based on structural characteristics and their known interactions with macromolecules, toxins isolated from snake venoms may represent a natural source of molecular scaffolds to obtain agents with anti-migratory and anti-invasive effects in cancer cells. In this review, we summarize recent evidence on the inhibitory effect of snake toxins on adhesion, migration, and invasion of cancer cells.

2. Snake Toxins as Inhibitors of Cancer Metastasis

There is ample literature showing that several isolated or recombinant snake venom toxins exhibit anti-cancer effects in vitro and in vivo preclinical models, inducing cell death via mitochondrial apoptotic pathway (intrinsic pathway) or necrosis [10–12]. In addition, certain toxins such as snake venom metalloproteases (SVMPs), disintegrins, phospholipases A2, C-type lectins (CLP), vascular apoptosis inducing proteins, and L-amino acid oxidases are able to inhibit angiogenesis [13–15] and activate the immune response during tumorigenesis [16]. While greater evidence on mechanisms of death induction of snake toxins on cancer cells have been reported, reduced information on the inhibitory mechanisms of adhesion, migration, and invasion of metastatic cancer cells is available. Despite the aforementioned information, it is possible distinguish three anti-metastatic mechanisms exhibited by at least six different snake toxin classes (Figure 1): involving (1) inhibition of extracellular matrix components (ECM)-dependent adhesion and migration, (2) inhibition of epithelial-mesenchymal transition, and (3) inhibition of migration by alterations in the actin/cytoskeleton network.

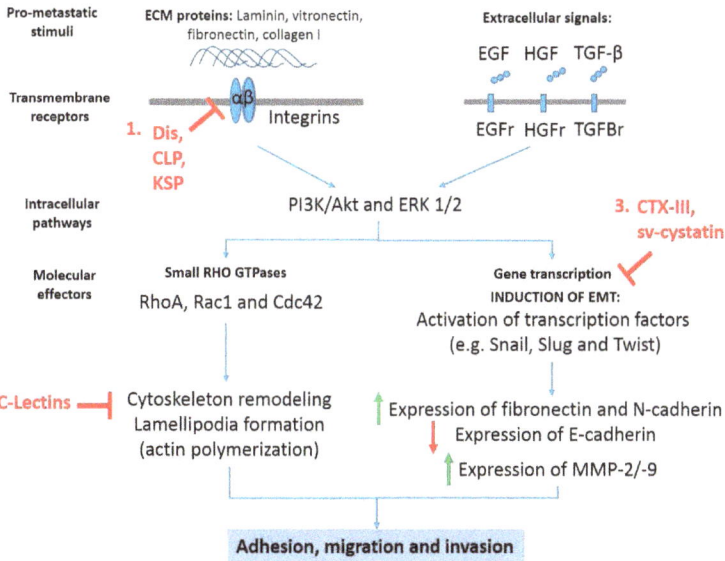

Figure 1. Anti-metastatic targets for snake toxins. Snake toxins inhibit pro-migratory and pro-invasive signals stimulated by extracellular matrix proteins and growth factors such as epidermal growth factor (EGF), hepatocyte growth factor (HGF), and transforming growth factor beta (TGF-β) though (1) inhibition of extracellular matrix (ECM) components-dependent adhesion and migration, (2) inhibition of migration by alterations in the actin/cytoskeleton network, and (3) inhibition of epithelial–mesenchymal transition (EMT). Dis: disintegrins; CLP: C-type lectin-like proteins; KSP: Kunitz-type serinprotease; C-Lectins: C-type lectins; CTX-III: cardiotoxin III; Sv-cystatin: snake venom cystatin.

3. Inhibition of Extracellular Matrix Component-Dependent Adhesion and Migration

During the initial steps of metastasis, it is required the interaction between ECM components and cancer cell, involving the ability of these cells to adhere to ECM components and migrate through them [17]. Integrins are the major receptor family present on the cell surface for adhesion to the ECM and include heterodimeric, transmembrane glycoproteins composed of α and β subunits [18], whose dimerization leads to 24 integrin pairs with distinct extracellular ligand-binding specificities [18]—such as collagen, laminin, vitronectin, and fibronectin—through the tripeptide motif Arg-Gly-Asp = RGD [19]. Abundant evidence has correlated the increased overexpression of certain integrins αvβ3, α5β1, and αvβ6 with cancer progression [20–22]. Integrins activate intracellular signaling that control cytoskeleton organization, cell polarity, and formation of leading edge of migrating cancer cells [22], being an attractive anti-cancer target for new antagonist molecules [23,24].

Three toxin classes (snake venom disintegrins, C-type lectin-like protein, and Kunit-like serinprotease inhibitor) have been reported with anti-migratory effect mediated by interaction with integrins in cancer cells, which are summarized in Table 1.

Snake venom disintegrins are small non-enzymatic proteins mostly derived from proteolytic processing of precursors that contain a metalloprotease domain, known as snake venom metalloproteases (SVMPs), which are phylogenetically related with ADAMs (a disintegrin and metalloprotease) [25–28]. This protein family, commonly found in the venoms of the Viperidae snakes [26] and some rear-fanged snakes [28–35], is classified according to their modular architecture with multiple non-catalytic domains in SVMP P-I, P-II, and P-III classes. Disintegrins are derived from proteolytic processing of P-II SMVP class and usually exhibit the canonical "RDG" integrin-recognition motif; however, non-canonical integrin-binding motif—such as "MLD", "KTS", and "VGD"—are exhibited in some snake venom disintegrins [36,37]. In addition, proteolysis from P-III SVMP class originates disintegrin-like proteins, which have covalently bound the "disintegrin-like" and "cysteine (Cys)-rich" domains [27]. Comprehensive classification and structural characteristics of SVMP are found in Takeda et al., 2012 [27] and Takeda, 2016 [38].

A disintegrin isolated from the venom of the Middle American rattlesnake (*Crotalus simus tzabcan*) named tzabcanin [39], which has 71 amino acids and contains the canonical RGD-binding domain, exhibits a weak or null cytotoxic effect on cancer cell lines [39], but remarkable inhibitory effect of fibronectin- and vitronectin-dependent cell adhesion. This toxin binds $\alpha v \beta 3$-integrins, which is the main receptor of the ECM protein vitronectin, inhibiting the adhesion and migration of melanoma and lung cancer cells [40].

Table 1. Snake toxins that inhibit the adhesion and migration of cancer cells by interaction with ECM components.

Toxin Name	Snake Species	Adhesive Motif	Integrin Target	ECM Ligand	Effect	Ref.
r-Cam-dis recombinant disintegrin	*Crotalus adamanteus*	RGD	$\alpha v \beta 3$	laminin-1	Inhibition of adhesion in pancreatic cancer cells	[41]
r-Colombistatins recombinant disintegrin-like domains from Class-III SVMP	*Bothrops colombiensis*	ECD	n.d.	collagen I	Inhibition of adhesion in SK-Mel-28 melanoma cells	[42]
DisBa-01, recombinant disintegrin	*Bothrops alternatus*	RGD	$\alpha v \beta 3$	fibronectin	Loss of cell directionality of migrating oral squamous carcinoma cells	[43]
r-mojastn-1, recombinant disintegrin	*Crotalus scutulatus scutulatus*	RGD	$\alpha v \beta 3$, $\alpha 3$, and $\beta 1$,	fibronectin and vitronectin	Inhibition of adhesion and migration of BXPC-3 pancreatic cancer cell line	[44,45]
r-viridistatin-2, recombinant disintigrin	*Crotalus viridis viridis*	RGD	$\alpha v \beta 3$	fibronectin and vitronectin	Inhibition of adhesion, migration and invasion of several cancer cell lines	[44,46]
Lebecin, C-type lectin-like protein	*Macrovipera lebetina*	-	$\alpha v \beta 3$	fibronectin and fibrinogen	Inhibition of adhesion and migration of MDA-MB-231 breast cancer cells	[47]
PIVL, Kunitz-type serin protease inhibitor	*Macrovipera lebetina transmediterranea*	RGN	$\alpha v \beta 3$	fibronectin and fibrinogen	Inhibition of adhesion, migration and invasion of human glioblastoma U87 cells	[48]

n.d.: not determined.

DisBa-01, a recombinant RGD-disintegrin produced from a cDNA venom gland library of *Bothrops alternatus*, inhibits in vivo angiogenesis and pulmonary metastasis [49]. In oral squamous carcinoma cells, DisBa-01 selectively decreases the migration speed and directionality of fibronectin-stimulated migration, increasing the adhesion area and rate of adhesion maturation. It lacks effects on migration of non-malignant cells such as fibroblasts. DisBa-01 exhibits a high affinity on fibronectin binding receptor $\alpha v \beta 3$ integrin [43]. Other recombinant disintegrins from Viperidae species have been reported such as $\alpha v \beta 3$ integrin antagonists, inhibiting the migration of cancer cells (Table 1). Additional disintegrins and disintegrin-like proteins from snake venoms reported with anti-cancer effect can be found in Selistre-de-Araujo et al., 2010 [50].

Interestingly, Lebecin, and PIVL isolated from *Macrovipera lebetina* venom, which belong two different toxin classes C-type lectin-like protein and Kunitz-type serin protease inhibitor, respectively, exhibit inhibitory effect on fibrinogen- and fibronectin-stimulated adhesion and migration.

Lebecin is a C-type lectin-like protein with α and β subunits of 129 and 131 amino acids, respectively [47]. In triple-negative breast cancer MDA-MB-231 cells, lebecin does not affect the viability. However, it inhibits the fibrinogen- and fibronectin-dependent adhesion and migration in a dose-dependent manner [47]. It has been described that lebecin interacts with αvβ3 integrin; but based on the high identity of its amino acid sequence with other C-type lectin-like protein previously reported from *Macrovipera lebetina* venom with inhibitory effect on adhesion, migration, and invasion of cancer cells [51,52], it has been suggested that lebecin can block other integrins such as α5β1 [47].

PIVL is a monomeric polypeptide chain bound by three disulfide linkages, which inhibits trypsin activity and lacks effects on the viability but blocks αvβ3 integrin-dependent migration, affecting the motility and cell directionality persistence of cancer cells [48]. PIVL also exhibits in vitro and in vivo anti-angiogenic effects [53].

4. Inhibition of Epithelial–Mesenchymal Transition

Epithelial–mesenchymal transition (EMT) is a process in which epithelial cells transdifferentiate into mesenchymal cells, losing their morphoinmunophenotypic characteristics. Interestingly, EMT occurs in normal and healthy tissues during angiogenesis and lymphangiogenesis; but in certain pathological conditions such as chronic inflammation, fibrosis and cancer is reactivated [6]. In tumors, EMT-like transitions involve the loss of components related with cell-cell interactions, apico-basal cell polarity and reorganization of cytoskeleton. Cancer cells with EMT have tumorigenic properties that non-EMT cells do not exhibit, such as a high migratory state that promote invasion and metastasis [4,5], lacking response to signals of oncogene-induced senescence [54] and resistance to anti-cancer drugs [55–57].

EMT can be induced by growth factors such as transforming growth factor beta (TGF-β), epidermal growth factor (EGF), hepatocyte growth factor (HGF), insulin-like growth factors 1 and 2 [40], activating RAS, Notch, and Wnt signalings which have been associated with poor prognosis and cancer progression [58,59]. During EMT, there is a reduction of the epithelial marker E-cadherin and an increase of the expression of mesenchymal markers vimentin, N-cadherin [60], as well as activation of transcription factors Snail, Slug, Twist, which act as repressor of E-cadherin [5,61].

Cardiotoxin III (CTX-III), a membrane toxin from Taiwan cobra (*Naja naja*) venom [62], inhibits the migration of cancer cells by reversion of EGF- and HGF-induced EMT. Previously, CTX-III has been described as a potent inductor of cell death in several human cancer cell lines [63–65] and a migration inhibitor of oral and breast cancer cells through activation of JNK and p38, without effect on ERK signaling, producing decreased metalloproteases-2 and -9 (MMP-2/-9) levels [66,67].

In breast cancer cells, the paracrine role of epidermal growth factor (EGF) and its receptor EGFR (ErbB-1) contribute to invasion, intravasation, and metastasis [68] through activation of extracellular signal-regulated kinase 1/2 (ERK1/2), STAT3, or PI3K/Akt signaling, promoting the EMT [69–71]. CTX-III inhibits the EGF-induced EMT in breast cancer cells, reducing EGFR phosphorylation and activation PI3K/Akt and ERK1/2. It reduces the MMP-9 levels [72] and the mesenchymal markers vimentin and N-cadherin and increases E-cadherin levels, inhibiting EGF-induced invasion and migration [72,73]. A similar effect of CTX-III on hepatocyte growth factor (HGF)-stimulated migration and invasion in breast cancer cells has been described [73–75].

Cancer cells can excrete cysteine-cathepsins, which are endopeptidases located intracellularly in endolysosomal vesicles [76] that are essential during the breakdown the ECM to promote the invasion and metastasis [77]. During EMT, cancer cells exhibit an increased extra- and intra-cellular proteolysis mediated by cathepsins, matrix metalloproteinases, urokinase-type plasminogen activator (uPA), and serinproteases such as kallikreins [78]. This proteolytic activity removes surface molecules involved in cell adhesion such as E-cadherin [79,80], limiting the cell–cell interaction and remodeling the extracellular matrix to uncover binding epitopes recognized by integrins and to form trials for cell migration [81]. Cysteine-cathepsins are regulated by natural inhibitors such as cystatins [82], which represent a group constituted by three types (type 1-stenfins, type-2 cystatins, type 3-kininogens) of cystatin domain containing proteins [83]. From *Naja naja atra* venom, it has been isolated a snake venom cystatin (Sv-cystatin) that exhibits a shorter sequence than other type-2 cystatins, such as cystatin M and cystatin C [84]. For this snake toxin, inhibitory effects on invasion and metastasis mediated by reduction of EMT markers has been described in MHCC97H liver cancer cells [85]. Sv-cystatin decreases the cathepsin B activity, MMP-2, and MMP-9 levels, increasing E-cadherin and decreasing EMT proteins N-cadherin and twist [85].

5. Alterations in the Actin/Cytoskeleton Network

During migration and invasion of cancer cells, the actin cytoskeleton is remodeled under extracellular stimuli, which is mediated by several receptors, including integrins [19]. Small GTPases Rho, Rac, and Cdc42 participate in the intracellular signaling involved in the control of the actin cytoskeleton architecture required for cell motility in individual and collective migration [86], which is a common signaling for normal and cancer cells [2]. The cell protrusion of a leading edge relies on Cdc42 and Rac activities, which are coupled to Rho activity-dependent contractility, supporting the movement of the cell body forward [87]. Consistent with the essential role of the cytoskeleton in promoting cancer migration, its deregulation may cause anti-adhesive and anti-migratory effects. Two snake venom calcium-dependent (C-type) lectins alter the actin/cytoskeleton network in cancer cells. C-type lectins identified from snake venoms are classified in two groups: C-type glycan-binding lectins; and C-type lectin-like proteins, which do not interact with sugars. The C-type glycan-binding lectins are homodimeric non-enzymatic proteins that contain a carbohydrate recognition domain (CRD), binding mainly with galactose [88].

Daboialectin, a low molecular weight C-type lectin isolated from *Daboia russelii* venom, produces morphological changes, including spindle-like shape with loss of cell–cell contacts in lung cancer cells A549 [89]. This snake toxin decreases the mRNA and protein levels of small GTPases Rho and Rac and increases the Cdc42 expression, which is in accordance with remarkable decrease of F-actin content, inhibition of migration and invasion observed in lung cancer cells treated with it [89].

BJcuL is a C-type lectin from *Bothrops jararacussu* venom composed by a disulfide-linked dimer with high affinity for glycoproteins containing β-D-galactosides [90]. BJcuL binds to cancer cells without affecting the adhesion of these cells to fibronectin, laminin, and type I collagen; however, it produces complete actin filament disorganization and disassembly in malignant cells [91]. This toxin does not block the integrin signaling [92], but it binds to cell surface with ECM glycoproteins, such as its substrate D-galactose, promoting the actin disassembles, an event that could accelerate cancer cell detachment from ECM, producing cell death [91].

6. Concluding Remarks

Given that malignant cells during metastasis exhibit molecular mechanisms different from those shown by non-metastatic and highly proliferative cancer cells, the conventional cytostatic drugs, which mainly target the cell proliferation, lack effects on the capacity to disseminate and grow in distant sites of metastatic cancer cells. This review highlights the need to search new anti-metastatic drugs. We identified three anti-metastatic mechanisms of action for at least six classes of toxins from snake venoms: (1) inhibition of ECM components-dependent adhesion and migration, (2) inhibition of EMT, and (3) inhibition of migration by alterations in the actin/cytoskeleton network.

These toxins may represent a natural source of molecular scaffolds to design new anti-migratory and anti-invasive agents by obtaining recombinant proteins or small molecules that act as antagonists of integrin signaling or inductors of actin disassembling by binding of cell surface glycoproteins. A selective inhibition of the signaling machinery involved in the cancer cell migration without affect those of migrating non-malignant cells is an important challenge for the new anti-metastatic drugs.

Interestingly, all anti-cancer evaluations on tumorigenic properties—such as proliferation, angiogenesis, invasion, and metastasis of malignant cells—have been performed with toxins isolated from front-fanged snake species, especially from *Viperidae* species; however, the potential therapeutic applications of toxins described from rear-fanged snake species—e.g., [28,30–32,93–95]—remain unexplored.

An extensive development and conjugation of drug delivery systems with some snake toxins, which has reduced the toxicity and improved the selectivity toward cancer cells [96,97], highlight their promising applications as direct anti-cancer agents or potential tools for the development of novel therapeutic strategies [16]. Finally, the in vivo validation of anti-metastatic effect described on in vitro cancer cell lines is a pending issue for drug discovery from snake toxins.

Acknowledgments: This work was supported by FONDECYT grant #1140753 (R.A.-M.), Programa de Investigación Asociativa en Cáncer Gástrico (PIA-CG, RU2107) (R.A.-M.) and FONDECYT postdoctoral fellowship #3170813 (F.A.U.).

Author Contributions: F.A.U. and R.A.-M. designed and contributed to the literature review, discussion, and writing of the manuscript.

Conflicts of Interest: The authors declare no conflict of interest.

References

1. Gandalovičová, A.; Rosel, D.; Fernandes, M.; Veselý, P.; Heneberg, P.; Čermák, V.; Petruželka, L.; Kumar, S.; Sanz-Moreno, V.; Brábek, J. Migrastatics-Anti-metastatic and anti-invasion drugs: Promises and challenges. *Trends Cancer* **2017**, *3*, 391–406. [CrossRef] [PubMed]

2. Riggi, N.; Aguet, M.; Stamenkovic, I. Cancer metastasis: A reappraisal of its underlying mechanisms and their relevance to treatment. *Annu. Rev. Pathol.* **2017**. [CrossRef] [PubMed]

3. Celià-Terrassa, T.; Kang, Y. Distinctive properties of metastasis-initiating cells. *Genes Dev.* **2016**, *30*, 892–908. [CrossRef] [PubMed]

4. Tomaskovic-Crook, E.; Thompson, E.; Thiery, J. Epithelial to mesenchymal transition and breast cancer. *Breast Cancer Res.* **2009**, *11*, 213. [CrossRef] [PubMed]

5. Lamouille, S.; Xu, J.; Derynck, R. Molecular mechanisms of epithelial-mesenchymal transition. *Nat. Rev. Mol. Cell Biol.* **2014**, *15*, 178–196. [CrossRef] [PubMed]

6. Karlsson, M.; Gonzalez, S.; Welin, J.; Fuxe, J. Epithelial-mesenchymal transition in cancer metastasis through the lymphatic system. *Mol. Oncol.* **2017**, *11*, 781–791. [CrossRef] [PubMed]

7. Lambert, A.; Pattabiraman, D.; Weinberg, R. Emerging biological principles of metastasis. *Cell* **2017**, *168*, 670–691. [CrossRef] [PubMed]

8. Sleeman, J.; Steeg, P. Cancer metastasis as a therapeutic target. *Eur. J. Cancer* **2010**, *46*, 1177–1180. [CrossRef] [PubMed]

9. Heath, A.S.; Yibin, K. Determinants of organotropic metastasis. *Annu. Rev. Cancer Biol.* **2017**, *1*, 403–423.

10. Aranda-Souza, M.; Rossato, F.; Costa, R.; Figueira, T.; Castilho, R.; Guarniere, M.; Nunes, E.; Coelho, L.; Correia, M.; Vercesi, A. A lectin from Bothrops leucurus snake venom raises cytosolic calcium levels and promotes B16-F10 melanoma necrotic cell death via mitochondrial permeability transition. *Toxicon* **2014**, *82*, 97–103. [CrossRef] [PubMed]

11. Ebrahim, K.; Shirazi, F.; Mirakabadi, A.; Vatanpour, H. Cobra venom cytotoxins; apoptotic or necrotic agents? *Toxicon* **2015**, *108*, 134–140. [CrossRef] [PubMed]

12. Prinholato da Silva, C.; Costa, T.; Paiva, R.; Cintra, A.; Menaldo, D.; Antunes, L.; Sampaio, S. Antitumor potential of the myotoxin BthTX-I from Bothrops jararacussu snake venom: Evaluation of cell cycle alterations and death mechanisms induced in tumor cell lines. *J. Venom. Anim. Toxins Incl. Trop. Dis.* **2015**, *21*. [CrossRef] [PubMed]

13. Guimarães, D.; Lopes, D.; Azevedo, F.; Gimenes, S.; Silva, M.; Aché, D.; Gomes, M.; Vecchi, L.; Goulart, L.; Yoneyama, K.; et al. In vitro antitumor and antiangiogenic effects of Bothropoidin, a metalloproteinase from *Bothrops pauloensis* snake venom. *Int. J. Biol. Macromol.* **2017**, *97*, 770–777. [CrossRef] [PubMed]

14. Dhananjaya, B.; Sivashankari, P. Snake venom derived molecules in tumor angiogenesis and its application in cancer therapy; an overview. *Curr. Top. Med. Chem.* **2015**, *15*, 649–657. [CrossRef] [PubMed]

15. Azevedo, F.; Lopes, D.; Cirilo-Gimenes, S.; Aché, D.; Vecchi, L.; Alves, P.; de Oliveira Guimarães, D.; Rodrigues, R.; Goulart, L.; de Melo Rodrigues, V.; et al. Human breast cancer cell death induced by BnSP-6, a Lys-49 PLA homologue from *Bothrops pauloensis* venom. *Int. J. Biol. Macromol.* **2016**, *82*, 671–677. [CrossRef] [PubMed]

16. Costa, T.; Menaldo, D.; Zoccal, K.; Burin, S.; Aissa, A.; Castro, F.; Faccioli, L.; Greggi-Antunes, L.; Sampaio, S. CR-LAAO, an L-amino acid oxidase from *Calloselasma rhodostoma* venom, as a potential tool for developing novel immunotherapeutic strategies against cancer. *Sci. Rep.* **2017**, *7*, 42673. [CrossRef] [PubMed]

17. Bartsch, J.; Staren, E.; Appert, H. Adhesion and migration of extracellular matrix-stimulated breast cancer. *J. Surg. Res.* **2003**, *110*, 287–294. [CrossRef]

18. Anderson, L.; Owens, T.; Naylor, M. Structural and mechanical functions of integrins. *Biophys. Rev.* **2014**, *6*, 203–213. [CrossRef] [PubMed]

19. Longmate, W.; DiPersio, C. Beyond adhesion: Emerging roles for integrins in control of the tumor microenvironment. *F1000Research* **2017**, *6*. [CrossRef] [PubMed]

20. Kwakwa, K.; Sterling, J. Integrin αvβ3 signaling in tumor-induced bone disease. *Cancers (Basel)* **2017**, *9*, 84. [CrossRef] [PubMed]

21. Niu, J.; Li, Z. The roles of integrin αvβ6 in cancer. *Cancer Lett.* **2017**, *403*, 128–137. [CrossRef] [PubMed]

22. Rathinam, R.; Alahari, S. Important role of integrins in the cancer biology. *Cancer Metastasis Rev.* **2010**, *29*, 223–237. [CrossRef] [PubMed]

23. Marelli, U.; Rechenmacher, F.; Sobahi, T.; Mas-Moruno, C.; Kessler, H. Tumor Targeting via Integrin Ligands. *Front. Oncol.* **2013**, *3*, 222. [CrossRef] [PubMed]

24. Kapp, T.; Rechenmacher, F.; Neubauer, S.; Maltsev, O.; Cavalcanti-Adam, E.; Zarka, R.; Reuning, U.; Notni, J.; Wester, H.; Mas-Moruno, C.; et al. A Comprehensive Evaluation of the Activity and Selectivity Profile of Ligands for RGD-binding Integrins. *Sci. Rep.* **2017**, *7*, 39805. [CrossRef] [PubMed]

25. Moura-da-Silva, A.; Theakston, R.; Crampton, J. Evolution of disintegrin cysteine-rich and mammalian matrix-degrading metalloproteinases: Gene duplication and divergence of a common ancestor rather than convergent evolution. *J. Mol. Evol.* **1996**, *43*, 263–269. [CrossRef] [PubMed]

26. Calvete, J.; Marcinkiewicz, C.; Monleón, D.; Esteve, V.; Celda, B.; Juárez, P.; Sanz, L. Snake venom disintegrins: Evolution of structure and function. *Toxicon* **2005**, *45*, 1063–1074. [CrossRef] [PubMed]

27. Takeda, S.; Takeya, H.; Iwanaga, S. Snake venom metalloproteinases: Structure, function and relevance to the mammalian ADAM/ADAMTS family proteins. *Biochim. Biophys. Acta* **2012**, *1824*, 164–176. [CrossRef] [PubMed]

28. Weldon, C.; Mackessy, S. Alsophinase, a new P-III metalloproteinase with α-fibrinogenolytic and hemorrhagic activity from the venom of the rear-fanged Puerto Rican Racer *Alsophis portoricensis* (Serpentes: Dipsadidae). *Biochimie* **2012**, *94*, 1189–1198. [CrossRef] [PubMed]

29. Peichoto, M.E.; Paes Leme, A.F.; Pauletti, B.A.; Batista, I.C.; Mackessy, S.P.; Acosta, O.; Santoro, M.L. Autolysis at the disintegrin domain of patagonfibrase, a metalloproteinase from *Philodryas patagoniensis* (Patagonia Green Racer; Dipsadidae) venom. *BBA Proteins Proteom.* **2010**, *1804*, 1937–1942. [CrossRef] [PubMed]

30. Ching, A.T.; Paes Leme, A.F.; Zelanis, A.; Rocha, M.M.; Furtado, M.D.F.; Silva, D.A.; Trugilho, M.R.; da Rocha, S.L.; Perales, J.; Ho, P.L.; et al. Venomics profiling of *Thamnodynastes strigatus* unveils matrix metalloproteinases and other novel proteins recruited to the toxin arsenal of rear-fanged snakes. *J. Proteome Res.* **2012**, *11*, 1152–1162. [CrossRef] [PubMed]

31. Ching, A.T.; Rocha, M.M.; Paes Leme, A.F.; Pimenta, D.C.; de Fatima, D.F.M.; Serrano, S.M.; Ho, P.L.; Junqueira-de-Azevedo, I.L. Some aspects of the venom proteome of the Colubridae snake *Philodryas olfersii* revealed from a Duvernoy's (venom) gland transcriptome. *FEBS Lett.* **2006**, *580*, 4417–4422. [CrossRef] [PubMed]

32. Urra, F.; Pulgar, R.; Gutiérrez, R.; Hodar, C.; Cambiazo, V.; Labra, A. Identification and molecular characterization of five putative toxins from the venom gland of the snake Philodryas chamissonis (Serpentes: Dipsadidae). *Toxicon* **2015**, *108*, 19–31. [CrossRef] [PubMed]

33. Kamiguti, A.; Theakston, R.; Sherman, N.; Fox, J. Mass spectrophotometric evidence for P-III/P-IV metalloproteinases in the venom of the Boomslang (*Dispholidus typus*). *Toxicon* **2000**, *38*, 1613–1620. [CrossRef]

34. Zhang, Z.; Zhang, X.; Hu, T.; Zhou, W.; Cui, Q.; Tian, J.; Zheng, Y.; Fan, Q. Discovery of toxin-encoding genes from the false viper *Macropisthodon rudis*, a rear-fanged snake, by transcriptome analysis of venom gland. *Toxicon* **2015**, *106*, 72–78. [CrossRef] [PubMed]

35. McGivern, J.; Wray, K.; Margres, M.; Couch, M.; Mackessy, S.; Rokyta, D. RNA-seq and high-definition mass spectrometry reveal the complex and divergent venoms of two rear-fanged colubrid snakes. *BMC Genom.* **2014**, *15*, 1061. [CrossRef] [PubMed]

36. Walsh, E.; Marcinkiewicz, C. Non-RGD-containing snake venom disintegrins, functional and structural relations. *Toxicon* **2011**, *58*, 355–362. [CrossRef] [PubMed]

37. Calvete, J. The continuing saga of snake venom disintegrins. *Toxicon* **2013**, *62*, 40–49. [CrossRef] [PubMed]

38. Takeda, S. ADAM and ADAMTS family proteins and snake venom metalloproteinases: A structural overview. *Toxins (Basel)* **2016**, *8*, E155. [CrossRef] [PubMed]

39. Saviola, A.; Modahl, C.; Mackessy, S. Disintegrins of *Crotalus simus* tzabcan venom: Isolation, characterization and evaluation of the cytotoxic and anti-adhesion activities of tzabcanin, a new RGD disintegrin. *Biochimie* **2015**, *116*, 92–102. [CrossRef] [PubMed]

40. Saviola, A.; Burns, P.; Mukherjee, A.; Mackessy, S. The disintegrin tzabcanin inhibits adhesion and migration in melanoma and lung cancer cells. *Int. J. Biol. Macromol.* **2016**, *88*, 457–464. [CrossRef] [PubMed]

41. Suntravat, M.; Barret, H.; Jurica, C.; Lucena, S.; Perez, J.; Sánchez, E. Recombinant disintegrin (r-Cam-dis) from Crotalus adamanteus inhibits adhesion of human pancreatic cancer cell lines to laminin-1 and vitronectin. *J. Venom Res.* **2015**, *6*, 1–10. [PubMed]

42. Suntravat, M.; Helmke, T.; Atphaisit, C.; Cuevas, E.; Lucena, S.; Uzcátegui, N.; Sánchez, E.; Rodriguez-Acosta, A. Expression, purification, and analysis of three recombinant ECD disintegrins (r-colombistatins) from P-III class snake venom metalloproteinases affecting platelet aggregation and SK-MEL-28 cell adhesion. *Toxicon* **2016**, *122*, 43–49. [CrossRef] [PubMed]

43. Montenegro, C.; Casali, B.; Lino, R.; Pachane, B.; Santos, P.; Horwitz, A.; Selistre-de-Araujo, H.; Lamers, M. Inhibition of αvβ3 integrin induces loss of cell directionality of oral squamous carcinoma cells (OSCC). *PLoS ONE* **2017**, *12*, e0176226. [CrossRef] [PubMed]

44. Lucena, S.; Castro, R.; Lundin, C.; Hofstetter, A.; Alaniz, A.; Suntravat, M.; Sánchez, E. Inhibition of pancreatic tumoral cells by snake venom disintegrins. *Toxicon* **2015**, *93*, 136–143. [CrossRef] [PubMed]

45. Lucena, S.; Sanchez, E.; Perez, J. Anti-metastatic activity of the recombinant disintegrin, r-mojastin 1, from the Mohave rattlesnake. *Toxicon* **2011**, *57*, 794–802. [CrossRef] [PubMed]

46. Lucena, S.; Jia, Y.; Soto, J.; Parral, J.; Cantu, E.; Brannon, J.; Lardner, K.; Ramos, C.; Seoane, A.; Sánchez, E. Anti-invasive and anti-adhesive activities of a recombinant disintegrin, r-viridistatin 2, derived from the Prairie rattlesnake (*Crotalus viridis viridis*). *Toxicon* **2012**, *60*, 31–39. [CrossRef] [PubMed]

47. Jebali, J.; Fakhfekh, E.; Morgen, M.; Srairi-Abid, N.; Majdoub, H.; Gargouri, A.; El Ayeb, M.; Luis, J.; Marrakchi, N.; Sarray, S. Lebecin, a new C-type lectin like protein from *Macrovipera lebetina* venom with anti-tumor activity against the breast cancer cell line MDA-MB231. *Toxicon* **2014**, *86*, 16–27. [CrossRef] [PubMed]

48. Morjen, M.; Kallech-Ziri, O.; Bazaa, A.; Othman, H.; Mabrouk, K.; Zouari-Kessentini, R.; Sanz, L.; Calvete, J.; Srairi-Abid, N.; El Ayeb, M.; et al. PIVL, a new serine protease inhibitor from *Macrovipera lebetina transmediterranea* venom, impairs motility of human glioblastoma cells. *Matrix Biol.* **2013**, *32*, 52–62. [CrossRef] [PubMed]

49. Ramos, O.; Kauskot, A.; Cominetti, M.; Bechyne, I.; Salla-Pontes, C.; Chareyre, F.; Manent, J.; Vassy, R.; Giovannini, M.; Legrand, C.; et al. A novel alpha(v)beta (3)-blocking disintegrin containing the RGD motive, DisBa-01, inhibits bFGF-induced angiogenesis and melanoma metastasis. *Clin. Exp. Metastasis* **2008**, *25*, 53–64. [CrossRef] [PubMed]

50. Selistre-de-Araujo, H.; Pontes, C.; Montenegro, C.; Martin, A. Snake venom disintegrins and cell migration. *Toxins (Basel)* **2010**, *2*, 2606–2621. [CrossRef] [PubMed]

51. Sarray, S.; Berthet, V.; Calvete, J.; Secchi, J.; Marvaldi, J.; El-Ayeb, M.; Marrakchi, N.; Luis, J. Lebectin, a novel C-type lectin from *Macrovipera lebetina* venom, inhibits integrin-mediated adhesion, migration and invasion of human tumour cells. *Lab. Investig.* **2004**, *84*, 573–581. [CrossRef] [PubMed]

52. Sarray, S.; Delamarre, E.; Marvaldi, J.; El Ayeb, M.; Marrakchi, N.; Luis, J. Lebectin and lebecetin, two C-type lectins from snake venom, inhibit alpha5beta1 and alphaV-containing integrins. *Matrix Biol.* **2007**, *26*, 306–313. [CrossRef] [PubMed]

53. Morjen, M.; Honoré, S.; Bazaa, A.; Abdelkafi-Koubaa, Z.; Ellafi, A.; Mabrouk, K.; Kovacic, H.; El Ayeb, M.; Marrakchi, N.; Luis, J. PIVL, a snake venom Kunitz-type serine protease inhibitor, inhibits in vitro and in vivo angiogenesis. *Microvasc. Res.* **2014**, *95*, 149–156. [CrossRef] [PubMed]

54. Ansieau, S.; Bastid, J.; Doreau, A.; Morel, A.; Bouchet, B.; Thomas, C.; Fauvet, F.; Puisieux, I.; Doglioni, C.; Piccinin, S.; et al. Induction of EMT by twist proteins as a collateral effect of tumor-promoting inactivation of premature senescence. *Cancer Cell* **2008**, *14*, 79–89. [CrossRef] [PubMed]

55. Fischer, K.; Durrans, A.; Lee, S.; Sheng, J.; Li, F.; Wong, S.; Choi, H.; El Rayes, T.; Ryu, S.; Troeger, J.; et al. Epithelial-to-mesenchymal transition is not required for lung metastasis but contributes to chemoresistance. *Nature* **2015**, *527*, 472–476. [CrossRef] [PubMed]

56. Kajiyama, H.; Shibata, K.; Terauchi, M.; Yamashita, M.; Ino, K.; Nawa, A.; Kikkawa, F. Chemoresistance to paclitaxel induces epithelial-mesenchymal transition and enhances metastatic potential for epithelial ovarian carcinoma cells. *Int. J. Oncol.* **2007**, *31*, 277–283. [CrossRef] [PubMed]

57. Shibue, T.; Weinberg, R. EMT, CSCs, and drug resistance: The mechanistic link and clinical implications. *Nat. Rev. Clin. Oncol.* **2017**, *14*, 611–629. [CrossRef] [PubMed]

58. Mulholland, D.; Kobayashi, N.; Ruscetti, M.; Zhi, A.; Tran, L.; Huang, J.; Gleave, M.; Wu, H. Pten loss and RAS/MAPK activation cooperate to promote EMT and metastasis initiated from prostate cancer stem/progenitor cells. *Cancer Res.* **2012**, *72*, 1878–1889. [CrossRef] [PubMed]

59. Bo, H.; Zhang, S.; Gao, L.; Chen, Y.; Zhang, J.; Chang, X.; Zhu, M. Upregulation of Wnt5a promotes epithelial-to-mesenchymal transition and metastasis of pancreatic cancer cells. *BMC Cancer* **2013**, *13*, 496. [CrossRef] [PubMed]

60. Foroni, C.; Broggini, M.; Generali, D.; Damia, G. Epithelial-mesenchymal transition and breast cancer: Role, molecular mechanisms and clinical impact. *Cancer Treat. Rev.* **2012**, *38*, 689–697. [CrossRef] [PubMed]

61. Nieto, M.; Cano, A. The epithelial-mesenchymal transition under control: Global programs to regulate epithelial plasticity. *Semin. Cancer Biol.* **2012**, *22*, 361–368. [CrossRef] [PubMed]

62. Bhaskaran, R.; Huang, C.; Chang, D.; Yu, C. Cardiotoxin III from the Taiwan cobra (*Naja naja atra*). Determination of structure in solution and comparison with short neurotoxins. *J. Mol. Biol.* **1994**, *235*, 1291–1301. [CrossRef] [PubMed]

63. Chien, C.; Chang, S.; Lin, K.; Chiu, C.; Chang, L.; Lin, S. Taiwan cobra cardiotoxin III inhibits Src kinase leading to apoptosis and cell cycle arrest of oral squamous cell carcinoma Ca9-22 cells. *Toxicon* **2010**, *56*, 508–520. [CrossRef] [PubMed]

64. Lin, K.; Su, J.; Chien, C.; Chuang, P.; Chang, L.; Lin, S. Down-regulation of the JAK2/PI3K-mediated signaling activation is involved in Taiwan cobra cardiotoxin III-induced apoptosis of human breast MDA-MB-231 cancer cells. *Toxicon* **2010**, *55*, 1263–1273. [CrossRef] [PubMed]

65. Chen, K.; Lin, S.; Chang, L. Involvement of mitochondrial alteration and reactive oxygen species generation in Taiwan cobra cardiotoxin-induced apoptotic death of human neuroblastoma SK-N-SH cells. *Toxicon* **2008**, *52*, 361–368. [CrossRef] [PubMed]

66. Yen, C.; Liang, S.; Han, L.; Chou, H.; Chou, C.; Lin, S.; Chiu, C. Cardiotoxin III inhibits proliferation and migration of oral cancer cells through MAPK and MMP signaling. *Sci. World J.* **2013**, *2013*, 650946. [CrossRef] [PubMed]

67. Lin, K.; Chien, C.; Hsieh, C.; Tsai, P.; Chang, L.; Lin, S. Antimetastatic potential of cardiotoxin III involves inactivation of PI3K/Akt and p38 MAPK signaling pathways in human breast cancer MDA-MB-231 cells. *Life Sci.* **2012**, *90*, 54–65. [CrossRef] [PubMed]

68. Chiang, S.; Cabrera, R.; Segall, J. Tumor cell intravasation. *Am. J. Physiol. Cell Physiol.* **2016**, *311*, C1–C14. [CrossRef] [PubMed]

69. Lo, H.; Hsu, S.; Xia, W.; Cao, X.; Shih, J.; Wei, Y.; Abbruzzese, J.; Hortobagyi, G.; Hung, M. Epidermal growth factor receptor cooperates with signal transducer and activator of transcription 3 to induce epithelial-mesenchymal transition in cancer cells via up-regulation of TWIST gene expression. *Cancer Res.* **2007**, *67*, 9066–9076. [CrossRef] [PubMed]

70. Balanis, N.; Carlin, C. Stress-induced EGF receptor signaling through STAT3 and tumor progression in triple-negative breast cancer. *Mol. Cell. Endocrinol.* **2017**, *451*, 24–30. [CrossRef] [PubMed]

71. Kim, J.; Kong, J.; Chang, H.; Kim, H.; Kim, A. EGF induces epithelial-mesenchymal transition through phospho-Smad2/3-Snail signaling pathway in breast cancer cells. *Oncotarget* **2016**, *7*, 85021–85032. [CrossRef] [PubMed]

72. Tsai, P.; Hsieh, C.; Chiu, C.; Wang, C.; Chang, L.; Lin, S. Cardiotoxin III suppresses MDA-MB-231 cell metastasis through the inhibition of EGF/EGFR-mediated signaling pathway. *Toxicon* **2012**, *60*, 734–743. [CrossRef] [PubMed]

73. Tsai, P.; Fu, Y.; Chang, L.; Lin, S. Taiwan cobra cardiotoxin III suppresses EGF/EGFR-mediated epithelial-to-mesenchymal transition and invasion of human breast cancer MDA-MB-231 cells. *Toxicon* **2016**, *111*, 108–120. [CrossRef] [PubMed]

74. Tsai, P.; Chu, C.; Chiu, C.; Chang, L.; Lin, S. Cardiotoxin III suppresses hepatocyte growth factor-stimulated migration and invasion of MDA-MB-231 cells. *Cell Biochem. Funct.* **2014**, *32*, 485–495. [CrossRef] [PubMed]

75. Tsai, P.; Fu, Y.; Chang, L.; Lin, S. Cardiotoxin III Inhibits Hepatocyte Growth Factor-Induced Epithelial-Mesenchymal Transition and Suppresses Invasion of MDA-MB-231 Cells. *J. Biochem. Mol. Toxicol.* **2016**, *30*, 12–21. [CrossRef] [PubMed]

76. Mohamed, M.; Sloane, B. Cysteine cathepsins: Multifunctional enzymes in cancer. *Nat. Rev. Cancer* **2006**, *6*, 764–775. [CrossRef] [PubMed]

77. Fonović, M.; Turk, B. Cysteine cathepsins and extracellular matrix degradation. *Biochim. Biophys. Acta* **2014**, *1840*, 2560–2570. [CrossRef] [PubMed]

78. Löser, R.; Pietzsch, J. Cysteine cathepsins: Their role in tumor progression and recent trends in the development of imaging probes. *Front. Chem.* **2015**, *3*, 37. [CrossRef] [PubMed]

79. Sobotič, B.; Vizovišek, M.; Vidmar, R.; Van Damme, P.; Gocheva, V.; Joyce, J.; Gevaert, K.; Turk, V.; Turk, B.; Fonović, M. Proteomic Identification of Cysteine Cathepsin Substrates Shed from the Surface of Cancer Cells. *Mol. Cell. Proteom.* **2015**, *14*, 2213–2228. [CrossRef] [PubMed]

80. Gocheva, V.; Zeng, W.; Ke, D.; Klimstra, D.; Reinheckel, T.; Peters, C.; Hanahan, D.; Joyce, J. Distinct roles for cysteine cathepsin genes in multistage tumorigenesis. *Genes Dev.* **2006**, *20*, 543–546. [CrossRef] [PubMed]

81. Friedl, P.; Alexander, S. Cancer invasion and the microenvironment: Plasticity and reciprocity. *Cell Biochem. Funct.* **2011**, *147*, 992–1009. [CrossRef] [PubMed]

82. Jedeszko, C.; Sloane, B. Cysteine cathepsins in human cancer. *Biol. Chem.* **2004**, *385*, 1017–1027. [CrossRef] [PubMed]

83. Shamsi, A.; Bano, B. Journey of cystatins from being mere thiol protease inhibitors to at heart of many pathological conditions. *Int. J. Biol. Macromol.* **2017**, *102*, 674–693. [CrossRef] [PubMed]

84. Brillard-Bourdet, M.; Nguyên, V.; Ferrer-di Martino, M.; Gauthier, F.; Moreau, T. Purification and characterization of a new cystatin inhibitor from Taiwan cobra (*Naja naja atra*) venom. *Biochem. J.* **1998**, *331*, 239–244. [CrossRef] [PubMed]

85. Tang, N.; Xie, Q.; Wang, X.; Li, X.; Chen, Y.; Lin, X.; Lin, J. Inhibition of invasion and metastasis of MHCC97H cells by expression of snake venom cystatin through reduction of proteinases activity and epithelial-mesenchymal transition. *Arch. Pharm. Res.* **2011**, *34*, 781–789. [CrossRef] [PubMed]

86. Mayor, R.; Etienne-Manneville, S. The front and rear of collective cell migration. *Nat. Rev. Mol. Cell Biol.* **2016**, *17*, 97–109. [CrossRef] [PubMed]

87. Zegers, M.; Friedl, P. Rho GTPases in collective cell migration. *Small GTPases* **2014**, *5*, e28997. [CrossRef] [PubMed]

88. Sartim, M.; Sampaio, S. Snake venom galactoside-binding lectins: A structural and functional overview. *J. Venom. Anim. Toxins Incl. Trop. Dis.* **2015**, *21*, 35. [CrossRef] [PubMed]

89. Pathan, J.; Mondal, S.; Sarkar, A.; Chakrabarty, D. Daboialectin, a C-type lectin from Russell's viper venom induces cytoskeletal damage and apoptosis in human lung cancer cells in vitro. *Toxicon* **2017**, *127*, 11–21. [CrossRef] [PubMed]

90. Carvalho, D.; Marangoni, S.; Oliveira, B.; Novello, J. Isolation and characterization of a new lectin from the venom of the snake *Bothrops jararacussu*. *IUBMB Life* **1998**, *44*, 933–938. [CrossRef]

91. Nolte, S.; de Castro Damasio, D.; Baréa, A.; Gomes, J.; Magalhães, A.; Mello Zischler, L.; Stuelp-Campelo, P.; Elífio-Esposito, S.; Roque-Barreira, M.; Reis, C.; et al. BJcuL, a lectin purified from *Bothrops jararacussu* venom, induces apoptosis in human gastric carcinoma cells accompanied by inhibition of cell adhesion and actin cytoskeleton disassembly. *Toxicon* **2012**, *59*, 81–85. [CrossRef] [PubMed]

92. De Carvalho, D.; Schmitmeier, S.; Novello, J.; Markland, F. Effect of BJcuL (a lectin from the venom of the snake *Bothrops jararacussu*) on adhesion and growth of tumor and endothelial cells. *Toxicon* **2001**, *39*, 1471–1476. [CrossRef]

93. Peichoto, M.E.; Teibler, P.; Mackessy, S.P.; Leiva, L.; Acosta, O.; Goncalves, L.R.; Tanaka-Azevedo, A.M.; Santoro, M.L. Purification and characterization of patagonfibrase, a metalloproteinase showing alpha-fibrinogenolytic and hemorrhagic activities, from *Philodryas patagoniensis* snake venom. *BBA Gen. Subj.* **2007**, *1770*, 810–819. [CrossRef] [PubMed]

94. Sánchez, M.N.; Timoniuk, A.; Maruñak, S.; Teibler, P.; Acosta, O.; Peichoto, M.E. Biochemical and biological analysis of *Philodryas baroni* (Baron's Green Racer; Dipsadidae) venom: Relevance to the findings of human risk assessment. *Hum. Exp. Toxicol.* **2014**, *33*, 22–31. [CrossRef] [PubMed]

95. Heyborne, W.H.; Mackessy, S.P. Identification and characterization of a taxon-specific three-finger toxin from the venom of the Green Vinesnake (*Oxybelis fulgidus*; family Colubridae). *Biochimie* **2013**, *95*, 1923–1932. [CrossRef] [PubMed]
96. Bhowmik, T.; Saha, P.; Sarkar, A.; Gomes, A. Evaluation of cytotoxicity of a purified venom protein from *Naja kaouthia* (NKCT1) using gold nanoparticles for targeted delivery to cancer cell. *Chem. Biol. Interact.* **2017**, *261*, 35–49. [CrossRef] [PubMed]
97. Badr, G.; Sayed, D.; Maximous, D.; Mohamed, A.; Gul, M. Increased susceptibility to apoptosis and growth arrest of human breast cancer cells treated by a snake venom-loaded silica nanoparticles. *Cell. Physiol. Biochem.* **2014**, *34*, 1640–1651. [CrossRef] [PubMed]

© 2017 by the authors. Licensee MDPI, Basel, Switzerland. This article is an open access article distributed under the terms and conditions of the Creative Commons Attribution (CC BY) license (http://creativecommons.org/licenses/by/4.0/).

toxins

Review

Therapeutic Potential of Cholera Toxin B Subunit for the Treatment of Inflammatory Diseases of the Mucosa

Joshua M. Royal [1,2,3] and Nobuyuki Matoba [1,2,3,*]

1   Department of Pharmacology and Toxicology, University of Louisville School of Medicine,
     Louisville, KY 40202, USA; joshua.royal@louisville.edu
2   Center for Predictive Medicine, University of Louisville, Louisville, KY 40202, USA
3   James Graham Brown Cancer Center, University of Louisville, Louisville, KY 40202, USA
*   Correspondence: n.matoba@louisville.edu; Tel.: +1-502-852-8412

Academic Editor: Steve Peigneur
Received: 18 October 2017; Accepted: 21 November 2017; Published: 23 November 2017

Abstract: Cholera toxin B subunit (CTB) is a mucosal immunomodulatory protein that induces robust mucosal and systemic antibody responses. This well-known biological activity has been exploited in cholera prevention (as a component of Dukoral® vaccine) and vaccine development for decades. On the other hand, several studies have investigated CTB's immunotherapeutic potential in the treatment of inflammatory diseases such as Crohn's disease and asthma. Furthermore, we recently found that a variant of CTB could induce colon epithelial wound healing in mouse colitis models. This review summarizes the possible mechanisms behind CTB's anti-inflammatory activity and discuss how the protein could impact mucosal inflammatory disease treatment.

Keywords: cholera toxin B subunit; mucosal immunity; immunomodulation; anti-inflammatory; retrograde trafficking; GM1 ganglioside

1. Introduction

Vibrio cholerae is a gram-negative bacterium that can colonize the gastrointestinal tract and cause life-threatening watery diarrhea. The principal virulence factor of *V. cholerae* is cholera toxin (CT), which consists of a catalytic A-subunit and a non-toxic homopentameric B-subunit (CTB) [1–3]. CTB binds cells through GM1 ganglioside receptors, which then mediates toxin entry into the cell. It has been previously shown that CTB can induce strong biological activities that can enhance or suppress immune effects under normal and various immunopathological conditions without the toxicity associated with the CTA subunit [4]. Consequently, CTB has been widely studied as a mucosal immunomodulatory agent.

In its most well-known immunostimulatory effects, CTB is used in the vaccine Dukoral®. Dukoral® is a WHO pre-qualified oral cholera vaccine which contains heat-killed whole cell *V. cholerae* and recombinant CTB (rCTB). Dukoral® stimulates the production of both antibacterial and antitoxin antibodies, including secretory immunoglobulin A (S-IgA) produced locally in the intestines [5]. CTB itself can induce potent mucosal and systemic antibody response upon mucosal administration in humans [6–8], which is largely due to the broad distribution of GM1 ganglioside on various cell types such as epithelial cells, macrophages, dendritic cells (DCs), B cells, T cells, and neurons [9–12]. Furthermore, the presence of GM1 ganglioside on the luminal surface of intestinal epithelial cells and antigen presenting cells (APCs) in the gut seems to be essential for CTB's strong mucosal immunostimulatory effects associated with MHC class II expression and local antigen enrichment [13]. In addition, CTB stimulates specific immunosuppressive effects against autoimmune disorders, excess inflammation, and allergic reactions [4,14–18]. We have recently shown that oral administration of a variant of CTB mitigates colitis in chemically-induced acute and chronic colitis mouse models [19]. Although the underlying mechanisms are not well understood, recent studies have shed some light on these immunosuppressive effects induced by CTB. Thus, this review will summarize published studies on CTB's impacts in mucosal inflammatory disease models, as well as the mechanisms associated with its therapeutic effect and the challenges that CTB faces as an immunomodulatory drug.

2. Cholera Toxin Structure and Mechanism in Gut Epithelial Cells

To reveal the mechanism of CTB-induced biological activity, we must first understand the molecule. CT is classified as an AB5 toxin family, which includes the toxins of *Shigella dysenteriae* and enterohaemorrhagic *Escherichia coli*. The toxins are usually composed of one A subunit and five B subunits (CTA and CTB, respectively, for CT). CTA consists of an enzymatically active 11-kDa N-terminal chain (CTA1) and a C-terminal chain (CTA2) that connects CTA to the central pore of CTB. CTB has the capacity to translocate the CTA across the plasma membrane, mediated by the binding of GM1 ganglioside, and then escort CTA from the plasma membrane into the endoplasmic reticulum (ER) [20,21]. The following summarizes CT's retrograde trafficking mechanism.

The five B-subunits form a central cylindrical pore lined by five amphipathic α-helices that help form a highly stable homopentamer. The pentamer contains five GM1 binding sites that lie on the outer edge of each B subunit [1,22]. Due to an avidity effect from the pentavalent binding capacity, CTB has a very strong affinity (K_D reported to be 5 pM to 1 nM) to GM1, which is mainly localized in lipid rafts on the plasma membranes of many cell types [9–12]. Once CT is bound to GM1 (up to five gangliosides at once), it is endocytosed by clathrin-dependent and independent mechanisms and trafficked via retrograde transport from the Golgi to the ER [21]. It is also known that CT can undergo transcytosis across epithelial cells from the apical to the basolateral surface. However, regardless of how the toxin enters the cell, CT travels to the trans-Golgi network via early endosomal vesicles, independent of the late endosome pathway. The C-terminus of CTA2 possesses a KDEL ER-retention signal for retrieval of CT from the cis-Golgi apparatus to the ER. Interestingly, the KDEL sequence is not vital for retrograde transport of CT to the ER. Mutations that alter the KDEL sequence on CT inhibit KDEL-dependent ER retrieval and decreased (albeit not completely) CT's toxification [23]. Thus, it is thought that CT's KDEL sequence—although not absolutely essential—improves the ER's retrieval of the dissociated CT from the Golgi apparatus and prolongs the time of retention within the ER [20,23,24]. Once in the ER, the CTA1-chain is dissociated from CTA2/CTB complex by protein disulfide isomerase (PDI). Subsequently, CTA1 enters the cytosol via the ER-associated degradation pathway and escapes proteasomal degradation [1,20]. On the other hand, the fate (and remaining function, if any) of CTA2/CTB after releasing CTA1 in the ER is not well documented. Meanwhile, CTA1 reaches the inner surface of the plasma membrane and catalyzes the ADP ribosylation of Gαs, thereby continuously activating adenylate cyclase to produce cAMP. Increased intracellular cAMP impairs sodium uptake and increases chloride outflow, leading to water secretion and diarrhea [20,25].

3. At the Cellular Level—What Is Known So Far

Although the virulence mechanism and intracellular trafficking of CT has been well studied, the anti-inflammatory mechanisms of CTB are much less studied and understood. After a comprehensive literature review, it seems that there are at least two separate modes of action induced by CTB to modulate inflammatory responses: one that is based on immune cell regulation, and another that is epithelial cell-mediated (Figure 1).

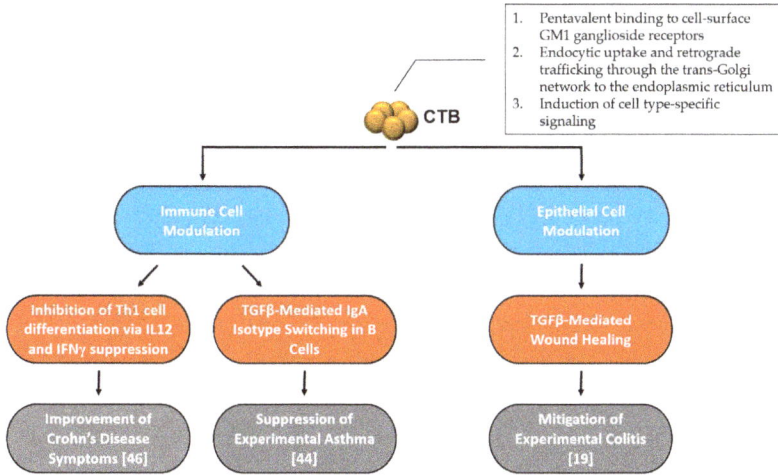

Figure 1. Summary of mechanisms involved in cholera toxin homopentameric B-subunit (CTB)'s inflammatory disease intervention.

In 1994, the immune suppressive effects of CTB were first reported by Sun et al. [26]. This report demonstrated that oral administration of mice with CTB conjugated with antigens (sheep red blood cells, horse red blood cells, and human γ-globulin) enhanced oral tolerance to the antigens, presumably through efficient presentation of antigens to immune cells in the gut-associated lymphoid tissue and the generation of regulatory cells. In a Commentary to this article, Weiner suggested that CTB could have enhanced tolerance by serving as a "selective mucosal adjuvant" and that this unique activity could be exploited to treat autoimmunity [27]. Subsequently, this seminal finding led to a new field of studies in which CTB-antigen conjugates were applied to induce tolerogenic reactions to the conjugated antigens in various immunopathological conditions (i.e., encephalomyelitis, autoimmune diabetes, autoimmune arthritis, uveitis) and IgE-mediated allergen hypersensitivity [14,16–18,28–38]. Through these studies, it became apparent there are two unique and distinct mechanisms of CTB responsible for the suppression of immunopathological reactions in allergy and autoimmune diseases: (1) to increase antigen uptake and presentation by different APCs through binding to their cell-surface GM1 ganglioside receptors and (2) to induce anti-inflammatory and immunoregulatory activities by directly or indirectly acting on specific immune cells. The latter mechanism points to the possibility that CTB by itself may act as an immunotherapeutic agent; however, only a handful of groups have actually proven that CTB alone—without co-administration or conjugation of antigens—can induce an anti-inflammatory response. Moreover, studies conducted with non-recombinant CTB (nrCTB, prepared by chemically dissociating CTA from CTB) can have significantly skewed experimental results due to trace amounts of CT and CTA [4,39,40]. For example, we have previously shown that picomolar concentrations (<10 ng/mL) of CT significantly inhibited lipopolysaccharide (LPS)-induced TNFα production in RAW264.7 cells, while recombinant (r)CTB failed to induce such an effect at a concentration as high as 10 μg/mL [4]. Thus, the use of rCTB is required to evaluate the effects unique to CTB.

3.1. Immune Cell Modulation

With regards to CTB's immune cell regulation, Kim et al. demonstrated in murine spleen B cells that rCTB dose-dependently increased IgA secretion and inhibited B cell growth [41]. In the presence of IL-2, rCTB significantly increased IgA isotype switching in LPS-activated B cells. These effects were reversed by the addition of an anti-TGFβ or soluble TGFβ1 receptor, which markedly inhibited rCTB-stimulated IgA response. Further analysis in the same report revealed that rCTB stimulated IgA2 B cells, upregulated TGFβ1 mRNA expression, and increased bioactive TGFβ1 levels, which is known to induce IgA isotype switching [41]. Thus, rCTB stimulated a TGFβ-mediated IgA response that was dependent on IL-2 as a cofactor. These findings have contributed to our understandings of how CTB stimulates B cell IgA production, and potentially oral tolerance as well (see below).

It is known that IgA antibodies help maintain mucosal homeostasis and play a role in immune protection [42,43]. Thus, it seems possible that rCTB administration could provide therapeutic effects in mucosal autoimmune disorders via IgA induction. For example, in an experimental mouse model of asthma, nrCTB suppressed the ability of DCs to prime for Th2 responses to inhaled allergen via an IgA-dependent manner [44]. In this study, co-administration of ovalbumin (OVA) and nrCTB suppressed classical features of asthma, including airway eosinophilia, Th2 cytokine synthesis, and bronchial hyperactivity in mice that were pre-sensitized with OVA-stimulated DCs in the lung. Furthermore, nrCTB treatment enhanced DCs' potential to induce Treg cells in vitro; however, these Treg cells did not provide protection when transferred into the airways of naïve mice that received OVA challenge. In contrast, the transfer of B cells from OVA+CTB-DCs-immunized mice to OVA-sensitized naïve mice significantly reduced eosinophilia and lymphocytosis. It was also found that nrCTB caused a TGFβ-dependent increase in antigen-specific IgA in the airway luminal secretion, and this was attributed to nrCTB's efficacy against the experimental asthma as the therapeutic effects were abrogated in mice lacking luminal IgA transporter (polymeric Ig receptor), which is necessary for the transport of dimeric IgA across the epithelium into the luminal mucosa [45].

Meanwhile, IgA may not be the sole factor contributing to CTB's ability to mitigate inflammatory diseases in the mucosa. For example, in the 2,4,6-trinitrobenzene sulfonic acid (TNBS)-induced mouse model of Crohn's disease, daily oral administration over a four-day period of 100 μg rCTB after the onset of TNBS-colitis immediately resolved weight loss and reduced inflammation [39]. In this case, the timing of mucosal restitution in regard to rCTB administration did not likely result in IgA production. In a similar TNBS-colitis study, rCTB administration reduced IL-12 and IFNγ secretion, inhibited STAT-4 and STAT-1 activation, and downregulated T-bet expression, indicating that rCTB inhibited mucosal Th1 cell signaling [46]. Moreover, these results were confirmed in a small multicenter, open-label, and nonrandomized clinical trial in which 15 patients with active CD received three oral doses of 5 mg rCTB per-week over 2 weeks (six doses total) and were examined 2, 4, 6, and 10 weeks after the start of the study. Of the 12 patients who finished the study per protocol, seven responded to treatment and five were in remission by week six and maintained remission through week 10 as defined by a CD activity index score $\leq$150 [47]. Of note, side effects seen in 33% of patients administered with CTB were mild (arthralgia, headache, and pruritus), and no safety concerns were raised throughout the trial [47].

Interestingly, rCTB did not reduce disease severity in an oxazolone-induced colitis model performed by the same group [39]. Oxazolone-induced colitis is mediated by IL-4 driven Th2 cells rather than IL-12/IFNγ-driven Th1 cells [39]. Thus, it appears that rCTB administration had a specific effect on specific T cell functions involved in TNBS-colitis [39]. Although the detailed mechanism by which rCTB inhibited Th1 cell was not elucidated, it is possible that the binding of CTB to GM1 ganglioside on immune cells resulted in a signaling cascade of events that led to Th1 inhibition, because non-GM1 binding CTB mutants do not modulate lymphocyte function [48]. In agreement with these findings, rCTB decreased monocyte-derived DC maturation and IL-12 production upon LPS stimulation in vitro [49]. Moreover, rCTB-pretreated, LPS-stimulated DCs induced low proliferating T cells that had enhanced production of IL-10 and reduced production of IFNγ. Rouquete-Jazdanian et al. additionally showed that the binding of rCTB to GM1 ganglioside directly prevented the activation and proliferation of CD4$^+$ T cells [50]. This effect was induced by rCTB-mediated sphingomyelinase activation that subsequently increased the production of ceramides, which are known cell cycle arrest inducers [51]. rCTB also inhibited protein kinase Cα, a pro-growth cellular regulator, which was linked to rCTB-induced lipid raft modifications and ceramide-mediated inactivation [52,53].

3.2. Epithelial Cell Modulation

Besides serving as a barrier lining the mucosal surface, epithelial cells have multiple functions associated with the maintenance of gut homeostasis and mucosal healing, and crosstalk between epithelial and immune cells is an important component of those complex biological processes [54,55]. Even though CTB first encounters epithelial cells in the gut, the CTB-mediated modulation of epithelial cells and its consequence to the mucosal immune system have largely been ignored in comparison to the protein's direct impacts on immune cells.

In one small study, CTB was shown to induce a dose-dependent increase of IL-10 mRNA levels in the colon epithelial cell-line T84 [56], hinting that CTB could induce epithelial cell-mediated immune modulation [57]. We have recently characterized CTB's global impacts on the gut to further our understanding of its unique biological activities. Using a plant-made recombinant CTB (CTBp) [58,59], we have shown that oral administration of the CTB variant significantly altered several immune cell populations in the colon lamina propria [19]. Two-weeks after two oral 30 µg CTBp administrations, Th2 and Treg cells increased in the colon lamina propria. This is not the first report of CTB-induced increase in these cell types [15,36,38,60,61]. For instance, it has been shown that oral administration of a CTB–insulin conjugate in NOD mice induced a shift from Th1 to Th2 profile while generating Treg cells [15]. Additionally, intraperitoneal administration of nrCTB to rats increased Treg cells in the peripheral blood 24–72 h after ischemia [60]. Besides the specific T helper cell subsets, our study has also revealed that innate immune cells—including dendritic cells, natural killer cells and macrophages (both M1 and M2)—populations were increased in the colon lamina propria two weeks after CTBp oral administration [19]. Furthermore, a global gene expression analysis revealed that CTBp had more pronounced impacts on the colon than the small intestine, with significant activation of TGFβ-mediated pathways in the colon mucosa [19]. Given that there is a strong link between epithelial-derived TGFβ and innate immune cells in wound healing [62–64], the results provided implications for the potential utility of CTBp to promote colonic mucosal health. Subsequently, we found that CTBp induced TGFβ-mediated wound healing in Caco2 colon epithelial cells. Furthermore, oral administration of CTBp in mice protected against colon mucosal damage in acute colitis induced by dextran sodium sulfate (DSS). Two oral doses of as low as 1 µg of CTBp mitigated clinical signs of disease (body weight loss, decreased histopathological scores, and blunted escalation of inflammatory cytokine levels) and upregulated wound healing-related genes [19]. Interestingly, CTBp administration prevented fibrosis associated with acute colitis in mice; hence, the protein did not appear to overstimulate TGFβ signaling. In fact, TGFβ gene expression levels were high during the early inflammatory phase and became lower in the recovery phase of the acute colitis model in CTBp-treated mice.

In contrast to TNBS-induced colitis, the DSS-colitis model closely approximates human ulcerative colitis (UC) [65–68]. Thus, the results point to the possibility that CTBp could be used to facilitate mucosal healing in the management of UC. Since the main driver of intestinal inflammation in the DSS model is the damage to the epithelial barrier lining the colon that allows intestinal microbiota into submucosal compartments [69], and since therapeutic effects were observed immediately upon CTBp administration, we concluded that CTBp's protective efficacy in the DSS colitis models were attained by the induction of TGFβ-mediated colonic epithelial wound healing. Given that UC poses an increased risk of developing colitis-associated colorectal cancer (CAC) [70,71], CTBp's effects were also examined in the azoxymethane (AOM)/DSS mouse model of CAC. Biweekly oral administration of CTBp over 9 weeks significantly reduced inflammation and tumorigenesis in this model, again highlighting its therapeutic potential in UC treatment [19].

It is of importance to point out that many of the effects observed in the aforementioned studies using CTBp may be unique to the plant-made variant, as it has a mutation at amino acid position 4 and an ER retention signal sequence at the C-terminus (N4S-CTB-SEKDEL; [58]). The ER-retention sequence was added to CTBp to improve production in planta, while Asn4→Ser mutation was introduced to avoid *N*-glycosylation [58,59]. The addition of the KDEL sequence to N4S-CTB significantly reduced

ER stress that otherwise caused poor production yield. It is thought that the KDEL sequence helped prolong CTBp's residence time in the ER to allow for proper folding and assembly.

The protein ER retention mechanism involving the KDEL receptor is highly conserved among eukaryotic organisms [72]. Thus, there is a possibility that the artificial KDEL sequence of CTBp may prolong the protein's residence in the epithelial cells upon binding to cell-surface GM1 ganglioside and retrograde transport into the ER, as has been demonstrated for CT [21,23]. Subsequently, this may induce a level of altered cell signaling. For example, interaction between CTBp's C-terminal KDEL sequence and KDEL receptors may have an impact on ER homeostasis [73,74]. The binding of proteins to the KDEL receptor and the induction of mild UPR have been linked to TGFβ activation, wound healing, colon epithelial cell prosurvival signaling, and protection from DSS-induced colitis [73,75–77]. Of note, CT is a known inducer of the UPR in epithelial cells due to the KDEL sequence on CTA [23,78,79], while CTB has no effect on the UPR or ER signaling [78].

Regardless of whether the ER retention signal had a significant contribution to the mucosal healing activity in the mouse colitis models, the study has provided evidence that CTB can exhibit a therapeutic effect against colitis in an epithelia-dependent manner, warranting further investigation of CTB's impacts on epithelial cells.

4. Conclusions—Challenges for the Use of CTB as an Immunomodulatory Drug

Although CTB has been administered in humans in the form of oral cholera vaccines over the past two decades, its development as an immunomodulatory drug will need to address unique issues associated with therapeutic use besides additional testing of safety and efficacy in specific disease indications. One of the principal questions is whether CTB's strong mucosal immunogenicity that induces a robust IgG and IgA immune response [4,58] is beneficial or dispensable to its anti-inflammatory/immunosuppressive effects. From a conventional biopharmaceuticals development standpoint, anti-drug antibodies constitute a theoretical risk because they may affect drug efficacy and pharmacokinetics, and potentially cause immunotoxicity [80,81]. However, induction of an antibody response—particularly that of IgA isotype—may play an important role in mitigating mucosal inflammation, as illustrated in the asthma study described in Section 3.1 [44]. The CD clinical trial showed an efficacy up to 10 weeks after repeated CTB administrations over 2 weeks [47]. Although not reported, the treatment regimen must have elicited high levels of anti-CTB antibodies in the gut and blood circulation. Thus, further investigation is necessary to address long-term efficacy following repeated CTB dosing. TGFβ seems to be a major denominator of CTB-induced immunomodulatory activities. TGFβ is a pleiotropic cytokine playing critical roles in cell differentiation and proliferation, as well as dynamic biological processes in wound healing and immune responses [82–84]. The cytokine is also involved in various pathological conditions. For example, elevated TGFβ levels have been correlated to the development of fibrosis following injury to the skin [85]. TGFβ mediates epithelial-to-mesenchymal transition (EMT) [86], and reduction of TGFβ1 levels in a mouse model of pulmonary fibrosis blunted fibrosis [87]. TGFβ signaling also has important implications in cancer. Although the cytokine functions as a suppressor of tumorigenesis at an early stage of tumor development, its expression is correlated with tumor progression and poor prognosis at late stages [84,88]. Collectively, the double-edged sword nature of TGFβ points to the importance of careful investigation of possible consequences upon long-term CTB dosing for the treatment of chronic inflammatory diseases. As mentioned in Section 3.2, CTBp treatment significantly mitigated gut inflammation and reduced tumor development in a model of CAC [19], providing a basis for further investigations of long-term therapeutic use of CTB for the treatment of IBD.

Of considerable interest may be population-based studies investigating potential association between the Dukoral® vaccine and gastrointestinal disorders involving mucosal inflammation. In a very recent study of patients who were diagnosed with colorectal cancer from July 2005 through December 2012 in Sweden, it was revealed that those who had previously received Dukoral® had a significantly reduced risk of death from colorectal cancer (CRC) [89]. Although the underlying mechanism is not clear at this point, the authors speculated that CTB might be associated with a risk reduction of CRC [89]. This observation warrants a comprehensive investigation on this subject.

In conclusion, even though CTB has been studied since the early 1970s [3], its immunomodulatory mechanisms appear to involve complex interplay between epithelial and immune cells that requires a systematic approach for comprehensive understanding. The studies highlighted herein strongly suggest CTB's potential as an effective mucosal anti-inflammatory agent with the potential to replace or supplement currently available therapies for the treatment of inflammatory disorders of the mucosa, such as anti-TNFα biologics used in IBD patients who are refractory to conventional medications. As anti-TNFα agents are administered systemically, these agents have limited efficacy for the induction of mucosal healing [90,91] and/or pose severe adverse reactions [92–94]. In contrast, CTB has few, if any, adverse effects (according to the CD clinical trial [47]), can directly heal lesions/ulcers, and blunt inflammation upon topical administration. Therefore, to aid in developing CTB-based therapeutic strategies against various mucosal immunopathological conditions, further research that delineates how CTB can modulate epithelial cell signaling and T cell functions simultaneously is warranted.

Acknowledgments: We thank Matthew Dent for critical reading of the manuscript. This manuscript is based on work supported by the Leona M. and Harry B. Helmsley Charitable Trust Fund.

Author Contributions: J.M.R. and N.M. conceived and designed the review idea and contents, and wrote the manuscript.

Conflicts of Interest: The authors declare no conflict of interest.

References

1. Sanchez, J.; Holmgren, J. Cholera toxin structure, gene regulation and pathophysiological and immunological aspects. *Cell. Mol. Life Sci.* **2008**, *65*, 1347–1360. [CrossRef] [PubMed]
2. Finkelstein, R.A.; LoSpalluto, J.J. Pathogenesis of experimental cholera. Preparation and isolation of choleragen and choleragenoid. *J. Exp. Med.* **1969**, *130*, 185–202. [CrossRef] [PubMed]
3. Lonnroth, I.; Holmgren, J. Subunit structure of cholera toxin. *J. Gen. Microbiol.* **1973**, *76*, 417–427. [CrossRef] [PubMed]
4. Baldauf, K.J.; Royal, J.M.; Hamorsky, K.T.; Matoba, N. Cholera toxin B: One subunit with many pharmaceutical applications. *Toxins* **2015**, *7*, 974–996. [CrossRef] [PubMed]
5. Cholera vaccines: WHO position paper. *Relev. Epidemiol. Hebd.* **2010**, *85*, 117–128.
6. Bergquist, C.; Johansson, E.L.; Lagergard, T.; Holmgren, J.; Rudin, A. Intranasal vaccination of humans with recombinant cholera toxin B subunit induces systemic and local antibody responses in the upper respiratory tract and the vagina. *Infect. Immun.* **1997**, *65*, 2676–2684. [PubMed]
7. Jertborn, M.; Nordstrom, I.; Kilander, A.; Czerkinsky, C.; Holmgren, J. Local and systemic immune responses to rectal administration of recombinant cholera toxin B subunit in humans. *Infect. Immun.* **2001**, *69*, 4125–4128. [CrossRef] [PubMed]
8. Kozlowski, P.A.; Cu-Uvin, S.; Neutra, M.R.; Flanigan, T.P. Comparison of the oral, rectal, and vaginal immunization routes for induction of antibodies in rectal and genital tract secretions of women. *Infect. Immun.* **1997**, *65*, 1387–1394. [PubMed]
9. Cuatrecasas, P. Interaction of Vibrio cholerae enterotoxin with cell membranes. *Biochemistry* **1973**, *12*, 3547–3558. [CrossRef] [PubMed]
10. Kuziemko, G.M.; Stroh, M.; Stevens, R.C. Cholera toxin binding affinity and specificity for gangliosides determined by surface plasmon resonance. *Biochemistry* **1996**, *35*, 6375–6384. [CrossRef] [PubMed]
11. MacKenzie, C.R.; Hirama, T.; Lee, K.K.; Altman, E.; Young, N.M. Quantitative analysis of bacterial toxin affinity and specificity for glycolipid receptors by surface plasmon resonance. *J. Biol. Chem.* **1997**, *272*, 5533–5538. [CrossRef] [PubMed]
12. Dawson, R.M. Characterization of the binding of cholera toxin to ganglioside GM1 immobilized onto microtitre plates. *J. Appl. Toxicol.* **2005**, *25*, 30–38. [CrossRef] [PubMed]
13. George-Chandy, A.; Eriksson, K.; Lebens, M.; Nordstrom, I.; Schon, E.; Holmgren, J. Cholera toxin B subunit as a carrier molecule promotes antigen presentation and increases CD40 and CD86 expression on antigen-presenting cells. *Infect. Immun.* **2001**, *69*, 5716–5725. [CrossRef] [PubMed]

14. Sun, J.B.; Rask, C.; Olsson, T.; Holmgren, J.; Czerkinsky, C. Treatment of experimental autoimmune encephalomyelitis by feeding myelin basic protein conjugated to cholera toxin B subunit. *Proc. Natl. Acad. Sci. USA* **1996**, *93*, 7196–7201. [CrossRef] [PubMed]

15. Ploix, C.; Bergerot, I.; Durand, A.; Czerkinsky, C.; Holmgren, J.; Thivolet, C. Oral administration of cholera toxin B-insulin conjugates protects NOD mice from autoimmune diabetes by inducing CD4+ regulatory T-cells. *Diabetes* **1999**, *48*, 2150–2156. [CrossRef] [PubMed]

16. Tarkowski, A.; Sun, J.B.; Holmdahl, R.; Holmgren, J.; Czerkinsky, C. Treatment of experimental autoimmune arthritis by nasal administration of a type II collagen-cholera toxoid conjugate vaccine. *Arthritis Rheum.* **1999**, *42*, 1628–1634. [CrossRef]

17. Rask, C.; Holmgren, J.; Fredriksson, M.; Lindblad, M.; Nordstrom, I.; Sun, J.B.; Czerkinsky, C. Prolonged oral treatment with low doses of allergen conjugated to cholera toxin B subunit suppresses immunoglobulin E antibody responses in sensitized mice. *Clin. Exp. Allergy* **2000**, *30*, 1024–1032. [CrossRef] [PubMed]

18. Stanford, M.; Whittall, T.; Bergmeier, L.A.; Lindblad, M.; Lundin, S.; Shinnick, T.; Mizushima, Y.; Holmgren, J.; Lehner, T. Oral tolerization with peptide 336–351 linked to cholera toxin B subunit in preventing relapses of uveitis in Behcet's disease. *Clin. Exp. Immunol.* **2004**, *137*, 201–208. [CrossRef] [PubMed]

19. Baldauf, K.J.; Royal, J.M.; Kouokam, J.C.; Haribabu, B.; Jala, V.R.; Yaddanapudi, K.; Hamorsky, K.T.; Dryden, G.W.; Matoba, N. Oral administration of a recombinant cholera toxin B subunit promotes mucosal healing in the colon. *Mucosal Immunol.* **2017**, *10*, 887–900. [CrossRef] [PubMed]

20. Wernick, N.L.; Chinnapen, D.J.; Cho, J.A.; Lencer, W.I. Cholera toxin: An intracellular journey into the cytosol by way of the endoplasmic reticulum. *Toxins* **2010**, *2*, 310–325. [CrossRef] [PubMed]

21. Chinnapen, D.J.; Chinnapen, H.; Saslowsky, D.; Lencer, W.I. Rafting with cholera toxin: Endocytosis and trafficking from plasma membrane to ER. *FEMS Microbiol. Lett.* **2007**, *266*, 129–137. [CrossRef] [PubMed]

22. Zhang, R.G.; Scott, D.L.; Westbrook, M.L.; Nance, S.; Spangler, B.D.; Shipley, G.G.; Westbrook, E.M. The three-dimensional crystal structure of cholera toxin. *J. Mol. Biol.* **1995**, *251*, 563–573. [CrossRef] [PubMed]

23. Lencer, W.I.; Constable, C.; Moe, S.; Jobling, M.G.; Webb, H.M.; Ruston, S.; Madara, J.L.; Hirst, T.R.; Holmes, R.K. Targeting of cholera toxin and Escherichia coli heat labile toxin in polarized epithelia: Role of COOH-terminal KDEL. *J. Cell Biol.* **1995**, *131*, 951–962. [CrossRef] [PubMed]

24. Fujinaga, Y.; Wolf, A.A.; Rodighiero, C.; Wheeler, H.; Tsai, B.; Allen, L.; Jobling, M.G.; Rapoport, T.; Holmes, R.K.; Lencer, W.I. Gangliosides that associate with lipid rafts mediate transport of cholera and related toxins from the plasma membrane to endoplasmic reticulm. *Mol. Biol. Cell* **2003**, *14*, 4783–4793. [CrossRef] [PubMed]

25. Sanchez, J.; Holmgren, J. Cholera toxin—A foe & a friend. *Indian J. Med. Res.* **2011**, *133*, 153–163. [PubMed]

26. Sun, J.B.; Holmgren, J.; Czerkinsky, C. Cholera toxin B subunit: An efficient transmucosal carrier-delivery system for induction of peripheral immunological tolerance. *Proc. Natl Acad. Sci. USA* **1994**, *91*, 10795–10799. [CrossRef] [PubMed]

27. Weiner, H.L. Oral tolerance. *Proc. Natl. Acad. Sci. USA* **1994**, *91*, 10762–10765. [CrossRef] [PubMed]

28. Bublin, M.; Hoflehner, E.; Wagner, B.; Radauer, C.; Wagner, S.; Hufnagl, K.; Allwardt, D.; Kundi, M.; Scheiner, O.; Wiedermann, U.; et al. Use of a genetic cholera toxin B subunit/allergen fusion molecule as mucosal delivery system with immunosuppressive activity against Th2 immune responses. *Vaccine* **2007**, *25*, 8395–8404. [CrossRef] [PubMed]

29. Ruhlman, T.; Ahangari, R.; Devine, A.; Samsam, M.; Daniell, H. Expression of cholera toxin B-proinsulin fusion protein in lettuce and tobacco chloroplasts–oral administration protects against development of insulitis in non-obese diabetic mice. *Plant Biotechnol. J.* **2007**, *5*, 495–510. [CrossRef] [PubMed]

30. Carter, J.E., III; Yu, J.; Choi, N.W.; Hough, J.; Henderson, D.; He, D.; Langridge, W.H. Bacterial and plant enterotoxin B subunit-autoantigen fusion proteins suppress diabetes insulitis. *Mol. Biotechnol.* **2006**, *32*, 1–15. [CrossRef]

31. Arakawa, T.; Yu, J.; Chong, D.K.; Hough, J.; Engen, P.C.; Langridge, W.H. A plant-based cholera toxin B subunit-insulin fusion protein protects against the development of autoimmune diabetes. *Nat. Biotechnol.* **1998**, *16*, 934–938. [CrossRef] [PubMed]

32. Sun, J.B.; Mielcarek, N.; Lakew, M.; Grzych, J.M.; Capron, A.; Holmgren, J.; Czerkinsky, C. Intranasal administration of a *Schistosoma mansoni* glutathione *S*-transferase-cholera toxoid conjugate vaccine evokes antiparasitic and antipathological immunity in mice. *J. Immunol. (Baltimore, Md: 1950)* **1999**, *163*, 1045–1052.

33. McSorley, S.J.; Rask, C.; Pichot, R.; Julia, V.; Czerkinsky, C.; Glaichenhaus, N. Selective tolerization of Th1-like cells after nasal administration of a cholera toxoid-LACK conjugate. *Eur. J. Immunol.* **1998**, *28*, 424–432. [CrossRef]

34. Czerkinsky, C.; Anjuere, F.; McGhee, J.R.; George-Chandy, A.; Holmgren, J.; Kieny, M.P.; Fujiyashi, K.; Mestecky, J.F.; Pierrefite-Carle, V.; Rask, C.; et al. Mucosal immunity and tolerance: Relevance to vaccine development. *Immunol. Rev.* **1999**, *170*, 197–222. [CrossRef] [PubMed]

35. Phipps, P.A.; Stanford, M.R.; Sun, J.B.; Xiao, B.G.; Holmgren, J.; Shinnick, T.; Hasan, A.; Mizushima, Y.; Lehner, T. Prevention of mucosally induced uveitis with a HSP60-derived peptide linked to cholera toxin B subunit. *Eur. J. Immunol.* **2003**, *33*, 224–232. [CrossRef] [PubMed]

36. Sun, J.B.; Xiao, B.G.; Lindblad, M.; Li, B.L.; Link, H.; Czerkinsky, C.; Holmgren, J. Oral administration of cholera toxin B subunit conjugated to myelin basic protein protects against experimental autoimmune encephalomyelitis by inducing transforming growth factor-beta-secreting cells and suppressing chemokine expression. *Int. Immunol.* **2000**, *12*, 1449–1457. [CrossRef] [PubMed]

37. Bergerot, I.; Ploix, C.; Petersen, J.; Moulin, V.; Rask, C.; Fabien, N.; Lindblad, M.; Mayer, A.; Czerkinsky, C.; Holmgren, J.; et al. A cholera toxoid-insulin conjugate as an oral vaccine against spontaneous autoimmune diabetes. *Proc. Natl. Acad. Sci. USA* **1997**, *94*, 4610–4614. [CrossRef] [PubMed]

38. Sun, J.B.; Czerkinsky, C.; Holmgren, J. Mucosally induced immunological tolerance, regulatory T cells and the adjuvant effect by cholera toxin B subunit. *Scand. J. Immunol.* **2010**, *71*, 1–11. [CrossRef] [PubMed]

39. Boirivant, M.; Fuss, I.J.; Ferroni, L.; De Pascale, M.; Strober, W. Oral administration of recombinant cholera toxin subunit B inhibits IL-12-mediated murine experimental (trinitrobenzene sulfonic acid) colitis. *J. Immunol. (Baltimore, Md: 1950)* **2001**, *166*, 3522–3532. [CrossRef]

40. Tamura, S.; Yamanaka, A.; Shimohara, M.; Tomita, T.; Komase, K.; Tsuda, Y.; Suzuki, Y.; Nagamine, T.; Kawahara, K.; Danbara, H.; et al. Synergistic action of cholera toxin B subunit (and Escherichia coli heat-labile toxin B subunit) and a trace amount of cholera whole toxin as an adjuvant for nasal influenza vaccine. *Vaccine* **1994**, *12*, 419–426. [CrossRef]

41. Kim, P.H.; Eckmann, L.; Lee, W.J.; Han, W.; Kagnoff, M.F. Cholera toxin and cholera toxin B subunit induce IgA switching through the action of TGF-beta 1. *J. Immunol. (Baltimore, Md: 1950)* **1998**, *160*, 1198–1203.

42. Reinholdt, J.; Husby, S. IgA and Mucosal Homeostasis. In *Madame Curie Bioscience Database*; Landes Bioscience: Austin, TX, USA, 2000–2013.

43. Corthesy, B. Role of secretory IgA in infection and maintenance of homeostasis. *Autoimmun. Rev.* **2013**, *12*, 661–665. [CrossRef] [PubMed]

44. Smits, H.H.; Gloudemans, A.K.; van Nimwegen, M.; Willart, M.A.; Soullie, T.; Muskens, F.; de Jong, E.C.; Boon, L.; Pilette, C.; Johansen, F.E.; et al. Cholera toxin B suppresses allergic inflammation through induction of secretory IgA. *Mucosal Immunol.* **2009**, *2*, 331–339. [CrossRef] [PubMed]

45. Neurath, M.F.; Finotto, S.; Glimcher, L.H. The role of Th1/Th2 polarization in mucosal immunity. *Nat. Med.* **2002**, *8*, 567–573. [CrossRef] [PubMed]

46. Coccia, E.M.; Remoli, M.E.; Di Giacinto, C.; Del Zotto, B.; Giacomini, E.; Monteleone, G.; Boirivant, M. Cholera toxin subunit B inhibits IL-12 and IFN-[61] production and signaling in experimental colitis and Crohn's disease. *Gut* **2005**, *54*, 1558–1564. [CrossRef] [PubMed]

47. Stal, P.; Befrits, R.; Ronnblom, A.; Danielsson, A.; Suhr, O.; Stahlberg, D.; Brinkberg Lapidus, A.; Lofberg, R. Clinical trial: The safety and short-term efficacy of recombinant cholera toxin B subunit in the treatment of active Crohn's disease. *Aliment. Pharmacol. Ther.* **2010**, *31*, 387–395. [CrossRef] [PubMed]

48. Aman, A.T.; Fraser, S.; Merritt, E.A.; Rodigherio, C.; Kenny, M.; Ahn, M.; Hol, W.G.; Williams, N.A.; Lencer, W.I.; Hirst, T.R. A mutant cholera toxin B subunit that binds GM1-ganglioside but lacks immunomodulatory or toxic activity. *Proc. Natl. Acad. Sci. USA* **2001**, *98*, 8536–8541. [CrossRef] [PubMed]

49. D'Ambrosio, A.; Colucci, M.; Pugliese, O.; Quintieri, F.; Boirivant, M. Cholera toxin B subunit promotes the induction of regulatory T cells by preventing human dendritic cell maturation. *J. Leukoc. Biol.* **2008**, *84*, 661–668. [CrossRef] [PubMed]

50. Rouquette-Jazdanian, A.K.; Foussat, A.; Lamy, L.; Pelassy, C.; Lagadec, P.; Breittmayer, J.P.; Aussel, C. Cholera toxin B-subunit prevents activation and proliferation of human CD4+ T cells by activation of a neutral sphingomyelinase in lipid rafts. *J. Immunol. (Baltimore, Md: 1950)* **2005**, *175*, 5637–5648. [CrossRef]

51. Dbaibo, G.S.; Pushkareva, M.Y.; Jayadev, S.; Schwarz, J.K.; Horowitz, J.M.; Obeid, L.M.; Hannun, Y.A. Retinoblastoma gene product as a downstream target for a ceramide-dependent pathway of growth arrest. *Proc. Natl. Acad. Sci. USA* **1995**, *92*, 1347–1351. [CrossRef] [PubMed]

52. Lee, J.Y.; Hannun, Y.A.; Obeid, L.M. Ceramide inactivates cellular protein kinase Calpha. *J. Biol. Chem.* **1996**, *271*, 13169–13174. [CrossRef] [PubMed]

53. Lee, J.Y.; Hannun, Y.A.; Obeid, L.M. Functional dichotomy of protein kinase C (PKC) in tumor necrosis factor-alpha (TNF-alpha) signal transduction in L929 cells. Translocation and inactivation of PKC by TNF-alpha. *J. Biol. Chem.* **2000**, *275*, 29290–29298. [CrossRef] [PubMed]

54. Leoni, G.; Neumann, P.A.; Sumagin, R.; Denning, T.L.; Nusrat, A. Wound repair: Role of immune-epithelial interactions. *Mucosal Immunol.* **2015**, *8*, 959–968. [CrossRef] [PubMed]

55. Kurashima, Y.; Kiyono, H. Mucosal Ecological Network of Epithelium and Immune Cells for Gut Homeostasis and Tissue Healing. *Annu. Rev. Immunol.* **2017**, *35*, 119–147. [CrossRef] [PubMed]

56. Ma, D.; Wolvers, D.; Stanisz, A.M.; Bienenstock, J. Interleukin-10 and nerve growth factor have reciprocal upregulatory effects on intestinal epithelial cells. *Am. J. Physiol. Regul. Integr. Comp. Physiol.* **2003**, *284*, R1323–R1329. [CrossRef] [PubMed]

57. Stordeur, P.; Goldman, M. Interleukin-10 as a regulatory cytokine induced by cellular stress: Molecular aspects. *Int. Rev. Immunol.* **1998**, *16*, 501–522. [CrossRef] [PubMed]

58. Hamorsky, K.T.; Kouokam, J.C.; Bennett, L.J.; Baldauf, K.J.; Kajiura, H.; Fujiyama, K.; Matoba, N. Rapid and scalable plant-based production of a cholera toxin B subunit variant to aid in mass vaccination against cholera outbreaks. *PLoS Negl. Trop. Dis.* **2013**, *7*, e2046. [CrossRef] [PubMed]

59. Hamorsky, K.T.; Kouokam, J.C.; Jurkiewicz, J.M.; Nelson, B.; Moore, L.J.; Husk, A.S.; Kajiura, H.; Fujiyama, K.; Matoba, N. N-glycosylation of cholera toxin B subunit in Nicotiana benthamiana: Impacts on host stress response, production yield and vaccine potential. *Sci. Rep.* **2015**, *5*, 8003. [CrossRef] [PubMed]

60. Zhang, L.; Huang, Y.; Lin, Y.; Shan, Y.; Tan, S.; Cai, W.; Li, H.; Zhang, B.; Men, X.; Lu, Z. Anti-inflammatory effect of cholera toxin B subunit in experimental stroke. *J. Neuroinflamm.* **2016**, *13*, 147. [CrossRef] [PubMed]

61. Aspord, C.; Czerkinsky, C.; Durand, A.; Stefanutti, A.; Thivolet, C. alpha4 integrins and L-selectin differently orchestrate T-cell activity during diabetes prevention following oral administration of CTB-insulin. *J. Autoimmun.* **2002**, *19*, 223–232. [CrossRef] [PubMed]

62. Wynn, T.A.; Vannella, K.M. Macrophages in Tissue Repair, Regeneration, and Fibrosis. *Immunity* **2016**, *44*, 450–462. [CrossRef] [PubMed]

63. Satoh, Y.; Ishiguro, Y.; Sakuraba, H.; Kawaguchi, S.; Hiraga, H.; Fukuda, S.; Nakane, A. Cyclosporine regulates intestinal epithelial apoptosis via TGF-beta-related signaling. *Am. J. Physiol. Gastrointest. Liver Physiol.* **2009**, *297*, G514–G519. [CrossRef] [PubMed]

64. Taverna, D.; Pollins, A.C.; Sindona, G.; Caprioli, R.M.; Nanney, L.B. Imaging mass spectrometry for assessing cutaneous wound healing: Analysis of pressure ulcers. *J. Proteome Res.* **2015**, *14*, 986–996. [CrossRef] [PubMed]

65. Kiesler, P.; Fuss, I.J.; Strober, W. Experimental Models of Inflammatory Bowel Diseases. *Cell. Mol. Gastroenterol. Hepatol.* **2015**, *1*, 154–170. [CrossRef] [PubMed]

66. Perse, M.; Cerar, A. Dextran sodium sulphate colitis mouse model: Traps and tricks. *J. Biomed. Biotechnol.* **2012**, *2012*, 718617. [CrossRef] [PubMed]

67. Clapper, M.L.; Cooper, H.S.; Chang, W.C. Dextran sulfate sodium-induced colitis-associated neoplasia: A promising model for the development of chemopreventive interventions. *Acta Pharmacol. Sin.* **2007**, *28*, 1450–1459. [CrossRef] [PubMed]

68. Jiminez, J.A.; Uwiera, T.C.; Douglas Inglis, G.; Uwiera, R.R. Animal models to study acute and chronic intestinal inflammation in mammals. *Gut Pathog.* **2015**, *7*, 29. [CrossRef] [PubMed]

69. Chassaing, B.; Aitken, J.D.; Malleshappa, M.; Vijay-Kumar, M. Dextran sulfate sodium (DSS)-induced colitis in mice. *Curr. Protoc. Immunol.* **2014**, *104*. [CrossRef]

70. Yashiro, M. Ulcerative colitis-associated colorectal cancer. *World J. Gastroenterol.* **2014**, *20*, 16389–16397. [CrossRef] [PubMed]

71. Grivennikov, S.I. Inflammation and colorectal cancer: Colitis-associated neoplasia. *Semin. Immunopathol.* **2013**, *35*, 229–244. [CrossRef] [PubMed]

72. Denecke, J.; De Rycke, R.; Botterman, J. Plant and mammalian sorting signals for protein retention in the endoplasmic reticulum contain a conserved epitope. *EMBO J.* **1992**, *11*, 2345–2355. [PubMed]

73. Yamamoto, K.; Hamada, H.; Shinkai, H.; Kohno, Y.; Koseki, H.; Aoe, T. The KDEL receptor modulates the endoplasmic reticulum stress response through mitogen-activated protein kinase signaling cascades. *J. Biol. Chem.* **2003**, *278*, 34525–34532. [CrossRef] [PubMed]

74. Cancino, J.; Capalbo, A.; Di Campli, A.; Giannotta, M.; Rizzo, R.; Jung, J.E.; Di Martino, R.; Persico, M.; Heinklein, P.; Sallese, M.; et al. Control systems of membrane transport at the interface between the endoplasmic reticulum and the Golgi. *Dev. Cell* **2014**, *30*, 280–294. [CrossRef] [PubMed]

75. Chusri, P.; Kumthip, K.; Hong, J.; Zhu, C.; Duan, X.; Jilg, N.; Fusco, D.N.; Brisac, C.; Schaefer, E.A.; Cai, D.; et al. HCV induces transforming growth factor beta1 through activation of endoplasmic reticulum stress and the unfolded protein response. *Sci. Rep.* **2016**, *6*, 22487. [CrossRef] [PubMed]

76. Matsuzaki, S.; Hiratsuka, T.; Taniguchi, M.; Shingaki, K.; Kubo, T.; Kiya, K.; Fujiwara, T.; Kanazawa, S.; Kanematsu, R.; Maeda, T.; et al. Physiological ER Stress Mediates the Differentiation of Fibroblasts. *PLoS ONE* **2015**, *10*, e0123578. [CrossRef] [PubMed]

77. Cao, S.S.; Song, B.; Kaufman, R.J. PKR protects colonic epithelium against colitis through the unfolded protein response and prosurvival signaling. *Inflamm. Bowel Dis.* **2012**, *18*, 1735–1742. [CrossRef] [PubMed]

78. Cho, J.A.; Lee, A.H.; Platzer, B.; Cross, B.C.; Gardner, B.M.; De Luca, H.; Luong, P.; Harding, H.P.; Glimcher, L.H.; Walter, P.; et al. The unfolded protein response element IRE1alpha senses bacterial proteins invading the ER to activate RIG-I and innate immune signaling. *Cell Host Microbe* **2013**, *13*, 558–569. [CrossRef] [PubMed]

79. Becker, B.; Blum, A.; Giesselmann, E.; Dausend, J.; Rammo, D.; Muller, N.C.; Tschacksch, E.; Steimer, M.; Spindler, J.; Becherer, U.; et al. H/KDEL receptors mediate host cell intoxication by a viral A/B toxin in yeast. *Sci. Rep.* **2016**, *6*, 31105. [CrossRef] [PubMed]

80. Yin, L.; Chen, X.; Vicini, P.; Rup, B.; Hickling, T.P. Therapeutic outcomes, assessments, risk factors and mitigation efforts of immunogenicity of therapeutic protein products. *Cell. Immunol.* **2015**, *295*, 118–126. [CrossRef] [PubMed]

81. Swanson, S.J.; Bussiere, J. Immunogenicity assessment in non-clinical studies. *Curr. Opin. Microbiol.* **2012**, *15*, 337–347. [CrossRef] [PubMed]

82. Biancheri, P.; Giuffrida, P.; Docena, G.H.; MacDonald, T.T.; Corazza, G.R.; Di Sabatino, A. The role of transforming growth factor (TGF)-beta in modulating the immune response and fibrogenesis in the gut. *Cytokine Growth Factor Rev.* **2014**, *25*, 45–55. [CrossRef] [PubMed]

83. Travis, M.A.; Sheppard, D. TGF-beta activation and function in immunity. *Annu. Rev. Immunol.* **2014**, *32*, 51–82. [CrossRef] [PubMed]

84. Morikawa, M.; Derynck, R.; Miyazono, K. TGF-beta and the TGF-beta Family: Context-Dependent Roles in Cell and Tissue Physiology. *Cold Spring Harb. Perspect. Biol.* **2016**, *8*. [CrossRef] [PubMed]

85. Penn, J.W.; Grobbelaar, A.O.; Rolfe, K.J. The role of the TGF-beta family in wound healing, burns and scarring: A review. *Int. J. Burns Trauma* **2012**, *2*, 18–28. [PubMed]

86. Kim, M.K.; Maeng, Y.I.; Sung, W.J.; Oh, H.K.; Park, J.B.; Yoon, G.S.; Cho, C.H.; Park, K.K. The differential expression of TGF-beta1, ILK and wnt signaling inducing epithelial to mesenchymal transition in human renal fibrogenesis: An immunohistochemical study. *Int. J. Clin. Exp. Pathol.* **2013**, *6*, 1747–1758. [PubMed]

87. Zhang, Y.Q.; Liu, Y.J.; Mao, Y.F.; Dong, W.W.; Zhu, X.Y.; Jiang, L. Resveratrol ameliorates lipopolysaccharide-induced epithelial mesenchymal transition and pulmonary fibrosis through suppression of oxidative stress and transforming growth factor-beta1 signaling. *Clin. Nutr. (Edinburgh, Scotland)* **2015**, *34*, 752–760. [CrossRef] [PubMed]

88. Colak, S.; Ten Dijke, P. Targeting TGF-beta Signaling in Cancer. *Trends Cancer* **2017**, *3*, 56–71. [CrossRef] [PubMed]

89. Ji, J.; Sundquist, J.; Sundquist, K. Cholera Vaccine Use Is Associated with a Reduced Risk of Death in Patients with Colorectal Cancer: A Population-based Study. *Gastroenterology* **2017**. [CrossRef] [PubMed]

90. Villanacci, V.; Antonelli, E.; Geboes, K.; Casella, G.; Bassotti, G. Histological healing in inflammatory bowel disease: A still unfulfilled promise. *World J. Gastroenterol.* **2013**, *19*, 968–978. [CrossRef] [PubMed]

91. Vaughn, B.P.; Shah, S.; Cheifetz, A.S. The role of mucosal healing in the treatment of patients with inflammatory bowel disease. *Curr. Treat. Options Gastroenterol.* **2014**, *12*, 103–117. [CrossRef] [PubMed]

92. Hansel, T.T.; Kropshofer, H.; Singer, T.; Mitchell, J.A.; George, A.J. The safety and side effects of monoclonal antibodies. *Nat. Rev. Drug Discov.* **2010**, *9*, 325–338. [CrossRef] [PubMed]

93. Murdaca, G.; Spano, F.; Contatore, M.; Guastalla, A.; Penza, E.; Magnani, O.; Puppo, F. Infection risk associated with anti-TNF-alpha agents: A review. *Expert Opin. Drug Saf.* **2015**, *14*, 571–582. [CrossRef] [PubMed]
94. Saleh, M.; Trinchieri, G. Innate immune mechanisms of colitis and colitis-associated colorectal cancer. *Nat. Rev. Immunol.* **2011**, *11*, 9–20. [CrossRef] [PubMed]

© 2017 by the authors. Licensee MDPI, Basel, Switzerland. This article is an open access article distributed under the terms and conditions of the Creative Commons Attribution (CC BY) license (http://creativecommons.org/licenses/by/4.0/).

Review

Animal Toxins Providing Insights into TRPV1 Activation Mechanism

Matan Geron †, Adina Hazan † and Avi Priel * 🔟

The Institute for Drug Research (IDR), School of Pharmacy, Faculty of Medicine, The Hebrew University of Jerusalem, Jerusalem 9112001, Israel; matan.geron@mail.huji.ac.il (M.G.); adina.hazan@mail.huji.ac.il (A.H.)
* Correspondence: avip@ekmd.huji.ac.il; Tel.: +972-2-675-7299
† These authors contributed equally to this work.

Academic Editor: Steve Peigneur
Received: 28 September 2017; Accepted: 13 October 2017; Published: 16 October 2017

Abstract: Beyond providing evolutionary advantages, venoms offer unique research tools, as they were developed to target functionally important proteins and pathways. As a key pain receptor in the nociceptive pathway, transient receptor potential vanilloid 1 (TRPV1) of the TRP superfamily has been shown to be a target for several toxins, as a way of producing pain to deter predators. Importantly, TRPV1 is involved in thermoregulation, inflammation, and acute nociception. As such, toxins provide tools to understand TRPV1 activation and modulation, a critical step in advancing pain research and the development of novel analgesics. Indeed, the phytotoxin capsaicin, which is the spicy chemical in chili peppers, was invaluable in the original cloning and characterization of TRPV1. The unique properties of each subsequently characterized toxin have continued to advance our understanding of functional, structural, and biophysical characteristics of TRPV1. By building on previous reviews, this work aims to provide a comprehensive summary of the advancements made in TRPV1 research in recent years by employing animal toxins, in particular DkTx, RhTx, BmP01, *Echis coloratus* toxins, APHCs and HCRG21. We examine each toxin's functional aspects, behavioral effects, and structural features, all of which have contributed to our current knowledge of TRPV1. We additionally discuss the key features of TRPV1's outer pore domain, which proves to be the target of the currently discussed toxins.

Keywords: TRPV1; outer pore domain; spider toxin; centipede toxin; scorpion toxin; snake toxin; sea anemone; nociception; venom; pain

---

1. Background

Noxious stimuli are detected by a wide array of peripherally located ion channel receptors, where their activation is transferred to the central nervous system mainly via small diameter, unmyelinated c fibers [1]. Due to the limited number of receptors charged with detecting innumerable stimuli, many of the ion channel receptors have the ability to detect and respond appropriately to a multitude of stimuli (polymodality) [2,3]. One such receptor, TRPV1, has the ability to detect internal and external noxious stimuli such as high temperature ($>42\ ^\circ$C), low pH, peptide toxins and capsaicin, the "hot" chemical in chili pepper [4–6]. TRPV1 activation, sensitization, and modulation have been indicated in many diseases, including irritable bowel syndrome, cancer, and diabetes [7–10].

Structurally, the TRPV1 receptor consists of four independent, identical protein subunits that assemble into a functional, non-selective cation channel (Figure 1) [11]. Each of the four subunits is composed of six transmembrane segments, with a pore-forming loop between segments five and six (pore helix) and intracellular *N*- and *C*- termini (Figure 1) [11,12]. Importantly, 25 amino acids in the S5 outer pore domain leading to the pore helix make up the pore turret, which has been proposed to be involved in heat-induced activation [13]. Over 112 known functional sites found along the sequence of each subunit are responsible for TRPV1's capacity to respond to a multitude of agonists, antagonists, and channel blockers [14]. One major binding site is the S3–S4 located vanilloid binding site, which is activated by endo-vanilloids such as anandamide, or exo-vanillolids such as capsaicin [15–18]. A second major binding domain has been described as the outer pore region, essential for proton-mediated TRPV1 activation [5]. Following the successfully constructed cryo-EM structure, we now understand that TRPV1 boasts a relatively broad outer pore domain, which has been suggested to enhance accessibility to ligands of variable size and charge [19]. Indeed, to date, all characterized animal toxins' binding sites were found to reside in the outer pore domain. During TRPV1 gating, considerable rearrangements occur in the outer pore, pore helix, and selectivity filter, but not in transmembrane segments 1–4 as would be expected when compared to the structurally-similar voltage gated channels [11,12,20]. In addition, TRPV1 activation is distinct in that it depends on the opening of two allosterically coupled gates. Large structural rearrangements in the outer pore domain affect the pore helix and selectivity filter, while a hydrophobic narrowing of the lower gate undergoes expansion during activation [11]. Functionally, it has been shown that activation of single subunit through the vanilloid binding site is sufficient to activate the entire channel [21]. Activation via the outer pore domain by protons requires four functional binding sites, indicating a discrepancy in the coupling mechanism between the two gates and the opening of the pore [21].

Due to its major role in the pain pathway, TRPV1 offers an attractive target for pharmacological manipulation in pain management [22]. There have been many attempts to design functional TRPV1 antagonists, which could potentially block pain transmission from the periphery. Even prior to the cloning of TRPV1, the potent and specific antagonist capsazepine was synthesized by modifying the structure of the naturally-occurring plant toxins, capsaicin and resiniferatoxin, in order to block nociceptive firing and potentially be used as an analgesic [23]. However, this molecule, along with many other molecules synthesized since then, has been found unsuitable for therapeutic use. The major stumbling blocks in successful pain therapy through TRPV1 inhibition are critical side effects such as hyperthermia and changes in core body temperature, or decreased noxious heat detection due to TRPV1's role in thermo-regulation and sensation [24]. A more successful approach, albeit imperfect, has been to take advantage of TRPV1 desensitization that occurs during strong activation by potent agonists such as capsaicin [25]. This approach is currently in use as a topical analgesic cream, but maintains side effects such as an initial burning sensation and potential damage to nociceptors [26]. Accordingly, more research is necessary to understand the function and regulation of TRPV1 in pain in order to develop molecules which affect only a single modality of the channel instead of blocking the channel's full array of functions.

Overall, an organism's nociceptive system serves as a beneficial protective mechanism to prevent or respond to tissue damage [2]. Venomous animals take advantage of this system and have developed highly specific, potent, and complex toxins to specifically target sensitive proteins in the pain system, such as TRPV1 [6]. As such, these molecules provide unique probes for understanding functionally important proteins in the peripheral pain pathway.

TRPV1's original cloning and characterization was made possible by the plant toxin capsaicin, used as a probe to search for a heat channel in the pain pathway [4]. Since then, capsaicin has become the "gold standard" of TRPV1 activation, and has contributed to critical developments over the last 20 years in understanding the role of TRPV1 in pruritis, cancer, weight loss, and in the cannabinoid system [8,9,27,28]. Resiniferatoxin (RTX), an ultra-potent phytotoxin agonist that is found in the plants *Euphorbia resinifera* and *Einhorbia poissonii* found in Morocco, has likewise become an invaluable research tool [29,30]. The successful elucidation of the TRPV1 cryo-EM structure in the open state was possible due to the extreme affinity and binding of RTX paired with the irreversible outer-pore binding Double knot Toxin (DkTx) [11]. Functionally, due to the strong activation and high affinity binding to TRPV1, RTX has been shown to selectively ablate TRPV1$^+$ nociceptors. Taking advantage of this, RTX administration is being clinically investigated as an analgesic in several ailments, such as severe burn subjects or bone cancer [31]. Additionally, vanillotoxins (VaTx1, VaTx2 and VaTx3), from the venom of the tarantula *Psalmopoeus cambridgei*, were the first described animal derived peptides that activate the somatosensory system, in this case specifically targeting TRPV1 (previously reviewed) [6,32]. This set of toxins, similar in structure to each other, contains three inhibitory cysteine knot (ICK) peptides [32]. ICK is a structural motif shared by toxins from venomous animals of many species, such as cone snails, spiders and scorpions [32]. These toxins have a unique feature of six cysteine residues that form sulfide bridges, producing a "knot-like" structure [33]. ICK toxins are most commonly involved in channel inhibition, such as Kv channels, causing paralysis and hyperexcitability [33]. Finally, some venomous animals (e.g, funnel web spider and *Heteractis crispa* sea anemone) were shown to produce toxins that inhibit TRPV1 rather than activating it. The purpose of these presumably analgesic toxins remain unknown [34].

Overall, toxins have provided unparalleled tools in understanding TRPV1 activation, regulation, and structure. In the pursuit to design more efficient and specific pain treatments through TRPV1 modulation, we further turn to toxins to elucidate the different binding sites of TRPV1 and its activation mechanism(s). To date, several animal toxins have been described that activate, modulate, or inhibit TRPV1. In this review, we focus on the recent animal toxins research that have contributed to our understanding of this receptor's unique activation profile and its inner workings as a receptor in the pain pathway.

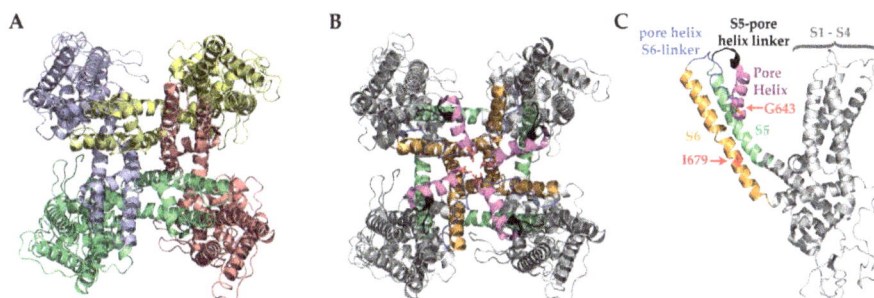

Figure 1. Pore-forming domains in the TRPV1 tetramer. (**A**) Top view of closed tetrameric TRPV1 channel showing transmembrane helices and emphasizing the intertwined subunits arranged around a central pore. Each subunit is color-coded individually. PDB ID: 5IRZ. (**B**) Color coded outer pore domain and pore-forming structures from a top-down view of the tetrameric channel. S5 (green) links with the pore helix (pink) via its linker (black). The pore helix connects to S6 (gold) via an outer pore linker (blue), which also harbors the upper selectivity filter Gly 643 (red). A lower selectivity filter (red) appears further down S6 at Ile 679 (red). S1–S4 are represented in grey. Note that the pore turret (23 AA) situated between S5 and the pore helix is omitted in this structure. PDB ID: 5IRZ. (**C**) Side view of a single TRPV1 subunit color coded as described in B. PDB ID: 5IRZ.

2. Double-Knot Toxin (DkTx)

2.1. Introduction

Vanillotoxins are ICK motif-containing spider proteins targeting TRPV1 [6,32]. These toxins activate the channel by binding to the extracellular pore domain [35]. One of these vanillotoxins, DkTx (8522 Da), has a unique bivalent structure: two tandemly repeated ICK motifs that are highly homologous (67% identity, Figure 2) [36,37]. This TRPV1-selective toxin is found in the venom of the Chinese bird spider (*Ornithoctonus huwena*), an old world aggressive tarantula [36,38]. Its bivalent structure allows DkTx to form an exceptionally stable complex with TRPV1's outer pore region resulting in persistent current conductance through the channel [36,37,39,40]. In contrast, the separated DkTx ICKs (knot 1, "K1"; knot 2, "K2") as well as the single knot vanillotoxins, VaTx1-3, produce reversible binding and activation of TRPV1 [36,39]. Additionally, DkTx's bivalency accounts for its increased potency and avidity of TRPV1 activation in comparison to other vanillotoxins [36,41].

2.2. Functional Aspects

An interesting feature of DkTx-induced TRPV1 current is its relatively slow kinetics during activation until maximal amplitude is reached [36]. However, co-application of DkTx and capsaicin results in increased binding and activation rates for the toxin as evident by faster onset of a stable, irreversible current typical to DkTx on both the single channel and whole cell levels [36]. Thus, it was suggested that DkTx preferentially binds, and subsequently locks, TRPV1 in the channel's open conformation [20,36,42]. Accordingly, in the absence of another activating stimulus, DkTx presumably binds TRPV1 during one of the channel's brief spontaneous transitions to the open state [20]. Further support for this notion can be derived from the apo structure of TRPV1 in which superimposition of DkTx clashes with the channel's side chains [20]. Thus, it is possible that channel activation by DkTx in low temperatures, when the open state excursions of TRPV1 are diminished, could be reduced. In addition, it could be expected that the toxin's ability to irreversibly lock and stabilize TRPV1 in its open state would yield maximal open probability, especially when compared to the flickering nature of channel activation by the reversible agonist capsaicin [36,43,44].

2.3. Effect on Nociception

Considering the toxin's unique activation profile of TRPV1, DkTx is expected to cause an excruciating and prolonged pain response [36]. Indeed, the bite of a Chinese-bird spider reportedly produces substantial pain and inflammation [38]. Nonetheless, experiments evaluating the specific effects of DkTx on pain and aversive responses have yet to be conducted. Additionally, the neuronal firing properties of TRPV1-positive nociceptors in response to DkTx application has yet to be characterized. Thus, whether prolonged TRPV1 activation by the toxin indeed leads to prolonged action potential firing on the neuronal level is still unknown.

2.4. Structural Features

A recent cryo-EM structure in near-atomic resolution of the DkTx-TRPV1 complex revealed that the two toxins' ICK motifs bind two adjacent subunits in the homo-tetrameric channel in an anti-parallel orientation [11,20,45]. Therefore, two DkTx molecules can fully occupy a single TRPV1 channel by binding to its outer pore domain [20,39,46]. DkTx's linker consists of seven amino acids, including two proline residues, which likely reduce its flexibility [36,39]. In addition, the toxin's linker adopts a tense and constrained conformation upon DkTx binding to TRPV1, which may reflect a linker's length that has evolved to match the distance between two adjacent ICK binding sites [20]. As different activators display a distinct activation stoichiometry of TRPV1, it is still not known how many knots bound to the channel are required in order to elicit channel gating [21]. In each TRPV1 subunit, the different toxin knots bind equivalent binding sites that are situated in the interface of two neighboring subunits [11,39]. Thus, when bound, a single knot interacts with the pore helix of one subunit and the pore loop preceding S6 in an adjacent subunit [20]. A previous alanine scan study has identified four TRPV1 residues in the S5-pore helix loop (I599; rTRPV1), S6-pore helix loop (F649; rTRPV1) and S6 (A657 and F659; rTRPV1) which are critical to channel activation by DkTx [36]. Additionally, molecular-dynamics simulations analysis suggested several TRPV1 residues that are in proximity to the bound DkTx and may be involved in toxin binding [39]. These include K535 and E536 (S4) (rTRPV1), Y631 (pore helix) (rTRPV1) and a stretch of residues between the pore helix and S6 (F649, T650, N652, D654, F655, K656, A657 and V658; rTRPV1). Thus, it was postulated that the more distant I599 and F659 (rTRPV1) contribute to DkTx-induced gating but not directly to the binding of the toxin [39]. It was further suggested that these two residues along with V595, F649 and T650 (rTRPV1) form a hydrophobic cluster which lies in the interface between the S5, S6, and the pore helix, and behind the selectivity filter [39]. Following DkTx binding, it is presumed that this cluster is disrupted and that residues in the cluster undergo substantial conformational changes which lead to ion permeation through the channel [39].

Although DkTx's ICK motifs are highly homologous and share the same binding site, they differ in binding orientation, leading to a higher potency observed in K2 than K1 [36,39]. Computational alanine scanning revealed that most residues involved in channel binding are conserved between the two ICKs as W11, G12, K14 and F27 from K1 as well as W53, G54, K56 and F67 from K2 were found to play an important role in TRPV1 activation (Figure 2) [39]. However, functional differences between the two knots could stem from other residues crucial for toxin binding which are variable between K1 and K2 such as K13 and S55, M25 and L65, and K35 and R75, respectively (Figure 2) [39]. Interestingly, these alternate residues are situated in a region that was shown to influence the knot's affinity to TRPV1 in a chimera study [39]. DkTx contains a large hydrophobic surface as each ICK contains two hydrophobic fingertips (K1: W11, M25, F27, and I28; K2: W53, L65, A66, F67, and I68) (Figure 2). However, these fingers are surrounded by acidic and basic residues (K1: E7, K13, K14, H30, and K32; K2: E47, E49, K56, E72, K73, and R75) (Figure 2) [20,39]. Similar amphipathic nature was previously shown to allow other ICK toxins to protrude into lipid environments [47–50]. Furthermore, tryptophan (Trp) fluorescence showed that DkTx is indeed able to undergo partitioning into lipid environments [39]. In addition, it was found that K1 permeates the membrane more efficiently than K2 [39]. Thus, it was speculated that in the course of DkTx binding, K1 initially partitions the membrane and then coordinates the binding of its more potent counterpart, K2, to TRPV1 [39]. However, there is still no evidence that this is indeed the case. An improved-resolution cryo-EM structure of the DkTx-bound TRPV1 shows that DkTx hydrophobic fingers penetrate 9 Å deep into the membrane thus causing local distortions in the lipid bilayer [20]. Moreover, this structure revealed that some membrane lipids form interactions with both DkTx fingers and TRPV1 residues thus producing a tripartite complex. For instance, DkTx's W11 (K1) and W53 (K2) form hydrophobic interactions with aliphatic chains whose polar head groups bind to R534 (S4; rTRPV1) in TRPV1 [20]. Other examples are F27 (K1) and F67 (K2) that similarly interact with triglycerides which are also in contact with S629 (pore helix; rTRPV1) as well as Y453 (S1; rTRPV1) [20]. These toxin-lipid-channel interactions along with DkTx-TRPV1 bonds likely compensate for the energetic penalty predicted for the interference of the organized lipid environment in this structure [20,39]. Overall, the interaction surface between DkTx and TRPV1 is quite limited as the bound toxin is also stabilized via its interaction with the lipid bilayer [20].

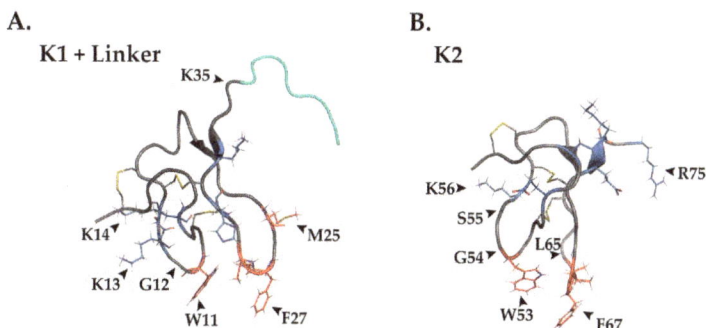

Figure 2. The amphipathic nature of the DkTx structure allows it to protrude the membrane bilayer. Individually visualized knots of DkTx, with the linker (cyan) between the two knots appearing on K1 (**A**) Hydrophobic residues are labeled in red and polar residues in blue. This amphipathic nature presumably enables DkTx to successfully protrude into the lipid environment of the cell membrane. Key amino acids indicated in TRPV1 binding according to computational scan studies have been labeled (). Disulfide bridges are labeled in yellow. K1, PDB ID2N9Z. (**B**) K2 knot of DkTx labeled as described in A. Amino acids in K2 are numbered according to the molecule in its entirety.PDB ID: 2NAJ.

3. RhTx

3.1. Introduction

RhTx is a peptide toxin (27 amino acids) that was identified in the venom of the Chinese red-headed centipede (*Scolopendra subspinipes mutilans*) [51]. This aggressive arthropod, which populates parts of eastern Asia and Australasia, can cause extreme localized pain upon envenomation [51,52]. It was found that RhTx is a selective TRPV1 activator, as it does not affect other TRPV channels (i.e., TRPV2-4). In addition, the voltage-gated potassium ion channel Kv2.1, which was previously shown to be inhibited by other peptide toxins targeting TRPV1 (i.e., VaTx1 and VaTx2), is neither inhibited nor activated by RhTx [32,51]. Similar to capsaicin-mediated TRPV1 activation, RhTx displays very rapid kinetics, in both channel opening and washout. This is in contrast to the slow-developing, slow washing DkTx described above [36,51].

3.2. Functional Aspects

RhTx activates TRPV1 with a comparable efficacy to capsaicin [51]. However, functional examinations revealed that TRPV1 activation by RhTx is highly temperature-dependent, as increased temperatures strongly potentiate the toxin activity [51,53]. In contrast to capsaicin, RhTx activity is significantly reduced at 20 °C compared to room temperature (RT) experiments and completely abolished at 10 °C [51]. Furthermore, 100 nM RhTx, 20% of the toxin's EC_{50} at RT, reduces heat activation threshold below body temperatures [51]. Thus, RhTx is a potent TRPV1 activator in mammals' physiological body temperatures. Single TRPV1 channel recordings revealed that RhTx evokes a near unity open probability while reduced current conductance was observed in negative holding membrane potentials in comparison to capsaicin [51]. Hence, additional experiments at negative holding potentials are required to understand the physiological relevance of the temperature-dependent activation by RhTx. In addition, RhTx desensitizes TRPV1's response to heat, whereas activation by capsaicin is unaffected. These findings imply that RhTx affects the heat activation machinery of the channel either directly or allosterically [51]. However, the changes in toxin potency observed could merely be the result of preferential RhTx binding to the open state of TRPV1. Thus, further evidence is required to establish the functional interaction between RhTx and heat, as well as with other TRPV1 activators such as protons and capsaicin.

3.3. Effect on Nociception

RhTx was shown to evoke an acute pain response when injected into mice [51]. However, longer term possible implications of RhTx application such as sensitivity to heat, cold and tactile stimuli were not tested. Firing properties and desensitization of TRPV1 positive neurons activated by this toxin are yet to be characterized as well. In addition, while other potent toxins (e.g., capsaicin) targeting TRPV1 eventually cause TRPV1+ fiber denervation, it is still not known whether this is the case for RhTx. The RhTx-induced pain response observed was comparable to the response elicited by the crude centipede venom [51]. Thus, RhTx could account for the excruciating pain evoked by this centipede's bite. Interestingly, a previously described toxin from the same venom, Ssm6a, was shown to selectively inhibit human and rodents Nav1.7 channels, causing analgesia [54]. Thus, the algogenic effect of this centipede's venom is surprising given the prominent role of Nav1.7 in nociception [55].

3.4. Structural Features

Structural analysis of RhTx revealed two disulfide bonds and a cluster of charged residues on one side of the molecule, endowing the toxin with polarity (Figure 3) [51,53]. Importantly, four charged residues (D20, K21, Q22 and E27) and one polar residue (R15) from the same structural domain were found to take part in RhTx-TRPV1 binding in a mutagenesis study, causing reduced or enhanced activation, respectively (Figure 3) [51]. These findings indicate that the toxin's charged surface forms electrostatic interactions with the channel [51]. No effect on RhTx activity was detected when hydrophobic residues were mutated, implying there is no significant interaction between the toxin and the lipid bilayer [51]. Further analysis of channel interaction with RhTx identified D602 (mTRPV1 turret), Y632 and T634 (mTRPV1 pore helix) as critical to TRPV1 activation by this toxin. Using Rosetta-based molecular docking, RhTx was predicted to bind at the interface of two TRPV1 subunits similar to DkTx [20,51]. Thus, it is possible that the binding sites of these two toxins are overlapping. Interestingly, another TRPV1 residue (L461 of the mouse TRPV1) located within the S1–S2 extracellular linker was found to facilitate RhTx-induced TRPV1 activation. This observation points to a possible role of this region, which was previously thought to be stationary, in the activation mechanism of TRPV1 [51,56].

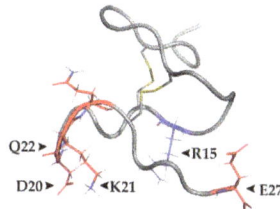

Figure 3. The polar RhTx binds TRPV1 through its charged surface. Structure of RhTx indicating the polarity of the molecule. Charged residues D20, K21, Q22 and E27 (red) have been indicated in TRPV1 binding, along with the polar residue R15. Two cysteine bridges are highlighted in yellow. PDB ID: 2MVA.

4. BmP01

4.1. Introduction

Scorpions are a source for diverse neurotoxins, which have contributed greatly to the study of their targeted ion channels [57]. BmP01, a 3178.6 Da (29 amino acids) protein, is the first scorpion toxin with TRPV1 activating properties to be described [58,59]. This toxin, found in the venom of *Mesobuthus martensii*, has a typical inhibitory cysteine knot (ICK) motif structure with three disulfide bonds (1–4, 2–5 and 3–6) stabilizing a compact and rigid protein fold (Figure 4) [58,59]. Within the vast ICK toxins family, BmP01 is further sub-classified as part of the a-KTX8 toxins subgroup with whom it shares similar topology and sequence features [60,61].

4.2. Functional Aspects

Concentration-response relationship experiments revealed that BmP01 dose-dependently activates the channel with comparable efficacy to capsaicin [59]. Accordingly, in single channel recordings it was shown that BmP01 produces conductance through TRPV1's pore similar to capsaicin, albeit with reduced open probability [61]. The kinetics of BmP01 activity and washout in electrophysiological recordings from TRPV1 expressing cells also resembles capsaicin [61]. The desensitizing properties of BmP01 on TRPV1 were not tested so far. BmP01 does not activate TRPV3, a closely related channel to TRPV1 [61]. The effect of this toxin on other prominent pain sensing TRP channels (e.g., TRPA1, TRPM8, and TRPV2) has yet to be determined. However, in addition to TRPV1, BmP01 was also found to modulate the activity of voltage-gated potassium channels by potently inhibiting mKv1.3, hKv1.3, and rKv1.1, but not mKv1.1, thus presenting species specificity [59]. The bi-functionality of this toxin might enable an enhanced nociceptive response to BmP01 as both activating TRPV1 and inhibiting Kv channels would result in hyperexcitable nociceptors [32,61–63]. Remarkably, BmP01 displays strong pH-dependent activity [61]. BmP01 displays low potency at neutral pH (EC$_{50}$ = 169.5 ± 12.3 µM) [59,61]. However, in acidic conditions the toxin's inhibitory effect in Kv1.3 is diminished, whereas its agonistic effect on TRPV1 is greatly potentiated (EC$_{50}$ = 3.76 ± 0.4 µM) [61]. This enhanced TRPV1 response is seen in pH values (~6.5) that were previously shown to sensitize the channel rather than activate it [5,61].

Thus, BmP01 and protons synergize to produce enhanced TRPV1 response [61]. Similar relations between BmP01 and heat (which similarly to protons modulates TRPV1 gating through the outer pore region), were yet to be reported. In addition, how BmP01 unique functional properties affect the neuronal response is yet to be characterized.

4.3. Effect on Nociception

Injection of 500 μM BmP01 evokes a pain response in wild-type mice but not in TRPV1 KO mice [59]. Therefore, this finding suggests that BmP01 does not activate other pain receptors in a physiologically significant manner. However, while BmP01-induced pain requires TRPV1, it is still not clear whether Kv1.3 contributes to this response. Nevertheless, it was shown that solely blocking this potassium channel does not evoke a pain response [59]. Furthermore, considering pH effect on both BmP01-dependent Kv inhibition and TRPV1 activation and that the pH of *Mesobuthus martensii* venom is acidic, TRPV1 is probably the main ion channel targeted by BmP01 to induce pain upon this scorpion's sting [61]. Thus, the pH-dependent sensitization of TRPV1 towards BmP01 reflects a strategy which enables physiologically relevant toxin concentrations to inflict pain following scorpion envenomation [61]. Indeed, injecting BmP01 in an acidic solution potentiated the response to this toxin in mice while such an effect was absent for capsaicin in low pH solutions [61]. Increased sensitivity to heat and tactile stimuli were not tested following BmP01 application.

4.4. Structural Features

ICK motifs are a common structural feature among many neurotoxins including the TRPV1 targeting toxins, DkTx and VaTxs, which bind the outer pore domain of the tetrameric TRPV1 [36,64,65]. Indeed, chimera studies along with docking calculations indicated that BmP01 also binds this channel domain [61]. Site directed mutagenesis screening identified three polar residues (E649, T651, and E652; hTRPV1) involved in TRPV1 activation by BmP01 which are in proximity to residues implicated in vanillotoxins-induced gating [36,61]. Interestingly, these residues are situated in the pore helix-S6 loop that was shown to influence TRPV1 gating [56,61]. Specifically, E649 was shown to undergo protonation under acidic conditions that leads to channel activation [5]. Alanine scanning and thermodynamic cycle analysis revealed that BmP01's K23 forms electrostatic interaction with E649 while the role of T651 and E652 as well as other possible interaction sites remain to be elucidated (Figure 4) [61]. Another site involved in proton-dependent TRPV1 gating is E601 (hTRPV1) which is protonated in slightly less acidic conditions than E649 [5]. Notwithstanding, while protonation at E601 only sensitizes TRPV1, protonation at both sites is required for robust channel activation [5,61]. Further functional tests showed that the protonation mimicking mutation, E601Q, potentiated BmP01 response in neutral pH while preventing protonation at this position by using the E601A mutation produced diminished toxin potency and pH dependence [61]. Thus, BmP01 takes advantage of the machinery mediating TRPV1 gating by protons to cause channel activation in pH values which otherwise only sensitize the channel [61].

K23 ▸

Figure 4. BmP01's K23 forms an electrostatic interaction with TRPV1 outer pore region. Structure of BmP01 indicating the typical ICK motif formed by three disulfide bonds (indicated in yellow). Red indicates key amino acid K23, which interacts with E649 of TRPV1, an important proton-binding site in TRPV1 channel activation. PDB ID: 1WM7.

5. *Echis coloratus* Toxins

5.1. Introduction

Pain is a hallmark of envenomation by most snakes [66]. However, little is known on how snake venoms produce this response, as only a few snake toxins targeting the pain pathway have been described so far [66–68]. Recently, peptides from the venom of the *Echis coloratus* viper were shown to activate TRPV1 [69]. This snake's bite produces an intense local burning pain along with local swelling and hemorrhagic disturbances [70]. In a screening of RP-HPLC fractions of *Echis coloratus* venom, three (F1, F7, F13) out of 24 fractions were found to produce TRPV1 response [69]. However, the structural and functional properties of F1 and F7 remain unknown, as these fractions were not further analyzed. Moreover, while F13 was shown to contain several proteins, this fraction was not subjected to further purification steps and the TRPV1-activating entity was not isolated [69]. Thus, the structure of the toxin targeting TRPV1 was not determined. Nonetheless, denaturing proteins in this fraction did not affect TRPV1 activation, indicating that the active compound is a low molecular weight peptide that presumably adopts a stable tight helical conformation similar to other heat resistant toxins previously described in snake venoms [69,71,72].

5.2. Functional Aspects

F13 evokes an outwardly rectifying TRPV1 current with extremely fast kinetics as F13 activation immediately terminates upon washout, similar to capsaicin and protons [69]. In addition, F13 evokes acute desensitization and tachyphylaxis of TRPV1 in the presence of calcium with typical kinetics [69]. F13 was shown to be a selective TRPV1 activator as no effect of this fraction was observed on other TRP channels expressed in nociceptors; TRPV2, TRPA1 and TRPM8 [69]. Modulation of voltage-gated ion channels by this fraction was not tested. In neurotropic assays, F13 presented NGF activity while SDS-PAGE analysis of this fraction revealed a protein with similar size to mouse NGF [69]. NGFs, which are commonly found in a variety of snake venoms, cause acute sensitization of TRPV1 indirectly through NGF receptors [73]. However, the TRPV1-activating component of this fraction was found to be independent of the NGF signaling pathway, as NGF receptor blockers do not alter TRPV1 response to F13 in a heterologous expression system [69]. These findings suggest that F13 contains both TRPV1 activating toxin and NGF, which likely synergize in vivo to produce an increased pain response mediated by TRPV1 [69].

5.3. Effect on Nociception

Envenomation by *Echis coloratus* snake results in hypotension, local swelling, necrosis and pain in humans [70,74,75]. While substances released from the damaged tissue are able to sensitize and activate nociceptors, this *Echis* snake was shown to produce also toxins that specifically target the pain pathway as reviewed here. However, as these TRPV1-targeting toxins were not isolated behavioral tests evaluating their specific effect *in vivo* are yet to be conducted.

5.4. Structural Features

The presumed peptide nature of the TRPV1-activating compound in F13 suggests that the binding site for this toxin resides in the extracellular domains of the channel. Indeed, mutations in the intracellular binding site for phytotoxins, the vanilloid binding site, that were previously shown to abolish capsaicin and resiniferatoxin activity do not influence F13-induced TRPV1 response [69]. Interestingly, the mutation A657P in the outer pore region of TRPV1, which renders the channel insensitive to DkTx and VaTx3, does not affect TRPV1 activation by F13 as well [69]. The possibility that F13 causes TRPV1 activation through the residues mediating protons-induced gating was not tested. Therefore, while this snake toxin likely binds the channel extracellularly, the exact binding site for the F13 toxin remains unknown. Additional mutagenesis and chimera studies along with structural and computational analysis of toxin structure and interactions with TRPV1 are required to locate this toxin's binding site.

6. Analgesic Polypeptide *Heteractis crispa* (APHC) Toxins

6.1. Introduction

While we can logically understand that toxins have evolved to specifically activate the pain system, accumulating evidence suggests that many toxins have components that possess analgesic properties. Such examples include mambalgins from the Black mamba snake or the previously described non-peptide toxin from the spider *Agelenopsis aperta*, shown to inhibit ASICs and TRPV1, respectively [76,77]. The toxins APHC1-3, derived from the sea anemone *Heteractis crispa* found in the Indio–Pacific, were the first polypeptide inhibitor of TRPV1 described [78].

6.2. Functional Aspects

All three toxins were shown to be partial, but potent, antagonists, with APHC3 displaying the highest level of inhibition (71% of 3 µM capsaicin), and the lowest IC_{50} value (18 nM) [78]. As is common with molecules presenting with partial antagonistic characteristics, a bimodal mechanism of action of APHC on TRPV1 activity was recently described [79]. In the presence of saturating concentrations of capsaicin (3 µM), APHC acts as an inhibitor, whereas current produced by very low concentrations (3 nM) of capsaicin are potentiated in the presence of the toxin [79]. Interestingly, APHC1 showed the highest levels of potentiation, showing an increase of 250% of the capsaicin - evoked current [79]. Furthermore, APHC3 potentiated TRPV1 activation in response to slightly acidic (pH 6.2) solution, whereas no inhibition was observed at highly acidic (pH 4.5) solutions [79]. Likewise, low concentrations of the synthetic molecule 2-aminoethoxydiphenyl borate (2APB) were subject to APHC potentiation, while high concentrations were unaffected [79].

6.3. Effect on Nociception

In vivo tests suggest that APHC1 and APHC3 both have the ability to produce analgesia in different behavioral pain models [80]. Both toxins display a dose-dependent inhibition of thermal nociception in doses up to 0.1 mg/kg injected intramuscularly, intraperitoneally, and intravenously [80]. Interestingly, APHC1 is a more potent inhibitor, as doses as low as 0.01 mg/kg are effective in increasing paw withdrawal latency from a hot plate, while the minimum effective dose of APHC3 was 0.05 mg/kg [80]. In comparison, potent serine protease inhibitors with similar structure and folding had no effect on thermal sensation, indicating that the basic structure alone is not enough to affect TRPV1 mediated thermal sensation [80]. When the toxins were tested for their effects in the formalin test for inflammation, the two candidates produced different effects: whereas APHC1 decreased both phases of the formalin test (acute and inflammatory pain), APHC3 attenuated only the second phase, indicating the inhibition of TRPV1 was effective only in modulating TRPV1 during inflammation [80]. The proposed mechanism is that APHC3 inhibits pH mediated TRPV1 activation, found during the inflammatory response, whereas APHC1 modulates additional TRPV1 modalities [80]. On the other hand, both toxins attenuated thermal hyperalgesia observed during complete Freund's adjuvant (CFA) injection in a dose dependent manner, although the effect of APHC1 was larger [80].

Similar to most molecules proposed to affect nociception through TRPV1, these molecules affect core body temperature [80]. Interestingly, unlike molecules that exclusively inhibit all of TRPV1's modalities, in vivo studies show that APHC1 and APHC3 do not cause hyperthermia [80]. Rather, these molecules decrease core body temperatures commonly observed during administration of TRPV1 selective agonists [80]. In accordance with their different inhibitory mechanisms, injection of APHC1 induced a sharp fall in body temperature of −0.8 °C within 30 min after administration, whereas APHC3 produced a slow decrease of −0.6 °C reached 60 min [80].

These toxins give an interesting insight into the possibility of TRPV1 modulation through an intermediate state. A potential aim for the development of novel analgesics through TRPV1 may be to avoid inhibiting TRPV as a whole and instead target a specific modality. These toxins provide an indication of the possible effects of partially inhibiting TRPV1.

6.4. Structural Features

The three highly homologous toxins have been characterized, differing in four of their 56 residues, and are estimated to be approximately 6187.0 Da each [78]. They contain unique features of protease inhibitors derived from bovine pancreas, such as a disulfide rich α/β structure common to BPTI/Kunitz-type enzymes [78]. Other molecules common to this group include SHPI-1, a trypsin inhibitor (85% homology with APHC), KAL-1, a potassium channel blocker, and calcicludine, a calcium channel blocker (52% APHC homology) [78].

Molecular modeling of the toxin on the background of cryo-EM structures of TRPV1 suggest that functionally important residues for toxin interactions are in the outer pore domain, involving the pore helix and two extracellular loops, and overlapping with the proton binding site [79]. Considering this, antagonism by the toxin at acidic pH may involve competitive inhibition rather than allosteric, in the case of other modalities [79]. Arginine residues in the toxin's C terminal mediate the strongest interactions between the toxin and TRPV1 outer pore domain [79]. Specifically, R48 of the toxin interacts with E648, E651, and Y653 (rTRPV1) of TRPV1's outer pore domain [79]. Similar to previously described toxins, APHC1 interacts simultaneously with two TRPV1 subunits as R51 of APHC1 interacts with E636 and Y627, and R55 of APHC1 interacts strongly with D646 of the adjacent subunit [79]. Moreover, two conformational states of the toxin, which depend on an open or closed TRPV1 channel, suggest that there is a two-way effect of conformational rearrangements of the channel affecting the toxin and the toxin affecting the channel [79]. Specifically, in the open state, the pore helix of TRPV1 pushes APHC1 about 2.5 Å in the external direction [79]. In this way, APHC toxins stabilize an intermediate state, producing a bimodal effect on TRPV1 gating [79].

7. *Heteractis crispa* RG 21 (HCRG21)

7.1. Introduction

Toxins with a kunitz-type domain act as protease inhibitors, yet some of these peptides have an additional neurotoxic activity as they also inhibit ion channels [81]. Such dual activity is observed in HCRG21, a kunitz-type peptide from the venom of *Heteractis crispa* [34]. This sea anemone toxin inhibits both trypsin and TRPV1 [34]. The structure of this bifunctional toxin contains three disulfide bonds and has a molecular mass of 6228 Da [34]. The HCRG21 sequence is highly homologous (82%–95%) to APHC1-3, TRPV1 inhibiting toxins from the same venom described above [34,78]. In addition to TRPV1 inhibitors, HCRG21 also shares high identity (84%) with ShPI-1, a potent inhibitor of several voltage-gated potassium channels [34]. However, HCGR21 does not modulate the activity of these channels [34].

7.2. Functional Aspects

TRPV1 inhibition by HCRG21 is dose-dependent with an $IC_{50} = 6.9 \pm 0.4$ μM when co-applied with 1 μM capsaicin [34]. The unclear evolutionary benefit of inhibiting TRPV1 combined with its low potency suggests that this toxin targets additional ion channels. Furthermore, it is not clear whether TRPV1 is inhibited by physiologically relevant toxin concentrations. However, unlike APHC1, HCRG21 is a full channel antagonist [34].

7.3. Effect on Nociception

Introduction of TRPV1 antagonists to the clinic as new analgesics was hampered as these inhibitors produced serious, adverse effects, namely hyperthermia and insensitivity to scalding heat [82–84].

Although these are on-target effects, it is thought that inhibiting one of the modalities activating TRPV1 would produce the desired analgesic effect while maintaining a favorable safety profile. However, HCRG21's ability to block TRPV1 activation by modalities other than capsaicin is yet to be reported.

7.4. Structural Features

Most residues that were found to be important in APHC1 binding to TRPV1 (E6, T14, E38, R48, and R51) are conserved in HCRG21, except for V31, which is substituted by proline [34]. This substitution could underlie the difference in efficacy between these two toxins. However, the significance of the aforementioned residues in HCRG21 activity is yet to be confirmed. Using homology modeling and molecular dynamics simulations, HCRG21 was hypothesized to interact with the regulatory domain in the intracellular TRPV1 C-terminal where PIP2 also binds the channel [34]. Nonetheless, both the ability of HCRG21 to penetrate the membrane and to block capsaicin currents through this site were not tested. A far likelier possibility, which was also suggested using computational analysis, is that HCRG21 binds the outer vestibule sitting directly on the TRPV1 channel pore thus presumably blocking the conductance of ions [34]. Conformational changes in the outer pore region and interaction with a residue in the channel's selectivity filter were also predicted upon toxin binding [34]. However, this computational data is not yet backed by any functional analysis experiments.

8. Toxins and the TRPV1 Outer Pore Domain

Toxins have evolved over thousands of years to target physiologically significant processes in order to exert a robust and acute effect in their victims [6]. To do so, toxins may modulate the activity of central ion channels by binding or affecting functionally important domains in these proteins [6,85]. Thus, toxins have been instrumental in understanding the structure and function of ion channels so far [41]. Common ion channel domains targeted by toxins include the voltage sensor, ion permeable pathway, and agonist binding sites [6]. Another channel site affected by toxins is the outer pore domain [6]. By binding to this region some toxins, like charybdotoxin, serve as a cork occluding ion conductance through the pore and directly inhibit channel activity [86]. However, other toxins either inhibit or activate channels by binding to sites in the outer pore domain that are involved in the channel's gating mechanism. Accordingly, as reviewed here, several toxins were shown to activate, modulate, or inhibit TRPV1 by binding to the outer pore domain of the channel (Figure 5 and Table 1) [34,36,51,61]. These findings signify the outer pore region of TRPV1 as a common domain for binding and gating of the channel by different peptide animal toxins, and highlight the importance of the outer pore domain in TRPV1 activation.

The outer pore region also plays a pivotal role in TRPV1 activation evoked by various stimuli other than toxins. For instance, extracellular protons potentiate and activate TRPV1 through glutamate residues (E600 and E648, respectively) found in this region [5,87]. Furthermore, protons were also found to perturb outer pore structure leading to sub-conductance TRPV1 currents [88]. Divalent cations were suggested to induce substantial conformational rearrangements in the outer pore region, which potentiates channel activity [89]. In addition, mutagenesis experiments revealed gain-of-function mutations within the extracellular pore domain [90]. Residues in this region were also found to be essential for heat activation of TRPV1 [91,92]. However, this does not necessarily mean that the channel's heat sensor lies within the outer pore domain, as other protein domains were also found to be equally involved in this process [93–95].

The proximity of so many binding sites and functionally important domains in the outer pore region raises the possibility that some toxins allosterically modulate TRPV1 activation by other stimuli. Indeed, RhTx was shown to reduce the threshold for heat activation by 6 °C [51]. In addition, BmP01 was found to affect the protons-induced gating machinery. The preferential binding of DkTx to the open TRPV1 conformation suggests that this toxin may also enhance the channel's response to other modalities in physiological settings [36,61].

Structural analyses of the TRPV1 channel in distinct conformations have confirmed that the outer pore domain undergoes substantial structural reorganization, which is associated with a shift in the pore helix relative position during the gating process [11,96]. In contrast, the S1–S4 domain remains static in different channel conformations and was suggested to serve as a scaffold for the S5-Pore-S6 domain [56]. TRPV1 and voltage gated ion channels share a similar topology and three-dimensional structure [46,97]. In voltage-gated ion channels, the outer pore region was shown to be relatively stationary during transitions between the apo and open states, while the S1–S4 domain where the voltage sensor is situated is highly dynamic [10,98]. Reflecting the differential gating mechanisms in these two channel types, VaTx3 inhibits Kv channels by binding to the channel's voltage sensor region (S3–S4) yet activates TRPV1 via the outer-pore domain to cause net hyper-excitability in effected neurons [32,36].

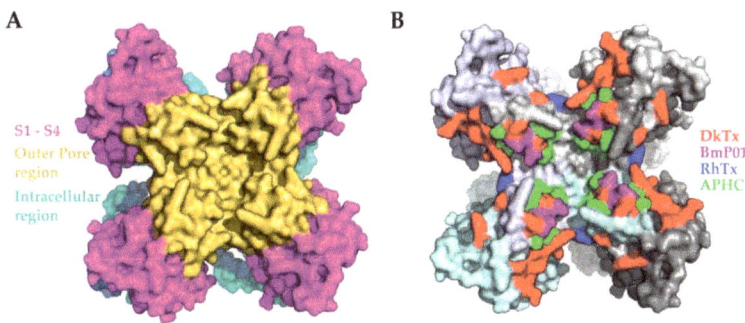

Figure 5. Significant amino acids for toxin-induced modulation of TRPV1 activity. (**A**) The outer pore region is collectively colored in gold, and S1–S4 are labeled in fuchsia. Intracellular structures are blue. PDB ID: 5IRZ. (**B**) Key amino acids indicated in channel activation by each individual toxin are labeled according to the toxin: DkTx, Red; RhTx, Blue; BmP01, Purple; APHC, Green. Whole subunits are lightly colored to visualize the interface between adjacent subunits. PDB ID: 5IRZ.

Table 1. Key toxin features and interactions with TRPV1.

Species	Toxin		
Chinese earth tiger tarantula *Chilobrachys guangxiensis*	Double-knot toxin (DkTx) [36]	**Key Amino Acids**	*TRPV1*: Y453, R534, K535, E536, I599, S629, Y631, F649, T650, A657, N652, D654, F655, K656, A657, V658, F659 *DkTx*: K1: W11, G12, K14, and F27 K2: W53, G54, K56 and F67
		Behavioral effects	Unknown
		Potency	EC$_{50}$ = 0.23 µM
Chinese red-headed centipede *Scolopendra subspinipes mutilans*	RhTx [51]	**Key Amino Acids**	*TRPV1*: D602, Y632, T634, Possibly L461 *RhTx*: D20, K21, Q22, R15, E27
		Behavioral effects	Acute pain response when injected into mice
		Potency	EC$_{50}$ = 521.5 ± 162.1 nM
Chinese Scorpion *Mesobuthus martensii*	BmP01 [59]	**Key Amino Acids**	*TRPV1*: E648, T651, E652
		Behavioral effects	Injection of 500 µM BmP01 evokes a pain response in wt mice but not in TRPV1 KO mice
		Potency	EC$_{50(Ph=6.5)}$ = 3.76 ± 0.4 µM EC$_{50(Ph=7.5)}$ = 169.5 ± 12.3 µM

Table 1. *Cont.*

Species	Toxin		
Palestine saw-scaled viper *Echis coloratus*	F13 [69]	Key Amino Acids	Unknown
		Behavioral effects	Unknown
		Potency	Unknown
Sebae anemone *Heteractis crispa*	Heteractis crispa RG 21 (HCRG21) [34]	Key Amino Acids	HCRG21: E6, T14, P31 E38, R48, R51
		Behavioral effects	Unknown
		Potency	$IC_{50} = 6.9 \pm 0.4 \mu M$
	Analgesic polypeptide Heteractis crispa (APHC1-3) [78]	Key Amino Acids	*TRPV1*: D648, E651, Y653, E636, Y627, D646 *APHC*: V31, R48, R51, R55
		Behavioral effects	• dose-dependent inhibition of thermal nociception • APHC1 decreases both phases of the formalin test (acute and inflammatory pain). APHC3 attenuates only inflammatory phase • attenuates thermal hyperalgesia observed during CFA injection • Injection of APHC1 decreased body temperature by -0.8 °C within 30 min after administration. Injection of APHC3 decreases body temperature by 0.6 °C 60 min after administration.
		Potency	APHC1: $IC_{50} = 6.9 \pm 0.4 \mu M$ APHC3: $IC_{50} = 18$ nM

9. Conclusions

In the past five years, many advances have been made in our understanding of TRPV1 activation, structure, and potential roles in pain management therapy, greatly due to the rich collection of animal toxins that target this ion channel [6,11,41]. In this review, we discuss the molecular traits of these toxins that have been pivotal to our understanding of TRPV1, with the aims of elucidating the detailed functioning of the pain system. It is only in this way that we can develop stronger, more precise tools to manipulate and control the nociceptive system.

Considering the discussed toxins, we find remarkable diversity in terms of structure, functional mechanism, and effect. However, the outer pore region of TRPV serves as a common and complex site in which toxins and other modalities (i.e., heat and protons) converge to modulate TRPV1 gating in distinct manners. For example, DkTx's unique structure produces a strong TRPV1 response, but does so through slow, prolonged TRPV1 activation by binding to the channel's open conformation [36]. In contrast, aversive behavior is also seen in RhTx, which produces strong, capsaicin-like TRPV1 activation [51]. Through RhTx, we also understand that conformational changes due to the ambient temperature may also affect the binding of the molecule [51]. Similarly, TRPV1 activation by BmP01 was shown to be potentiated by low pH values [59]. Thus, these toxins may take advantage of existing conditions at the site of injected venom (the presence of inflammatory mediators, protons, and body temperature) as a general strategy to enhance their potency and effect. Moreover, considering the recently characterized *Echis coloratus* activity on TRPV1, we can understand that, although all toxins thus far target the outer pore domain, key amino acids in TRPV1 activation remain to be described [69]. Other toxins, such as the APHC family or HCRG21 which display analgesic properties, have an unclear purpose for inhibiting TRPV1 [34,78]. Nonetheless, these toxins offer insight into the behavioral effects of inhibiting TRPV1.

Acknowledgments: This work was supported by the Israel Science Foundation (Grant 1444/16; to A.P.), the Brettler Center and David R. Bloom Center, School of Pharmacy (The Hebrew University of Jerusalem; to A.P.), a Jerusalem Brain Community (JBC) doctoral Fellowship (to A.H.), and Paula Goldberg Scholarship (to M.G.).

Author Contributions: M.G., A.H. and A.P. wrote the paper.

Conflicts of Interest: The authors declare no conflict of interest.

References

1.   Basbaum, A.I.; Bautista, D.M.; Scherrer, G.; Julius, D. Cellular and molecular mechanisms of pain. *Cell* **2009**, *139*, 267–284. [CrossRef] [PubMed]
2.   Dubin, A.E.; Patapoutian, A. Nociceptors: The sensors of the pain pathway. *J. Clin. Investig.* **2010**, *120*, 3760–3772. [CrossRef] [PubMed]
3.   Patapoutian, A.; Tate, S.; Woolf, C.J. Transient receptor potential channels: Targeting pain at the source. *Nat. Rev. Drug Discov.* **2009**, *8*, 55–68. [CrossRef] [PubMed]
4.   Caterina, M.J.; Schumacher, M.A.; Tominaga, M.; Rosen, T.A.; Levine, J.D.; Julius, D. The capsaicin receptor: A heat-activated ion channel in the pain pathway. *Nature* **1997**, *389*, 816–824. [CrossRef] [PubMed]
5.   Jordt, S.E.; Tominaga, M.; Julius, D. Acid potentiation of the capsaicin receptor determined by a key extracellular site. *Proc. Natl. Acad. Sci. USA* **2000**, *97*, 8134–8139. [CrossRef] [PubMed]
6.   Bohlen, C.J.; Julius, D. Receptor-targeting mechanisms of pain-causing toxins: How ow? *Toxicon* **2012**, *60*, 254–264. [CrossRef] [PubMed]
7.   Yiangou, Y.; Facer, P.; Dyer, N.; Chan, C.; Knowles, C.; Williams, N.; Anand, P. Vanilloid receptor 1 immunoreactivity in inflamed human bowel. *Lancet* **2001**, *357*, 1338–1339. [CrossRef]
8.   Brown, D.C.; Iadarola, M.J.; Perkowski, S.Z.; Erin, H.; Shofer, F.; Laszlo, K.J.; Olah, Z.; Mannes, A.J. Physiologic and antinociceptive effects of intrathecal resiniferatoxin in a canine bone cancer model. *Anesthesiology* **2005**, *103*, 1052–1059. [CrossRef] [PubMed]
9.   Suri, A.; Szallasi, A. The emerging role of TRPV1 in diabetes and obesity. *Trends Pharmacol. Sci.* **2008**, *29*, 29–36. [CrossRef] [PubMed]
10. Cao, E.; Cordero-Morales, J.F.; Liu, B.; Qin, F.; Julius, D. TRPV1 channels are intrinsically heat sensitive and negatively regulated by phosphoinositide lipids. *Neuron* **2013**, *77*, 667–679. [CrossRef] [PubMed]
11. Cao, E.; Liao, M.; Cheng, Y.; Julius, D. TRPV1 structures in distinct conformations reveal activation mechanisms. *Nature* **2013**, *504*, 113–118. [CrossRef] [PubMed]
12. Zheng, J. Molecular Mechanism of TRP Channels. In *Comprehensive Physiology*; John Wiley & Sons, Inc.: Hoboken, NJ, USA, 2013; Volume 3, pp. 221–242.
13. Cui, Y.; Yang, F.; Cao, X.; Yarov-Yarovoy, V.; Wang, K.; Zheng, J. Selective disruption of high sensitivity heat activation but not capsaicin activation of TRPV1 channels by pore turret mutations. *J. Gen. Physiol.* **2012**, *139*, 273–283. [CrossRef] [PubMed]
14. Winter, Z.; Buhala, A.; Ötvös, F.; Jósvay, K.; Vizler, C.; Dombi, G.; Szakonyi, G.; Oláh, Z. Functionally important amino acid residues in the transient receptor potential vanilloid 1 (TRPV1) ion channel—An overview of the current mutational data. *Mol. Pain* **2013**, *9*, 30. [CrossRef] [PubMed]
15. Kumar, R.; Hazan, A.; Basu, A.; Zalcman, N.; Matzner, H.; Priel, A. Tyrosine residue in the TRPV1 vanilloid binding pocket regulates deactivation kinetics. *J. Biol. Chem.* **2016**, *291*, 13855–13863. [CrossRef] [PubMed]
16. Jordt, S.E.; Julius, D. Molecular basis for species-specific sensitivity to "hot" chili peppers. *Cell* **2002**, *108*, 421–430. [CrossRef]
17. Kumar, R.; Hazan, A.; Geron, M.; Steinberg, R.; Livni, L.; Matzner, H.; Priel, A. Activation of transient receptor potential vanilloid 1 by lipoxygenase metabolites depends on PKC phosphorylation. *FASEB J.* **2017**, *31*, 1238–1247. [CrossRef] [PubMed]
18. Yang, F.; Xiao, X.; Cheng, W.; Yang, W.; Yu, P.; Song, Z.; Yarov-Yarovoy, V.; Zheng, J. Structural mechanism underlying capsaicin binding and activation of the TRPV1 ion channel. *Nat. Chem. Biol.* **2015**, *11*, 518–524. [CrossRef] [PubMed]
19. Liao, M.; Cao, E.; Julius, D.; Cheng, Y. Structure of the TRPV1 ion channel determined by electron cryo-microscopy. *Nature* **2013**, *504*, 107–112. [CrossRef] [PubMed]
20. Gao, Y.; Cao, E.; Julius, D.; Cheng, Y. TRPV1 structures in nanodiscs reveal mechanisms of ligand and lipid action. *Nature* **2016**, *534*, 347–351. [CrossRef] [PubMed]
21. Hazan, A.; Kumar, R.; Matzner, H.; Priel, A. The pain receptor TRPV1 displays agonist-dependent activation stoichiometry. *Sci. Rep.* **2015**, *5*, 12278. [CrossRef] [PubMed]
22. Mickle, A.D.; Shepherd, A.J.; Mohapatra, D.P. Sensory TRP Channels. In *Progress in Molecular Biology and Translational Science*; Elsevier, Inc.: Amsterdam, The Netherlands, 2015; Volume 131, pp. 73–118.

23. Bevan, S.; Hothi, S.; Hughes, G.; James, I.F.; Rang, H.P.; Shah, K.; Walpole, C.S.J.; Yeats, J.C. Capsazepine: A competitive antagonist of the sensory neurone excitant capsaicin. *Br. J. Pharmacol.* **1992**, *107*, 544–552. [CrossRef] [PubMed]

24. Gavva, N.R.; Bannon, A.W.; Hovland, D.N.; Lehto, S.G.; Klionsky, L.; Surapaneni, S.; Immke, D.C.; Henley, C.; Arik, L.; Bak, A.; et al. Repeated administration of vanilloid receptor TRPV1 antagonists attenuates hyperthermia elicited by TRPV1 blockade. *J. Pharmacol. Exp. Ther.* **2007**, *323*, 128–137. [CrossRef] [PubMed]

25. Anand, P.; Bley, K. Topical capsaicin for pain management: Therapeutic potential and mechanisms of action of the new high-concentration capsaicin 8% patch. *Br. J. Anaesth.* **2011**, *107*, 490–502. [CrossRef] [PubMed]

26. Bertrand, H.; Kyriazis, M.; Reeves, K.D.; Lyftogt, J.; Rabago, D. Topical mannitol reduces capsaicin-induced pain: Results of a pilot-level, double-blind, randomized controlled trial. *PM&R* **2015**, *7*, 1111–1117. [CrossRef]

27. Van der Stelt, M.; Di Marzo, V. Endovanilloids. Putative endogenous ligands of transient receptor potential vanilloid 1 channels. *Eur. J. Biochem.* **2004**, *271*, 1827–1834. [CrossRef] [PubMed]

28. Kittaka, H.; Uchida, K.; Fukuta, N.; Tominaga, M. Lysophosphatidic acid-induced itch is mediated by signalling of LPA 5 receptor, phospholipase D and TRPA1/TRPV1. *J. Physiol.* **2017**, *595*, 2681–2698. [CrossRef] [PubMed]

29. Chou, M.Z.; Mtui, T.; Gao, Y.-D.; Kohler, M.; Middleton, R.E. Resiniferatoxin binds to the capsaicin receptor (TRPV1) near the extracellular side of the S4 transmembrane domain. *Biochemistry* **2004**, *43*, 2501–2511. [CrossRef] [PubMed]

30. Raisinghani, M.; Pabbidi, R.M.; Premkumar, L.S. Activation of transient receptor potential vanilloid 1 (TRPV1) by resiniferatoxin. *J. Physiol.* **2005**, *567*, 771–786. [CrossRef] [PubMed]

31. Elokely, K.; Velisetty, P.; Delemotte, L.; Palovcak, E.; Klein, M.L.; Rohacs, T.; Carnevale, V. Understanding TRPV1 activation by ligands: Insights from the binding modes of capsaicin and resiniferatoxin. *Proc. Natl. Acad. Sci. USA* **2016**, *113*, E137–E145. [CrossRef] [PubMed]

32. Siemens, J.; Zhou, S.; Piskorowski, R.; Nikai, T.; Lumpkin, E.A.; Basbaum, A.I.; King, D.; Julius, D. Spider toxins activate the capsaicin receptor to produce inflammatory pain. *Nature* **2006**, *444*, 208–212. [CrossRef] [PubMed]

33. Craik, D.J.; Daly, N.L.; Waine, C. The cystine knot motif in toxins and implications for drug design. *Toxicon* **2001**, *39*, 43–60. [CrossRef]

34. Monastyrnaya, M.; Peigneur, S.; Zelepuga, E.; Sintsova, O.; Gladkikh, I.; Leychenko, E.; Isaeva, M.; Tytgat, J.; Kozlovskaya, E. Kunitz-type peptide HCRG21 from the sea anemone Heteractis crispa is a full antagonist of the TRPV1 receptor. *Mar. Drugs* **2016**, *14*, 229. [CrossRef] [PubMed]

35. Min, J.-W.; Liu, W.-H.; He, X.-H.; Peng, B.-W. Different types of toxins targeting TRPV1 in pain. *Toxicon* **2013**, *71*, 66–75. [CrossRef] [PubMed]

36. Bohlen, C.J.; Priel, A.; Zhou, S.; King, D.; Siemens, J.; Julius, D. A bivalent tarantula toxin activates the capsaicin receptor, TRPV1, by targeting the outer pore domain. *Cell* **2010**, *141*, 834–845. [CrossRef] [PubMed]

37. Bae, C.; Kalia, J.; Song, I.; Yu, J.; Kim, H.H.; Swartz, K.J.; Kim, J.I. High yield production and refolding of the double-knot toxin, an activator of TRPV1 channels. *PLoS ONE* **2012**, *7*, e51516. [CrossRef] [PubMed]

38. Liang, S. An overview of peptide toxins from the venom of the Chinese bird spider Selenocosmia huwena Wang [=Ornithoctonus huwena (Wang)]. *Toxicon* **2004**, *43*, 575–585. [CrossRef] [PubMed]

39. Bae, C.; Anselmi, C.; Kalia, J.; Jara-Oseguera, A.; Schwieters, C.D.; Krepkiy, D.; Won Lee, C.; Kim, E.-H.; Kim, J.I.; Faraldo-Gómez, J.D.; et al. Structural insights into the mechanism of activation of the TRPV1 channel by a membrane-bound tarantula toxin. *eLife* **2016**, *5*, e11273. [CrossRef] [PubMed]

40. Julius, D. TRP Channels and Pain. *Annu. Rev. Cell Dev. Biol.* **2013**, *29*, 355–384. [CrossRef] [PubMed]

41. Kalia, J.; Milescu, M.; Salvatierra, J.; Wagner, J.; Klint, J.K.; King, G.F.; Olivera, B.M.; Bosmans, F. From foe to friend: Using animal toxins to investigate ion channel function. *J. Mol. Biol.* **2015**, *427*, 158–175. [CrossRef] [PubMed]

42. Dilly, S.; Lamy, C.; Marrion, N.V.; Liégeois, J.-F.; Seutin, V. Ion-channel modulators: More diversity than previously thought. *ChemBioChem* **2011**, *12*, 1808–1812. [CrossRef] [PubMed]

43. Liu, B.; Hui, K.; Qin, F. Thermodynamics of heat activation of single capsaicin ion channels VR1. *Biophys. J.* **2003**, *85*, 2988–3006. [CrossRef]

44. Hui, K.; Liu, B.; Qin, F. Capsaicin activation of the pain receptor, VR1: Multiple open states from both partial and full binding. *Biophys. J.* **2003**, *84*, 2957–2968. [CrossRef]

45. Liao, M.; Cao, E.; Julius, D.; Cheng, Y. Single particle electron cryo-microscopy of a mammalian ion channel. *Curr. Opin. Struct. Biol.* **2014**, *27*, 1–7. [CrossRef] [PubMed]
46. Kalia, J.; Swartz, K.J. Exploring structure-function relationships between TRP and Kv channels. *Sci. Rep.* **2013**, *3*, 1523. [CrossRef] [PubMed]
47. Lee, S.-Y.; MacKinnon, R. A membrane-access mechanism of ion channel inhibition by voltage sensor toxins from spider venom. *Nature* **2004**, *430*, 232–235. [CrossRef] [PubMed]
48. Gupta, K.; Zamanian, M.; Bae, C.; Milescu, M.; Krepkiy, D.; Tilley, D.C.; Sack, J.T.; Vladimir, Y.-Y.; Kim, J.I.; Swartz, K.J. Tarantula toxins use common surfaces for interacting with Kv and ASIC ion channels. *eLife* **2015**, *4*, e06774. [CrossRef] [PubMed]
49. Jung, H.J.; Lee, J.Y.; Kim, S.H.; Eu, Y.-J.; Shin, S.Y.; Milescu, M.; Swartz, K.J.; Kim, J.I. Solution structure and lipid membrane partitioning of VSTx1, an inhibitor of the KvAP potassium channel. *Biochemistry* **2005**, *44*, 6015–6023. [CrossRef] [PubMed]
50. Takahashi, H.; Kim, J.I.; Min, H.J.; Sato, K.; Swartz, K.J.; Shimada, I. Solution structure of hanatoxin1, a gating modifier of voltage-dependent K+ channels: Common surface features of gating modifier toxins. *J. Mol. Biol.* **2000**, *297*, 771–780. [CrossRef] [PubMed]
51. Yang, S.; Yang, F.; Wei, N.; Hong, J.; Li, B.; Luo, L.; Rong, M.; Yarov-Yarovoy, V.; Zheng, J.; Wang, K.; et al. A pain-inducing centipede toxin targets the heat activation machinery of nociceptor TRPV1. *Nat. Commun.* **2015**, *6*, 8297. [CrossRef] [PubMed]
52. Chen, M.; Li, J.; Zhang, F.; Liu, Z. Isolation and characterization of SsmTx-I, a specific Kv2.1 blocker from the venom of the centipede Scolopendra Subspinipes Mutilans L. Koch. *J. Pept. Sci.* **2014**, *20*, 159–164. [CrossRef] [PubMed]
53. Undheim, E.A.B.; Jenner, R.A.; King, G.F. Centipede venoms as a source of drug leads. *Expert Opin. Drug Discov.* **2016**, *11*, 1139–1149. [CrossRef] [PubMed]
54. Yang, S.; Xiao, Y.; Kang, D.; Liu, J.; Li, Y.; Undheim, E.A.B.; Klint, J.K.; Rong, M.; Lai, R.; King, G.F. Discovery of a selective NaV1.7 inhibitor from centipede venom with analgesic efficacy exceeding morphine in rodent pain models. *Proc. Natl. Acad. Sci. USA* **2013**, *110*, 17534–17539. [CrossRef] [PubMed]
55. Nassar, M.A.; Stirling, L.C.; Forlani, G.; Baker, M.D.; Matthews, E.A.; Dickenson, A.H.; Wood, J.N. Nociceptor-specific gene deletion reveals a major role for Nav1.7 (PN1) in acute and inflammatory pain. *Proc. Natl. Acad. Sci. USA* **2004**, *101*, 12706–12711. [CrossRef] [PubMed]
56. Zheng, J.; Ma, L. Structure and function of the ThermoTRP channel pore. In *Current Topics in Membranes*; Elsevier Inc.: Amsterdam, The Netherlands, 2014; Volume 74, pp. 233–257.
57. Smith, J.J.; Hill, J.M.; Little, M.J.; Nicholson, G.M.; King, G.F.; Alewood, P.F. Unique scorpion toxin with a putative ancestral fold provides insight into evolution of the inhibitor cystine knot motif. *Proc. Natl. Acad. Sci. USA* **2011**, *108*, 10478–10483. [CrossRef] [PubMed]
58. Wu, G.; Li, Y.; Wei, D.; He, F.; Jiang, S.; Hu, G.; Wu, H. Solution structure of BmP01 from the venom of scorpion Buthus martensii Karsch. *Biochem. Biophys. Res. Commun.* **2000**, *276*, 1148–1154. [CrossRef] [PubMed]
59. Hakim, M.; Jiang, W.; Luo, L.; Li, B.; Yang, S.; Song, Y.; Lai, R. Scorpion toxin, BmP01, induces pain by targeting TRPV1 channel. *Toxins (Basel)* **2015**, *7*, 3671–3687. [CrossRef] [PubMed]
60. Zhu, S.; Peigneur, S.; Gao, B.; Luo, L.; Jin, D.; Zhao, Y.; Tytgat, J. Molecular diversity and functional evolution of scorpion potassium channel toxins. *Mol. Cell. Proteom.* **2011**, *10*, M110.002832. [CrossRef] [PubMed]
61. Yang, S.; Yang, F.; Zhang, B.; Lee, B.H.; Li, B.; Luo, L.; Zheng, J.; Lai, R. A bimodal activation mechanism underlies scorpion toxin–induced pain. *Sci. Adv.* **2017**, *3*, e1700810. [CrossRef] [PubMed]
62. Swartz, K.J.; MacKinnon, R. An inhibitor of the Kv2.1 potassium channel isolated from the venom of a Chilean tarantula. *Neuron* **1995**, *15*, 941–949. [CrossRef]
63. Bautista, D.M.; Sigal, Y.M.; Milstein, A.D.; Garrison, J.L.; Zorn, J.A.; Tsuruda, P.R.; Nicoll, R.A.; Julius, D. Pungent agents from Szechuan peppers excite sensory neurons by inhibiting two-pore potassium channels. *Nat. Neurosci.* **2008**, *11*, 772–779. [CrossRef] [PubMed]
64. Daly, N.L.; Craik, D.J. Bioactive cystine knot proteins. *Curr. Opin. Chem. Biol.* **2011**, *15*, 362–368. [CrossRef] [PubMed]
65. Zhu, S.; Darbon, H.; Dyason, K.; Verdonck, F.; Tytgat, J. Evolutionary origin of inhibitor cystine knot peptides. *FASEB J.* **2003**, *17*, 1765–1767. [CrossRef] [PubMed]

66. Chacur, M.; Gutiérrez, J.M.; Milligan, E.D.; Wieseler-Frank, J.; Britto, L.R.G.; Maier, S.F.; Watkins, L.R.; Cury, Y. Snake venom components enhance pain upon subcutaneous injection: An initial examination of spinal cord mediators. *Pain* **2004**, *111*, 65–76. [CrossRef] [PubMed]

67. Zhang, C.; Medzihradszky, K.F.; Sánchez, E.E.; Basbaum, A.I.; Julius, D. Lys49 myotoxin from the Brazilian lancehead pit viper elicits pain through regulated ATP release. *Proc. Natl. Acad. Sci. USA* **2017**, *114*, E2524–E2532. [CrossRef] [PubMed]

68. Bohlen, C.J.; Chesler, A.T.; Sharif-Naeini, R.; Medzihradszky, K.F.; Zhou, S.; King, D.; Sánchez, E.E.; Burlingame, A.L.; Basbaum, A.I.; Julius, D. A heteromeric Texas coral snake toxin targets acid-sensing ion channels to produce pain. *Nature* **2011**, *479*, 410–414. [CrossRef] [PubMed]

69. Geron, M.; Kumar, R.; Matzner, H.; Lahiani, A.; Gincberg, G.; Cohen, G.; Lazarovici, P.; Priel, A. Protein toxins of the Echis coloratus viper venom directly activate TRPV1. *Biochim. Biophys. Acta Gen. Subj.* **2017**, *1861*, 615–623. [CrossRef] [PubMed]

70. Benbassat, J.; Shalev, O. Envenomation by Echis coloratus (Mid-East saw-scaled viper): A review of the literature and indications for treatment. *Isr. J. Med. Sci.* **1993**, *29*, 239–250. [PubMed]

71. Gomes, A.; Choudhury, S.R.; Saha, A.; Mishra, R.; Giri, B.; Biswas, A.K.; Debnath, A.; Gomes, A. A heat stable protein toxin (drCT-I) from the Indian Viper (Daboia russelli russelli) venom having antiproliferative, cytotoxic and apoptotic activities. *Toxicon* **2007**, *49*, 46–56. [CrossRef] [PubMed]

72. Argos, P.; Rossmann, M.G.; Grau, U.M.; Zuber, H.; Frank, G.; Tratschin, J.D. Thermal stability and protein structure. *Biochemistry* **1979**, *18*, 5698–5703. [CrossRef] [PubMed]

73. Zhang, X.; Huang, J.; McNaughton, P.A. NGF rapidly increases membrane expression of TRPV1 heat-gated ion channels. *EMBO J.* **2005**, *24*, 4211–4223. [CrossRef] [PubMed]

74. Gilon, D.; Shalev, O.; Benbassat, J. Treatment of envenomation by Echis coloratus (mid-east saw scaled viper): A decision tree. *Toxicon* **1989**, *27*, 1105–1112. [CrossRef]

75. Fainaru, M.; Eisenberg, S.; Manny, N.; Hershko, C. The natural course of defibrination syndrome caused by Echis colorata venom in man. *Thromb. Diath. Haemorrh.* **1974**, *31*, 420–428. [PubMed]

76. Diochot, S.; Baron, A.; Salinas, M.; Douguet, D.; Scarzello, S.; Dabert-Gay, A.-S.; Debayle, D.; Friend, V.; Alloui, A.; Lazdunski, M.; et al. Black mamba venom peptides target acid-sensing ion channels to abolish pain. *Nature* **2012**, *490*, 552–555. [CrossRef] [PubMed]

77. Kitaguchi, T.; Swartz, K.J. An inhibitor of TRPV1 channels isolated from funnel Web spider venom. *Biochemistry* **2005**, *44*, 15544–15549. [CrossRef] [PubMed]

78. Andreev, Y.A.; Kozlov, S.A.; Koshelev, S.G.; Ivanova, E.A.; Monastyrnaya, M.M.; Kozlovskaya, E.P.; Grishin, E.V. Analgesic compound from sea anemone Heteractis crispa is the first polypeptide inhibitor of vanilloid receptor 1 (TRPV1). *J. Biol. Chem.* **2008**, *283*, 23914–23921. [CrossRef] [PubMed]

79. Nikolaev, M.V.; Dorofeeva, N.A.; Komarova, M.S.; Korolkova, Y.V.; Andreev, Y.A.; Mosharova, I.V.; Grishin, E.V.; Tikhonov, D.B.; Kozlov, S.A. TRPV1 activation power can switch an action mode for its polypeptide ligands. *PLoS ONE* **2017**, *12*, e0177077. [CrossRef] [PubMed]

80. Andreev, Y.; Kozlov, S.; Korolkova, Y.; Dyachenko, I.; Bondarenko, D.; Skobtsov, D.; Murashev, A.; Kotova, P.; Rogachevskaja, O.; Kabanova, N.; et al. Polypeptide modulators of TRPV1 produce analgesia without hyperthermia. *Mar. Drugs* **2013**, *11*, 5100–5115. [CrossRef] [PubMed]

81. Ranasinghe, S.; McManus, D.P. Structure and function of invertebrate Kunitz serine protease inhibitors. *Dev. Comp. Immunol.* **2013**, *39*, 219–227. [CrossRef] [PubMed]

82. Wong, G.Y.; Gavva, N.R. Therapeutic potential of vanilloid receptor TRPV1 agonists and antagonists as analgesics: Recent advances and setbacks. *Brain Res. Rev.* **2009**, *60*, 267–277. [CrossRef] [PubMed]

83. Kaneko, Y.; Szallasi, A. Transient receptor potential (TRP) channels: A clinical perspective. *Br. J. Pharmacol.* **2014**, *171*, 2474–2507. [CrossRef] [PubMed]

84. Khairatkar-Joshi, N.; Szallasi, A. TRPV1 antagonists: The challenges for therapeutic targeting. *Trends Mol. Med.* **2009**, *15*, 14–22. [CrossRef] [PubMed]

85. Trim, S.A.; Trim, C.M. Venom: The sharp end of pain therapeutics. *Br. J. Pain* **2013**, *7*, 179–188. [CrossRef] [PubMed]

86. Miller, C.; Moczydlowski, E.; Latorre, R.; Phillips, M. Charybdotoxin, a protein inhibitor of single Ca2+-activated K+ channels from mammalian skeletal muscle. *Nature* **1985**, *313*, 316–318. [CrossRef] [PubMed]

87. Cromer, B.A.; McIntyre, P. Painful toxins acting at TRPV1. *Toxicon* **2008**, *51*, 163–173. [CrossRef] [PubMed]

88. Liu, B.; Yao, J.; Wang, Y.; Li, H.; Qin, F. Proton inhibition of unitary currents of vanilloid receptors. *J. Gen. Physiol.* **2009**, *134*, 243–258. [CrossRef] [PubMed]
89. Yang, F.; Ma, L.; Cao, X.; Wang, K.; Zheng, J. Divalent cations activate TRPV1 through promoting conformational change of the extracellular region. *J. Gen. Physiol.* **2014**, *143*, 91–103. [CrossRef] [PubMed]
90. Myers, B.R.; Bohlen, C.J.; Julius, D. A yeast genetic screen reveals a critical role for the pore helix domain in TRP channel gating. *Neuron* **2008**, *58*, 362–373. [CrossRef] [PubMed]
91. Grandl, J.; Kim, S.E.; Uzzell, V.; Bursulaya, B.; Petrus, M.; Bandell, M.; Patapoutian, A. Temperature-induced opening of TRPV1 ion channel is stabilized by the pore domain. *Nat. Neurosci.* **2010**, *13*, 708–714. [CrossRef] [PubMed]
92. Yang, F.; Cui, Y.; Wang, K.; Zheng, J. Thermosensitive TRP channel pore turret is part of the temperature activation pathway. *Proc. Natl. Acad. Sci. USA* **2010**, *107*, 7083–7088. [CrossRef] [PubMed]
93. Hilton, J.K.; Rath, P.; Helsell, C.V.M.; Beckstein, O.; Van Horn, W.D. Understanding thermosensitive transient receptor potential channels as versatile polymodal cellular sensors. *Biochemistry* **2015**, *54*, 2401–2413. [CrossRef] [PubMed]
94. Clapham, D.E.; Miller, C. A thermodynamic framework for understanding temperature sensing by transient receptor potential (TRP) channels. *Proc. Natl. Acad. Sci. USA* **2011**, *108*, 19492–19497. [CrossRef] [PubMed]
95. Yao, J.; Liu, B.; Qin, F. Modular thermal sensors in temperature-gated transient receptor potential (TRP) channels. *Proc. Natl. Acad. Sci. USA* **2011**, *108*, 11109–11114. [CrossRef] [PubMed]
96. Steinberg, X.; Lespay-Rebolledo, C.; Brauchi, S. A structural view of ligand-dependent activation in thermoTRP channels. *Front. Physiol.* **2014**, *5*, 171. [CrossRef] [PubMed]
97. Fernández-Ballester, G.; Ferrer-Montiel, A. Molecular modeling of the full-length human TRPV1 channel in closed and desensitized states. *J. Membr. Biol.* **2008**, *223*, 161–172. [CrossRef] [PubMed]
98. Long, S.B.; Campbell, E.B.; Mackinnon, R. Voltage sensor of Kv1.2: Structural basis of electromechanical coupling. *Science* **2005**, *309*, 903–908. [CrossRef] [PubMed]

© 2017 by the authors. Licensee MDPI, Basel, Switzerland. This article is an open access article distributed under the terms and conditions of the Creative Commons Attribution (CC BY) license (http://creativecommons.org/licenses/by/4.0/).

toxins

MDPI

Review

Botulinum Toxin for the Treatment of Neuropathic Pain

JungHyun Park [1] and Hue Jung Park [2,*]

[1] Department of Anaesthesiology & Pain Medicine, Incheon St. Mary's Hospital, College of Medicine,
The Catholic University of Korea, Incheon 21431, Korea; happyjj@catholic.ac.kr
[2] Department of Anaesthesiology & Pain Medicine, Seoul St. Mary's Hospital, College of Medicine,
The Catholic University of Korea, Seoul 06591, Korea
* Correspondence: huejung@catholic.ac.kr.; Tel.: +82-2258-2236 or +82-2258-6157

Academic Editor: Steve Peigneur
Received: 10 August 2017; Accepted: 21 August 2017; Published: 24 August 2017

Abstract: Botulinum toxin (BoNT) has been used as a treatment for excessive muscle stiffness, spasticity, and dystonia. BoNT for approximately 40 years, and has recently been used to treat various types of neuropathic pain. The mechanism by which BoNT acts on neuropathic pain involves inhibiting the release of inflammatory mediators and peripheral neurotransmitters from sensory nerves. Recent journals have demonstrated that BoNT is effective for neuropathic pain, such as postherpetic neuralgia, trigeminal neuralgia, and peripheral neuralgia. The purpose of this review is to summarize the experimental and clinical evidence of the mechanism by which BoNT acts on various types of neuropathic pain and describe why BoNT can be applied as treatment. The PubMed database was searched from 1988 to May 2017. Recent studies have demonstrated that BoNT injections are effective treatments for post-herpetic neuralgia, diabetic neuropathy, trigeminal neuralgia, and intractable neuropathic pain, such as poststroke pain and spinal cord injury.

Keywords: botulinum toxin; neuropathic pain; neuropathic pain treatment

---

1. Introduction

Botulinum toxin (BoNT) has been used for decades in the treatment of diseases, such as dystonia or seizures, and cosmetic treatments. BoNT is useful in conditions such as strabismus because it causes long lasting but reversible paralysis via the administration of small amounts locally [1,2]. As BoNT purification technology develops, the range of use of this drug has been expanded, and the number of Food and Drug Administration (FDA)-approval diseases has also increased. Common to these applications is the fact that BoNT is absorbed from the neuromuscular junction or parasympathetic axon terminal to the motor neuron terminal because the toxin is responsible for the release of acetylcholine. It is important to note that these effects are not systematically redistributed but only localized. Numerous reports suggest that local administration of BoNT has a significant effect on neuropathic pain.

For a long time, the analgesic effect of Botulinum toxin A (BoNT-A) was considered to be due to the effect of muscle relaxation [3–5]. However, recent studies using BoNT in neuropathic pain models have demonstrated that BoNT has an analgesic effect independent of muscle relaxation by demonstrating dissociation of the duration of muscle relaxation and duration of pain relief [6].

In this paper, we investigate the mechanism of BoNT in neuropathic pain by examining the effects of the drug for intractable neuropathic pains, such as postherpetic neuralgia, diabetic neuropathy, complex regional pain syndrome, trigeminal neuralgia, phantom limb pain, spinal cord injury-induced neuropathic pain, and central poststroke pain.

2. Structure of Botulinum Toxin

BoNT is protein group produced by anaerobic bacteria called *Clostridium botulinum*, which has approximately 40 subtypes. However, seven serotypes are typically noted based on antigen specificity. Botulinum toxin A (BoNT-A) and B (BoNT-B) are the most commonly used drugs. Particularly, BoNT-A type has a molecular weight of approximately 900,000. BoNT-A is a double chain protein. The light chain (LC) is active, whereas the heavy chain (HC) is not active. BoNT binds to the acceptor at the nerve end and enters the nerve ending by receptor-mediated endocytosis. LC binds to the exogenous protein involved in exocytosis and breaks down the peptide bond of the protein transporter to block exocytosis and acetylcholine secretion. The C-terminal receptor-binding domain, which constitutes the heavy chain of BoNT, binds to ganglioside receptors and specific proteins on the cell membrane. This binding induces endocytosis of HC-LC. In general, acetylcholine binding to the acetylcholine receptor of the motor endplate is necessary for muscle contraction. At this time, the acetylcholine exocytosis process is necessary in presynaptic membrane. The normal acetylcholine exocytosis process requires three proteins: the synaptosomal associated protein-25 kDa (SNAP-25), syntaxin, and the vesicle-associated membrane protein (VAMP)/synaptobrevin in the presynaptic membrane. These proteins are called soluble N-ethylmaleimide (SNARE) proteins. As a zinc-dependent endoprotease, the LC of BoNT cleaves intracellular SNARE. This cleavage interferes with SNARE-mediated protein transport and transmitter release, blocking muscle innervation at the neuromuscular junction and resulting in flaccid paralysis [7,8]. This effect of BoNT LC is dependent on the serotype, but it persists from days to months [9,10].

BoNT-A and BoNT-B are effective in neuropathic pain. Mice can be treated with nerve ligation to induce mononeuropathy and cisplatin to induce polyneuropathy. BoNT-B improves allodynia and hyperalgesia [11]. A clinically reported case report demonstrates that BoNT-B improves pain and symptoms in complex regional pain syndrome (CRPS) patients with a lumbar sympathetic block [12].

3. Mechanism of Action of Botulinum Toxin for Neuropathic Pain (Experimental Study)

BoNT also reduces and alters neuropathic pain in several animal models via the following mechanisms. BoNT inhibits the secretion of pain mediators (substance P, glutamate, and calcitonin gene related protein (CGRP)) from the nerve endings and dorsal root ganglions (DRG), reduces local inflammation around the nerve endings, deactivates the sodium channel, and exhibits axonal transport. We will review the various mechanisms by which BoNT reduces neuropathic pain (Figure 1).

3.1. BoNT Inhibits the Release of Pain Mediators from the Peripheral Nerve Terminal, DRG, and Spinal Cord Neuron

The effect of BoNT on the secretion of sensory neurotransmitters has been documented in several animal models. BoNT reduces normal CGRP release and capsaicin-induced DOA secretion and has additional effects on the TRPV1 pathway [13]. According to Meng et al., in a rat trigeminal ganglion sensory peptidergic neuron cell culture model, BoNT cleaves neuronal SNARE and blocks neurotransmitter secretion [14]. Durham et al. also reported a prophylactic advantage in migraine headaches via blocking the release of neuropeptides, such as CGRP from the trigeminal ganglion neuronal culture [15].

Fan et al. demonstrated that BoNT significantly reduces TRPV1 protein levels. Several studies demonstrated that TRPV1 plays a crucial role in arthritis pain, and this article examined the causal relationship between the antinociceptive effect of BoNT and the expression of TRPV1 in DRG of rats with arthritic pain. No significant changes in TRPV1 mRNA levels were observed via RT-PCR performed with different BoNT doses (1, 3, and 10 U); However, BoNT or TRPV1 protein levels were significantly decreased. This paper demonstrates the antinociceptive mechanism of BoNT by reducing TRPV1 expression by inhibiting plasma membrane trafficking after intra-articular administration [16].

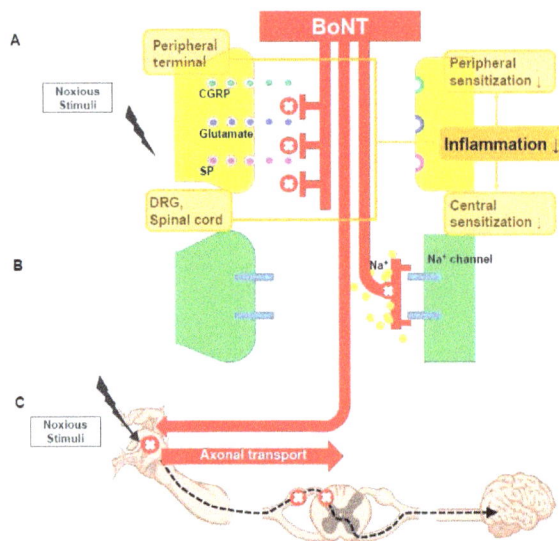

Figure 1. (**A**) Noxious stimuli cause inflammation through the release of neuropeptides and inflammatory mediators, which can cause peripheral sensitization. This action also occurs in DRG, dorsal horn of spinal cord and can lead to central sensitization. Botulinum toxin (BoNT) inhibits the release of pain mediators in peripheral nerve terminal, DRG, and spinal cord neuron, thereby reducing the inflammatory response and preventing the development of peripheral and central sensitization. Symbols; SP, substance P; CGRP, calcitonin gene related protein; DRG, dorsal root ganglion; (**B**) The hyperexcitability and spontaneous action potential mediated by the Na channel in peripheral sensory neuron contribute to the pathophysiology of neuropathic pain. BoNT can control neuropathic pain by blocking the Na channel; (**C**) Some of the BoNT appear to retrograde transport along the axons. SNAP-25 is cleaved in the dorsal horn of the spinal cord and central nuclei after a small amount of BoNT is administered to the periphery, thereby boosting the retrograde transport of BoNT.

3.2. BoNT Reduces Inflammation

Cyclophosphamide (CYP) was injected into the bladder of rats to induce CYP-induced cystitis, and HCL was injected into the bladder to induce acute injury. The bladder was harvested and compared with the Sham group. The cells were cultured in a solution containing BoNT to compare neurotransmitters. CGRP and substance P were significantly increased in the acute injury group compared with the control group, and substance P was significantly increased in the CYP-induced cystitis group. After exposure to BoNT, neurotransmitter levels were significantly reduced. In this article, we found that BoNT has an anti-inflammatory effect on chronic inflammation and acute injury [17]. In a chronic arthritis dog model, intraarticular BoNT injections are effective for up to 12 weeks [18,19]. The anti-inflammatory effect of BoNT reduces the release of peripheral neurotransmitters and inflammatory mediators.

However, the effects are debated. Rojecky et al. injected carrageenan and capsaicin into the hindpaw of the rat, and rats were treated with BoNT five days before injection. No significant differences in edema and plasma protein extravasation were noted between the group injected with BoNT and the group without BoNT [20]. In addition, Sycha et al. reported that the BoNT group and the control group had no direct effect on acute, noninflammatory pain in the group treated with BoNT upon skin exposure to Ultraviolet B [21]. Chuang et al. measured cyclooxygenase-2 (COX-2) levels in the capsaicin-induced prostatitis model. COX-2 is a key enzyme that is an important mediator of inflammation and pain. COX-2 expression was induced as assessed by Western blotting or immunostaining. Inflammation was induced upon injection of capsaicin into the prostate of an adult male rat. Another group was pretreated with 20 U BoNT one week before injection of capsaicin. The expression of COX-2 was reduced in spinal sensory and motor neurons and the prostate in the pretreatment group [22].

BoNT also decreases local inflammation around the nerve terminal. According to the report of Cui et al., BoNT was administered to the footpads in formalin-inflammatory pain model rats. The antinociceptive effect started 5 h after BoNT treatment and persisted for greater than 12 days. In addition, edema was reduced, but no localized muscle weakness was observed. Formalin-induced glutamate release was also significantly reduced. This finding demonstrates that local inflammation around the nerve endings is reduced in the absence of obvious muscle weakness [23].

3.3. BoNT Deactivates Sodium Channels

BoNT also deactivates the sodium channel. Na current stimulates numerous cellular functions, such as transmission, secretion, contraction, and sensation. BoNT-A changes the Na current of a neuronal excitable membrane, which is different from that of local anesthetics, tetrodotoxin, and antiepileptic drugs that completely control the Na current via a concentration-dependent manner [24].

3.4. BoNT Exhibits Axonal Transport

BoNT exhibits axonal transport function from the periphery to the CNS, and administering BoNT to the facial and trigeminal nerve causes SNAP-25 cleavage in the central nuclei. In addition, a small amount of BoNT was injected into the hind limb, confirming the cleavage of the SNAP-25 in the ventral horn and the dorsal horn of the ipsilateral spinal cord, thereby demonstrating the retrograde axonal transport function of BoNT [25]. In addition, the BoNT effect on both sides has been reported after injecting BoNT on one side [26–28]. In animal studies, the anti-nociceptive effect of BoNT was studied in paclitaxel-induced peripheral neuropathy. The withdrawal nociceptive reflex was reduced after paclitaxel injection into the hind paw of the rat. BoNT was injected into one side, but the analgesic effect was observed on both sides. Diffusion into blood circulation may affect the central nervous system, but the dose was too low to cause systemic side effects. BoNT is also too large to pass the BBB, so the theory that BoNT is transmitted from the periphery to the central nervous system through the axon is possible [28]. To prevent retrograde axonal transport, Rojecky et al. confirmed the antinociceptive effect of unilateral transport of the axonic transport blocker colchicine in the ipsilateral sciatic nerve [26], which also demonstrated the retrograde axonal transport of BoNT.

However, this notion is controversial. Tang et al. injected ^{125}I-radiolabeled free BoNT into the gastrocnemius muscle of rats and rabbit eyelids and observed BoNT in various tissues at different time points. In both rabbits and rats, systemic effects were absent, and most of the toxins remained in the injection site. The authors concluded that most of the BoNT remained near the injection site and did not cause systemic toxicity [29].

Whether BoNT is transported retrograde from the injection site remains controversial. However, retrograde axonal transport has been demonstrated in numerous papers. Marinelli et al. analyzed the expression of cl-SNAP-25 from the nerve endings of the hind paw to the spinal cord after applying BoNT to the periphery. Immunostained cl-SNAP-25 was detected in all tissues. Additional experiments were performed to assess whether the growth state of Schwann cells interacts with BoNTs. As a result, BoNT regulated the proliferation of Schwann cells to inhibit acetylcholine release. This result demonstrates retrograde trafficking of BoNT [30].

4. Clinical Study of Botulinum Toxin for Neuropathic Pain

4.1. Trigeminal Neuralgia

A review of the efficacy of Botulinum toxin (BoNT) on trigeminal neuralgia (TN) has been reported in approximately 11 cases, including three RCT papers. This review includes the largest number of clinical trials for neuropathic pain for BoNT. In a randomized, double-blind, placebo-controlled study of 42 patients, Wu et al. performed a parallel design with intradermal or submucosal injection of 75 U of BoNT-A in 22 patients. Twenty patients in the control group received 1.5 mL saline. In the BoNT group, 68.8% of patients had a visual analog scale (VAS) reduction of greater than 50%. In the control group, a VAS reduction of greater than 50% was noted in 15% of the patients [31]. In addition, a randomized, double-blind, placebo-controlled study was performed in 84 adults with TN by Zhang et al. 28 control subjects were treated with saline, 27 with 25 U BoNT-A, and 29 with 75 U BoNT-A. The response rates in the 25 U and 75 U groups were 70.4% and 86.2%, respectively, which were significantly different from the control group (32.1%). However, no significant differences were noted between the two groups [32]. According to Zuniga et al., 20 patients received 50 U of BoNT-A, and 16 controls received the same dose of saline. VAS was 4.9 vs. 6.63 at two months follow-up. No significant differences were noted between the two groups. At three months, there was a significant difference at 4.75 vs. 6.94 [33].

Prospective, open, and case series for trigeminal neuralgia are reported in three studies. According to Bohluli et al., 15 TN patients were administered 50–100 U of BoNT-A in the trigger zone without any special injection mode. All patients reported a reduction in pain frequency and VAS score [34]. Zuniga et al. reported 12 trigeminal neuralgia patients who underwent subcutaneous injection in the trigger zone, and a reduction in VAS lasting greater than two months was noted in 10 patients [35]. Turk et al. also reported that injection of 50 U BoNT-A at 1.5–2 cm depth around the zygomatic arch was performed in eight patients, and the incidence of pain and VAS were reduced in all patients [36]. The above papers are summarized in Table 1.

Table 1. Botulinum toxin for trigeminal neuralgia.

Study Design	Number of Patients	Method of Injection (Total Volume)	Result	Reference
Randomized double-blind, placebo-controlled	42	Intradermal, submucosal (75 U/saline 1.5 mL)	50% VAS reduction 68.8% (Botulinum toxin (BoNT) group) 15% (Control)	[31]
Randomized, double-blind, placebo-controlled	84 (27 BoNT 25 U, 29 BoNT 75 U, 28 control)	Intradermal, submucosal (25 U/75 U/saline 1 mL)	Visual analog scale (VAS) reduction 70.4% (25 U) vs. 86.2% (75 U) vs. 32.1% (Control)	[32]
Randomized, double-blind, placebo-controlled	36 (20 BoNT, 16 control)	Intramuscular (50 U/saline 1 mL)	VAS (BoNT vs. Control) 4.9 vs. 6.63 (2 months) 4.75 vs. 6.94 (3 months)	[33]
Prospective, open, case series	15	Injected at the trigger zones (50–100 U)	All patients improved frequency and severity of pain attacks	[34]
Prospective, open, case series	12	Subcutaneous (20–50 U)	VAS reduced lasting more than 2 months in 10 patients.	[35]
Prospective, open, case series	8	Around zygomatic arch, 1.5–2 cm depth (50 U per point, total 100 U)	Incidence of pain and VAS were reduced in all patients.	[36]

4.2. Postherpetic Neuralgia

Two BoNT RCTs for postherpetic neuralgia (PHN) have been reported. Xiao et al. performed a randomized, double-blind, placebo-controlled study of 60 patients with PHN. The following study groups were included: the BoNT group, 0.5% lidocaine group, and 0.9% saline group. These patients were treated 5 U/mL BoNT-A, 0.5% lidocaine and 0.9% saline in the affected dermatome, respectively. Follow-up was performed at one day, seven days, and three months after drug administration. The BoNT group exhibited significantly improved VAS and sleep quality compared with the other two groups [37]. In addition, Apalla et al. performed a randomized, double-blind, placebo-controlled study on 30 adults with PHN, and the affected sites were divided into a chessboard of 5 U BoNT-A per injection. Thirteen of the 15 patients had a VAS reduction of at least 50% lasting approximately 16 weeks and a significantly reduced the sleep score [38]. Previously, there were reports on the antinociceptic effect of BoNT. Liu et al. reported that the VAS decreased from 10 to 1 after BoNT-A injection into the lesion, and the effect persisted for 52 days [39]. Sotiriou et al. reported assessed a case series of three patients. The affected site was divided into a chessboard form using a total of 100 U BoNT-A with 5 U injected at each point. The VAS started to decrease in three days and continued to decrease for greater than two months [40]. These papers are summarized in Table 2.

Table 2. Botulinum toxin for postherpetic neuralgia.

Study Design	Number of Patients	Method of Injection (Total Volume)	Result	Reference
Randomized, double-blind, placebo-controlled	60	Subcutaneous BoNT 5 U, 0.5% lidocaine, 0.9% saline per site	Significantly VAS pain score was decreased and sleep time improved	[37]
Randomized, double-blind, placebo-controlled	30	Divided into chessboard 5 U per site	50% VAS reduction of 13 patients	[38]
Case report	1	Fan pattern injection 100 U	VAS decrease from 10 to 1 Lasted for 52 days	[39]
Case series	3	Divided into chessboard 5 U per site (100 U)	VAS decrease and continued to 2 months	[40]

4.3. Post-Surgical Neuralgia

Four reports on the efficacy of BoNT on post-surgical neuralgia, including RCT articles, have been published. RCT articles include post-herpetic neuralgia and post-traumatic neuralgia. According to Ranoux et al., 29 patients with focal painful neuropathy and mechanical allodynia were included in a randomized, double-blind, placebo-controlled study. Up to 20–190 U BoNT-A was injected into the pain site intradermally. The injections reduced VAS, burning sensation, and allodynic brush sensitivity and improved QOL [41]. Layeeque et al. also observed postoperative pain. In 48 breast cancer patients subject to mastectomy, 22 patients were treated with BoNT-A in the pectoralis major, serratus anterior, and rectus abdominis muscle, and 26 control group patients were not treated. The group treated with BoNT reported improved post-operative pain, and post-operative analgesic use was significantly reduced. In addition, the tissue expander was removed from one patient in the BoNT group and five patients in the control group. The BoNT group did not complain of any particular complications [42]. A case report described satisfactory results from subcutaneous injection of BoNT-A in a 67-year-old patient with post-thoracotomy pain for more than two years postoperatively. The pain site was divided into 1-square centimeter. Then, 2.5 U of BoNT-A was injected into the middle, and 100 U BoNT-A was administered in total. The patient reported improved pain after five days, and pain relief persisted for up to 12 weeks [43]. According to Rostami et al., eight cancer patients with persistent focal pain were treated with surgery or radiotherapy. BoNT-A was injected intramuscularly or subcutaneously into the localized pain area. All patients reported significant VAS improvement, and a significant improvement in QOL was also noted [44]. The above studies are described in Table 3.

Table 3. Botulinum toxin for post-surgical neuralgia.

Study Design	Number of Patients	Method of Injection (Total Volume)	Result	Reference
Randomized, double-blind, placebo-controlled	29 (4 Postherpetic neuralgia, 25 Post-traumatic, post-surgical neuropathy)	Intradermal (20–190 U)	Decrease VAS, neuropathic nature pain and improve in quality of life	[41]
Prospective, non-randomized, placebo-controlled	48 (22 BoNT, 26 control)	Intramuscular (100 U)	Post-operative pain and analgesic use was reduced	[42]
Case report	1	Subcutaneous Affected zone was drawn with divisions of approximately 1 cm², 2.5 U per site (100 U)	Improvement in pain was about 50% as measured on the VAS and persisted at 12 weeks	[43]
Pilot, prospective	8	Intramuscular, subcutaneous (100 U)	All patients had VAS improvement	[44]

4.4. Diabetic Neuropathy

Two randomized, double-blind, placebo-controlled studies used BoNT for pain control of diabetic neuropathy (DN). In a study of 20 DN patients, Yuan et al. reported that 4 U of BoNT-A per site (total 50 U) was administered to the dorsum of foot, and 44% of patients had a clear reduction in VAS lasting three months and improved sleep quality [45]. Ghasemi et al. conducted a study similar to the previous paper, except that the BoNT dose was 8–10 U per site in 40 DN patients. A decrease in neuropathic pain score (NPS) and Douleur Neuropathique 4 (DN4) scores were reported in that study [46]. A meta-analysis of these two articles concluded that DN has a significant association between BoNT and pain relief [47]. The above papers are described in Table 4.

Table 4. Botulinum toxin for diabetic neuropathy.

Study Design	No. of Patients	Method of Injection (Total Volume)	Result	Reference
Randomized, double-blind, placebo-controlled, cross-over trial	20	Intradermal 4 U per site at dorsum of foot (50 U per each foot)	44.4% of the BoNT group experienced a reduction of VAS within 3 months.	[45]
Randomized, double-blind, placebo-controlled	40	Intradermal, dorsum of the foot, in a grid distribution pattern, total 12 sites 8–10 U per site	Decrease in neuropathic pain score and Douleur Neuropathique 4	[46]

4.5. Occipital Neuralgia

Kapural et al. retrospectively analyzed six patients injected with 50 U BoNT-A in the occipital nerve and found that the VAS was significantly reduced. Five patients exhibited pain relief lasting greater than four weeks [48]. Taylor et al. reported that 100 U of BoNT-A was administered to the occipital protuberance in the prospective, open, and case series. Improvement in sharp/shooting pain was noted, but no definite improvement in dull/aching pain was indicated [49]. Occipital neuralgia has been assessed in only two case series without an RCT article, so these studies are insufficient to prove the effectiveness of BoNT. The above papers are also described in Table 5.

Table 5. Botulinum toxin for occipital neuralgia.

Study Design	No. of Patients	Method of Injection (Total Volume)	Result	Reference
Case series	6	Occipital nerve block 50 U for each block (100 U)	Significant VAS reduction and pain relief lasting >4 weeks	[48]
Prospective, open, case series	6	Greater and lesser occipital nerve block (100 U)	Improvement in sharp/shooting pain, no definite improvement in dull/aching pain	[49]

4.6. Carpal Tunnel Syndrome

Breuer et al. conducted a randomized, double-blind, placebo-controlled study of 20 patients. In this study, 2,500 U of BoNT-B or saline was injected into hypothena muscle and tentorium associated with carpal tunnel. Tingling sensation, pain, and pain related to improved sleep were noted, but there was no significant difference compared with the control group [50]. In a prospective, open, pilot study of five patients, a total of 30 U of BoNT-A was injected intracarpally. Of the five patients, three reported insignificant pain relief, and none had electrophysiological changes [51]. These results suggest that the use of BoNT in carpal tunnel syndrome is not effective. These papers are described in Table 6.

Table 6. Botulinum toxin for carpal tunnel syndrome.

Study Design	No. of Patients	Method of Injection (Total Volume)	Result	Reference
Randomized, double-blind, placebo-controlled	20	Intramuscular, hypothena muscle, tentorium (2500 U)	No significant difference compared to the control group	[50]
Prospective, open, pilot	5	Intracapal 30 U for each carpal tunnel (60 U)	Three patients insignificant reduced pain, none had electrophysiological change.	[51]

4.7. CRPS

Safarpour et al. reported that two patients with CRPS had a reduction of CRPS and myofascial pain with the intramuscular administration of 20 U BoNT-A per site and trigger point injection [52]. They also performed randomized, prospective, double-blind, placebo-controlled, open-label extension studies of BoNT in CRPS patients. Fourteen patients with CRPS were divided into the BoNT group ($n = 8$) and control group ($n = 6$). A total of 40–200 U (5 U per point) BoNT was administered to the affected area with allodynia. No difference was found between the interventional group and the placebo group, and this study was terminated early due to the intolerance of BoNT [53]. In another study, lumbar sympathetic block was performed in a randomized, double-blind, placebo-controlled crossover study. Patients received standard LSGB on one side, and 10 mL of 0.5% bupivacaine was used. The same patient was injected with a crossover (another side) injection of 75 U BoNT-A in 10 mL of 0.5% bupivacaine. The control group has a median of 10 days, whereas the BoNT group has a median of 71 days [54]. In a case series published by Choi et al., two patients who experienced short-term effects on the lumbar sympathetic block were injected with 5000 U of BoNT-B in 0.25% levobupivacaine with a lumbar sympathetic block. VAS, allodynia, edema, coldness, and analgesic drug usage were reduced [12]. In a prospective, open case series of 11 patients with CRPS symptoms in upper limb girdle muscles, a total of 300 U of BoNT-A was administered to the pain-related muscles at 25–50 U. All patients exhibited improved VAS, allodynia, hyperalgesia, and skin color after 6–12 weeks [55]. In a retrospective, uncontrolled, unblended study of 37 patients, as a result of administering a total of 100 U of BoNT-A (10–20 U per pain site), 97% of patients reported pain reduction, and the average pain score decreased by 43% [56]. Except for one negative study, positive results have been published.

However, these studies include a low class papers, and the effect of BoNT in CRPS patients has not been proven. These papers are summarized in Table 7.

Table 7. Botulinum toxin for complex regional pain syndrome (CRPS).

Study Design	Number of Patients	Method of Injection (Total Volume)	Result	Reference
Case series	2	Intramuscular Trigger point 20 U per site	Reduction of CRPS pain and myofascial pain	[52]
Randomized, prospective, double-blind, placebo-controlled, and open-label extension	14 (8 BoNT group, 6 control group)	Intradermal, subcutaneous Allodynia area 5 U per site (40–200 U)	No difference between BoNT group and placebo group, terminated study early.	[53]
Randomized, double-blind, placebo-controlled crossover	9 (18 cases)	Lumbar sympathetic block 75 U BoNT + 0.5% bupivacaine/0.5% bupivacaine	Longer duration of pain reduction (BoNT vs. control/71 days vs. 10 days)	[54]
Case series	2	Lumbar sympathetic block 5000 U BoNT-B + 0.25% levobupivacaine	VAS and CRPS symptoms were reduced.	[12]
Prospective, open case series	11	Affected site, 25–50 U per site (300 U)	All patients had improved VAS, allodynia, hyperalgesia, and skin color after 6 to 12 weeks	[55]
Retrospective, uncontrolled, unblended	37	Affected site, 10–20 U per site (100 U)	The 97% patients reduced pain. (average pain reduction of 43%)	[56]

4.8. Phantom Limb Pain

In a prospective, randomized, double-blind pilot study, 14 patients with phantom limb pain were treated with 50 U per site for a total of 250–300 U BoNT-A. In addition, a lidocaine and depomedrol mixture was administered at the focal tender point. VAS was assessed monthly in patients before and six months after treatment. Both groups reported improved pain. The BoNT group had an advantage over pain control during the 3–6 months, but phantom limb pain was not completely alleviated [57]. There is a case report in which the effect of BoNT was effective in reducing phantom limb pain for greater than 12 months. In total, 25 U of BoNT-A was injected into the trigger point of the stump at four sites, and the patient was able to reduce the pain medication given that the pain was significantly eliminated [58]. The effect of BoNT on phantom limb pain cannot be verified because only low-grade studies on phantom limb pain have been reported. The above papers are also listed in Table 8.

Table 8. Botulinum toxin for phantom limb pain.

Study Design	No. of Patients	Method of Injection (Total Volume)	Result	Reference
Prospective, randomized, double-blind, pilot	14	Intramuscular/cutaneous/subcutaneous/ neuroma (EMG guidance) 50 U per site (250–300 U)	Both groups improved pain and BoNT group had an advantage over pain control during 3–6 months but could not completely change phantom limb pain.	[57]
Case series	3	EMG guidance into points with strong fasciculation (500 U)	Phantom pain, pain medication could be reduced, the gait became more stable and the artificial limb was better tolerated.	[58]

4.9. Spinal Cord Injury-Induced Neuropathic Pain

In a study of 40 patients with spinal cord injury-induced neuropathic pain, a randomized, double-blind, placebo-controlled design was used. In the BoNT group, 200 U BoNT-A was divided into 40 sites, and 4 mL of saline was administered to the control group in a similar manner. Pain intensities were assessed using VAS, the Korean version of the short-form McGill Pain Questionnaire

(SF-MPQ), and the Korean version of the World Health Organization Quality of Life (WHOQOL-BREF) questionnaire. The same procedure was performed at baseline and four and eight weeks. The BoNT group exhibited a statistically significant decrease in VAS at four and eight weeks compared with the placebo group, and SF-MPQ was also significantly reduced compared with the placebo group. However, there was no significant difference between the control group and the BoNT group in the Korean version of the WHOQOL-BREF, which assesses physical health, psychological social relationship, and environmental domains [59]. A similar paper was published in 2017, and a randomized, double-blind, placebo-controlled study was performed in 44 patients with spinal cord injury-induced neuropathic pain. The BoNT group received 200 U of BoNT-A at the pain site, and the control group received the same amount of saline at the pain site. Unlike the above paper, patients received the same treatment once daily for eight weeks. The primary outcome of pain was measured on a VAS scale, and the secondary outcome was measured by the SF-MPQ and the WHOQOL-BREF questionnaire. At four and eight weeks, both primary and secondary outcomes were measured and evaluated. No adverse effect was noted in both groups. VAS and SF-MPQ were significantly decreased in the BoNT group compared with placebo group at four and eight weeks, respectively. The difference from the above paper is that the WHOQOL-BREF also exhibited a statistically significant decrease compared with the placebo group [60].

In addition, there have been several case reports of neuropathic pain associated with spinal cord injury. Jabbari et al. reported that two patients who had burning pain and allodynia after spinal cord injury injected with 5 U of BoNT-A at 16–20 sites in the pain site maintained significant VAS reduction for greater than three months [61]. Han et al. mentioned that 20 U of BoNT-A was injected into 10 painful areas in patients with spinal cord injuries, and VAS was decreased from 96 to 23 [62]. The use of BoNT for spinal cord injury is considered to be effective based on a statistically significant RCT journal report. These papers are listed in Table 9.

Table 9. Botulinum toxin for spinal cord injury-induced neuropathic pain.

Study Design	Number of Patients	Method of Injection (Total Volume)	Result	Reference
Randomized, double-blind, placebo-controlled	40	Subcutaneous (200 U)	Significantly VAS was decreased at 4 and 8 weeks.	[59]
Randomized, double-blind, placebo-controlled	44	Subcutaneous (200 U) Once daily for 8 weeks	Significantly VAS was decreased at 4 and 8 weeks.	[60]
Case	2	Subcutaneous 5 U of BoNT at 16–20 sites	Significant VAS reduction for more than 3 months	[61]
Case	1	Subcutaneous 20 U of BoNT at 10 sites	VAS decreased from 96 to 23.	[62]

4.10. Central Poststroke Pain

Poststroke patients often use BoNT due to poststroke spasticity. However, some recent reports have reported that BoNT is used for central poststroke pain control. Shippen et al. injected BoNT in patients with elbow flexor spasticity with central poststroke pain. The patients had severe neuropathic pain at the site of the spasticity and received 100 U BoNT-A of Biceps Brachii, 75 U Brachialis and 25 U Brachioradialis. After the second day, the pain was reduced, and the spasticity was improved one week after administration. The patients repeat BoNT every three months to control pain [63]. Barbosa et al. also published a case report in which an analgesic effect was obtained using BoNT-A in patients with central poststroke pain. In two patients with stroke, injection of BoNT-A 200 U into the affected area under EMG guidance resulted in a decrease in NRS after a 3-month follow-up [64]. A randomized, double-blind, placebo-controlled trial of 273 patients with poststroke spasticity was performed. In total, 74.3% of the patients had stroke-related pain, and 47.3% were suffering from greater than NRS 4. Patients were divided into two groups: BoNT-A and standard care vs. placebo and standard care. The degree of pain was compared 12 weeks from the baseline, and the BoNT group reported significantly less pain compared with the placebo group. The reduction in pain persisted for up to 52 weeks [65]. This is the first RCT assessing the control of neuropathic pain with BoNT in patients with poststroke spasticity. Therefore, BoNT may be effective in patients with central poststroke pain. The above papers are summarized in Table 10.

Table 10. Botulinum toxin for central poststroke pain.

Study Design	Number of Patients	Method of Injection (Total Volume)	Result	Reference
Case	1	Intramuscular Biceps Brachii 100 U, Brachialis 75 U and Brachioradialis 25 U	Pain was reduced after 2 days, spasticity was improved after 1 week.	[63]
Case	2	Intramuscular Affected muscle (200 U)	NRS reduction for more than 3 months	[64]
Randomized, double-blind, placebo-controlled	273 (139 BoNT, 134 control)	Intramuscular Dosing was determined by investigator, second injection was performed with an open label and at least 12 weeks after the first injection	Significantly VAS was decreased at 12 weeks and reductions in pain were sustained through Week 52.	[65]

5. Adverse Effects

BoNT-A has minimal irreversible medical adverse effect. Regarding the use of BoNT in cervical dystonia, side effects, including neck muscle weakness, dysphagia, pain during swallowing, and flu-like symptoms, are rarely reported. The use of BoNT in blepharospasm and cerebral palsy is associated with unilateral or bilateral ptosis, hematoma, and lower limb weakness and pain. When BoNT is used in neuropathic pain, relatively minor complications, such as antibody formation and immune-related complications, are reported when a small amount of BoNT-A enters the circulatory system [66]. BoNT-B can also be used to obtain effective results when neutralizing antibodies are present in BoNT-A, and the effect is reduced. [67,68].

6. Conclusions

Before beginning BoNT therapy, patients with neuropathic pain require a careful assessment of functional limitations, goals, and expected outcomes. The guidelines of the American Academy of Neurology recommend the use of BoNT-A in neuropathic pain as follows. In postherpetic neuralgia, trigeminal neuralgia, and spinal cord injury-induced neuropathic pain, BoNT is effective (Level A) and BoNT is probably effective in post-surgical neuralgia, diabetic neuropathy, and central poststroke pain (Level B). In neuropathic pain, such as occipital neuralgia, CRPS, and phantom limb pain, a large and well-designed blinded and randomized controlled trial is needed to evaluate the effect of BoNT. The route of administration of BoNT is different for each article. There are no clinical guidelines for administration of BoNT for neuropathic pain. Most treatments are subcutaneous or intradermal, and BoNT is also injected intramuscularly or into the surrounding tissues. In some papers, BoNT is injected into the skin as a chessboard. In other studies, BoNT is directly injected into the nerve. In particular, the development of ultrasound technology can accurately inject drugs near the nerve, and BoNT injection near the nerve is emerging as an alternative method [69].

There is a need for comparative studies on whether these methods are effective and safe or which methods are more effective than others. In addition, studies should be carried out to compare the minimum doses that are effective. Large, well-designed clinical trials are needed to address these problems.

Conflicts of Interest: The authors declare no conflict of interest.

References

1. Thenganatt, M.A.; Fahn, S. Botulinum toxin for the treatment of movement disorders. *Curr. Neurol. Neurosci. Rep.* **2012**, *12*, 399–409. [CrossRef] [PubMed]
2. Pellizzari, R.; Rossetto, O.; Schiavo, G.; Montecucco, C. Tetanus and botulinum neurotoxins: Mechanism of action and therapeutic uses. *Philos. Trans. R. Soc. Lond. B Biol. Sci.* **1999**, *354*, 259–268. [CrossRef] [PubMed]
3. Porta, M. A comparative trial of botulinum toxin type A and methylprednisolone for the treatment of myofascial pain syndrome and pain from chronic muscle spasm. *Pain* **2000**, *85*, 101–105. [CrossRef]
4. Foster, L.; Clapp, L.; Erickson, M.; Jabbari, B. Botulinum toxin A and chronic low back pain. A randomized, double-blind study. *Neurology* **2001**, *56*, 1290–1293. [CrossRef] [PubMed]
5. Fishman, L.M.; Anderson, C.; Rosner, B. BOTOX and physical therapy in the treatment of piriformis syndrome. *Am. J. Phys. Med. Rehabil.* **2002**, *81*, 936–942. [CrossRef] [PubMed]
6. Park, H.J.; Lee, Y.; Lee, J.; Park, C.; Moon, D.E. The effects of botulinum toxin A on mechanical and cold allodynia in a rat model of neuropathic pain. *Can. J. Anaesth.* **2006**, *53*, 470–477. [CrossRef] [PubMed]
7. Johnson, E.A.; Montecucco, C. BOTULISM. In *Handbook of Clinical Neurology*; Andrew, G.E., Ed.; Elsevier: Amsterdam, The Netherlands, 2008; pp. 333–368.
8. Montecucco, C.; Schiavo, G. Mechanism of action of tetanus and botulinum neurotoxins. *Mol. Microbiol.* **1994**, *13*, 1–8. [CrossRef] [PubMed]
9. Pantano, S.; Montecucco, C. The blockade of the neurotransmitter release apparatus by botulinum neurotoxins. *Cell. Mol. Life Sci.* **2014**, *71*, 793–811. [CrossRef] [PubMed]
10. Whitemarsh, R.C.; Tepp, W.H.; Johnson, E.A.; Pellett, S. Persistence of botulinum neurotoxin a subtypes 1–5 in primary rat spinal cord cells. *PLoS ONE* **2014**, *9*, e90252. [CrossRef] [PubMed]
11. Park, H.J.; Marino, M.J.; Rondon, E.S.; Xu, Q.; Yaksh, T.L. The effects of intraplantar and intrathecal botulinum toxin type B on tactile allodynia in mono and polyneuropathy in the mouse. *Anesth. Analg.* **2015**, *121*, 229–238. [CrossRef] [PubMed]
12. Choi, E.; Cho, C.W.; Kim, H.Y.; Lee, P.B.; Nahm, F.S. Lumbar sympathetic block with botulinum toxin type B for complex regional pain syndrome: A case study. *Pain Phys.* **2015**, *18*, 911–916.
13. Kharatmal, S.B.; Singh, J.N.; Sharma, S.S. Voltage-gated sodium channels as therapeutic targets for treatment of painful diabetic neuropathy. *Mini Rev. Med. Chem.* **2015**, *15*, 1134–1147. [CrossRef] [PubMed]
14. Meng, J.; Wang, J.; Lawrence, G.; Dolly, J.O. Synaptobrevin I mediates exocytosis of cgrp from sensory neurons and inhibition by botulinum toxins reflects their anti-nociceptive potential. *J. Cell Sci.* **2007**, *120*, 2864–2874. [CrossRef] [PubMed]
15. Durham, P.L.; Cady, R. Insights into the mechanism of onabotulinumtoxinA in chronic migraine. *Headache* **2011**, *51*, 1573–1577. [CrossRef] [PubMed]
16. Fan, C.; Chu, X.; Wang, L.; Shi, H.; Li, T. Botulinum toxin type A reduces TRPV1 expression in the dorsal root ganglion in rats with adjuvant-arthritis pain. *Toxicon* **2017**, *133*, 116–122. [CrossRef] [PubMed]
17. Lucioni, A.; Bales, G.T.; Lotan, T.L.; McGehee, D.S.; Cook, S.P.; Rapp, D.E. Botulinum toxin type A inhibits sensory neuropeptide release in rat bladder models of acute injury and chronic inflammation. *BJU Int.* **2008**, *101*, 366–370. [CrossRef] [PubMed]
18. Heikkila, H.M.; Hielm-Bjorkman, A.K.; Morelius, M.; Larsen, S.; Honkavaara, J.; Innes, J.F.; Laitinen-Vapaavuori, O.M. Intra-articular botulinum toxin A for the treatment of osteoarthritic joint pain in dogs: A randomized, double-blinded, placebo-controlled clinical trial. *Vet. J.* **2014**, *200*, 162–169. [CrossRef] [PubMed]
19. Hadley, H.S.; Wheeler, J.L.; Petersen, S.W. Effects of intra-articular botulinum toxin type A (Botox®) in dogs with chronic osteoarthritis. *Vet. Comp. Orthop. Traumatol.* **2010**, *23*, 254–258. [CrossRef] [PubMed]
20. Bach-Rojecky, L.; Dominis, M.; Lackovic, Z. Lack of anti-inflammatory effects of botulinum toxin A in experimental models of inflammation. *Fundam. Clin. Pharmacol.* **2008**, *22*, 503–509. [CrossRef] [PubMed]
21. Sycha, T.; Samal, D.; Chizh, B.; Lehr, S.; Gustorff, B.; Schnider, P.; Auff, E. A lack of antinociceptive or antiinflammatory effect of botulinum toxin A in an inflammatory human pain model. *Anesth. Analg.* **2006**, *102*, 509–516. [CrossRef] [PubMed]
22. Chuang, Y.C.; Yoshimura, N.; Huang, C.C.; Wu, M.; Chiang, P.H.; Chancellor, M.B. Intraprostatic botulinum toxin A injection inhibits cyclooxygenase-2 expression and suppresses prostatic pain on capsaicin induced prostatitis model in rat. *J. Urol.* **2008**, *180*, 742–748. [CrossRef] [PubMed]

23. Cui, M.; Khanijou, S.; Rubino, J.; Aoki, K.R. Subcutaneous administration of botulinum toxin A reduces formalin-induced pain. *Pain* **2004**, *107*, 125–133. [CrossRef] [PubMed]
24. Shin, M.C.; Wakita, M.; Xie, D.J. Inhibition of membrane Na+ channels by A type botulinum toxin at femtomolar concentrations in central and peripheral neurons. *J. Pharmacol. Sci.* **2012**, *118*, 33–42. [CrossRef] [PubMed]
25. Matak, I.; Riederer, P.; Lacković, Z. Botulinum toxin's axonal transport from periphery to the spinal cord. *Neurochem. Int.* **2012**, *61*, 236–239. [CrossRef] [PubMed]
26. Bach-Rojecky, L.; Lackovic, Z. Central origin of the antinociceptive action of botulinum toxin type A. *Parmacol. Biochem. Behav.* **2009**, *94*, 234–238. [CrossRef] [PubMed]
27. Bach-Rojecky, L.; Salkovic-Petrisic, M.; Lackovic, Z. Botulinum toxin type A reduces pain supersensitivity in experimental diabetic neuropathy: Bilateral effects after unilateral injection. *Eur. J. Pharmacol.* **2010**, *633*, 10–14. [CrossRef] [PubMed]
28. Favre-Guilmard, C.; Auguet, M.; Chabrier, P.E. Different antinociceptive effects of botulinum toxin type A in inflammatory and peripheral polyneuropathic rat models. *Eur. J. Pharmacol.* **2009**, *617*, 48–53. [CrossRef] [PubMed]
29. Tang-Liu, D.D.; Aoki, K.R.; Dolly, J.O.; de Paiva, A.; Houchen, T.L.; Chasseaud, L.F.; Webber, C. Intramuscular injection of 125I-botulinum neurotoxin-complex versus 125I-botulinum-free neurotoxin: Time course of tissue distribution. *Toxicon* **2003**, *42*, 461–469. [CrossRef]
30. Marinelli, S.; Vacca, V.; Ricordy, R.; Uggenti, C.; Tata, A.M.; Luvisetto, S.; Pavone, F. The analgesic effect on neuropathic pain of retrogradely transported botulinumneurotoxin A involves Schwann cells and astrocytes. *PLoS ONE* **2012**, *7*, e47977. [CrossRef] [PubMed]
31. Wu, C.J.; Lian, Y.J.; Zheng, Y.K.; Zhang, H.F.; Chen, Y.; Xie, N.C.; Wang, L.J. Botulinum toxin type A for the treatment of trigeminal neuralgias: Results from a randomized, double-blind, placebo-controlled trial. *Cephalgia* **2012**, *32*, 443–450. [CrossRef] [PubMed]
32. Zhang, H.; Lian, Y.; Ma, Y.; Chen, Y.; He, C.; Xie, N.; Wu, C. Two doses of botulinum toxin type A for the treatment of trigeminal neuralgia: Observation of therapeutic effect from a randomized, double-blind, placebo-controlled trial. *J. Headache Pain.* **2014**, *15*, 65. [CrossRef] [PubMed]
33. Zuniga, C.; Piedimonte, F.; Diaz, S.; Micheli, F. Acute treatment of trigeminal neuralgia with onabotulinum toxin A. *Clin. Neuropharmacol.* **2013**, *36*, 146–150. [CrossRef] [PubMed]
34. Bohluli, B.; Motamedi, M.H.; Bagheri, S.C.; Bayat, M.; Lassemi, E.; Navi, F.; Moharamnejad, N. Use of botulinum toxin A for drug-refractory trigeminal neuralgia: Preliminary report. *Oral Surg. Oral Med. Oral Pathol. Oral Radiol. Endod.* **2011**, *111*, 47–50. [CrossRef] [PubMed]
35. Zuniga, C.; Diaz, S.; Piedimonte, F.; Micheli, F. Beneficial effects of botulinum toxin type A in trigeminal neuralgia. *Arq. Neuropsiquiatr.* **2008**, *66*, 500–503. [CrossRef] [PubMed]
36. Turk, U.; Ilhan, S.; Alp, R.; Sur, H. Botulinum toxin and intractable trigeminal neuralgia. *Clin. Neuropharmacol.* **2005**, *28*, 161–162. [PubMed]
37. Xiao, L.; Mackey, S.; Hui, H.; Xong, D.; Zhang, Q.; Zhang, D. Subcutaneous injection of botulinum toxin A is beneficial in postherpetic neuralgia. *Pain Med.* **2010**, *11*, 1827–1833. [CrossRef] [PubMed]
38. Apalla, Z.; Sotiriou, E.; Lallas, A.; Lazaridou, E.; Ioannides, D. Botulinum toxin A in postherpetic neuralgia: A parallel, randomized, double-blind, single-dose, placebo-controlled trial. *Clin. J. Pain* **2013**, *29*, 857–864. [CrossRef] [PubMed]
39. Liu, H.T.; Tsai, S.K.; Kao, M.C.; Hu, J.S. Botulinum toxin A relieved neuropathic pain in a case of post-herpetic neuralgia. *Pain Med.* **2006**, *7*, 89–91. [CrossRef] [PubMed]
40. Sotiriu, E.; Apalla, Z.; Panagiotidou, D.; Ioannidis, D. Severe post-herpetic neuralgia successfully treated with botulinum toxin A: Three case reports. *Acta Derm.-Venereol.* **2009**, *89*, 214–215.
41. Ranoux, D.; Attal, N.; Morain, F.; Bouhassira, D. Botulinum toxin type A induces direct analgesic effects in chronic neuropathic pain. *Ann. Neurol.* **2008**, *64*, 274–283. [CrossRef] [PubMed]
42. Layeeque, R.; Hochberg, J.; Siegel, E.; Kunkel, K.; Kepple, J.; Henry-Tillman, R.S.; Dunlap, M.; Seibert, J.; Klimberg, V.S. Botulinum toxin infiltration for pain control after mastectomy and expander reconstruction. *Ann. Surg.* **2004**, *240*, 608–613. [CrossRef] [PubMed]
43. Fabregat, G.; Asensio-Samper, J.M.; Palmisani, S.; Villanueva-Pérez, V.L.; De Andrés, J. Subcutaneous botulinum toxin for chronic post-thoracotomy pain. *Pain Pract.* **2013**, *13*, 231–234. [CrossRef] [PubMed]

44. Rostami, R.; Mittal, S.O.; Radmand, R.; Jabbari, B. Incobotulinum toxin-A improves post-surgical and post-radiation pain in cancer patients. *Toxins* **2016**, *8*, 22. [CrossRef] [PubMed]
45. Yuan, R.Y.; Sheu, J.J.; Yu, J.M.; Chen, W.T.; Tseng, I.J.; Change, H.H.; Hu, C.J. Botulinum toxin for diabetic neuropathic pain: A randomized double-blind crossover trial. *Neurology* **2009**, *72*, 1473–1478. [CrossRef] [PubMed]
46. Ghasemi, M.; Ansari, M.; Basiri, K.; Shaigannejad, V. The effects of intradermal botulinum toxin type A injections on pain symptoms of patients with diabetic neuropathy. *J. Res. Med. Sci.* **2014**, *19*, 106–111. [PubMed]
47. Lakhan, S.E.; Velasco, D.N.; Tepper, D. Botulinum toxin-A for painful diabetic neuropathy: A meta-analysis. *Pain Med.* **2015**, *16*, 1773–1780. [CrossRef] [PubMed]
48. Kapural, L.; Stillman, M.; Kapural, M.; Mclntyre, P.; Guirgius, M.; Mekhail, N. Botulinum toxin occipital nerve block for the treatment of severe occipital neuralgia: A case series. *Pain Pract.* **2007**, *7*, 337–340. [CrossRef] [PubMed]
49. Taylor, M.; Silva, S.; Cottrell, C. Botulinum toxin type A in the treatment of occipital neuralgia: A pilot study. *Headache* **2008**, *48*, 1476–1481. [CrossRef] [PubMed]
50. Breuer, B.; Sperber, K.; Wallenstein, S.; Kiprovski, K.; Calapa, A.; Snow, B.; Pappagallo, M. Clinically significant placebo analgesic response in a pilot trial of botulinum B in patients with hand pain and carpel tunnel syndrome. *Pain Med.* **2006**, *7*, 16–24. [CrossRef] [PubMed]
51. Tsai, C.P.; Liu, C.Y.; Lin, K.P.; Wnag, K.C. Efficacy of botulinum toxin type A in the relief of carpal tunnel syndrome. *Clin. Drug Investig.* **2006**, *26*, 511–515. [CrossRef] [PubMed]
52. Safarpour, D.; Jabbari, B. Botulinum toxin a (Botox) for treatment of proximal myofascial pain in complex regional pain syndrome: Two cases. *Pain Med.* **2010**, *11*, 1415–1418. [CrossRef] [PubMed]
53. Safarpour, D.; Salardini, A.; Richardson, D.; Jabbari, B. Botulinum toxin A for treatment of allodynia of complex regional pain syndrome: A pilot study. *Pain Med.* **2010**, *11*, 1411–1414. [CrossRef] [PubMed]
54. Carroll, I.; Clark, J.D.; Mackey, S. Sympathetic block with botulinum toxin to treat complex regional pain syndrome. *Ann. Neurol.* **2009**, *65*, 348–351. [CrossRef] [PubMed]
55. Argoff, C.E. Botulinum toxin type A treatment of myofascial pain in patients with CRPS Type 1 (reflex sympathetic dystrophy): A pilot study. In Proceedings of the World Pain Congress (IASP) Meeting, Vienna, Austria, 22–27 August 1999.
56. Kharkar, S.; Ambady, P.; Venkatesh, Y.; Schwartzman, R.J. Intramuscular botulinum toxin in complex regional pain syndrome: Case series and literature review. *Pain Phys.* **2011**, *14*, 419–424.
57. Wu, H.; Sultana, R.; Taylor, K.B.; Szabo, A. A prospective randomized double-blinded pilot study to examine the effect of botulinum toxin type A injection vs. Lidocaine/Depomedrol injection on residual and phantom limb pain. *Clin. J. Pain* **2012**, *28*, 108–112. [CrossRef] [PubMed]
58. Kern, U.; Martin, C.; Scheicher, S.; Muler, H. Long-term treatment of phantom and stump pain with Botulinum toxin type A over 12 months. A first clinical observation. *Nervenarzt* **2004**, *75*, 336–340. [CrossRef] [PubMed]
59. Han, Z.A.; Song, D.H.; Oh, H.M.; Chung, M.E. Botulinum toxin type A for neuropathic pain in patients with spinal cord injury. *Ann. Neurol.* **2016**, *79*, 569–578. [CrossRef] [PubMed]
60. Li, G.; Lv, C.A.; Tian, L.; Jin, L.J.; Sun, P.; Zhao, W. A randomized controlled trial of botulinum toxin A for treating neuropathic pain in patients with spinal cord injury. *Medicine* **2017**, *96*, e6919. [CrossRef] [PubMed]
61. Jabbari, B.; Maher, N.; Difazio, M.P. Botulinum toxin A improved burning pain and allodynia in two patients with spinal cord pathology. *Pain Med.* **2003**, *4*, 206–210. [CrossRef] [PubMed]
62. Han, Z.A.; Song, D.H.; Chung, M.E. Effect of subcutaneous injection of botulinum toxin A on spinal cord injury-associated neuropathic pain. *Spinal Cord* **2014**, *52*, S5–S6. [CrossRef] [PubMed]
63. Shippen, C.; Bavikatte, G.; Mackarel, D. The benefit of botulinum toxin A in the management of central post stroke pain: A case report. *J. Neurol. Stroke* **2017**, *6*, 00218. [CrossRef]
64. Camoes-Barbosa, A.; Neves, A.F. The analgesic effect of abobotulinum and incobotulinum toxins type A in central poststroke pain: Two case reports. *PM&R* **2016**, *8*, 384–387.
65. Wissel, J.; Ganapathy, V.; Ward, A.B.; Borg, J.; Ertzgaard, P.; Herrmann, C.; Haggstrom, A.; Sakel, M.; Ma, J.; Dimitrova, R.; et al. Onabotulinum toxin A improves pain in patients with post-stroke spasticity: Findings from a randomized, double-blind, placebo-controlled trial. *J. Pain Symptom Manag.* **2016**, *52*, 17–26. [CrossRef] [PubMed]

66. Sławek, J.; Madaliński, M.H.; Maciag-Tymecka, I.; Duzyński, W. Frequency of side effects after botulinum toxin A injections in neurology, rehabilitation and gastroenterology. *Pol. Merkur. Lek.* **2005**, *18*, 298–302.

67. Kessler, K.R.; Skutta, M.; Benecke, R. Longterm treatment of cervical dystonia with botulinum toxin A: Efficacy, safety, and antibody frequency. German Dystonia Study Group. *J. Neurol.* **1999**, *246*, 265–274. [CrossRef] [PubMed]

68. Greene, P.; Fahn, S.; Diamond, B. Development of resistance to botulinum toxin type A in patients with torticollis. *Mov. Disord.* **1994**, *9*, 213–217. [CrossRef] [PubMed]

69. Moon, Y.E.; Choi, J.H.; Park, H.J.; Park, J.H.; Kim, J.H. Ultrasound-guided nerve block with botulinum toxin type A for intractable neuropathic pain. *Toxins* **2016**, *8*. [CrossRef] [PubMed]

© 2017 by the authors. Licensee MDPI, Basel, Switzerland. This article is an open access article distributed under the terms and conditions of the Creative Commons Attribution (CC BY) license (http://creativecommons.org/licenses/by/4.0/).

MDPI

St. Alban-Anlage 66

4052 Basel

Switzerland

Tel. +41 61 683 77 34

Fax +41 61 302 89 18

www.mdpi.com

Toxins Editorial Office

E-mail: toxins@mdpi.com

www.mdpi.com/journal/toxins

www.ingramcontent.com/pod-product-compliance
Lightning Source LLC
Chambersburg PA
CBHW051715210326
41597CB00032B/5494